Marketing/Technical/Regulatory Sessions

of the Composites Institute's
International Composites EXPO '97

January 27–29, 1997

SESSION 1

COMPOSITES BASICS: AN EDUCATIONAL SEMINAR FOR CIVIL INFRASTRUCTURE PROFESSIONALS

Presenter: John P. Busel, *Composites Institute*

NOTE: No formal papers were submitted for this seminar. A handbook entitled *Introduction to Composites* was distributed to attendees at this session.

MONDAY MORNING—JANUARY 27

SESSION 2

INTRODUCTION TO COMPOSITES: AN EDUCATIONAL SEMINAR

NOTE: No formal papers were submitted for this seminar. A handbook entitled *Introduction to Composites* was distributed to attendees at this session.

SESSION 3

COMPOSITE BRIDGES: TRANSITIONING FROM THE TRADITIONAL TO THE NEW IN CIVIL ENGINEERING TECHNOLOGY

Session Manager: John P. Busel, *Composites Institute*
Moderator: Mark Greenwood, *Owens Corning*
Vice Moderator: Dr. N. Raghupathi, *PPG Industries*

3-A "Two Federal Agencies Experiment with Composite Bridges: Case Studies on the Acceptance of Composite Materials," Glen Mandigo, *GHI. Inc.*

3-B "Design and Construction of a Lightweight FRP Work Platform for Use at Oyster Creek Nuclear Power Plant," Roy J. Wilson, G. Eric Johansen, Scott Ribble and Adrian Fogle, *E. T. Techtonics, Inc.*

3-C "Fiber Reinforced Composite Decks for Infrastructure Renewal—Results and Issues," Vistasp M. Karbhari, Frieder Seible, Gilbert A. Hegemier and Lei Zhao, *University of California, San Diego*

3-D "The Carbon Shell System for Modular Short and Medium Span Bridges," Frieder Seible, Gilbert Hegemier, Vistasp Karbhari, Rigoberto Burgueño and Andrew Davol, *University of California, San Diego*

3-E "Design and Evaluation of a Modular FRP Bridge Deck," Roberto Lopez-Anido, Hota V. S. GangaRao, Venkata Vedam and Nikki Overby, *West Virginia University*

3-F "Design and Construction of FRP Pedestrian Bridges: 'Reopening the Point Bonita Lighthouse Trail'," G. Eric Johansen, Roy J. Wilson, Frederic Roll, P. Garret Gaudini and Kerryn Gray, *E. T. Techtonics, Inc.*

SESSION 4

BRAIDING AND FABRICS: ADVANCED REINFORCEMENT ARCHITECTURE OPENS NEW HORIZONS

Session Manager: Andrew Head, *A&P Technology*

4-A "Improvement of Lateral Compressive Strength in Braided Pipe—New Braided Fabrication," H. Hamada, A. Nakai, M. Masui and M. Sakaguchi, *Kyoto Institute of Technology*

4-B "Braided Reinforcements: A Versatile, Cost-Effective Alternative to Traditional Reinforcement," Jeff Martin and Andrew A. Head, *A&P Technology, Inc.*

4-C "Fabrics with Enhanced Surface Chemistry Increase Laminate Properties, Improve Design Confidence and Improve Cost Performance for Scrimp, HLU-SU, RTM, and Pultruded Composites," Gordon L. Brown, Jr. and Buddy Creech, *Clark-Schwebel, Inc.*

4-D "State-of-the-Art Developments Yield Braided Reinforcement Designs that Meet Unique Composite Product Requirements," Andrew A. Head and Jeff Martin, *A&P Technology*

4-E "A Cost-Effective Approach to High-Strength Pultruded Composites," Michael W. Klett, *PPG Industries, Inc.*

SESSION 5

SIX MAJOR TESTING TECHNIQUES FOR COMPOSITES: CHOOSING AND USING THE RIGHT ONE

Session Manager/
Moderator: Reynaldo De La Rosa,
Interplastic Corporation

Vice Moderator: Justin B. Berman, *U.S. Army CERL*

5-A **"Nondestructive Testing Techniques Used to Evaluate the Progression in Damage of a Swirl Mat Composite,"** D. C. Worley II, P. K. Liaw and R. S. Benson, *University of Tennessee;* W. A. Simpson, Jr. and J. M. Corum, *Oak Ridge National Laboratory*

5-B **"Comparison of Three-Point and Four-Point Flexural Bending Tests,"** Denise Theobald, Jack McClurg and James G. Vaughan, *University of Mississippi*

5-C **"Progressive Fracture and Damage Tolerance of Composite Pressure Vessels,"** Christos C. Chamis and Pascal K. Gotsis, *NASA Lewis Research Center;* Levon Minnetyan, *Clarkson University*

5-D **"Development of Accelerated Test Methods to Determine the Durability of Composites Subject to Environmental Loading,"** T. Russell Gentry, Lawrence C. Bank, Aaron Barkatt, Luca Prian and Feng Wang, *The Catholic University of America*

5-E **"Residual Structural Integrity of a Polymer Matrix Composite Structural Component,"** Albert H. Cardon, Michel Bruggeman, Yang Qin and Soumalia Harou-Kouka, *University Brussels*

SESSION 6

ENHANCING THE PROPERTIES OF CONCRETE COMPOSITES: AN EFFECTIVE R_x FOR A TRADITIONAL MATERIAL

Session Manager/
Vice Moderator: John P. Busel, *Composites Institute*
Moderator: Dr. Salem Faza,
    Marshall Industries Composites Inc.

6-A   **"An FRP Grid for Bridge Deck Reinforcement,"** Habib Rahman and Charles Kingsley, *National Research Council Canada*; John Crimi, *Autocon Composites Inc.*

6-B   **"Constructability Assessment of FRP Rebars,"** David H. Deitz and Issam E. Harik, *University of Kentucky*

6-C   **"Yield Line Study of Concrete Slabs with FRP Rebars,"** Zia Razzaq and Ram Prabhakaran, *Old Dominion University*; Mohamed Adnan and El-wani Dabbagh, *University of Aleppo*; Mike M. Sirjani, *Norfolk State University*

6-D   **"Aging of Glass Fiber Composite Reinforcement,"** Max L. Porter, *Iowa State University*; Jacob Mehus, *REINERTSEN Engineering*; Kurt A. Young, *Walter P. Moore & Associates, Inc.*; Ed O'Neil, *Army Corps of Engineers*; Bruce A. Barnes, *Black & Veatch*

6-E   **"FRP Grids for Reinforced Concrete: An Investigation of Fiber Architecture,"** Renata S. Engel, Charles E. Bakis, Antonio Nanni and Michael Croyle, *The Pennsylvania State University*

6-F   **"McKinleyville Bridge: Construction of the Concrete Deck Reinforced with FRP Rebars,"** Sanjeev V. Kumar, Hemanth K. Thippeswamy and Hota V. S. GangaRao, *West Virginia University*

SESSION 7

UNDERSTANDING THE REGULATORY GRAB-BAG

Session Manager: John Schweitzer, *Composites Institute*

7-A   **"Rats and Mice, TRI and PEL: Status of Assorted SIRC Programs,"** Jack Snyder, *SIRC*

7-B   **"Tote that Drum: NFPA 30 and Storage of Polyester Resin,"** Nancy Dehmlow, *GLS Corporation*

7-C   **"TCLP, RCRA, Subpart CC, LDR and Other Servings of Alphabet Soup,"** John Schweitzer, *Composites Institute*

7-D   **"CFA Open Molding Emissions Study,"** Robert Lacovara, *Composites Fabricators Association*

7-E   **"Filament Winding Styrene Emissions Study,"** L. Craigie and G. Webster, *Dow Chemical Company*

7-F   **"Title V Review and Update: Will I Ever Get My Permit?"** Rosemarie Kelley, *Keller & Heckman*

NOTE: No formal papers were submitted for this session.

SESSION 8

COMPOSITES IN ACTION

The format of this session was a hands-on demonstration; therefore, there were no formal papers.

SESSION 9

REBARS, REINFORCEMENTS AND REPAIRS: COMPOSITES TO THE RESCUE

Session Manager/
Vice Moderator: John P. Busel, *Composites Institute*
Moderator: Rick Pauer, *Reichhold Chemicals Inc.*

9-A "**Strengthening of Concrete Structures—State of the Art and Future Needs**," Peter H. Emmons, Alexander M. Vaysburd, Jay Thomas and Miroslav Vadovic, *Structural Preservation Systems, Inc.*

9-B "**Experimental and Analytical Studies of Adhesive Property for Rehabilitation of Concrete Structures by Using Composite Materials**," Hiroyuki Hamada, Asami Nakai and Shigeo Urai, *Kyoto Institute of Technology*; Atsushi Yokoyama, *Mie University*

9-C "**Tensile Reinforcement by FRP Sheets Applied to RC**," Antonio Nanni, Charles E. Bakis, Thomas E. Boothby, Elizabeth M. Frigo and Young-Joo Lee, *The Pennsylvania State University*

9-D "**Evaluation of the Hybridization of Wood Crossties with Glass Fiber-Reinforced Composites (GFRC)**," H. V. S. GangaRao, S. S. Sonti and M. C. Superfesky, *West Virginia University*; N. Raghupathi and E. A. Martine, *PPG Industries*; T. H. Dailey, *Indspec Chemical Corporation*

9-E "**The Tom's Creek Bridge Rehabilitation and Field Durability Study**," John L. Lesko, Richard E. Weyers, John C. Duke, Michael D. Hayes and Joe N. Howard, *Virginia Tech*; Daniel E. Witcher and Glenn Barefoot, *Morrison Molded Fiber Glass*; Randy Formica, *Town of Blacksburg, Virginia*; Jose Gomez, *Virginia Transportation Research Council*; Ernesto Villalba and Julius F. J. Volgyi, *Virginia Department of Transportation*

SESSION 10

AN OVERVIEW OF REGULATIONS GOVERNING EMISSIONS FROM COMPOSITES MANUFACTURING

Session Manager: John Schweitzer, *Composites Institute*

10-A "**Overview of Regulations of Emissions from Composite Manufacturing**," John Schweitzer, *Composites Institute*

10-B "**Status of EPA MACT for Composites**," Madeline Strum, *EPA*

10-C "**MACT for Boatbuilding**," John McKnight, *NMMA*

10-D "**PolyAdd Control Technology for Composite Manufacturing: A Case Study**," R. Goltz, *Dow Chemical Company*

10-E "**Economic and Non-Economic Impact of Incineration of Emissions from Composite Manufacturing**," R. Haberlein, *EECS*

NOTE: No formal papers were submitted for this session.

SESSION 11

DEALING WITH THE CHARACTERISTICS OF SMC/BMC: FLOW, SHEAR, SHRINKAGE, SINK MARKS, AND PAINTING

Session Manager: Hamid Kia, *General Motors*
Moderator: Mike Gruskiewicz, *Premix, Inc.*
Vice Moderator: Donald Robertson, *3M Company*

11-A "**Predicting Sink-Mark Formation in Ribbed SMC Parts**," Sari K. Christensen, Esther M. Sun and Tim A. Osswald, *University of Wisconsin-Madison*

11-B "**Analysis of Cure and Flow Behavior of SMC During Compression Molding**," Hiroyuki Hamada and Keigo Futamata, *Kyoto Institute of Technology*; Hajime Naito, *Seksui Chemical Co., Ltd.*

11-C "**Study of In-Mold Flow Behavior of SMC**," Kenichi Hamada, Tetsuya Harada, Takashi Tomiyama and Hirokazu Yamada, *Dainippon Ink and Chemicals, Inc.*

11-D "**Iosipescu Shear Testing of Sheet Molding Composites (SMC)**," E. M. Odom, *University of Idaho*; T. A. Kvanvig, T. P. VanHyfte and T. H. Grentzer, *Ashland Chemical Company*

11-E "**Shrinkage Control of Low-Profile Unsaturated Resins Cured at Low Temperature**," Wen Li and L. James Lee, *The Ohio State University*

11-F "**An Investigation of Post-Fill Curing Behavior of Compression Molded SMC Parts**," H. U. Akay, O. Selcuk and O. Gurdogan, *Technalysis, Inc.*; M. Revellino and L. Saggese, *Iveco, SpA*

TUESDAY MORNING—JANUARY 28 (CONTINUED)

SESSION 12

RESINS, COATINGS, FIBERS AND FOAMS—PARTNERS IN COMPOSITES FORMULATION

Session Manager: Dr. Louis Ross, *Interplastic Corporation*

12-A "Viscosity Increase of High Molecular Weight Unsaturated Polyester Resin through Use of Filler and/or Thickener," Eiichiro Takiyama, Yoshitaka Hatano and Fumio Matsui, *Showa High Polymer Co., Ltd.*

12-B "High-Performance Synthetic Foams," Harry S. Katz and Radha Agarwal, *Utility Development Corp.*

12-C "Influence of Surface Treatment of Filler on Hydrothermal Aging of Particle-Filled Polymer Composite," Tohru Morii and Nobuo Ikuta, *Shonan Institute of Technology;* Hiroyuki Hamada, *Kyoto Institute of Technology*

12-D "The Role of Constituent Materials in the Development of Syntactic Foam" Thomas J. Murray and Noel J. Tessier, *Emerson and Cuming Composite Materials, Inc.*

12-E "Newly Developed Flexible Vinyl Ester Resin with Air-Drying," Tomoaki Aoki, Kazuyuki Tanaka, Kouji Arakawa and Kanemasa Nomaguchi, *Hitachi Chemical Company Ltd*

SESSION 13

COMPOSITES—LEADING THE OFFENSIVE IN CORROSION-RESISTANT DUTY FOR LARGE INDUSTRIAL VESSELS

Session Manager: Ben Bogner, *Amoco Chemical Company*
Moderator: Frank Cassis, *FAC Associates*

13-A "Large Diameter FRP Applications in FGD Systems," John C. McKenna and A. Herbert, *Ershigs, Inc.*

13-B "The Suitability of Isopolyester FRP for Water Treatment and Sewer Applications," H. R. Dan Edwards and Ben Bogner, *Amoco Chemical Company*

13-C "Case Histories of ASME RTP-1 Stamped Vessels Designed by Subpart A and Subpart B," Alfred L. Newberry, *FEMech Consulting*

13-D "The 0° and 90° Laminate—An Advanced Fabrication Method for FRP Corrosion Resistant Vessels and Equipment," Joe Lassandro, *Viatec, Inc.*

13-E "Dual Laminate Piping," Gary Alan Glein, *Norcore Plastics*

NOTE: No formal papers were submitted for this session.

SESSION 14

DESIGN AND TESTING OF PULTRUDED COMPOSITE STRUCTURES

Session Manager/
Moderator: Jeffrey D. Martin, *Martin Pultrusions Group, Inc.*

Vice Moderator: Dawn Watkins, *Martin Pultrusions Group, Inc.*

14-A "ASCE/PIC Standard for Design of Pultruded Composite Structures," Richard E. Chambers, *Chambers Engineering P.C.;* Max L. Porter and Ashvin A. Shah, *Iowa State University*

14-B "Design Considerations for Pultruded Composite Beam-to-Column Connections Subjected to Cyclic and Sustained Loading Conditions," Ayman S. Mosallam, *California State University*

14-C "Simulation of Progressive Failure of Pultruded Composite Beams in Three Point Bending Using LS-DYNA3D," David W. Palmer, Lawrence C. Bank and T. Russell Gentry, *The Catholic University of America*

14-D "An Investigation of the Influence of Tightening Torque and Seawater on Bolted Pultruded Composite Joints," R. Prabhakaran, Satish Devara and Zia Razzaq, *Old Dominion University*

14-E "An Experimental Investigation of Bolt-Preload Relaxation in Pultruded Composite Joints," R. Prabhakaran, Ravi Srinivas and Zia Razzaq, *Old Dominion University*

14-F "The Effect of Environmental Exposure on the Behavior of Pultruded Mechanical Connections," Robert L. Yuan, *University of Texas;* Shane E. Weyant, *Creative Pultrusions, Inc.*

SESSION 15

MATERIAL SUBSTITUTION WITH THERMOPLASTIC COMPOSITES

Session Manager: Cliff Watkins, *PPG Industries, Inc.*
Moderator: Larry Ferguson, *PPG Industries, Inc.*

15-A **"Replacing Metal with Engineering Composites,"** Dan Voture, *LNP Engineering Plastics*

15-B **"High-Performance Nylon Composites in Automotive,"** Clint Christian, *DuPont Automotive*

15-C **"The Northstar Air Intake Manifold—Conversion from Metal to Plastics,"** Ken Baraw, *BASF Corporation*

15-D **"Polyphenylene Sulfide Composites: Applications in Automotive Fuel Systems,"** Jonas Angus, *Hoechst Technical Polymers*

15-E **"Advanced Thermoplastic Composites for Office Seating,"** James Hurley, *BASF Corporation*

15-F **"Advances in *in-situ* Fiber Placement of Thermoplastic Matrix Composites for Commercial Applications,"** Michael J. Pasanen, *Automated Dynamics Corporation*

15-G **"Optimum Beam Design for Automotive Pedal Application,"** Chul S. Lee, *Allied Signal, Inc.*; Brian Irwin, *University of Michigan*

NOTE: No formal papers were submitted for this session.

TUESDAY, AFTERNOON—JANUARY 28

SESSION 16

WAYS TO ACCELERATE COMPOSITES' ACCEPTANCE IN THE CIVIL INFRASTRUCTURE MARKET: HOW YOUR COMPANY CAN BENEFIT FROM CURRENT PROJECTS

Session Manager/
Vice Moderator: John P. Busel, *Composites Institute*
Moderator: Doug Barno, *Composites Institute*

Part 1

16-A "Composite Repair/Upgrade of Concrete Civil Engineering Structures" Orange S. Marshall, Jr. and Pamalee A. Brady, *U.S. Army Construction Engineering Research Laboratories (CERL)*; John P. Busel, *Composites Institute*

16-B "Advancements in Smart-Tagged Composites for Infrastructure Applications," Justin R. Berman, Robert F. Quattrone and John P. Voyles, *U.S. Army Construction Engineering Research Laboratories (CERL)*; John P. Busel, *Composites Institute*

16-C "Design and Development of FRP Composite Piling Systems," Richard Lampo, *U.S. Army Corps of Engineers*; Ali Maher, *Rutgers University*, John P Busel, *Composites Institute*; Robert Odello, *U.S. Naval Facilities Engineering Service Center*

16-D "Development and Demonstration of a Modular FRP Deck for Bridge Construction and Replacement," Roberto Lopez-Anido and Hota V. S. GangaRao, *West Virginia University*; Jonathan Trovillion, *U.S. Army Corps of Engineers*; John Busel, *Composites Institute*

Part 2

16-E "Composite Elements for Navy Waterfront Facilities," Robert J. Odello, George E. Warren and Daniel Hoy, *U.S. Naval Facilities Engineering Service Center*; John P. Busel, *Composites Institute*

Session Manager/
Moderator: Antonio Nanni,
Pennsylvania State University

A panel discussion was held on the progress of the International Research on Advanced Composites in Construction (IRACC), a collaborative effort of seventy experts from Europe, Japan and the Americas. The mission of the IRACC panel is to seek new forms of international collaboration, to address the need for international standardization and coding, and to facilitate national priorities in light of international achievements. The following five areas were addressed: non-prestressed reinforcement (rebar) for concrete, prestressed reinforcement (cables, tendons) for concrete, structural shapes, structural FRP systems, and concrete repair systems.

SESSION 17

NEW DEVELOPMENTS IN SMC/BMC MATERIALS: FROM LOW DENSITY TO LOW PRESSURE, WITH A FEW STOPS IN BETWEEN

Session Manager: Hamid Kia, *General Motors*
Moderator: Lewis Perkey, *Molding Products Division, Interplastic*
Vice Moderator: Donald Robertson, *3M Company*

17-A "New Advances in Low Pressure Sheet Molding Compound," Louis Dodyk, Burr Leach and Andrew Ratermann, *Cambridge Industries*

17-B "The Use of Divinylbenzene Monomer for Improved Fiber-Reinforced Composites," K. E. Atkins, R. C. Gandy, C. G. Reid and R. L. Seats, *Union Carbide Corporation*

17-C "A New Structural SMC Resin," Steven P. Hardebeck and Clark B. Wade, *Alpha/Owens Corning*

17-D "High Performance Vinyl Ester Molding Compound," Koichi Akiyama, Kenichi Morita, Hideki Terada, Hiroya Okumura and Takashi Shibata, *Takeda Chemical Industries, Ltd.*

17-E "Development of Low Density SMC Formulations in the Transportation Market," James R. Lemke and Michael Sommer, *BYK-Chemie USA*

17-F "Compounding and Molding Parameters that Lead to Edge Paint Pops in SMC Composite Panels," Michael W. Klett and Steven J. Morris, *PPG Industries, Inc.*

SESSION 18

COMPOSITES IN ACTION

The format of this session was a hands-on demonstration; therefore, there were no formal papers.

SESSION 19

SMART MATERIAL SYSTEMS: NEW DEVELOPMENTS IN THE MONITORING OF STRUCTURAL COMPOSITES

Session Manager: John P. Busel, *Composites Institute*

Moderator: Dr. Craig A. Rogers,
University of South Carolina

Vice Moderators: Dr. James Sirkis, *University of Maryland;*
Dr. Victor Giurgiutiu,
University of South Carolina

Image having the ability to test the structural integrity of a large composite system at any point during its manufacture, installation and extended service life. This is just one of the "near future" realities to be explored during this comprehensive session and workshop.

19-A "**Damage Detection in Composite Rotorcraft Flexbeams Using the De-Reverberated Response,**" K. A. Lakshmanan and Darryll J. Pines, *University of Maryland*

19-B "**Embedded Fiber Optic Sensor for Filament Wound Composite Structures,**" Chia-Chen Chang and Jim Sirkis, *University of Maryland;* Rich Foedinger, *Technology Development Associates, Inc.;* Terry Vandiver, *Redstone Arsenal*

19-C "**Improved Active Tagging, Non-Destructive Evaluation Techniques for Full-Scale Structural Composite Elements,**" Zao Chen, *Virginia Tech;* Victor Giurgiutiu and Craig A. Rogers, *University of South Carolina;* Robert Quattrone and Justin Berman, *U.S. Army Corps of Engineers*

19-D "**Tagged Composites Mechanical Properties: Investigation and Correlation with Fractographic Examination,**" Victor Giurgiutiu, *University of South Carolina;* Jerone Gagliano, *Virginia Polytechnic Institute and State University;* Robert Quattrone, Justin Berman and John Voyles, *U.S. Army Corps of Engineers*

SESSION 20

ADDRESSING FIRE PERFORMANCE STANDARDS: CREATING A DOORWAY TO CIVIL ENGINEERING MARKETS—A PANEL DISCUSSION

Session Manager: John Schweitzer, *Composites Institute*

A panel of experts lead a discussion on the perceptions held by civil infrastructure professionals regarding composites' performance in a fire. Is the lack of fire standards a barrier to the use of composites in infrastructure applications? Topics included the NIST program on fire safety, performance-based fire protection standards, and a case study of fire testing for composites.

SESSION 21

PROCESSING AND RECYCLING: TWO SIDES OF THE SMC/BMC COIN

Session Manager: Hamid Kia, *General Motors*

Moderator: William Mellian, *Owens Corning*

Vice Moderator: Loren Larson, *3M Company*

Part 1

21-A "**Supercritical Fluid (SCF) Application of SMC Primers: Balancing Transfer Efficiency and Appearance,**" Jeffrey D. Goad and James Hansen, *Union Carbide Corporation*

21-B "**Acrylic BMC for Artificial Marble,**" Hiroshi Kato, Kazunori Kobayashi, Takuji Koyanagi and Ryozou Amano, *Inax Corp.*

21-C "**Warpage in Injection Molded FRP: Establishing Causes and Cures Using Numerical Analysis,**" Stefan Kukula, Makoto Saito, Naoki Kikuchi, Tadaaki Shimeno and Akio Muranaka, *Kobe Steel Ltd.*

Part 2

Recyclability of SMC/BMC materials is still subject to question in many academic and industrial circles. This portion of the session brought together speakers who addressed issues such as the technologies, the economics and the infrastructure for SMC/BMC recycling. Included in the topics were updates on the most recent usage of recyclates in European and North American automotive applications.

WEDNESDAY MORNING—JANUARY 29 (CONTINUED)

SESSION 22

RESINS AND ADDITIVES: COMPONENTS FOR ACHIEVING HIGH PERFORMANCE COMPOSITES

Session Manager: Dr. Louis Ross, *Interplastic Corporation*

22-A "**High-Performance Modified-Phenolic Piping System,**" Joie L. Folkers and Ralph S. Friedrich, *Ameron International*

22-B "**Novel, Thermally Triggered Catalyst/Promoter Additives for Unsaturated Polyester, Vinyl Ester and Epoxy Resin Cure,**" Steven P. Bitler, Mark A. Wanthal, David A. Kamp, Paul A. Meyers and David D. Taft, *Landec Corporation*

22-C "**A New Development in Low Profile Additives,**" Norisha Ujikawa and Masumi Takamura, *NOF Corporation;* F. B. Laurent and Clive B. Bucknall, *Cranfield University*

22-D "**Development of High Molecular Weight Unsaturated Polyester Resin,**" Eiichiro Takiyama, Yoshitaka Hatano and Fumio Matsui, *Showa High Polymer Co., Ltd.*

22-E "**Introduction to Phenolic Resin—Application, Processability and Behaviour,**" Peter Falandysz, *Composites By Design*

22-F "**Formulation Optimization of an Acrylic Modified Polyester Resin for Use in Resin Transfer Molding,**" Louis Ross and Matthew Kastl, *Interplastic Corporation*

SESSION 23

INNOVATIONS IN PULTRUSION DESIGN, PROCESSING AND CONTROL

Session Manager/
Moderator: Jeffrey D. Martin,
Martin Pultrusions Group, Inc.

Vice Moderator: Vic Leonino, *LCM Manufacturing, Inc.*

23-A "**Evaluation of the Quality and Consistency of Open-Bath Wet-Out Techniques for Pultruded Composites,**" Ellen Lackey and James G. Vaughan, *University of Mississippi*

23-B "**Low Cost Methods for VOC Abatement in the Pultrusion Plant and for Compliance with the Clean Air Act,**" Jeff Martin, *Martin Pultrusions Group;* Douglas S. Barno, *VVK Weege of North America Inc.*

23-C "**A Material/Processing Approach to Density Reduction and Cost Savings for Pultruded Rods and Profiles,**" Steven G. Katz and Darryl A. Payne, *ISORCA, Inc.*

23-D "**A Mass Produced, Trans-Tibial Prosthesis Made of Pultruded FRP,**" Makoto Saito, Hiroki Nakayama and Shuuji Hagiwara, *Kobe Steel Ltd.;* Seishi Sawamura, Mikio Yuki and Ichiro Kitayama, *Hyogo Rehabilitation Center;* Richard Reed and Bill Carroll, *Glastic Corporation*

23-E "**Cure Systems for Pultrusion: Multiple Peroxide vs. Promoted,**" Ted M. Pettijohn, *Witco Corporation*

SESSION 24

METHODS, MATERIALS, TESTING AND EMISSIONS CONTROL FOR OPEN MOLDING

Session Manager: Richard C. Adams, *Amoco Chemicals*

24-A "**Reaction Spray Molding Using Polyurethane and Polyurethane Technology,**" Rodney D. Jarbor, *Futura Coatings, Inc.*

24-B "**Exploration of Innovative Composite Joining Methods Using Abrasive Waterjet Machining,**" David G. Taggart, Joel A. Kahn, Thomas J. Kim and Madhusarathi Nanduri, *University of Rhode Island*

24-C "**Residual Compressive Strength of Hand Lay-Up Glass Fiber-Reinforced Plastic Laminates Due to Low Velocity Impact,**" Ivan Volpoet, Chihdar Yang and Su-Seng Pang, *Louisiana State University*

24-D "**90% Reduction in Styrene Evaporation from White Gelcoat,**" Karl Rodland and Roar Mork, *Jotun Polymer AS*

24-E "**A Method for Fabrication of Structural Adhesive Joints,**" Ronnal P. Reichard, *Structural Composites, Inc.*

SESSION 25

NEW TECHNOLOGIES FOR EXPANDING INFRASTRUCTURE APPLICATIONS: DAMS, HIGHWAY SYSTEMS AND RAILROADS

Session Manager/
Vice Moderator: John P. Busel, *Composites Institute*
Moderator: Rick Pauer, *Reichhold Chemicals Inc.*

25-A "Using the Highway Innovative Technology Evaluation Center to Bring New Proprietary Products into the Highway Market," David A. Reynaud, *HITEC*

25-B "Conceptual Design, Optimization, Development, and Prototype Evaluation of Composite-Reinforced Wood Railroad Crossties," Julio F. Davalos and Pizhong Qiao, *West Virginia University;* Michael G. Zipfel, *Lockheed Martin Astronautics;* Kira L. Kaleps, *Industrial Fiberglass Specialties, Inc.*

25-C "Composite Wicket Gate Development at the Olmsted Prototype Dam," Mostafiz Chowdhury, Robert Hall and Byron McClellan, *U.S. Army Corps of Engineers;* Peter Hoffman and James Mundloch, *McDonnell Douglas Corporation*

25-D "FRP Composite Rebars Come of Age: A Rational Strategy for Commercialization," Charles R. McClaskey, *Reichhold Chemicals, Inc.*

25-E "Innovation of Pultrusion Structural Shapes for Infrastructure," Glenn Barefoot, Dave Sitton, Clint Smith and Dan Witcher, *Morrison Molded Fiber Glass Company*

SESSION 26

COMPOSITES IN ACTION

The format of this session was a hands-on demonstration; therefore, there were no formal papers.

SESSION MANAGERS

RICHARD ADAMS
Amoco Chemical Company

SESSION MANAGER FOR:
Methods, Materials, Testing and Emissions Control for
  Open Molding

BEN BOGNER
Amoco Chemical Company

SESSION MANAGER FOR:
Composites—Leading the Offensive in
  Corrosion-Resistant Duty for Large Industrial Vessels

JOHN P. BUSEL
Composites Institute

SESSION MANAGER FOR:
Composite Bridges: Transitioning from the Traditional to
  the New in Civil Engineering Technology
Enhancing the Properties of Concrete Composites: An
  Effective R_x for a Traditional Material
Rebars, Reinforcements and Repairs: Composites to the
  Rescue
Ways to Accelerate Composites' Acceptance in the Civil
  Infrastructure Market: How Your Company Can
  Benefit from Current Projects
SMART Material Systems: New Developments in the
  Monitoring of Structural Composites
New Technologies for Expanding Infrastructure
  Applications: Dams, Highway Systems and Railroads

REYNALDO DE LA ROSE
Interplastic Corporation

SESSION MANAGER FOR:
Six Major Testing Techniques for Composites: Choosing
  and Using the Right One

ANDREW HEAD
A&P Technology

SESSION MANAGER FOR:
Braiding and Fabrics: Advanced Reinforcement
  Architecture Opens New Horizons

HAMID KIA
General Motors

SESSION MANAGER FOR:
Dealing with the Characteristics of SMC/BMC: Flow,
  Shear, Shrinkage, Sink Marks, and Painting
New Developments in SMC/BMC Materials: From Low
  Density to Low Pressure, With a Few Stops in Between
Processing and Recycling: Two Sides of the SMC/BMC
  Coin

JEFFREY MARTIN
Martin Pultrusions Group, Inc.

SESSION MANAGER FOR:
Design and Testing of Pultruded Composite Structures
Innovations in Pultrusion Design, Processing and
  Control

DR. LOUIS ROSS
Interplastic Corporation

SESSION MANAGER FOR:
Resins, Coatings, Fibers and Foams—Partners in
  Composites Formulation
Resins and Additives: Components for Achieving High
  Performance Composites

JOHN SCHWEITZER
Composites Institute

SESSION MANAGER FOR:
Understanding the Regulatory Grab-Bag
An Overview of Regulations Governing Emissions from
  Composites Manufacturing
Addressing Fire Performance Standards: Creating a
  Doorway to Civil Engineering Markets—A Panel
  Discussion

CLIFF WATKINS
PPG Industries, Inc.

SESSION MANAGER FOR:
Material Substitution with Thermoplastic Composites

Composites Basics: An Educational Seminar for Civil Infrastructure Professionals

NOTE: No formal papers were submitted for this seminar.

Introduction to Composites: An Educational Seminar

NOTE: No formal papers were submitted for this seminar.

Composite Bridges: Transitioning from the Traditional to the New in Civil Engineering Technology

Two Federal Agencies Experiment with Composite Bridges: Case Studies on the Acceptance of Composite Materials

GLEN MANDIGO

ABSTRACT

The U.S. Federal Highway Administration (FHWA) and the U.S. Forest Service (USFS) both recently began experimenting with composite pedestrian bridges.

The Federal Lands Highway Office (FLHO) of the FHWA initiated in June 1996 an effort to assess composite materials for bridge design and construction in an effort to promote the use of innovative technology and materials. A new pedestrian span along the George Washington Parkway in Washington, DC was selected for a demonstration project because of its relative low risk, small size, and overall low cost. A principal FLHO objective for the project was gaining an internal design capability for composite structures to better service client agencies, such as the National Park Service, that are already using composite bridges.

The USFS experimental effort also resulted from increased interest in composite bridges by USFS trail managers. Rather than selecting a flagship bridge project, the USFS initiated a broader based research effort to evaluate composite structures, including those already in service. The USFS emphasis was again on developing an internal capability to work with the new materials.

Certain common elements from these applications case studies offer lessons to the composites industry: 1) bridge engineers respond best to their own customers; 2) composite suppliers must equip agency engineers to work with composite materials (e.g. application focused design guides, software); 3) low-risk structures present a near-term mechanism to introduce composite materials.

INTRODUCTION

Composite materials have rapidly gained greater visibility in the civil infrastructure community. Engineers, facility owners, designers, and researchers have expressed interest in applying composite materials to address particular problems. In response to this growing interest level, two U.S. government agencies initiated demonstration and research projects to evaluate composite structures.

The Federal Lands Highway Office (FLHO) of the U.S. Federal Highway Administration (FHWA) and the U.S. Forest Service (USFS) both recently began experimenting with composite bridges. The objectives for each agency's effort and the original motivations for looking at composite materials were similar and offer insight for firms and organizations promoting composite structures.

The USFS manages over 500,000 miles of roads and 500,000 miles of hiking trails, including thousands of trail bridges. Annual USFS budgets for bridge engineering and maintenance average around $8 million. USFS engineering authority is distributed into nine

Glen Mandigo, GHL Incorporated, 1020 19th Street, N.W., Suite 520, Washington, D.C. 20036, (202) 429-9714

regions of the country. USFS region engineers are supported by engineers at the national forests and by national forest trail managers. USFS engineers also work closely with FLHO engineers on roadway projects that cross national forests.

The FLHO functions as a highway department for federally owned lands. FLHO's customers therefore include the U.S. Forest Service, the National Park Service, the Bureau of Indian Affairs, and other agencies managing federal lands. FLHO provides these agencies with services similar to those provided by a state highway department (e.g., engineering, construction contracting support, maintenance contracting, bridge inspection, technology development).

FLHO manages over 30,000 miles of roads, including 3,000 bridges. Annual FLHO budgets for bridge engineering and maintenance average around $135 million. Engineering authority is distributed into three division across the country.

The experimental composite bridge efforts were initiated and managed by the bridge engineering sections of each agency. This paper describes these efforts and highlights certain common elements from these initiatives that offer lessons to the composites industry: 1) bridge engineers respond best to their own customers; 2) composite suppliers must equip agency engineers to work with composite materials (e.g. application focused design guides, software); 3) low-risk structures present a near-term mechanism to introduce composite materials.

U.S. FOREST SERVICE

The National Park Service along with the U.S. Forest Service (USFS) pioneered the use of composite trail bridges. The USFS installed two composite bridges and the National Park Service installed over six composite spans. Composite materials offered advantages over conventional metal and wood trail bridge materials for many of USFS' unique bridging requirements. Composite bridges also proved to be cost competitive with conventional building materials, particularly when construction/installation costs were included for remote locations. Trail managers correspondingly expressed interest in purchasing more and more composite trail bridges.

However, to better manage the introduction of these new materials into the USFS bridge inventory, USFS bridge engineers in the fall of 1995 placed an informal hold on additional purchases and installations of composite bridges. Bridge engineers needed time to gather more data and experience with the composite bridge materials. Two technical topics were identified about which more information was needed: procedures and guidelines associated with ownership of the composite structures (e.g., repair, maintenance, inspection), and quality control or standards for design and installation (e.g., design guidelines, standards for installation).

To address these concerns, during the spring and summer of 1996 USFS engineers gathered input from various composite material vendors, academic experts, other owners of composite trail bridges (e.g. National Park Service), and USFS' own technical and engineering resources. That information and experience which was not available in written form was collected through meetings, conferences, and forums involving USFS engineers and composites experts.

As might be expected, the USFS found it difficult to obtain enough information and data to fully address its primary concerns. Information on particular issues was often unavailable (e.g., inspection and damage criteria, crack propagation under pedestrian load spectrums, creep data, AASHTO standards) and on other issues, that information which was available was incomplete or only partially relevant to USFS interests (e.g., connections, environmental exposure, computer analysis packages, design guides).

The USFS considered the National Science Foundation testing program on composite materials and on composite bridge structures to be the most reliable and comprehensive source of useful data and experience. Vendor information and direct experiences from other owners of composite structures were also valuable information resources.

Having found information to address only a limited number of its concerns, the USFS initiated its own testing program to evaluate composite bridges. Having established a test and evaluation program, and wishing to balance the growing interest of the USFS trail mangers in deploying composite bridges, USFS engineers correspondingly lifted their informal hold on purchasing and installing composite bridge structures with the following conditions: 1) conventional materials and construction alternatives are determined to be unacceptable, 2) significant potential cost savings can be demonstrated, 3) bridges are deployed in a low-risk location (i.e., spans under 50 feet, average pedestrian loading), 4) installed composite bridges must be enrolled in the USFS composite bridge testing program, 5) composite bridges receive approval from the USFS regional bridge engineer.

The conditional deployment of short-span composite structures was a recognition by USFS engineers of the relative low-risk associated with short-span structures. From available information on composite structures, USFS engineers concluded that short-span composite bridges presented acceptable financial, structural, and environmental risk. Unfortunately, data on composite spans over 50 feet was not sufficient to support a similar conclusion.

The experimental deployment of composite bridges was also a recognition of the growing popularity of composite bridges among USFS trail managers. Short-span composite bridges were competitive on cost with traditional materials, and lightweight composite structures were often easier to install. USFS engineers wanted to be responsive to their customers while minimizing risks.

The USFS testing program began in November 1996 and was expected to continue for the next three to five years. The USFS Missoula Technical Development Center in Missoula, Montana was charged with administering the program. The program will be coordinated with the U.S. Park Service, the manufacturer(s) of composite bridges, and the U.S. Federal Highway Administration Office of Recreational Trails. Initial funding was targeted at around $50,000.

The goal of the Missoula testing program is to determine the stability and durability of the composite materials, the integrity of the design and connections. If the USFS deems it necessary and appropriate this program may also develop design and construction guidelines for composite trail bridges. Preliminary results are expected by the end of September 1997.

FEDERAL LANDS HIGHWAYS OFFICE

Federal Lands Highways' (FLHO) interest in composite materials resulted from its commitment to customer support. As the engineering authority for roads crossing federally owned lands, FLHO received a growing number of inquiries about composite bridges.

Recent emphasis on composite materials by the Federal Highway Administration provided additional motivation to FLHO to examine composite structures. For example, the U.S. Department of Transportation (DoT) ranked composite materials as one of its highest research priority areas for FY 1997.

To be better able to provide assistance to its customers in this new technology area, FLHO engineers determined they needed greater experience with composite materials and a better understanding of how to design structures using composites. Specific FLHO research objectives were similar to USFS concerns about composite materials including, design standards and tools (e.g., computer structural design tools, cost estimating techniques, materials data) and data on performance (e.g., environmental exposure, damage/durability).

FLHO proposed a demonstration project as a means of gaining familiarity with and understanding of composite. The pedestrian span over Spout Run in Washington, DC, shown in figure 1, was selected as a leading candidate demonstration structure. The bridge was considered a good initial project because it was already scheduled for construction, the pedestrian span was considered a low-risk application, and the structure was accessible for exhibiting the new technology. A major, and ultimately fatal, disadvantage of the proposed bridge was that the basic design and configuration shown in figure 1 was already approved and therefore locked in.

Figure 1: Proposed Composite Bridge Over Spout Run Along
George Washington Parkway in Washington, DC.

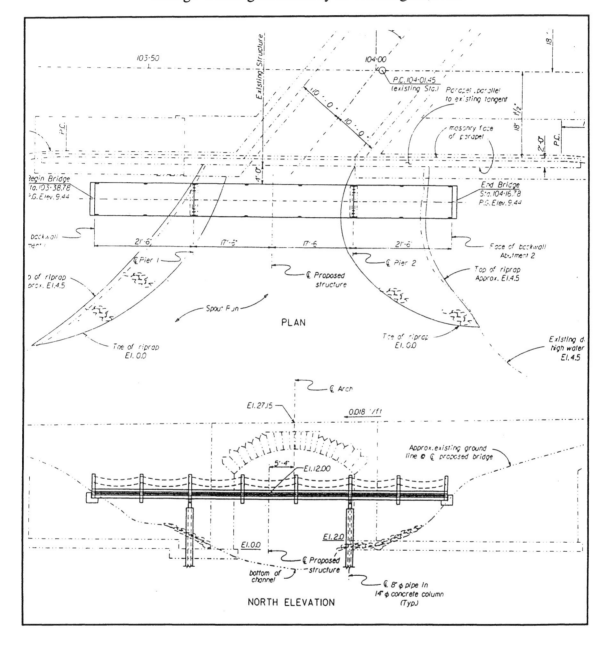

As part of the demonstration project planning exercise, FLHO engineers gathered information on composites throughout the summer of 1996 and developed a plan that would allow them to use the bridge project as a composites teaching site for the engineering department. Additionally, matching research and deployment funds from FHWA were identified to support the training aspect of the project.

FLHO gathered input from various composite bridge vendors and academic experts to prepare a proposal to the National Park Service officials who managed the George Washington Parkway, along which the bridge would be built. Approval by this customer agency was necessary before the project could go forward.

Items of particular interest to FLHO included cost of the structure, criteria for selecting materials, design guidelines, durability, and connections. These items were addressed to the preliminary satisfaction of FLHO engineers through interactive sessions with composite manufacturers, FHWA researchers and other composites experts.

Ultimately, this FLHO demonstration project was put on hold in the fall of 1996 because the appearance of the proposed composite bridge was determined to be architecturally inconsistent with the character of the George Washington Parkway. The restrictions resulting from the preapproval of the basic bridge design and George Washington Parkway landscape architects' preference for stone and wood on all Parkway structures conspired to eliminate composites as a building material for the proposed site. Finally, flooding at the site was also determined to be unacceptably high for purposes of a demonstration project site (Hurricane Fran produced flood waters at least 4 feet over the bridge deck level).

FLHO engineers determined to continue their search for another demonstration site where they can deploy a composite structure. At the time this paper was submitted, no site had been identified. FLHO continues to gather information on composites and plans to participate in a new composites initiative developed under the FHWA Headquarters Bridge Office.

CONCLUSIONS

These case studies illustrate that opportunities to deploy composite structures continue to expand. Increased awareness of composites has resulted in approval from various engineering organizations to experiment with the new materials. Certain common elements from these composite applications case studies offer lessons to firms and organizations marketing composites for bridge applications.

1) Bridge engineers respond best to their own customers

Marketing efforts need to target end users/owners of composite structures in addition to bridge engineers. Both USFS engineers and the FLHO engineers were sensitive to their customers' interest in composite bridges. Without such interest, it is doubtful that any number of vendor calls would have resulted in any demonstration projects or test programs.

This lesson is particularly relevant in determining how to best approach state highway departments. Local interest in composites technology would go a long way to support present industry interactions at the state engineering department levels.

2) Composite suppliers must equip agency engineers to work with composite materials (e.g. application focused design guides, software)

USFS engineers expressed concern about their ability to maintain composite bridges once they were built. What happens three years down the line when an engineer needs to determine if a front-end loader can safely cross a composite pedestrian bridge? Before composite structures will be deployed in larger numbers, engineers must be provided the tools to work with the materials.

Developing design guidelines and validating computer analysis tools is a principal objective of both the USFS and FLHO efforts. The need for such tools comes up repeatedly in every discussion on the application of composites outside the aerospace industry. Technical topics of particular interest include connections/joints, environmental effects on material properties, durability and damage tolerant design, and design of repairs.

3) Low-risk structures present a near-term mechanism to introduce composite materials.

Low-risk is a leading criteria for both the USFS and FLHO composite deployment efforts. With over 60 composite pedestrian bridges in service in the U.S., composite materials are still considered experimental because they are new to many engineering departments that must ultimately approve their use. Thus, conservatism drives the identification of low-risk deployment opportunities when working with new users of composite materials.

Pedestrian bridge structures have emerged as the leading candidate for low-risk composite bridge deployment. Several development and demonstration projects have been initiated in 1995 and 1996, with many more pedestrian projects pending proposal acceptance. These structures represent a potential to create a robust database and an opportunity to address many of the design and materials concerns expressed by engineering offices.

RECOMMENDATIONS

Certain recommendations to the composites industry follow from the composite experimentation efforts underway at the USFS and FLHO. These recommendations could expand the number of efforts to deploy composite structures and ultimately speed the process of introducing composites as generally accepted building materials.

1) Target to a greater extent owners/users of structures

Owners and users of structures exert great influence over designers and engineers in selecting materials and designs. Beyond qualifying composites as a new building material, composite industry marketing efforts need to target to a greater extent owners/users. Without interest from their customers, engineers and designers have less incentive to use composites.

This recommendation applies specifically to public sector transportation construction, where the incentives and paths to technology deployment are not always clear. Public ownership implies a distributed owner/user group, however, persons associated with financing public construction are always key owner/user targets, including state legislatures, transportation comptrollers, state transportation chief administrative officers (CAOs), directors of roadway maintenance, etc.

Fortunately, a number of successful deployments of composite materials are already in place to support a marketing effort targeting public sector owners/users. Also, the benefits of using composite materials are often very consistent with the general interests of public sector owners/users. For example, composite repair of a freeway overpass in Florida is an excellent illustration of how composites can cost-effectively stretch the useful life of existing infrastructure.

Some example themes to use in marketing composite materials to owners/users include:

- Technology innovations can ease budget pressures.
- Composite technologies have been proven, are being used, and are available for immediate deployment to enhance services.
- State and federal programs are available to support deployment of composite technologies.

2) Develop application-specific design guides

Design guides are essential for expanding the use of composite materials in civil construction. Sufficient data now exists for application-specific design guides, particularly for pedestrian structures. The only issue, then, is who will pay for the development and distribution of the design guides.

Engineers and designers in civil construction are accustomed to receiving "free" technical materials support in the form of industry-supplied design guides and through professional committees. The USFS and FLHO efforts to develop pedestrian bridge design guides for composite structures can be considered unique. These projects, therefore, represent a significant opportunity for collaboration between the composite industry and construction engineers.

The composites industry should also consider establishing a mechanism through which to share information about the many experimental pedestrian bridge projects and to help fund the development of a comprehensive design guide. The risk in not participating in projects such as the USFS and FLHO composite deployment efforts is that resulting design guides will be diverse and some may not contain complete information sets. Without access to important composite industry data, different and incomplete design guides could delay the broader use of composites.

BIOGRAPHY

Glen Mandigo is an Associate with GHL Incorporated, a Washington, DC-based government relations and marketing firm specializing in advanced materials. Mr. Mandigo represents clients to the U.S. Congress and the Administration to expand the use of advanced technologies for infrastructure and surface transportation. Mr. Mandigo is also Co-Chair of the Suppliers of Advanced Composite Materials Association Market Expansion Working Group on Civil and Marine Infrastructure. Mr. Mandigo received his B.S. in Aerospace Engineering and was a composites structural engineer with Lockheed Aeronautical Systems Company and Advanced Aerodynamics and Structures, Inc. Mr. Mandigo's Washington experience includes a position with the U.S. House of Representatives Subcommittee on Space and consulting relationships with a number of federal agencies including the National Aeronautics and Space Administration, the U.S. Department of Transportation, and the Transportation Research Board.

Design and Construction of a Lightweight FRP Work Platform for Use at Oyster Creek Nuclear Power Plant

ROY J. WILSON, G. ERIC JOHANSEN, SCOTT RIBBLE AND ADRIAN FOGLE

ABSTRACT

This paper examines the design and fabrication of a motorized fiberglass reinforced plastic (**FRP**) platform span to be used to facilitate the refueling process in a nuclear power plant. The pultruded structural system is mounted on motorized steel trucks and incorporates additional mechanical and electrical subsystems, for use by workers servicing the reactor core.

SITE CONSTRAINTS

The refueling process in a modern nuclear power generating station is a highly complex and expensive undertaking. In addition to the reactor servicing costs, loss of generating power while the reactor is off-line translates into a considerable expense for the utility company. Consequently, any reduction in servicing time during the refueling process can result in substantial savings for the utility.

Typically, the design of a nuclear reactor features a deep, water-filled well that contains the active fuel rods, along with adjacent water-filled wells which hold the spent fuel rods and equipment being used in the refueling process. Most of the refueling work is done from above, using specialized tools to reach deep into the reactor well. Typically, trained personnel use a massive, multilevel steel platform, positioned above the pool, as a staging area from which to work. This platform is mounted on motorized wheels which move it along heavy rails imbedded in a trench in the floor. The steel platform, weighing approximately 21,150 kg, (45,000 pounds), along with an overhead traveling crane, are permanently installed at the plant, and represent the principal material handling equipment used during refueling.

Equipment access to the reactor's upper level refueling floor is limited to a service hatch, which at Oyster Creek measured 6.2 m (20'-3") by 5.5 m (18'-0). Equipment too large to fit through this opening must be broken down into subassemblies, and then re-assembled by personnel that have been trained to work in a potentially radioactive environment. Generally, on site assembly that requires unusual materials such as epoxies, etc., would not be acceptable, as only strictly approved substances are allowed on the reactor floor.

Roy J. Wilson, G. Eric Johansen, Scott Ribble and Adrian Fogle, E. T. Techtonics, Inc., P.O. Box 156, 4006 Butler Pike, Plymouth Meeting, PA 19462.

DESIGN CRITERIA

E. T. Techtonics, Inc. was asked by General Electric Nuclear Energy Division to design and fabricate a smaller FRP version of the existing motorized steel platform, to be used by GE's servicing crews during the refueling for Oyster Creek Nuclear Power Generating Station in New Jersey. The purpose of this platform was not to replace the larger steel structure, but to work directly alongside it in performing the numerous primary and secondary operations required. In doing so, the "feel" of the span needed to be stiff, so as not to induce any uneasiness in workers performing delicate operations while on the platform. Further, our FRP design was expected to compete favorably, both on a cost and performance basis, against existing aluminum platforms, which had already been in use at other plants.

Our platform was expected to satisfy the following basic design criteria:
- Overall length of 13.7 m (45'-0) and decking width of 1.2 m (4'-0).
- Uniform live loading of 100 psf, with a deflection not to exceed L/360.
- Concentrated loading of 1,000 lbs. on a 1 square foot area anywhere on the deck.
- Factor of safety (based on Ultimate Strength of FRP material) of 2.5.
- Complete structural analysis, using STAAD 3, including Category 1 seismic analysis.
- Full scale deflection load test, to confirm results of computer analysis.

In addition to these basic structural requirements, the platform needed to integrate the following design objectives for efficient use by GE's work crews:
- A permanently attached motorized wheel base to allow the platform to roll on the existing steel rail system used by the larger platform.
- A removable auxiliary wheel base with resilient wheels and an independent framing system, capable of allowing the entire span to be manually rolled on the existing concrete floor, while being lifted to a height of 1.2 m (4'-0) off the floor to clear existing handrail obstructions.
- A complete electrical fit-out to provide power to both the motorized wheel base and for 18 utility outlets on multiple circuits for powering equipment used by work crews.
- A fully enclosed deck area to a height of .46 m (1'-6"), to prevent anything accidentally falling from the deck area into the reactor well.
- All type 304 stainless steel fasteners and locknuts, and an outboard tie-off rail for use by workers lowering equipment into the reactor well.

STRUCTURAL SYSTEM

The platform was constructed as a Pratt truss, with which E. T. Techtonics, Inc., has substantial previous experience in designing, based on our lightweight FRP structural system (**LONGSPAN PRESTEK**). To build the truss we used fire-retardant, isophthalic polyester resin standard structural channels and square tubes pultruded by Bedford Reinforced Plastics Inc., Bedford, PA. Channel profiles employed in the design included 203.2 mm (8") in the double beam top and bottom chords and 152.4 mm (6") lateral crosspieces. Truss compression posts and tension diagonals were designed using 50.8 mm (2") tubes. Horizontal bracing (not shown in the photographs) was designed using 50.8 mm (2") tubes and provides lateral support to the lower chord. Ladders, end handrails, and kickplates were designed using standard industry safety yellow, fire-retardant tubes and shapes. Truss connections generally employ 19 mm (3/4") stainless steel bolts and locknuts. The system was designed with a 25.2 mm (1") precamber at midspan to allow for fabrication tolerances and initial dead load. FRP fabrication was carried out by Structural Fiberglass Inc. of Bedford, PA. Design, analysis, shop drawings, and platform assembly was provided by E. T. Techtonics, Philadelphia, PA.

We placed the FRP decking (also supplied by Bedford Plastics, which utilizes an interlocking panel to panel design) to run parallel to the length of the span, supported on crosspieces that were in turn supported by the bottom chord of the truss. This facilitated our ability to ensure a complete seal of the deck surface, using an overlapping arrangement of FRP angles, toeplates, and plastic side panels, to satisfy the client's concern over work related items falling from the platform deck into the reactor pool. (Even loss of an item as small as a washer into the reactor pool is an unacceptable condition.) As a walking surface, the deckboard presented a design problem in that standard slip resistant grit surfacing could not be applied. The client's need for frequent washdown to remove any potential radioactivity during the cleanup process, as well as the concern that over time grit material might become dislodged and fall into the reactor pool, made a grit surface an unacceptable condition. However, as FRP surfaces lacking grit tend to be slippery when wet, it was decided that the plant would apply a slip resistant mat to the smooth deckboard once it was installed, which could be removed later for cleaning.

To further complicate the design, the structural and secondary systems all needed to be spliced into three pre-assembled sections, so that the span sections could be hoisted up into the reactor servicing area through the existing access hatch. This required close attention to the design and assembly of the secondary systems, so that the platform could be reassembled on the plant floor with minimal effort by GE's work crew, using standard hand tools.

MECHANICAL SYSTEM

Providing a motorized wheelbase for the FRP structural system presented a significant design challenge. Not only was it necessary to address the ability of the FRP span to withstand these unusual loading conditions, but numerous site obstructions had to be cleared, requiring fabrication of a specially designed drive system.

In the traveling crane/hoist industry, the flanged wheel systems which move the system are called "trucks". These trucks typically travel on overhead rail systems, and are controlled by a hanging push-button "pendant" with directional controls. This keeps all the moving parts of the system off the plant floor and clear of any obstructions. In most small size truck systems of the type we used, the drive system utilizes a direct drive motor reducer, mounted horizontally off the end of the truck.

Our design and the site obstructions required us to redesign the drive system to utilize a chain and sprocket power transfer from the motor reducer (mounted on top of the trucks) to the wheels. We were able to use a relatively lightweight steel truck assembly, as the dead load of the FRP platform was relatively low, as was the concentrated load, compared to a typical overhead crane system. The control pendant, mounted on a stainless traveller wire attached to the top chord of the truss, allows the operator to operate the span from various work locations. An electronic "soft start" device dampens the initial sensation of motion during repositioning of the platform. The platform moves at a speed 50 fpm. (Integrating the electrical wiring for the "soft start" control into the wiring harness encountered some difficulty and confusion. Soft start control is not a standard item in industrial overhead crane design, and required an electrician familiar with the use of such a device.)

The wheelbase of the truck system we used was shorter than that recommended in standards developed by the Crane Manufacturer's Association (CMA). Those standards determine wheelbase as a function of overall span length to prevent cribbing or racking of the crane wheels on the rail system. However, the client needed a minimum wheelbase to allow the FRP platform to be brought in as close as possible to the steel platform, to facilitate refueling work. We used a dual motor system (one motor reducer mounted on each truck) to minimize any tendency for cribbing. Safety items incorporated on the truck system include disc brakes, which remain in locked position except when power is supplied. To prevent overturning during a seismic event, steel rail grabbers were provided which enclose the rail head and prevent the truck from being lifted or thrown off the rail.

Our auxiliary truck system (see photo # 2), was designed to mount independently of the motorized truck system, to allow the platform to be used to service an adjacent equipment pool which is not connected to the rail system. These auxiliary trucks were completely pre-assembled and could be quickly installed by hoisting the platform off the rails (using the plant's overhead crane) and slipping the auxiliary trucks posts and kneebrace between the lower chord double beams. As before, this assembly is attached to the platform via 19 mm (3/4") bolts. Once attached to the auxiliary trucks, the platform is at a height of 1.2 m (4'-0) off the floor, and can be rolled laterally over the floor as required while clearing existing safety handrails. The platform while mounted on the auxiliary trucks is not seismically qualified, as it is not used over the reactor vessel. This truck system has .3048 m (12") diameter wheels and manually operated floor locks to prevent rolling once positioned.

TESTING

Computer analysis of the FRP system indicated the span met all the required structural and seismic objectives. After fabrication of the span, deflection was successfully tested by applying a partial load to the span (55 psf concrete block) and measuring deflection at midspan. Full operation of the motorized truck system was tested using a gas-powered generator to supply 460/3 phase/60 current to the motors. Lateral bracing of the FRP span was tested by creating a locked wheel condition on one end of the span, and applying motor power to the other end of the span, inducing possible twist into the structure. Finally, the structure was tested for manual rolling with the platform attached to the auxiliary trucks at a height of 1.2 m (4'-0) off the floor. It was found that one man at midspan or two persons at the ends, could easily roll the structure across a flat concrete floor. The estimated total weight of the span was 2,268 kg (5,000 pounds).

CONCLUSIONS

The span has performed very well as designed. During shop testing, the span exhibited excellent vertical stiffness at midspan under multiple pedestrian loads. Lateral stiffness of the

FRP structure under moving load, a primary concern during the design process, was found to be very good. The complete platform was light enough for crews to easily hoist the subassemblies through the equipment hatch for splicing, and light enough to allow manual positioning of the platform both on the plant floor and on the steel rails. Indeed, the light weight, ease of disassembly, and easy to clean surface of the FRP platform would potentially allow it to be transported to other power plants for use in the infrequent refueling process, although there are presently no plans to do this. Whether this is possible would first depend on whether the structure becomes contaminated during use, and the degree to which any such contamination could be successfully removed to allow the platform to be taken out of the containment area.

We were unable to view the platform in use during the actual refueling process, due to strict regulations in effect when the reactor containment is open. However, GE has been sufficiently pleased with its performance to be considering use of fiberglass spans in other plants, including several requiring an even more stringent seismic analysis. The only unexpected use condition noted by work crews was their report of experiencing occasional static electricity charges while walking on the span. As we have never experienced such a phenomenon while walking on FRP decking, this was attributed to possible interaction between workers protective footgear and the non-slip decking material applied to the deck by the plant. However, this phenomenon, if attributable to the smooth FRP decking surface, might bear further investigation if platform or catwalk spans are to be used in potentially explosive environments such as grain elevators. The long-term effect of a potentially radioactive environment on the stability of resins used in the pultrusion is presently unknown, due to lack of sufficient research data in this area. Therefore, this span is classified as a "commercial grade" structure. We expect the platform to be serviceable for the life of the plant.

REFERENCES

Johansen, G. Eric, Roll, Dr. Frederick, Wilson, Roy J, 1993, "Creep and Relaxation of PRESTEK Structural Systems", ANTEC '93, New Orleans, LA..

Johansen, G. Eric, Roll, Dr. Frederick, Wilson, Roy J, 1994, "Spanning Staircase Rapids with a Prestressed RP Truss Structural System", 4th Int. Conf. on Short & Medium Span Bridges, Halifax, Nova Scotia, Canada..

Johansen, G. Eric, Roll, Dr. Frederick, Wilson, Roy J, 1995, "Strength/Stiffness Characteristics of a Prestressed RP Truss Structural System", SPI Composites Institute 50th Annual Conference, Cincinnati, OH.

Johansen, G. Eric, Roll, Dr. Frederick, Wilson, Roy J, Gaudini, P. Garrett, 1996, "Design and Construction of Two FRP Pedestrian Bridges in Haleakala National Park, Maui, Hawaii", Advanced Composite Materials in Bridges and Structures, 2nd International Conference, Montreal, Quebec, Canada.

Photo #1: FRP platform mounted on motorized trucks.

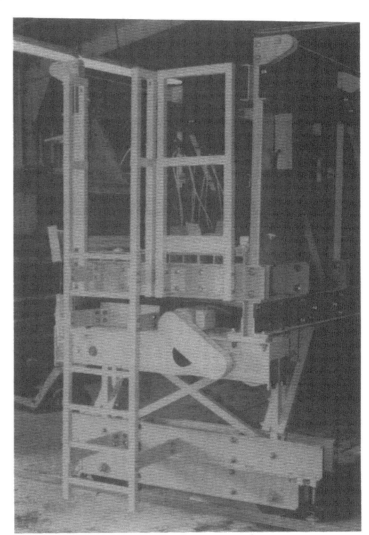

Photo #2: FRP platform mounted on auxiliary manually operated trucks.

Fiber Reinforced Composite Decks for Infrastructure Renewal—Results and Issues

VISTASP M. KARBHARI, FRIEDER SEIBLE,
GILBERT A. HEGEMIER AND LEI ZHAO

ABSTRACT

Based on their high strength- and stiffness-to-weight ratios, corrosion resistance, environmental durability and inherent tailorability, fiber reinforced polymer composites are being increasingly considered for use in infrastructure renewal. One area of potential application of these materials is in the fabrication of light weight bridge decks that can be deployed in uses ranging from the replacement of deteriorating decks, to the erection of completely new superstructure. This paper discusses test results from an ongoing investigation related to the use of composite bridge deck components. Issues related to manufacturing, overall durability, and joining are also discussed.

INTRODUCTION

Over the past decade, the issue of deteriorating infrastructure and lifelines has become a topic of critical importance in the United States, and to an equal extent in Europe. The deterioration of decks, superstructure elements and columns can be traced to reasons ranging from aging and environmentally induced degradation to poor initial construction and lack of maintenance. Although the aspect of seismic retrofit of columns has received the most emphasis over the past few years, the issue related to the need for renewal of bridge decks is also critical. It is estimated that most bridges in the US on an average last 68 years, whereas their decks last 35 years, and that "during the 1990's, 40% of the total highway deck area in the US will become 35 years old, statistically ready for replacement" (Bettigole, 1990). Some estimates from the mid west and from regions where there is extensive use of road salt, put the life of a conventional bridge deck at about 10 years, thereby requiring extensive triage, or expensive replacement within a short time period. States such as Wisconsin report that with the exclusion of painting, bridge deck repair and replacement accounts for between 75 and 90% of annual maintenance costs associated with the structure (BIRL, 1995). Added to the problems of deterioration are the issues related to the need for higher load ratings (HS15 to HS20, for example) and increased number of lanes to accommodate the ever increasing traffic flow on the major arteries. Beyond the costs and visible consequences associated with continuous retrofit and repair of such structural components, are the real consequences related to losses in productivity and overall economies related to time and resources caused by delays and detours.

The high strength-to-weight and stiffness-to-weight ratios, corrosion and fatigue

Vistasp M. Karbhari, Frieder Seible, Gilbert A. Hegemier and Lei Zhao; Division of Structural Engineering; University of California, San Diego; 9500 Gilman Drive; La Jolla, CA 92093-0085

resistance of fiber reinforced composites, in addition to their tailorability makes them attractive for use in replacement bridge decks or in new bridge systems. Besides the potentially lower overall life-cycle costs (due to decreased maintenance requirements), such decks would be significantly lighter, thereby affecting savings in substructure costs, enabling the use of higher live load levels in the cases of replacement decks, and bringing forth the potential of longer unsupported spans and enhanced seismic resistance. There has been considerable activity over the past decade in the area of composite reinforcement for concrete bridges with the reinforcement ranging from composite rebar and grids to composite cables for external and internal post-tensioning. The current application, however, emphasizes the use of fiber reinforced composites for the entire deck or as part of the actual superstructure system itself. Such a concept, in general, is not new, having been used previously in the Miyun bridge in China and the Aberfeldy Footbridge in Scotland, among others. The focus of this paper will be on developments in the area of replacement bridge decks capable of being placed on pre-existing concrete and steel girders, as well as being used in new bridge systems.

DEVELOPMENT APPROACH

The overall test program was undertaken at the University of California, San Diego (UCSD) under funding from the Federal Highway Administration (FHWA) and the Advanced Research Projects Agency (ARPA) as part of a University-Industry Consortium, and was aimed at the development of lightweight, degradation resistant fiber reinforced composite decks for primary use in replacement. The overall criteria that were used to guide the development of these decks included: (i) the development of stiffnesses through the appropriate use of face sheets and internal core configurations that would fall in the range between the uncracked and cracked stiffness of existing reinforced concrete decks, (ii) the development of equivalent energy levels at acceptable displacement levels as a means of building in a safety factor due to the elastic behavior of composite sections, and (iii) the development of processing methods that would be cost-effective and which ensured repeatability and uniformity.

CONFIGURATION	COMPONENT SCALE			
	3' - 4'	6' - 8'	14'	15' x 7.5' Panels
Balsa Core (Dupont)	●			
Foam Filled Boxes (Dupont)	●		●	
Foam Filled Truss (Dupont)	●	●	●	●
Foam Filled Hat Sections (Dupont)			●	●
Pultruded Profiles With Face Sheets (Lockheed-Martin)	●	●	●	
Hybrids (Lockheed-Martin)				●
Corrugated Core (Core-Kraft)		●		
"Egg Crate" Core (Northrop-Grumman)			●	

Figure 1: Matrix of Test Specimens

A building block approach was used in this program with tests being conducted at the subcomponent, component and field-size levels (Figure 1). Components were fabricated by a number of industrial partners (as listed in Figure 1) using Wet-Layup, Pultrusion and Resin-Infusion processes for manufacturing. Deck specimens fabricated using the pultrusion process incorporated pultruded profiles (hollow boxes and I sections) with face sheets that were fabricated using the wet layup/sprayup process. Two different versions of resin infusion were used and in each case foam cores were used as tooling aids rather than for structural purposes. In all cases except one, the decks were fabricated using E-glass fibers in nonwoven fabric form using either a vinylester of polyester resin system. The depth of the specimens was restricted to nine inches so as to enable direct replacement with existing concrete decks.

TEST RESULTS

The overall enveloping load-displacement profiles of a number of subcomponents are shown in Figure 2. Each of these used a different core configuration in order to assess the effectiveness of various strategies. The truss core configuration consisted of triangular foam cores wrapped with fabric, which were assembled within face sheets to form the structure. In order to assess the effectiveness of additional reinforcement at the nodes formed by the adjacent triangles, the second configuration used a woven fabric insert that strengthened the node area. The effect of this on the overall ductility and performance of the system is clearly apparent, as response appears to shift from linear elastic to "pseudo-ductile" with the ductility being afforded by matrix cracking and gradual separation between fabric layers at internal nodes and along the inclined webs.

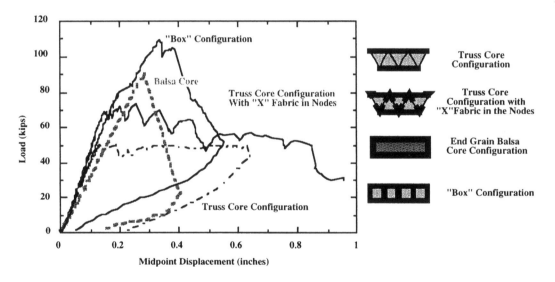

Figure 2: Load-displacement response of subcomponents tested in three-point bend mode

Tests were also conducted on slightly larger specimens (truss core type using the Resin Infusion processing method and corrugated core type using the wet layup process) to ascertain repeatability and uniformity, both of which were found to be satisfactory. Larger component level tests were conducted on 14 ft long beams using a test setup that consisted of two high strength concrete abutments that were set at 8 ft centers in a hydrostone bed and post-tensioned to the test floor. Continuity was provided on the cantilever end of the beam outside the supported span through the use of a structural steel yoke placed at the theoretical point of inflection of the adjacent span. The yoke was tied through the test floor. Each specimen was extensively strain gaged and deflection profiles were monitored using linear potentiometers.

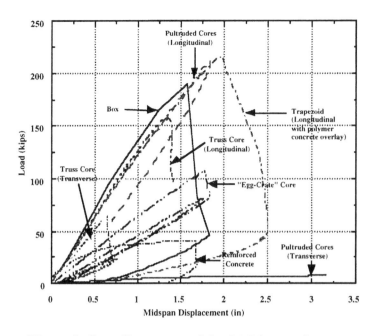

Figure 3: Overall response of the 14 ft beam elements

The fiber reinforced composite deck specimens show failure loads far in excess of that shown by the reinforced concrete specimen with initial structural stiffnesses that are comparable. The "box" configuration and the "trapezoid" configuration also show significantly enhanced energy levels as can be seen from Figure 3. Obviously further optimization is necessary in order to achieve the appropriate performance levels at the minimum cost. Costs are directly related to the thicknesses of composite used in the skins and in the webs formed with the core configuration. A concern often voiced about the use of fiber reinforced composites for primary structure is that related to brittle behavior and catastrophic failure. In most cases, the components showed considerable load capacity and continued to carry load, albeit at significantly lower levels, even after substantial cracking and fracture were seen. The strain profiles recorded at the top and bottom surfaces of the box type component are depicted in Figures 4 a and b respectively. As can be seen in Figure 5, for a specimen with a trapezoidal core configuration with the cores spanning the length of the component, displacement increases are gradual and predictable, even at very high load levels.

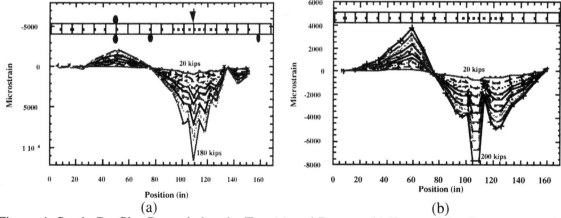

(a)   (b)

Figure 4: Strain Profiles Recorded at the Top (a) and Bottom (b) Faces of the Component with the "Box" Configuration Depicted in Figure 3

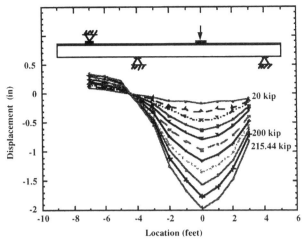

Figure 5: Displacement Profile for a Trapezoidal Core Component (at 20 kip load intervals)

Field scale deck panels of size 15' x 7.5' x 9" were also tested for a limited number of concepts. These panels were fabricated in pairs, one each to be tested in the laboratory for assessment of stiffness and load-deflection response, and the other to be placed in a road test-bed for monitoring. Deck panels were fabricated following preset design criteria based on the achievement of a stiffness level between that of a cracked and uncracked concrete deck, and with the appropriate energy level to ensure safety. A panel using a glass/carbon hybrid construction was also tested, with and without a 0.75" polymer concrete overlay. All panels were approximately 8.25" thick, leaving 0.75" for the polymer concrete overlay. Panels were tested in the laboratory under simply supported end conditions with loads being introduced into the deck through two pads 6' apart, and a spreader beam to approximate AASHTO HS loading conditions. Test results are shown in Figure 6.

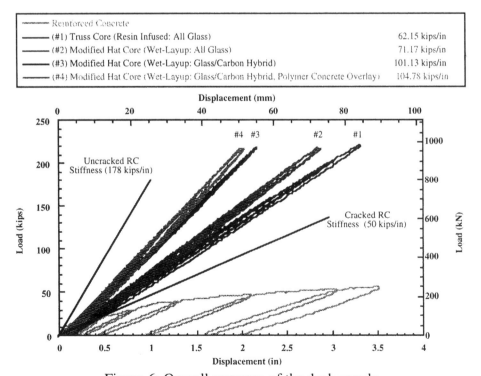

Figure 6: Overall response of the deck panels

In each case, the composite decks were tested to a load of about 220 kips after which the test was terminated. It should be noted that the composite decks were 3 to 4 times lighter than the reinforced concrete deck panel which weighed about 12,500 lbs.

4 deck panels (3 composite and one reinforced concrete) have been placed in a test site at UCSD in the road bed simply supported between two abutments 14 feet apart on centers. Deck-to-deck connections were made through the use of shear keys. The site is on a heavily traveled road and hence affords the opportunity for field monitoring. The decks are instrumented with strain gages and linear potentiometers and response is monitored at regular intervals. It is envisioned that monitoring will provide critical information on the actual behavior of these decks under actual traffic conditions.

ISSUES AND FUTURE DIRECTIONS

The tests conducted to date have shown that fiber reinforced composite decks can be fabricated using a variety of processes with performance levels based on quasi-static testing that are acceptable for use in replacement strategies. A number of issues still need to be investigated before the actual implementation of such decks in the field. Some of these issues relate to response under dynamic loads, long-term durability of materials, connections between the decks and supporting girders, and connections between the decks and barriers and side rails. Due to the plethora of choices possible in fiber and resin types, fabric architectures and processing methods, it is critical that specifications be developed based on performance standards, not just at the materials level, but also at the component and systems level. These standards must necessarily include recommendations on partial safety factors to accommodate variations due to differences in raw materials, quality achievable through different manufacturing processes, durability based on material choices and bridge environment, structural response and damage. It is highly probable that such deck systems will interact with pre-existing or new structural components fabricated from conventional materials and it is critical that interface issues as related to coefficients of thermal expansion, residual stresses and moduli differences are resolved without adversely affecting overall systems response. Issues such as cost-effectiveness and overall manufacturability must be addressed in the near future if such schemes are to compete with conventional strategies. It must be noted that overall cost-effectiveness is intrinsically tied to the degree of efficiency in materials conversion through the choice of a manufacturing process or processes. It is conceivable that future directions will emphasize the use of hybrid construction, encompassing not just hybrids of different reinforcing fibers, but also hybrids combining fiber reinforced composites with conventional materials such as steel and concrete. Future development in deck systems will have to consider a systems level approach based on the specifics of the application under consideration.

ACKNOWLEDGMENTS

The support of ARPA, Caltrans and FHWA is gratefully acknowledged.

REFERENCES

Bettigole, N.H. (1990), "Replacing Bridge Decks," *Civil Engineering*, 9, pp. 76-77.

BIRL (1995), "Definition of Infrastructure Specific Markets for Composite Materials," *BIRL Report.*

The Carbon Shell System for Modular Short and Medium Span Bridges

FRIEDER SEIBLE, GILBERT HEGEMIER, VISTASP, KARBHARI,
RIGOBERTO BURGUEÑO AND ANDREW DAVOL

ABSTRACT

Advanced composite materials, originally developed for aerospace and defense applications, have become of great interest to the civil engineering and construction industry, because of their unique characteristics related to mechanical response and environmental durability. This paper summarizes the recent research developments of a new structural concept for the design of new short and medium span bridge systems. The concept consists of a framing system of premanufactured carbon shells/tubes filled on-site fully or partially with concrete, where the carbon shells have the dual function of stay-in-place formwork and reinforcement for the structural elements. The paper will discuss the development of this modular carbon shell bridge concept documented by analytical and experimental research results, and will discuss the opportunities and limitations for this new construction system.

INTRODUCTION

At the University of California, San Diego (UCSD), an advanced composite Carbon Shell System (CSS) concept for new structural systems is being developed under an ARPA (Advanced Research Projects Agency) TRP (Technology Reinvestment Project) program and a FHWA (Federal Highway Administration) project (Seible, et al., 1995b).

The Carbon Shell System emerges from the unique combination of conventional civil construction materials and new polymer matrix composites. A premanufactured advanced composite carbon tube is filled fully or partially with concrete depending on the strength, stiffness, and stability requirements for the structural component. The carbon tubes are manufactured by the wet filament winding procedure, which is a proven, reliable and cost effective process for the construction of high quality advanced composite structural components. In this new concept, the concrete provides the compression force transfer and the carbon shell

---

Frieder Seible, Professor of Structural Engineering
Gilbert Hegemier, Professor of Applied Mechanics
Vistasp Karbhari, Assistant Professor of Structural Engineering
Rigoberto Burgueño, and Andrew Davol, Graduate Research Assistants
Division of Structural Engineering, University of California, San Diego, La Jolla, California 92093-0085

the functions of formwork for the concrete, reinforcement for the tension force transfer in bending and shear, and, most importantly, a high level of confinement of the concrete core. This system combination takes advantage of the excellent tailorable mechanical characteristics of advanced composite materials and the full utilization of the compressive strength of the well confined concrete core (Seible et al., 1995a).

The advantages offered by this system can only be fully utilized when appropriate connection systems and concepts can be developed and proven. Due to the linear elastic nature of the carbon fiber reinforced composite, the focus is on the development of a design philosophy addressing both ductility and strength connection principles. The research program aims therefore to systematically develop the basis of the CSS concept for new structural components made of filament wound carbon shells and filled on-site with concrete. The research focuses on three areas namely, (1) the analytical modeling and characterization of the concrete filled carbon shell system and the proposed connections, (2) the development of appropriate design models, and (3) the experimental large scale validation of components and assemblages to fully characterize the structural response.

STRUCTURAL CHARACTERIZATION OF THE CARBON SHELL SYSTEM

The analytical modeling and characterization of the concrete filled carbon shell system and the proposed connections, together with the development of appropriate design models are two of the primary research objectives. A number of assumptions were made regarding the interaction between the longitudinal flexural stress, the circumferential hoop tension stress resulting from the confinement of the concrete, and the shear stress carried by the shell. This enables a simple but rational approach to determine analytical models to characterize the response of the system, which with corresponding experimental investigation will provide the basis for further more rigorous analytical modeling. Design and assessment models developed at UCSD for retrofit measures of existing concrete columns constitute the basis for the analytical modeling. Most of these analytical tools are considered applicable to the carbon shell system, with simple modifications being proposed where appropriate (Seible et al., 1995a).

For analysis purposes the equivalent properties of the laminated shell are used to describe the behavior of the system through the use of classical lamination theory. A biaxial stress state is assumed to exist in the shell taking into account the interaction between flexure, shear and confinement effects considering full composite action between the shell and the concrete core. The concrete model employed neglects the concrete tensile carrying capability and considers the compression response to be dependent on the confining pressure. The carbon shell works to provide confining pressure by limiting the expansion of the concrete core. A confinement model which accounts for the passive nature of the confining action due to the carbon shell is used.

Pilot lateral load tests on two concrete filled circular carbon shell bridge columns have demonstrated the potential of this innovative application for the development of new framing systems (Burgueño, et al., in progress). A ductile design concept, shown in Figure 1a, consisting in simple starter bar connections penetrating from the column footing and encased with a concrete filled carbon shell has resulted in a very ductile overall system as seen in Figure 2a. The system displayed large inelastic deformation capacities in a well confined plastic hinge matching almost exactly the response of a conventionally reinforced concrete column reference test. Furthermore, it was verified that only short development lengths of these splice bars are required due to the high confinement from the circular carbon tube. Thus, ductile connection details utilizing mild reinforcement splice bars in the concrete core across the joint can be envisaged. A strength design concept, shown in Figure 1b, relying on the capacities of the carbon shell system without the use of steel for anchorage was successfully tested displaying a linear elastic response up to failure as seen in Figure 2b. Connection design concepts in the form of premanufactured threaded and/or bonded inserts for the design of structural members where either simple strength design is used, or where parts of the structure are to be protected from inelastic action can thus be conceived. The experimental evaluation of a third pilot test column consisting of an adhesively bonded inner sleeve joint, shown in Figure 1c, is currently in progress. Even though the joint concept is applied here for column members, it can be envisioned that the test unit represents one half of a symmetrical longitudinal joint detail for both column and beam elements.

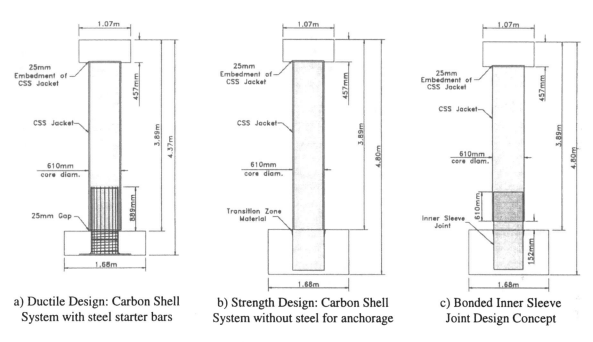

a) Ductile Design: Carbon Shell System with steel starter bars

b) Strength Design: Carbon Shell System without steel for anchorage

c) Bonded Inner Sleeve Joint Design Concept

Figure 1. Carbon Shell System Pilot Test Units

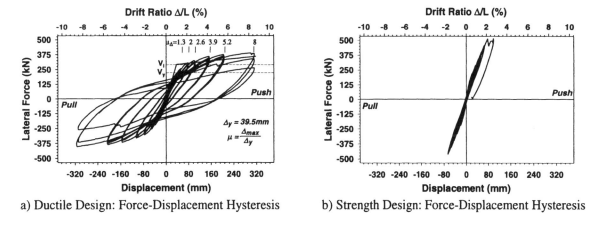

a) Ductile Design: Force-Displacement Hysteresis

b) Strength Design: Force-Displacement Hysteresis

Figure 2. Force-Displacement Response of Carbon Shell Systems

MODULAR CARBON SHELL BRIDGE SYSTEMS

The concept of concrete filled carbon shells has been extended beyond bridge columns to complete structural systems for modular short to medium span bridge systems. It is envisaged that complete bridge systems can be composed of linear segments of carbon shell tubes assembled together by means of longitudinal and off-angle joints with appropriate mechanisms depending on the response requirements for the structural component. With the development of the previously mentioned analytical tools and proper design details, the concrete filled carbon shell system can provide a viable design alternative for new bridge structures. A brief description of the development of complete bridge systems based on the carbon shell concept follows.

The concept of a concrete filled advanced composite carbon shell system was applied to the conceptual design of the *CERF, Columbus Indiana Bridge* (Seible et al., 1995c, 1995d). Two solutions were proposed for the full clearance of the 61 m span bridge without any intermediate piers. Figure 3a depicts a solution consisting of a three dimensional space truss supporting a precast prestressed concrete deck. The space truss system is composed of a lower chord suspended from the deck system by means of two inclined Warren-type trusses. An alternate concept for the same 61 m span bridge site in the form of a dual tied arch bridge is shown in Figure 3b. This bridge system consists of free standing twin tied-arches. The tie beams are provided by postensioned concrete filled carbon tubes. The bridge deck system is composed of transverse beams connected to the tie beams in a link-and-log system, and supported from both arches by means of inclined hangers.

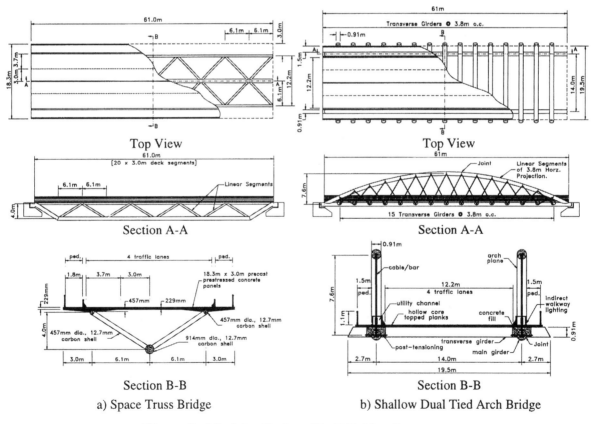

Figure 3. Modular Carbon Shell Bridge Systems

Figure 4. Carbon Shell Cable-Stayed Bridge Concept

The design concept has been extended to the 137 m long 4-lane cable stayed *I-5/Gilman* vehicular traffic bridge, shown in Figure 4, based on the concrete filled carbon shell concept for the two edge girders which provide well confined anchorages and compression force transfer for the inclined cable stays. The deck concept consists of 230 mm deep E-glass panels spanning longitudinally 3.8 m over transverse carbon tubes (Seible et al., 1996).

The concrete filled carbon shell system can also be applied to modular short span bridge systems such as the *Kings Stormwater Channel* bridge, shown in Figure 5. The bridge system consists of a continuous two span structure over an intermediate pier. The spaced beam-and-slab superstructure system is composed of longitudinal carbon shell girders connected across their tops with a thin continuous advanced composite deck. Preliminary parametric studies have indicated that different span lengths can be accommodated through this concept by using modular carbon shells with different diameters and wall thicknesses. The typical span to depth ratios for simpy supported concrete filled carbon shell bridge systems (as shown in Figure 5) range from 17 to 20 including the depth of the 150 mm bridge deck. Longer spans and higher span to depth ratios can be achieved with postensioning for the concrete filled carbon tubes as well as continuous girder action.

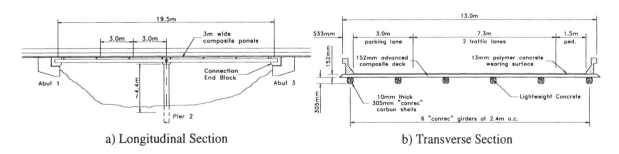

a) Longitudinal Section   b) Transverse Section

Figure 5. Modular Short Span Carbon Shell Bridge

CONCLUSIONS

Quickly expanding research and application of advanced composite materials in civil engineering construction world-wide, demonstrate both the need for new construction materials, and the technical advantages provided by these synthetic fibers and appropriate matrix systems in conjunction with conventional structural materials

The carbon shell system represents a new concept which combines the superior mechanical characteristics of directional strength in tension in the direction of the composite fibers with the dominant characteristics of concrete in compression, and steel for inelastic deformation capacity. With the development and validation of strength and ductile design approaches together with the development of appropriate joint details, the carbon shell system has demonstrated the potential to become a viable framing system alternative. Furthermore, through several examples it has been shown that the concept constitutes a promising new construction system for short and medium span bridges.

The demonstrated technical advantages to-date clearly show that advanced composites will be applied in future civil engineering projects. The extent of these applications will depend on (1) the resolution of outstanding issues such as repairability, fire, durability and environmental concerns, (2) the extent to which automation in the manufacturing process can reduce cost, (3) the availability of validated codes, standards and guidelines which can be used as design references and tools by the civil engineering community, and (4) the degree of quality control and quality assurance can be developed and provided during the manufacturing/installation phase utilizing unskilled general construction labor.

REFERENCES

Burgueño, R., Seible, F., and Hegemier, G., in progress. "Concrete Filled Carbon Shell Bridge Piers Under Simulated Seismic Loads - Experimental Studies," Advanced Composites Technology Transfer Consortium, Report No. ACTT-95/12, University of California, San Diego, La Jolla, California.

Seible, F., Burgueño, R., Abdallah, M.G., Nuismer, R. 1995a. "Advanced Composite Carbon Shell Systems for Bridge Columns under Seismic Loads", *Proceedings from the National Seismic Conference on Bridges and Highways*, San Diego, California.

Seible, F., Hegemier, G.A., and Karbhari, V. 1995b. "Advanced Composites for Bridge Infrastructure Renewal," *Fourth International Bridge Engineering Conference Volume I*, San Francisco, California.

Seible, F., Hegemier, G.A., Karbhari, V., Burgueño, R., and Davol, A. 1995c. "Shallow Dual Tied Carbon Shell Arch Bridge," *CERF 1996 Innovation Awards Program*, University of California, San Diego, La Jolla, California.

Seible, F., Hegemier, G.A., Karbhari, V., Davol, A., and Burgueño, R. 1995d. "Carbon Shell Space Truss Bridge," *CERF 1996 Innovation Awards Program, Innovative Concepts Award*, University of California, San Diego, La Jolla, California.

Seibe, F., Nagy, G., Lockheed Martin Corporation, Trans-Science Corporation, and University of California, San Diego. 1996. "Design Study of an All Advanced Composite Cable Stayed Bridge - I-5/Gilman Bridge in San Diego, Vol. IV - Design Alternates," Advanced Composites Technology Transfer Consortium, Report No. ACTT-96/06, University of California, San Diego, La Jolla, California.

Design and Evaluation of a Modular FRP Bridge Deck

ROBERTO LOPEZ-ANIDO,[1] HOTA V. S. GANGARAO,[2]
VENKATA VEDAM[3] AND NIKKI OVERBY[3]

ABSTRACT

This article presents the characterization of a modular fiber reinforced polymer (FRP) composite bridge deck that was designed with hexagonal cross-sectional shape and multi-axial fiber architecture. The FRP H-deck was modeled and designed using an engineering approach. Static and fatigue test results of FRP H-decks are reported. A finite element model of the FRP deck is correlated with experimental data. The design limit states of FRP decks for highway bridges are presented. The design of FRP H-decks is controlled by deflection limit states.

INTRODUCTION

The FRP deck cross-section and fiber architecture were designed for use in highway bridges. A cross-section made of full-depth hexagons and half-depth trapezoids has shown enhanced performance for bridge deck loads. The fiber architecture is made of E-glass multi-axial stitched fabrics with binderless chopped strand mats. These fabrics were manufactured by Brunswick Technologies Inc. (BTI). The flanges and web elements have fiber reinforcements tailored to provide optimal structural performance. The matrix is a vinylester resin with good weatherability and resistance to harsh environments. The resulting product is called H-deck. The H-deck modules are placed transversely to the traffic direction and are supported by longitudinal beams. The supporting beams (steel or FRP wide flange sections) can be spaced up to 9 ft. (2.74 m) apart. The H-deck system was designed to meet AASHTO (1994) highway bridge design requirements. H-deck test prototypes were fabricated using Vacuum Assisted Resin Transfer Molding (VARTM) by HardCore-DuPont, L.L.C. Some aspects of this fabrication process are patented as SCRIMP™ (Seeman Composites Resin Infusion Molding Process). The FRP deck is evaluated for stiffness, strength and fatigue resistance. The objective of this work is to characterize through experimental evaluations and numerical analysis a modular FRP deck system, and to present a design approach for using the deck in highway bridges.

---

[1] Research Assistant Professor, Department of Civil and Environmental Engineering (CEE), and Constructed Facilities Center (CFC), West Virginia University (WVU), Morgantown, WV 26506-6103.
[2] Professor, CEE, and Director of CFC, WVU.
[3] Graduate Research Assistant, CEE, CFC, WVU.

DESCRIPTION OF H-DECK

The cross-section dimensions of the H-deck are shown in Figure 1, where the depth was constrained to 8 in (203 mm) to utilize the modular system for concrete deck replacement.

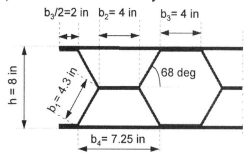

Figure 1. CROSS-SECTION DIMENSIONS OF H-DECK

The fiber architecture made of bi-, tri- and quadri-axial fabrics was tailored to develop fiber continuity between the deck elements (web and flanges), and to provide adequate fiber reinforcement along main stress paths (See Figure 2).

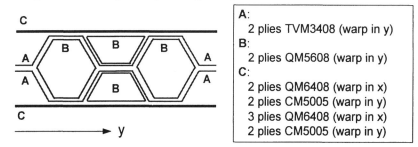

Figure 2. FIBER ARCHITECTURE OF H-DECK

MODELING OF FRP DECKS

A material property pre-processor was used to predict the strength and stiffness properties of deck elements. For each multi-axial fabrics listed in Figure 2, laminated plates were fabricated by VARTM. Stiffness and strength properties for these plates were tested at two different labs, and the average values are utilized in this work. Alternatively, the mechanical properties of a unidirectional composite layer can be evaluated based on properties of the constituent materials (fibers and resins) by utilizing micro-mechanics models. For flange and web elements, the mechanical properties of the composite laminate (e.g., bending stiffness, and elastic moduli) are evaluated based on macro-mechanics, i.e., lamination theory. Finally, the cross-sectional stiffness coefficients of the deck are computed. For example, the major orthotropic bending stiffness results in $D_{xx} = 7.45 \cdot 10^4 \, kip \cdot in = 8.42 \cdot 10^3 \, kN \cdot m$. The computation of properties for the H-deck was implemented in spread-sheet format.

DESIGN LIMIT STATES

Highway bridges designed with FRP H-deck modules need to meet or exceed the AASHTO (1984) design criteria during their service life. The basic design relationship is:

$$\left(\text{Load Effects} \times \text{Load Factor}\right) \leq \left(\text{Nominal Strength} \times \text{Resistance Factor}\right) \qquad (1)$$

The load effects (deformations, stresses or stress resultants) and the corresponding load factors are determined based on the AASHTO (1984) design specifications for highway bridges. Load factors need to be considered for dead load, vehicular load, earthquake, wind, creep, and temperature gradient, among others. Several load combinations shall be satisfied for strength, extreme event, serviceability and fatigue limit states. The vehicular live load consists of a standard design truck (or tandem) or a design lane load. In addition, a dynamic load allowance, IM, is applied as an increment to the static wheel load to account for wheel load impact from moving vehicles.

The live load utilized in this article is based on the AASHTO standard HS20 design truck. West Virginia Division of Highways specified an HS25 design truck that is obtained by multiplying the standard HS20 truck load by a factor 1.25. The resulting HS25 wheel load is 20000 lb. (89.0 kN). In the test specimens, an equivalent wheel load is applied on a rectangular patch (See Figure 3).

The nominal strength in Eq. (1) is based on the deck dimensions as shown on the plans and on permissible stresses, strains, deformations or specified strength of the FRP materials. The nominal strength is reduced by statistically-based structural and material factors. Structural resistance factors are utilized for bending, shear, bearing, tension and compression. Material resistance factors are introduced to account for manufacturing effects and for the long-term properties of composites. Main manufacturing material factors account for process repeatability, tooling quality, dimensional tolerances, curing control, void content, and fiber misalignment. Main long-term performance material factors account for physical and chemical aging, moisture absorption, UV exposure, and freeze-thaw effects.

EXPERIMENTAL SET-UP

H-deck specimens, 120 x 45 in (3.05 x 1.14 m) fabricated by VARTM, were tested under static and fatigue loads along the major stiffness direction. Two types of loading were applied: (1) transverse line load with a width of 11.5 in (292 mm) to establish longitudinal stiffness properties under cylindrical bending; and (2) rectangular patch load, 10 x 20 in. (250 x 500 mm), with the larger dimension perpendicular to traffic flow to simulate the action of a wheel load (See Figure 3).

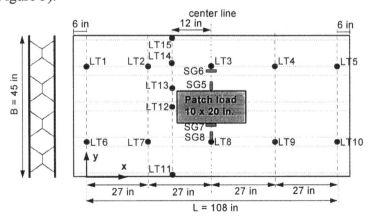

Figure 3. LOCATION OF LVDTs AND STRAIN GAGES ON TOP FACE OF DECK

Strain gages were placed on the outer faces of top and bottom flanges in both longitudinal and transverse directions. LVDTs (Linear Variable Displacement Transducer) were mounted on the deck top and bottom flanges to measure vertical deflections (See Figures 3 and 4).

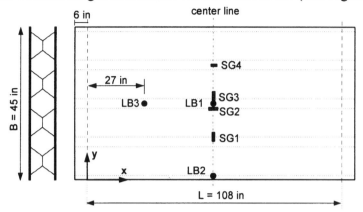

Figure 4. LOCATION OF LVDTs AND STRAIN GAGES ON BOTTOM FACE OF DECK

EVALUATION OF DEFLECTIONS

The H-deck was tested under static loads for three different span lengths: 5 ft. (1.52 m), 7 ft. (2.13 m), and 9 ft. (2.74 m). The maximum load applied was 90 kips (400.3 kN) due to limitations of the test frame. In all the tests, the deck exhibited a linear elastic response. The live load portion of load combination Service I including impact, IM, is used for deflection control (AASHTO 1994). Excessive deformation can cause premature deterioration of the wearing surface and affect the performance of fasteners. For orthotropic plate decks the deflection limit for vehicular load is Span/300.

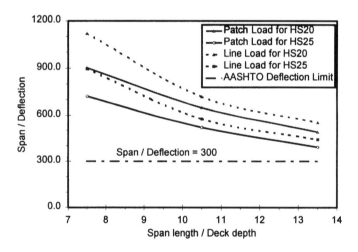

Figure 5. EXPERIMENTAL SPAN-DEFLECTION RATIO FOR DESIGN LOADS

In Figure 5, the experimental span/deflection ratios for a simply-supported module with a width of 45 in (1.14m) are related to the AASHTO deflection limit. The data points were computed as the average of 10 test repetitions. For each test, midspan deflections were corrected from support settlements and a linear regression analysis was conducted. It shall be noticed that in a bridge deck continuously supported on beams, the distance between inflection points results smaller than the actual beam spacing.

FINITE ELEMENT ANALYSIS

A finite element model (FEM) was created for the H-deck using ANSYS™ 5.0. FRP deck elements are modeled using eight-node layered shell elements (SHELL99). Each composite layer represents a multi-axial fabric with elastic properties that were obtained experimentally. A simply supported deck module with a length of 108 in (2.74 m) under a central patch load (See Figure 3) was modeled using 368 shell elements with 1026 nodes. The FEM deflection-load ratios (flexibility coefficients) are correlated with the experimental results in TABLE I.

TABLE I. EXPERIMENTAL-FEM CORRELATIONS FOR DEFLECTION-LOAD RATIO

	Bottom	Top	
	LB1	LT3	LT8
	(milli in/kip)	(milli in/kip)	(milli in/kip)
Experimental	10.52	8.45	7.81
Finite Element M.	11.09	8.14	8.56

Experimental and numerical deflection curves for a unit patch load are shown in Figure 6. The transverse variation of deflection represents the wheel load distribution capacity of the H-deck in the minor stiffness direction. Experimental strain evaluations per unit load are correlated with the FEM predictions in TABLE II.

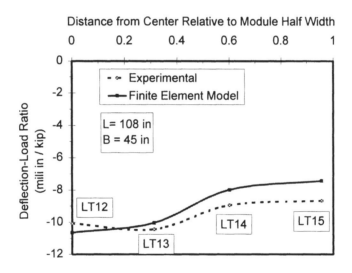

Figure 6. EXPERIMENTAL AND FEM CURVES FOR DEFLECTION-LOAD RATIO

TABLE II. EXPERIMENTAL-FEM CORRELATIONS FOR STRAIN-LOAD RATIO

	Longitudinal (x) (με/kip)		Transverse (y) (με/kip)		
	Top	Bottom	Top		Bottom
	SG7	SG2	SG5	SG8	SG3
Experimental	-27.5	36.7	-33.2	20.8	4.8
Finite Element M.	-34.1	37.7	-32.4	30.2	3.2

Transverse tensile and compressive strains are developed on the deck top face. The strain levels for an HS-25 wheel load with an impact factor of 1.33 are below 1000 μe showing that the deck has a huge reserve of strength capacity. Nevertheless, tensile strains on the top face need to be considered for the serviceability limit state of polymer concrete overlay cracking.

FATIGUE TESTS

Fatigue tests on the H-deck were conducted at U.S. Army Corps of Engineers, Construction Engineering Research Laboratory (CERL), in Champaign, Illinois. A deck module (See Figure 3) was tested for a load cycle ranging from 2000 lb. (8.9 kN) to 50000 lb. (222.4 kN). A static test was conducted before the fatigue cycles, and then after every 500,000 cycles. A total of 2,000,000 cycles were applied with a load frequency of 3 Hz. The fatigue load range, 48000 lb. (213.5 kN), represents an HS25 wheel load of 20000 lb. (89.0 kN) times a deck joint impact factor of 1.75, and times a conservatism factor for extended service life of 1.35. The flexibility ratio after incremental fatigue cycling was measured at various locations on the top and bottom face of the deck without noticing any degradation in deck stiffness (See TABLE III).

TABLE III. FLEXIBILITY RATIO AFTER FATIGUE CYCLING

	Fatigue	Top (milli-in / kip)					Bottom (milli-in / kip)		
Test	Cycles	LT11	LT12	LT13	LT14	LT15	LB1	LB2	LB3
1	0	9.12	10.07	10.43	8.95	8.67	10.52	9.31	7.19
2	500,000	9.37	10.71	10.81	9.24	8.96	10.90	9.51	7.37
3	1,000,000	9.23	10.75	10.20	8.93	8.75	10.53	9.08	7.28
4	1,500,000	9.03	10.49	9.91	8.64	8.50	10.56	8.76	7.07
5	2,000,000	8.94	10.45	9.89	8.63	8.49	10.55	8.70	7.13
6	2,000,000	9.03	10.59	10.00	8.74	8.61	10.62	8.74	7.17

CONCLUSION

Experimental and numerical evaluations of deflections and strains showed good agreement. Tensile stresses on the deck top face control the serviceability of the overlay. The deck did not experience stiffness degradation due to fatigue. The design of the FRP H-deck is controlled by deflection limit states. The H-deck met AASHTO requirements for highway bridges.

ACKNOWLEDGMENT

Part of the work presented in this article was sponsored by the CPAR Program of the U.S. Army Corps of Engineers through a partnership among CERL, WVU-CFC, and the Composites Institute. Another part of the work presented herein was sponsored by the PTP Program of Federal Highway Administration and West Virginia Department of Transportation, Division of Highways. The financial support is gratefully acknowledged.

REFERENCES

American Association of State Highway and Transportation Officials (1994), "LRFD Bridge Design Specifications," Washington, DC.

Design and Construction of FRP Pedestrian Bridges: "Reopening the Point Bonita Lighthouse Trail"

G. ERIC JOHANSEN, ROY J. WILSON, FREDERIC ROLL,
P. GARRETT GAUDINI AND KERRYN GRAY

ABSTRACT

The construction of two fiberglass reinforced plastic **(FRP)** truss bridges in Golden Gate National Recreation Area, San Francisco, CA have been investigated in this study. The two bridges, 10.7 m (35'-0") and 21.4 m (70'-0") in length, were constructed using E.T. Techtonics, Inc.'s new lightweight **FRP** building system **(LONGSPAN PRESTEK).** The system is constructed using fiberglass/isophthalic polyester resin channels and tubes. The case study will address the structural concerns involving seismic issues and extreme wind conditions, the use of camber and X-bracing in the design to increase overall strength and stiffness, difficult site constraints which had to be overcome in the erection of the bridges, and the overall advantages derived in using the **LONGSPAN PRESTEK** System in comparison to traditional structural systems such as wood, steel and concrete in a marine environment of this type. It will also evaluate the aesthetic considerations in the overall design of the bridges, i.e. the use of camber and the color white for the bridges, and the resulting visual image in the Marin Headlands landscape.

INTRODUCTION

In the summer of 1995, E.T. Techtonics, Inc. was contacted by the Golden Gate National Park Association concerning the design and construction of two lightweight **FRP** bridges for the Point Bonita Lighthouse Trail in the Marin Headlands near San Francisco, CA. The lighthouse trail which overlooks the Golden Gate Bridge is the most popular attraction in the Marin Headlands area and is used by more than 250,000 visitors a year. It is one of the most memorable walks in the Park Service as the 1-1/2 mile + trail leads one along sheer rocky cliffs which drop 120'-0" to the Pacific Ocean below. This route takes the visitor through a long hand-chiseled approach tunnel, a marvelous piece of mid 19th century engineering which has remained relatively intact even with recent seismic events. Walking along the cliffs, one is struck by the relentless force of the Pacific Ocean which is "crashing" around you. One is even more impressed by the strength of the wind as there is this constant reminder in your face as one moves along the trail. Below, one can see a number of seals

Dr. G. Eric Johansen, Roy J. Wilson, Dr. Frederic Roll, P. Garrett Gaudini and Kerryn Gray, E. T. Techtonics, Inc., 213 Monroe St., Philadelphia, PA 19147.

bobbing in the water searching for food with the Golden Gate Bridge as a backdrop. Needless to say, the landscape is very dramatic.   Finally, one reaches the 150'-0" long swaying suspension bridge which leads out to the point and the lighthouse.   The vertical and horizontal stability of the "swinging" bridge does not instill a great deal of structural confidence in the visitor, but having made it this far there is not turning back.   One crosses the bridge and arrives at the lighthouse (See Fig. 1).

The history of the Point Bonita Lighthouse is an interesting account.   The lighthouse was originally constructed in 1855 high above Point Bonita to prevent cargo ships from crashing against Golden Gate's rocky shores.   Unfortunately, it was found that the lighthouse was sited too far above the water.   In dense fog (which is a common occurrence in this area), approaching ships could not see the lighthouse beacon (a Fresnel lens made in Paris which was the finest lens available at that time).   It was decided that a new lighthouse would be built at Point Bonita's tip on a a narrow spit of land jutting out into the Pacific Ocean.   This site proved difficult as it was on what builders referred to as "rotten rock" which was unstable sandstone, chert and shale cliffs.   Historical records indicate that construction of the lighthouse was hindered as this so-called "rotten rock" would frequently (and unexpectedly) slide into the sea.   Finally in 1877, a foundation was secured and completed, and the lighthouse as it now stands was constructed.

Unfortunately this "rotten rock" problem has created numerous trail problems for the National Park Service.   In December of 1992, a six foot wall of rock that had served as a natural guard rail along the trail unexpectedly slid into the ocean narrowing the trail from eight feet in width to six feet.   The Park Service closed the lighthouse trail due to safety concerns. After a slope stability assessment of the site by a local engineering firm, it was decided to construct 2 bridges over areas of the trail which were determined to be unstable.   The proposed span lengths along the cliffs were 10.7 m (35'-0") and 21.4 m (70'-0").   Both bridges posed construction problems due to the slope stability issues. The site was also inaccessible by road (drop-off point for trucks was approximately 1-1/2 miles from the construction site).   If the bridges were constructed using traditional materials (steel, wood and concrete) , the structures would require an airlift to the site.   Further, the Park Service had been having maintenance problems with traditional wood, steel and concrete structures in this rigorous marine environment of the Marin Headlands.   Severe corrosion and rot required constant attention and replacement.   It was also important that aesthetically the "image" of the bridges structures responded to the nature of the existing suspension bridge structure and the lighthouse.   Given the above issues, the lightweight, maintenance-free, aesthetically pleasing characteristics of **FRP** components were an excellent structural alternative for this particular bridge project.

DESCRIPTION OF THE PROPOSED STRUCTURES

It was determined that the shorter span 10.7 m (35'-0") bridge would be constructed using the standard **FRP** Pratt truss system developed by E.T. Techtonics, Inc. for use in pedestrian and equestrian type applications.   This approach uses a standard double beam top and bottom chord connected with single vertical compression members and diagonal tension members for the truss members.   For the proposed bridge, the trusses were connected side to side with lateral and horizontal bracing members.   Standard 203.2 mm (8") channels were used for the top and bottom truss chords with 50.8 mm (2") hollow square tubes for the vertical and diagonal members which connected the chords together. 203.2 mm (8") channels were used for the horizontal cross pieces. 50.8 mm (2") hollow square tubes were required for the horizontal bracing. Standard 3"x12" pressure treated southern yellow pine was used for the decking system. 304 SS 19 mm (3/4") dia. steel bolts were used for the primary truss connections. The bridge was also cambered 101.6 mm (4") at the request of the client for aesthetic purposes (See Fig. 2).

The longer span 21.4 m(70'-0") bridge required a different design. The proposed clear span length using the double beam system with pin/roller end conditions was the longest ever attempted by E.T. Techtonics, Inc. From our previous completed projects, it had been determined that issues such as lateral buckling of the compression chord, location of splices, hole tear-out in the diagonal members, sidesway, and overall vibration characteristics of the structure become important design considerations for spans over 15.4 m (50'-0"). This warranted using a larger channel, i.e. 203.2 mm (8") rather than our standard design which uses 152.4 mm(6"), and X bracing (tension and compression diagonals) to strengthen and stiffen the overall design. Further, the client wanted the structure cambered at least 304.8 mm (12") for aesthetic reasons. Cambering the system would increase the strength and stiffness of the overall structure, but would create more involved fabrication issues.

The double beam chord system was constructed with 203.2 mm (8") channels with 50.8 mm x 50.8 mm x 6.4 mm (2" x 2"x 1/4") hollow square tube vertical posts and diagonal X bracing connecting the chords together. 203.2 mm (8") channels were used for the horizontal cross pieces which extended out approximately .91 m (3'-0") every 3.08 m (10'-0") bay of the truss system. 50.8 mm x 50.8 mm x 6.4 mm (2"x 2"x1/4") hollow square tube outrigger posts connected to these channel extensions to provide increased lateral support for the compression chord of the double beam truss. 50.8 mm x 50.8 mm x 6.4 mm (2" x2"x1/4") hollow square tubes were used for the horizontal bracing. One 203.2 mm (8") channel stringer ran on top of the horizontal cross pieces to provide additional support for the decking system. The system was spliced in two locations on each truss due to shipping issues. All connections were bolted using 304SS 19 mm (3/4") dia. steel bolts for the primary truss system. All **FRP** shapes (both bridges) were pultruded by Creative Pultrusions, Inc. of Alumbank, PA (See Fig. 3).

Outrigging the structural system addressed the lateral buckling issue of the top compression chord of the truss by increasing the overall lateral stiffness of the chord. Using this with the 203.2 mm (8") channel which has greater lateral stiffness than a 152.4 (6") channel resulted in a structure which has excellent side to side characteristics. The use of X bracing(tension and compression diagonals) decreased the potential for hole tearout and/or bearing failures in these members. The overall strength/stiffness of the system was also increased allowing for heavier loads on the longer span. Vibration characteristics were also improved by the increased stiffness of the structure. X bracing also reduced overall joint slippage therefore minimizing deflection due to dead load and fabrication tolerances. Cambering the system increased the strength and stiffness of the structure.

Computer models of the proposed designs were developed using STAAD 3 to evaluate the bridges for dead load only, dead load + live load, wind load (125 mph and 40 psf uplift), Seismic Zone 4, temperature variations of 75 F., and natural frequency. The pedestrian bridges were designed for 85 psf live load in accordance with specifications determined by the Golden Gate National Park Association (GGNPA). The structures were analyzed for pin/roller and pin/pin end conditions. The **FRP** (fiberglass/isopolyester resin) design specifications used were as follows:

FRP Ultimate Strength (E.T. Techtonics Laboratory Tests)

Compression: 25,000 psi (F.S. = 3, 8,333 psi allowable)
Tension: 25,000 psi (F.S. = 3, 8,333 psi allowable)
Bending: 25,000 psi (F.S. = 3, 8,333 psi allowable)

FRP Young's Modulus (E.T. Techtonics Laboratory Tests)

203.2 mm (8") Channel: 3250 ksi
50.8 mm x 50.8 mm x 6.4 mm (2"x2"x1/4") Square Tube: 2500 ksi

The serviceability criteria established for the project was as follows:

Live load deflection: L/360 (Bridge would be initially cambered 12" to eliminate any initial deflection due to dead load and joint slippage)

Minimum Frequency: 5 cycles/sec.(Design for pedestrian loading)

Based on this criteria, it was determined from the three dimensional computer analysis that the proposed bridge designs would meet the natural frequency requirements established by GGNPA. From our experience, FRP bridge structures over 15.4 m (50'-0") are susceptible to vibration issues ("liveliness") unless properly designed. The long span 21.4 m (70'-0") bridge could not be "springy" given its location on the cliffs and approximately 120'-0" above the ocean. The analysis determined that the frequency of the system was in the range of 5.0 cycles/sec. without the camber and 5.5 - 6.0 cycles/sec. with the camber. It is interesting to note the increase in overall stiffness in the bridge structure due to camber.

CONSTRUCTION

Due to the inaccessibility of the site in the Marin Headlands, the National Park Service specified the maximum length of **FRP** components 10.7 m (35'-0"). This required the splicing of the 21.4 m (70'-0") bridge in two places resulting in member lengths of 6.9 m (22'-6"), 7.7 m (25'-0"), and 6.9 m (22'-6"). Further, the system was shipped unassembled because of handling issues. Fabrication was done by Structural Fiberglass, Inc. of Bedford, PA (See Fig. 4).

Erection of the two bridges presented some dangerous site constraints due to the eroding cliffs. It was decided after careful evaluation and the Coast Guard's offer to airlift the bridges in by helicopter at no cost that air lifting the bridges was the best solution. The lifting capacity of the helicopter was approximately 2350 kg (5000 lbs.). The 10.7 m (35'-0") bridge weighed approximately 650 kg. (1700 lbs.) without the wood deck so this posed no problem. It was estimated that the 21.4 m (70'-0") bridge weighed approximately 2350 kg. (5000 lbs.) without the wood deck. This could pose an issue, but given the overload capacity of the helicopter was not a problem. Assembly of the bridges took place in a parking area 3km (2 miles) from the site. The field was easily accessible to the helicopter. The bridges were airlifted out over the water and into place along the cliffs. Total erection time was approximately 1-1/2 hrs.

Design, fabrication and shipping of the bridges to the site had taken approximately 60 days. Site work, pouring of concrete for the foundations, placement of anchor bolts, and erection of the bridges took an additional 30 days. The entire project, start to finish, was completed in 90 days.

AESTHETIC CONSIDERATIONS

One of the major concerns of E.T. Techtonics, Inc. in the design of **FRP** bridges is aesthetics. **FRP** offers unique aesthetic characteristics in comparison to traditional materials such as wood, steel, and concrete. **FRP** can be specified in any shape and color and provided with a variety of textures. In the development of the **PRESTEK** system, there has been major attention given to the aesthetic nature of the designs, in particular the longspan bridges. In this particular design for the Golden Gate National Recreation Area, the client requested that white be used for the color of the bridges in order to match the color of the existing suspension bridge and lighthouse which are white. This seemed to be an excellent choice. White is a major color used in the marine environment. It "reads" very well in this type of landscape. Our only concern is how the color white will weather over time as it is known to take on a yellowish tint. Color retention in pultrusions is a major concern and must be further addressed by the industry.

CONCLUSIONS

The above case study demonstrates the feasibility of constructing economical longspan bridges on inaccessible sites with **FRP**. The unique strength/stiffness characteristics of the **LONGSPAN PRESTEK** System combined with its lightweight and non-corrosive nature provide distinct advantages over traditional construction materials such as wood, steel, and concrete. Besides being easy to assemble and erect, **PRESTEK** is aesthetically pleasing as well as low maintenance. It is a particularly cost effective solution in those applications which have difficult site constraints such as the Pont Bonita Lighthouse Trail.

This investigation is limited to span lengths up to 21.4 m (70'-0"). From laboratory tests, computer analysis, and field observations, pedestrian bridges up to 33 m (100'-0") can be constructed with the same components. The **FRP** components, wood decking and metal fasteners for this project weighed approximately 5000 kg (11,000 lbs.). The 12.3 (35'-0") span bridge weighed approximately 1363 kg (3,000 lbs.). The 24.6 (70'-0") span bridge weighed approximately 3636 kg (8,000 lbs.). The total cost of the project including design, fabrication and shipping (excluding erection) was $45,000. Completion time for the job was 90 days from start of design to end of construction.

ACKNOWLEDGMENTS

The investigation is based on the work of Robert LeRicolais, a pioneer in the development of lightweight structural configurations.

REFERENCES

Johansen, G. Eric, Roll, Dr. Frederic, and Wilson, R.W., "Spanning Devil's Pool with a Prestressed Kevlar Cable/FRP Tube Structural System", Advanced Composite Materials in Bridges and Structures, ACMBS, pp.435-445.

Johansen, G. Eric, Roll, Dr. Frederic, Wilson, R.W., and Erki, Dr. M.A., " Creep Relaxation Characteristics of a Prestressed Kevlar Cable/FRP Tube Structural System", Advanced Composite Materials in Bridges and Structures, ACMBS, pp. 445-455.

Johansen, G. Eric, Roll, Dr. Frederic, and Wilson, R.W., "Ultimate Strength Characteristics of an FRP/Kevlar Cable Structural System", SPI Composites Institute 48th SPI Conference Proceedings, Cincinnati, OH.

Johansen, G. Eric, Roll, Dr. Frederic, and Wilson, R.W., "Creep and Relaxation of PRESTEK Structural Systems", ANTEC '93, New Orleans, LA 1993.

Johansen, G. Eric, Roll, Dr. Frederic, and Wilson, R.W. , " Spanning Staircase Rapids with a Prestressed RP Truss Structural System", 4th International Conference on Short and Medium Span Bridges, Halifax, Nova Scotia, Canada, 1994.

Johansen, G. Eric, Roll, Dr. Frederic, and Wilson, R.W. , "Strength/Stiffness Characteristics of a Prestressed RP Truss Structural System", SPI Composites Institute 50th Annual Conference, Cincinnati, OH 1995.

Johansen, G. Eric, Roll, Dr. Frederic, and Wilson, R.W., "Design Applications of a Prestressed FRP/Kevlar Cable Structural System", Conference on NSF Research Transformed into Practice", Arlington, VA 1995.

Johansen, G. Eric, Roll, Dr. Frederic, Wilson, R.W., Olson, Noah, Millman, Martin, and Silvey, Michael, " Advanced Composite Material Support Frames: An Evaluation of the Bovee Meadow Bridge at Lake crescent, WA", SPI Composites Institute 51st Annual Conference, Cincinnati, OH 1996.

Johansen, G. Eric, Roll, Dr. Frederic, Wilson, R.W., Gaudini, P. Garrett, "Design and Construction of Two FRP Pedestrian Bridges in Haleakala National Park, Maui, Hawaii", Advanced Composite Materials in Bridges and Structures, 2nd International Conference, Montreal, Quebec, Canada, 1996.

Johansen, G. Eric, Roll, Dr. Frederic, Wilson, R.W., Gaudini, P. Garrett, "Indiana Jones and the Temple of Doom Revisited", Wilson Forum -East, Alexandria, VA, 1996

Creative Pultrusion Design Guide, 1987 CPI.

FIGURE 1. Point Bonita Lighthouse and "Swaying" Suspension Bridge

FIGURE 2. 35'-0" Span Bridge

FIGURE 3. 70'-0" Span Bridge

FIGURE 4. Shop Assembly by E.T. Techtonics, Inc.

Braiding and Fabrics: Advanced Reinforcement Architecture Opens New Horizons

Improvement of Lateral Compressive Strength in Braided Pipe—New Braided Fabrication

H. HAMADA, A. NAKAI, M. MASUI AND M. SAKAGUCHI

ABSTRACT

The purpose of this study is to fabricate the braided composite tubes which have high lateral compressive strength without interlaminar delamination.   The numerical analysis method was applied to the analyses for simulating the mechanical behaviors of multi-layered braided composite tube.   From the analytical results, the validity of fiber bundles which were oriented in the interlamina was confirmed.   According to the analytical results, new braiding system were proposed in order to fabricate the braided composite tube with high strength and thickness.   The proposed new braided composite tube possesses higher lateral compressive properties than that of laminated braided composite tube.

1   INTRODUCTION

The high specific strength and good corrosion resistance of a fiber reinforced plastic (FRP) pipe has led to increasing uses in the chemical and other industries in recent years.   FRP pipes have been remarkably applied to the second structural parts in the moving components such as automobile, airplane and helicopter, and so on.   For FRP pipes, main fabrication methods were filament winding or sheet laminating methods.   When these tubular members are used in structural applications, high modulus and high strength equivalent to those of metallic material are required.   In order to obtain sufficient mechanical properties, thick tubular members are made with multiple layers.   However, laminated composite tubes have limited application, due to occurrence of delamination.

In order to overcome the limitation of laminated composite tubes, it is necessary to fabricate the composite tube with thickness and strength in through-the-thickness direction.   Recently various fabrication methods for FRP pipes have been developed.   One of these method is "Braiding technique".   We have already proposed a new fabrication system for three dimensional braided composites.   Also, this braiding technique can fabricate three-dimensional joint pipes.

The purpose of this study is to fabricate the braided composite tubes which have high lateral compressive strength without interlaminar delamination.   Firstly, the analytical method for simulating the mechanical behaviors of the braided composite was introduced and the analytical method was applied to the analyses of multi-layered braided composite tube.   From the analytical results, the validity of fiber bundles which were oriented in the interlamina was investigated. According to the analytical results, new braiding system were proposed in order to fabricate the braided composite tube with high strength and thickness.   Particularly, braided tubes which have high lateral compressive strength were fabricated.   Lateral compressive tests were carried out and it was confirmed that proposed system could be used for high strength and thick composite tube.

2   ANALYTICAL APPROACH FOR BRAIDED COMPOSITE TUBE WITH THICKNESS AND STRENGTH

2-1 CONCEPT OF MODELING

Hiroyuki Hamada, Asami Nakai, Mikio Masui and Masahide Sakaguchi, Kyoto Institute of Technology, Matsugasaki, Sakyo-ku, Kyoto 606, JAPAN

In the case of textile composites, fiber orientation angle, crimp, continuity of fiber and stress transmission system between fibers greatly affect the mechanical behavior. So, the analytical model has to represent the weaving structure faithfully. Moreover, various range of fractures from microscopic fracture to macroscopic fracture in the composite have to be considered. It might be difficult to represent all ranges of fracture in a single model.

Figure 1 show the five modeling steps for simulating the mechanical behaviors of the braided composite precisely. Basically, results obtained from micro model are used in the analyses of macro model. For example, elastic modulus and strength in the micro model are used in the macro model as material constants. According to this system, macroscopic analysis can be performed considering microscopic phenomena. This analytical method can be applied to predict the various mechanical properties of braided composite, by selecting and combining these models according to the analytical objects.

Inherently interlamina delamination is fatal

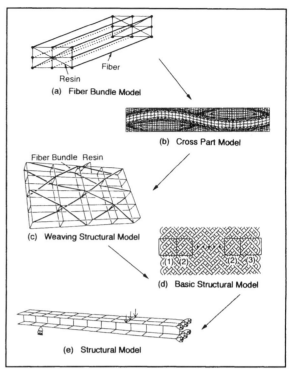

failures occurring in normal laminated pipe under lateral compressive load. In this study, the analytical method to represent the interlamina of

Fig.1　Five modeling steps for simulating the mechanical behaviors of the braided composite

multi-layered braided composites was proposed and the elastic modulus and initial interlaminar delamination stress of multi-layered braided composite tube under lateral compressive load were examined.

2-2 ANALYTICAL METHOD

In this study, the analytical method of the connection between 'Weaving structural model' and 'Structural model' is described, so that 'Weaving structural model' can be regarded as the micro model and 'Structural model' as the macro model. The analytical method consisted of two steps of calculations. Elastic modulus, shear modulus and Poisson's ratio of repeating unit were obtained by using the micro model, and then the material properties were fed into the macro model. Both micro model and macro model were analyzed by using the finite element method.

The finite element division for the micro model is shown in Figure 1(c). This analytical model is a part of the braided composites in which repeating unit is used to express faithfully the fiber orientation state. The weaving structure of the braided composite is expressed by connecting beam elements. The thick lines and the fine lines express the fiber bundle and resin in the braided composite tube, respectively. The resin elements are set up on both surfaces of the model, since the resin exists at the surface of the braided composite. Also, resin existing between fiber bundle crossings is expressed by using beam element in order to consider the transmission of force at crossing point between fiber bundles through resin. In this analysis, the micro model with fiber orientation angle of 60 degree was prepared.

The macro model used in this analysis, namely the finite element division for the 3-layered braided composite tube, is shown in

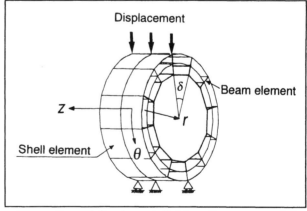

Figure.2　Finite element divisions for macro model

Figure 2. Here, the local coordinate system is used, in which $r-$, $\theta-$ and $z-$ axes indicate the radial, circumference and axial directions respectively. The model is constructed of anisotropic shell and beam elements. The shell elements represent the braided composite and have the material constants obtained by the micro model. The beam elements represent the interlaminar region of the 3-layered braided composite tube. In this study, two types of macro model were prepared and its lateral compressive properties were examined. One is the macro model for laminated tube with three-layer (Laminated model) in which the beam elements have resin properties. The other is the macro model for self-reinforced tube with through-the-thickness fiber in the interlamina (Self-reinforced model) and a part of beam elements have material properties of braided composite in order to represent the through-the-thickness fiber in the interlamina. The loading condition and boundary condition were also shown in Figure 2.

2-3 ANALYTICAL RESULTS

Firstly, the normal stress σ_{rr} and shear stress $\tau_{r\theta}$, τ_{rz} components in beam elements of laminated model were calculated. The results are shown in Figure 3. In this figure, abscissa denotes the beam element position by using the center angle δ, between the beam element position and the initial supported point before loading. In Figure 3, it is found that all stress components in the circumferential direction changed as the sine curve with period π. However the amplitude in $\tau_{r\theta}$ was much larger than that in σ_{rr} and τ_{rz}. This results indicate that the initial interlaminar delamination is mainly caused by shear stress $\tau_{r\theta}$. From these results, the initial interlaminar delamination occur at center angle of $\pm$ 150 degree of cross section due to shear stress $\tau_{r\theta}$. Table I shows the calculated elastic modulus and initial fracture stress of the macro model. The elastic modulus and initial fracture stress of self-reinforced model

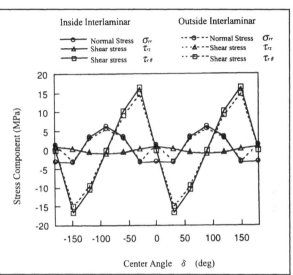

Fig.3 Stress components in beam elements plotted against center angle δ

were larger than that of laminated model. This results indicate that the occurrence of the initial interlaminar delamination can be restrained by the through-the-thickness fibers.

Moreover, in order to investigate the behavior of 3-layered braided composite tube after initial fracture, the beam element in which initial fracture occurred was removed from the initial macro model and the elastic modulus and initial fracture stress of the macro model with a defect were calculated. Here, for convenience, the elastic modulus and initial fracture stress were refereed as second modulus and second initial fracture stress respectively. The calculated second modulus and second initial fracture stress were also shown in Table I and both value of self-reinforced model were larger than that of laminated model. It was considered that the analytical results indicated the difficulty of propagation of interlaminar delamination on self-reinforced tube with through-the-thickness fiber.

From these analytical results, the validity of fiber bundles which were oriented in the interlamina was confirmed.

TABLE I - Lateral compressive modulus and Initial fracture stress of macro model

	Lateral Compressive Modulus (GPa)	Initial Fracture Stress (MPa)	Second Lateral Compressive Modulus (GPa)	Second Initial Fracture Stress (MPa)
Laminated Tube	5.10	557	4.77	520
Multi-Reciprocal Tube	5.51	567	5.14	531

3 FABRICATION OF BRAIDED COMPOSITE TUBE WITH THICKNESS AND STRENGTH

Figure 4(a) shows the conventional braiding mechanism. The spindles hold the bobbins and supply the yarn. The spindle movement makes the intertwining of the yarn, so that the braiding structure is formed. The braiding mechanism consists of taking up mechanism and spindle movement. Spindles move along the circle, so that the spindle movement can be regarded to a one-dimensional movement. The formed fabric is taken up in the vertical direction. Apparently the taking up of the braided fabric is also a one-dimensional movement. Simple shapes such as tubular braid and flat braid can be fabricated by combination of these two one-dimensional movements.

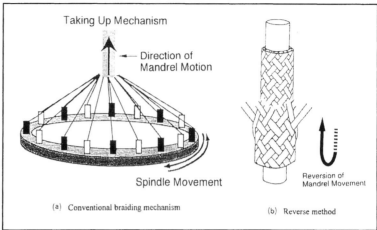

Fig.4 Multi-Reciprocal Braiding Process

In order to orient the fibers in the interlamina of laminated composite tube according to the above analytical results, reversible movements of the mandrel are added to the conventional braiding process as shown in Figure 4(b). This braiding process is called "Multi-Reciprocal Braiding Process". This process initially developed with the aim of fabricating thick braided composite tube. The basic concept of this process is that mandrel, taking up movement, moves reciprocally in spite of circular spindle movement. In the braiding operations, the mandrel goes up and reverses. This reciprocal movement is repeated until the required thickness can be obtained. The fibers continuously orient without cutting them.

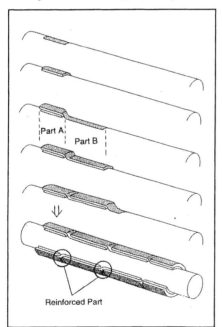

Fig.5 Fabrication method of braided composite tube with through-the-thickness fabric

The braided composite tube with through-the-thickness fabric can be fabricated by the technique as shown in Figure 5. During the braiding process, the reversible points changed. Firstly, three layers are fabricated by normal reciprocal process at part A and then the third layer comes to the first layer at part B. Therefore, during this process, braiding fiber results in orientation at interlamina region between part A and B. This part which fiber orients at interlaminar is called "Reinforced Part". It is expected that through-the-thickness fabric of the braided composite tube would suppress growth of delamination in a thick composite tube. This is called "Multi-Reciprocal Tube".

In this study, lateral compressive behaviors of multi-reciprocal tube were examined. Laminated tube with three-layer was also prepared in order to compare with the multi-reciprocal tube. Reinforcement fiber used for fabrication of the specimens was glass fiber (E-glass Roving, 1100tex : Nitto Boseki Co., Ltd.). As matrix, epoxy resin (EPICLON830 : DAINIPPON INK AND CHEMICAL, Inc.) was used.

4. LATERAL COMPRESSIVE PROPERTIES

4-1 STATIC LATERAL COMPRESSIVE TEST

The inside diameter of the specimens was 17.5mm and the outside diameter was 21.5mm. The specimen length was 25mm and the braiding angle was 60 degree. The specimen was placed between the upper and lower plate and was deformed by pushing down the upper plate. Lateral compressive tests were carried out by using Instron Universal Testing Machine (Type 4206) at testing speed of 2mm/min.

Figure 6 shows relation between lateral compressive load and displacement on the multi-reciprocal tube and laminated tube. Both specimen exhibited linear increase of load at small displacement. After the load reached the maximum value, sudden reductions in load were observed in case of laminated tube. On the other hand, the load reached the maximum value and kept same value in case of multi-reciprocal tube and ultimate displacement of multi-reciprocal tube was higher than that of laminated tube. This is because the difference of fracture propagation. From the observation during the lateral compressive tests with the eye, in the case of laminated tube, the interlaminar delamination was occurred at one of the both width edges and propagated without a stop from the edge to the other edge rapidly. Whereas, in the case of multi-reciprocal tube, the propagation of interlaminar delamination which occurred at one of the both edges was impeded at the reversible points. Therefore, the sudden reduction in load were not observed in multi-reciprocal tube.

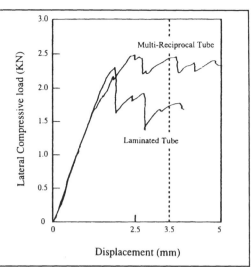

Fig.6   Relation between load and displacement

TABLE II - Lateral compressive modulus and strength

	Lateral Compressive Modulus (GPa)	Lateral Compressive Strength (MPa)	Energy Absorption (J)
Laminated Tube	5.19	374	4.96
Multi-Reciprocal Tube	5.00	410	5.61

Table II lists lateral compressive modulus and strength. Lateral compressive modulus of multi-reciprocal tube nearly equal to that of laminated tube. Whereas lateral compressive strength of multi-reciprocal tube is higher than that of laminated tube, because the interlamina was reinforced with through-the-thickness fabric and the interlaminar strength was enhanced. Moreover, the energy absorption to a displacement of 3.5mm was 5.61J in the case of multi-reciprocal tube, whereas the energy absorption of laminated tube was 4.96J. These results indicate the validity of fiber bundles which are oriented through the interlamina.

Moreover, the lateral compressive properties of self-reinforced tubes with different number of reinforced part were examined. Length of the specimen was 50mm and the number of reinforced part in each specimens ranged from 1 to 4. Figure 7 shows relation between lateral compressive load and displacement. As the number of reinforced part increased, sudden reduction in load after the load reached the maximum value was impeded. Maximum load increased as the number of reinforced part increased. These results indicate that the lateral

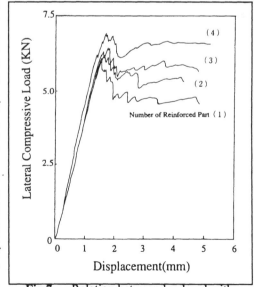

Fig.7   Relation between load and with different number of reinforced part

compressive properties were enhanced by increasing the number of reinforced part.

4-2 DYNAMIC LATERAL COMPRESSIVE TEST

There is possibility that multi-reciprocal tube can be used as high energy absorption parts to impact, since the reduction of load after maximum load is impeded in multi-reciprocal tube. Therefore, energy absorption to impact was examined by dynamic lateral compressive test. Dynamic lateral compressive behaviors of multi-reciprocal tubes with 2 and 4 reinforced part and laminated tube were compared. The length of the specimen was 50mm. Dynamic lateral compressive test was carried out at 15km/h with the drop weight impact machine. Figure 8 shows relation between lateral compressive load and displacement. In multi-reciprocal tubes, reduction of load after maximum load was smaller than laminated tube and sudden reduction of load was not observed.

Maximum loads were 5.40KN, 7.30KN, 8.58KN in order of number of reinforced part, 0, 2, 4, and increased as number of reinforced part increased. Moreover, the energy absorption to displacement of 10mm was 37.6J, 54.8J, 61.0J in order of number of reinforced part, 0, 2, 4, and also

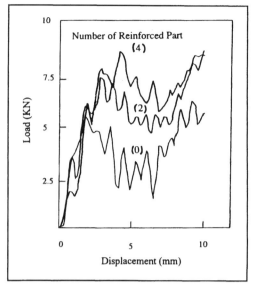

Fig.8 Relation between load and displacement in impact test

increased as number of reinforced part increased. It is considered that this reason is that the propagation of interlaminar delamination is impeded at the reinforced parts and the sudden reduction of load is further impeded by increasing the number of reinforced parts. From these results, it turned out that the multi-reciprocal tube could be a high energy absorption part.

5   CONCLUSION

In this study, the analytical method for simulating the mechanical behaviors of the braided composite was proposed and the analytical method was applied to the analyses of multi-layered braided composite tube. From the analytical results, the validity of fiber bundles which were oriented in the interlamina was confirmed. According to the analytical results, the braided composite tube with through-the-thickness fabric was fabricated and the lateral compressive properties were measured. In the case of multi-reciprocal tube, the propagation of interlaminar delamination was impeded at the reinforced part. Lateral compressive properties were enhanced by fiber bundles which are oriented in the interlamina. It can be expected that the composite tube with thickness and strength in through-the-thickness direction can be fabricated by multi-reciprocal braiding process.

BIOGRAPHIES

Hiroyuki Hamada
Dr. Hiroyuki Hamada is presently an associate professor of Faculty of Textile Science at Kyoto Institute of Technology. He received his B. Eng. degree from Doshisha University in 1978, M. Eng. degree from Doshisha University in 1980 and Dr. Eng. degree from Doshisha University in 1985.

Asami Nakai
Miss. Asami Nakai is presently a graduate student at The University of Tokyo. She received his B. Eng. degree from Kyoto Institute of Technology in 1994 and M. Eng. degree from Kyoto Institute of Technology in 1996.

Mikio Masui
Mr. Mikio Masui is presently a graduate student at Kyoto Institute of Technology. He received his B. Eng. degree from Kyoto Institute of Technology in 1996.

Masahide Sakaguchi
Mr. Masahide Sakaguchi is presently a graduate student at Kyoto Institute of Technology.

Braided Reinforcements: A Versatile Cost-Effective Alternative to Traditional Reinforcement Approaches

JEFF MARTIN AND ANDREW A. HEAD

ABSTRACT

Braided reinforcements are an alternative reinforcement architecture that many processors do not consider when developing a new composite product. In fact, braided reinforcements are perhaps the most misunderstood and under-rated reinforcement architecture available to our industry. Misconceptions about braid abound, yet braided reinforcements are very efficient in delivering unique and optimum properties in composite laminates.

This paper will explore the misconceptions that exist about braid. It will address the type of properties that are possible using braided reinforcements. The versatility of the braiding process demonstrates how flexible this form of architecture can be in meeting mechanical properties in all composite processes. The paper will conclude by identifying braided reinforcements as a cost-effective alternative to traditional reinforcements.

INTRODUCTION

Although the Composites Institute Proceedings are about the newest high tech processes and materials, this paper will discuss an ancient art which when executed with modern reinforcement materials becomes an alternate material, perhaps the ultimate reinforcement architecture which just maybe is the most cost-effective choice for composite processors. This ancient art is coming to the forefront as a preferred way to add strength, durability and economy to composite manufacturing.

The subject is braided reinforcements.

The word braid has many connotations, from hirsute decoration to the "boondoggle" gifts the kids bring home from camp, to some ubiquitous products we take for granted such as shoelaces, drapery cords, and candle wicks.

Braiding is one of the most ancient of fabric arts and it actually predates the evolution of man. Nature has used braids for millions of years, using the technique to fashion the skin of the shark, and to strengthen the stems of tuber plants.

Man may have taken his cue from these natural uses, for braiding has been a part of life since neolithic times. Braid has been used by the Egyptians over 3,000 years ago for necklaces and in hair styles. The Japanese used braids for the famous Karakumi belts they wore.

Jeff Martin and Andrew A. Head, A&P Technology, Inc.

Our own American cowboys depended on plaited leather lariats to capture horses and cattle. And long before Edison, artificial light was enhanced by braided gas lamp mantles.

Braids are important in today's industry too. Sleeving for electrical wiring, reinforcements for rubber and pressurized hose.

And braids are equally important to medical science. Braided pre-pregs are used for prosthetic devices, splints and orthotic devices. Stents and catheters used in heart surgery procedures use braid as well.

Most important to Composites Institute people, braids, though ancient in concept, are the latest in high tech ways to improve composites. From golf clubs to cargo trailers, industrial rollers to high performance aircraft engines, from tennis racquets to insitu conduit linings, braided reinforcement fibers are part the future of the composites industry.

BRAIDED REINFORCEMENTS - AN IMPORTANT PART OF YOUR COMPOSITE FUTURE

We are "introducing" braids to the Composites Institute because braided reinforcements, are not only a "well kept secret" but they may be the most under-rated, misunderstood reinforcement style available. Braids have tremendous potential to add mechanical properties, while reducing lay-up times and molding cycles and helping to reduce component weights and production costs.

If you are not considering braided reinforcements when designing a composite structure, you are overlooking a versatile efficient problem solver.

To whet your appetite, here are a few of the property improvements you can effect with braids:

Hockey Sticks   Aircraft Ducting   Snowboard   Knee Brace

- Improved shear properties, as in the floor of a cargo container,
- Higher impact strength, as in a hockey stick ,
- Superior screw retention, as in a container floor beam,
- Improved fatigue as in several aircraft components,
- Better damage tolerance, as in a container corner beam,
- Better torsional strength, as in a knee brace,
- Thinner wall sections, as in a snowboard,

- Improved ductility when used in wall reinforcements in seismic zones,
- Better torsional and dimensional stability, which can translate into thinner walls, lighter weight, and lower cost in this new generation wake board.

MISCONCEPTIONS ABOUT BRAID

While braided reinforcements are beginning to find their way into composite applications, there are a number of misconceptions and misunderstandings about the product.

Many feel that braided materials are only available in a tubular form. While tubular braids, know as sleevings, are a common form and easy to produce, braids are also available as tapes with finished edges, as slit tapes, in overbraids on your component, as net-shape preforms, cored braids and in rope form. And most recently you can get any of these forms produced as an epoxy pre-preg. This variety of styles should cover virtually all the requirements of the composites industry.

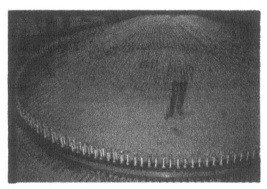

A second misconception is that braids are only available in narrow widths. While many companies do only produce narrow width braids on braiders with a limit of 144 carriers, one firm manufactures braids on machines with up to 600 carriers. This permits sleeving diameters up to 52 inches and tape widths up to 36 inches. Depending on the angle required wider products are also possible. In addition, a braider with 800 carriers is planned to be installed in 1997.

Another misconception is that there are big property compromises due to fiber crimp in braids. The braiding process does take fibers out of their tensile mode. One supplier of stitched reinforcements acknowledged that while some discuss tensile property reductions of 7 to 20 percent, a more realistic range is 10-15 percent. However, when braided reinforcements are pretensioned in the mold these percentages drop considerably to a point of being statistically insignificant. While those percentages refer to tensile strengths, braid is unique in that it gives other unique properties not possible with other reinforcement styles. Some of these property pluses are due to the interlocked nature of the fibers that keeps them from moving under load. These are torsional stability, impact or toughness, shear strength and screw retention.

A braided reinforcement improved shear strength in floor panel by 40% over stitched materials.
Photo Courtesy: Stoughton Composites, Inc.

In marine containers, braided reinforcements improved shear and screw retention by 60% over a mat/roving composite and 40% over a stitched fabric reinforced composite. Braid also offers dampening and ductility properties that are just not as easily obtainable or even possible with other reinforcement styles.

Many believe that braided materials are expensive. In the past the braiding process was slow and most work was performed on small machines. Like many material processes, the costs of the braiding process are heavily influenced by the number of pounds per hour that can be processed. The more pounds processed per hour, the lower the manufacturing cost per pound, and ultimately the lower the selling price. With better processing controls higher processing speeds are now possible. With machines with more carriers, larger fabrics can be executed, and more pounds per hour can be processed. The result of these actions is a much lower cost per pound for all braided materials, making it very competitive with other reinforcement styles.

Probably the biggest misnomer is that braided reinforcements are only for "performance driven applications". Braids certainly do enhance performance - in all composite laminates! But due to the lower costs per pound, braid can now be used in more generic laminates. With so many unique properties and the efficiency with which it delivers all properties, thinner laminates with equal properties are also possible, at a lower total cost to traditionally reinforced laminates!

VARIATIONS IN FIBER AND BRAIDED REINFORCEMENT ARCHITECTURE

There is a tendency, when braided reinforcements are suggested, to reject them unless a tubular shape is needed. This attitude overlooks the versatility in fiber choices, fiber architecture and fiber content offered by the advanced braiding technology in use in 1997.

In addition to standard tubular biaxial braids, triaxial braids with added longitudinal fibers are routinely available. An almost unlimited mix of reinforcing fibers can be incorporated in hybrid braids. The modern braiding machinery can be adapted to produce special braid designs to meet specific needs.

Tubes, which are only part of the story, are available in diameters from .05" to 36".

Many applications need flat reinforcements. Contrary to some popular views, flat tapes are routinely available in widths from a fraction of an inch to 36" wide.

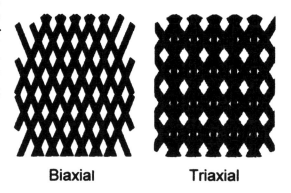

Biaxial   Triaxial

To better understand what a braid is and is not, it is helpful to define braids. Braiding is "a method of intertwining three or more threads in such a way that they cross one another and are laid together, without twisting, in diagonal formation". In tubular braids, there are two sets of continuous yarns, overlaid in a diagonal pattern. In flat braids, one set of yarns is used, and each yarn is interwoven with every other yarn in a zigzag pattern from edge to edge.

Recent developments in braiding machines allow us to do much more than expected by that simple definition. A braid today may have as many as 900 "threads" or "yarns". Braid angles can be infinitely varied from 15° to 75° from the machine direction and the braid angle can be changed (customized) during the braiding process. Extra longitudinal fibers can be introduced, and fiber materials can be mixed. Any conventional reinforcement fiber can be braided as long as it is reasonably flexible. Fiberglass, carbon, aramid, polyester, ceramic, polyethylene, and other synthetic fibers are routinely braided.

Of all the ways of introducing fibers to strengthen polymer composites, braid may be the most versatile. Let's look at what braids can do compared to more conventional ways of arranging fibers.

• compared to woven fabrics, braids offer drapability and can be tailored to give more isotropic properties. In multi-layer composites, braided fibers tend to nest and this helps reduce interlaminar failures. In tubular composites, woven and stitched fabrics must be cut and overlapped; braids are continuous in nature with continuous fibers running from end to end.

• compared to filament winding, braids offer tailored changes in fiber angle and provide properties similar to filament wound structures. Braids eliminate the delamination worries of multi-layered wound structures and can conform to bends and changes is cross-section, such as flanges, that baffle the filament winder. Braided fabrics can mimic the properties of a filament wound structure without the capital investment in winding equipment

• compared to unidirectional fabrics, braids have these same advantages, and can add to the properties of conventional unidirectional structures such as pultruded profiles.

• compared to a normal preform, fibers can be braided into a net shape preform, adding the extra strength of continuous fibers and better handling characteristics.

While braids may substitute for other forms of reinforcements, they can perform equally important supporting roles, being used in conjunction with other reinforcements, or being added as a selective reinforcement in critical component locations that may get subsequent fabrication.

The fiber angle on a simple sleeve can be tailored, reducing the labor in making a golf shaft while adding zip to the drive. With excellent fatigue properties braid also adds longer life to the club.

Adding a braided tape made with texturized fiberglass yarn to a pultruded beam section helped Stoughton Composites add strength just where it was needed for the screw retention required when fastening down the floor of a shipping container.

BRAID PROVIDES UNIQUE CHARACTERISTICS

Conformity is one of braid's principal virtues; a braided reinforcement can conform to compound complex shapes without losing its integrity. An example of this conformity feature can be shown in aircraft ducting and the fittings for pipes where flanges are molded with a single braided sleeving providing complete flange coverage.

This conformability extends itself beyond complex or compound shapes to tapered tubular structures. In either its normal form or in pre-

Single ply braided sleeving can conform to the body and flange of compound, complex parts.

impregnated form braided sleeving is easy to apply and tension over tapered mandrels. Multiple plys are, likewise easy to handle and apply. Lighting poles and golf club shafts are examples of products that take advantage of braids conformability to these structures.

While braided reinforcements provide excellent torsional strength in these applications the load transfer characteristics of braid make it excellent in tensile applications. In fact, in the tensile mode, braid also exhibits another of its unusual characteristics, ductility. As the product is loaded in tensile the fibers begin to pick up the load and distribute it from fiber to fiber until the full braid has evenly supported the complete load. So long as the resin matrix has the elongation necessary the composite will efficiently operate in a ductile fashion until fully loaded in tensile. Only this fiber architecture operates in this manner. Braid makes an ideal material for laminating over the walls of buildings in seismic regions due to its ductile behavior.

Braided reinforcements conform to tapered structures

When it is necessary to make an opening in composite components, cutting reinforcing fibers weakens the structure. Using braids by simply spreading the braid around the desired opening allows the hole to be reinforced without any fibers being cut. This not only allows the opening, it adds strength by densifying the fibers around the opening rather than reducing strength by cutting out fibers. The handhole in lighting poles is a great place to take advantage of this selective reinforcing feature possible with braided reinforcement.

In all these unique characteristics braid is able to perform so efficiently because the fibers run continuously throughout the composite. In woven or stitched fabrics the fibers are cut and applied in an overlapping manner that cannot give the strength found with continuous fiber reinforced braided reinforcements.

BRAID FINDS A PLACE IN ALL COMPOSITE PROCESSES

Another way braid demonstrates its versatility is in the fact that it is adaptable to virtually all the composite processes, from RTM and SCRIMP® to hand lay-up, autoclave molding, pultrusion, filament winding and compression molding.

Open mold processes like hand lay-up and SCRIMP molding are ideal for braid due to its conformability over compound shapes. RTM and autoclave molding also find considerable advantage in the use of braided reinforcement plys. In fact, many times the laminate can be made thinner because the continuous fiber nature of braids, along with its conformability allow more efficient laminates that perform better because there are no overlaps or seams in the laminate. Bladder molding is also a good process for braided reinforcements. It has an additional advantage when braid is used, because when the bladder is filled with air the braided fiber architecture is stretched putting the fibers in the tensile mode where we can utilize 100 percent of their strength value.

Pultrusion and filament winding also use braid, but generally more for the selective reinforcement of a specific area within the part where shear or screw retention is required or in areas where fabrication will subsequently take place. Pultruded tubular shapes are possible using a braid insertion device (BID) that allows the tubular braid to be put over the mandrel and pulled over it through the die.

CONCLUSION

The proven successes of braided reinforcements suggest some future possibilities.

What is often said about many products is particularly true of braids, "The possibilities are limited only by the imagination!" The challenge to the designer is to scrutinize the critical areas in a composite structure. Using braids, the need for highly stressed areas does not mean that the entire component must be similarly strong. Adding braids can strengthen the critical areas while allowing less material to be used in non-critical areas, saving weight and reducing cost.

When the need for stiffness or torsional strength may have caused an overdesign to compensate for severely non-isotropic properties, braids can help. By substituting or adding braids, thinner sections may provide the required stiffness or torsional properties with lighter weight and lower cost. The new designs for skis, wake boards and snow boards demonstrate how effectively braid performs in these functions.

Sections that will encounter high shear stresses may survive better if the potential for interlaminar shear is diminished by using braids, whose third dimension characteristic provides a valuable linkage between two layers of reinforcement.

More often than not, braids can provide a solution to the problem of how to achieve optimum properties. Braids are versatile and adaptable to imaginative solutions. Braid should be viewed at the beginning of a project as an alternative material, perhaps as the ultimate reinforcement and THE cost effective choice - a combination of virtues that are not only worth considering but should hardly be ignored in today's competitive business environment!

BIOGRAPHIES

Jeff Martin, is a sales and marketing consultant with 30 years of experience in composite materials, processes and markets. His background includes nearly ten years with Owens Corning Fiberglas in reinforcement marketing. He assists A & P Technology, Inc. with their marketing and sales efforts to customers in all composite processes.

Jeff is a member of the Society of the Plastics Industry's (SPI) Composites Institute (CI) division, where he has served as a member of the Board of Directors for four years and Conference Chairman for their Annual Conference. He has managed the Pultrusion Sessions of that conference for the past 20 years. He is also a member of the European Pultrusion Technology Association (EPTA). Jeff has also lectured extensively on composite materials and processes, world-wide.

Andrew A. Head, is the president of A & P Technology, Inc., originally part of Atkins & Pearce, Inc., the world's largest braiding company. Atkins & Pearce has been in business since 1817. A & P Technology was established in 1995 to address the industry's needs for braided reinforcements and braided materials for advanced technology applications.

Andy is a 1986 graduate of Dartmouth College with a degree in economics. He is a member of SPI and numerous other trade associations.

Fabrics with Enhanced Surface Chemistry Increase Laminate Properties, Improve Design Confidence and Improve Cost Performance for Scrimp, HLU-SU, RTM, and Pultruded Composites

GORDON L. BROWN, JR. AND BUDDY CREECH

ABSTRACT

Through chemistry and finishing expertise, Clark-Schwebel, Inc. has produced fabrics that give significantly higher laminate mechanical properties in glass fiber reinforced thermoset laminates than traditional fabrics that use glass fiber rovings or yarns which have a "resin compatible sizing" applied by the glass fiber manufacturer. These "enhanced" fabrics will allow engineers/designers to design with significantly increased laminate mechanical properties. Compared with the weaving of many individual yarns (or rovings) each sized individually by the glass fiber producer, the finishing of a woven fabric simultaneously coats all yarns (or rovings) in the fabric, thus providing uniform finish application. The uniform finish application increases consistency thus promoting enhanced designer confidence leading to lower safety factors, in many applications, and thus reducing the cost of the finished composite to the end user.

The following processes will be able to take the most advantage of these increased properties and increase in design confidence: 1) vacuum assisted transfer molding, 2) pultrusion, 3) lamination to the surface of balsa and other cores, 4) hand lay-up, 5) RTM and vacuum bag molding, and 6) others.

INTRODUCTION

SUPERFABRIC™ brand of woven and non-woven fabrics (patent pending) from Clark-Schwebel, Inc. features enhanced surface chemistry which allows fabricators the ability to produce more cost effective parts and designers to decrease safety factors making composites more cost effective versus traditional structures made of wood, steel and aluminum. These new fabrics improve significantly the interface region involving the glass fiber, sizing on the glass fiber, and the thermoset resin matrix as evidenced by significantly enhanced laminate mechanical properties both before and after water boil exposure.

In the manufacture of these new surface enhanced woven and non-woven fabrics, glass fiber rovings and yarn reinforcements were used which contain a resin compatible sizing applied by the glass fiber manufacturer. The processing of the fabric may include multiple

---

Gordon L. Brown, Jr. and Buddy Creech, Clark-Schwebel, Inc., P.O. Box 2627, Anderson, SC 29622.

steps, but the tailored finish is always applied to the surface of the existing sizing system to produce the enhanced surface that improves the interface between the glass fiber reinforcement and the resin matrix as evidenced by significantly increased laminate mechanical properties.

It is well known that the producers of glass fiber reinforcements apply a sizing to the surface of glass fibers to perform a variety of functions. The primary purpose in the design of the sizing is to give the customers of the glass fiber producers a product which will not only process excellently in their operation, but give good laminate mechanical properties when combined with a thermoset resin matrix.

But there are chemicals within the sizing system that are placed there in order that the glass fiber manufacturers may successfully process the glass fiber reinforcements within their own facility. Some functions of the sizing system within a glass fiber manufacturers operations are:

1) Reduce or eliminate glass fiber filament breakage.
2) Allow proper package build.
3) Allow for proper curing with controlled binder migration of the sizing system as water is removed during the drying process.
4) Give the proper level of integrity to the glass fiber strand.

The result of the above is that the properties of the end use product as made with these glass fiber rovings or yarns are compromised to a degree by the incorporation of selected chemicals within the sizing system to accommodate the needs of the glass fiber manufacturer.

The surface enhancing chemistry creates a synergy between the sizing system as supplied by the glass manufacturer and the particular thermoset resin matrix system. This modification to the laminate interface contributes to a "better" laminate. Designers and engineers will be able to design using higher laminate properties and their confidence level with composites should be increased allowing them to reduce their safety factors which will allow composite material systems to be more cost competitive to traditional materials such as steel, wood and aluminum.

INDUSTRY DRIVEN TECHNOLOGY

The Market Development Alliance (MDA) of the Composites Institute has many aggressive programs to increase the usage of composite material systems across a broad spectrum of end use applications. The "Infrastructure" applications have been identified as being ideal applications where the many benefits of composite material systems will be realized.

The "Infrastructure" community has a long history of conservatism. Many of the structures in our infrastructure place "life and limb" as a primary consideration in their design and as such, the engineers and designers must be convinced of the long term durability and the "design considerations" which go into the use of a new material system. One of the primary functions of the Market Development Alliance is to build this increase in confidence by partnering with such a agencies as the U.S. Army's Core of Engineers in designing demonstration projects where composites with their associated benefits seem to have a high probability of success but where the industry needs to present a united front in demonstrating our new materials systems to these new end users.

In most composite applications, composites on a first installed basis are more expensive than the traditional material systems consisting of wood, steel, and aluminum. Composite manufacturers have argued for years that their material systems should and must be evaluated on a "life cycle cost". One of the primary features of a composite is its corrosion resistance and in many applications this results in a significantly longer life span than traditional materials and as such, the economics are very favorable toward composites if "life cycle costing" is considered.

But the "harsh reality of life" has taught us all that the people providing the money for capital expenditures are definitely interested in minimizing the "initial cost".

Therefore, it is absolutely necessary for composite material systems to be designed in the most cost effective fashion as possible in order to present the most favorable "initial cost picture". One of the best ways to improve the cost competitiveness of composite material systems in structural applications is to have the confidence of the designers/engineers so that the safety factors do not penalize composite materials. Today, with many engineers/designers just becoming comfortable with composite material systems, they are reluctant to use the safety factors being used with traditional material systems.

One of the reasons for the conservative nature of designers/engineers is the variability in laminate mechanical properties as a result of the many different potential combinations of reinforcing materials, a wide variety of resin systems combined with the different processes for combining the reinforcements in the resin matrices.

The focus of this paper is directed specifically to building increased confidence in the minds of the designers/engineers. By using continuous forms of reinforcements in a more cost effective manner, the fabricator can supply components at a lower cost. By showing the designer/engineer that the materials that are being recommended do contribute significantly to improving the bond between the resin matrix and the surface of the reinforcement, the confidence in the long term performance of the composite material systems can and will be enhanced particularly as compared to composites made with continuous forms of reinforcements in which the surface of the reinforcement has not been enhanced.

LABORATORY LAMINATE PREPARATION AND TESTING

Clark-Schwebel, Inc. made and tested laminates in its laboratory. The general guidelines for laminate preparation follow. Results are shown in Table I and Table II. Vinyl ester resins (Dow Chemical's 8084) were used for wet lay-up laminates and cured with cobalt napthenate/MEK peroxide. All laminate testing done in crosswise direction.

- Laminates were approximately 15" x 16".

- Approximate thickness of laminates was 0.150".

- Fabrics tested were uni-weft with approximately 90% of the reinforcement in the weft direction.

- Approximate fiber volume of the finished laminate in the weft direction was 45%.

- Laminates are formed by placing the resin and fabric on a release film which is formed into a bag.

   The laminate is rolled to remove air.

   It is then cured in a hydraulic press at 85°F (for consistency) using 50 psi and 150 mil stops.

   It is post-cured in an oven for 8 hours at 140°F.

- Phenolic resin is Borden's Durite SC-1008.

The fabric was prepregged (not a wet lay-up).

They were stacked and pressed in a hydraulic press at 300 psi and 325°F for 1 hour.

TABLE I - SURFACE ENHANCED FABRIC PERFORMANCE WITH VINYL ESTER RESIN

	Laminate A Style 14542 (Non-Enhanced)	Laminate B Style 14542 (Surface Enhanced) Version A	Laminate C Style 14542 (Surface Enhanced) Version B	Laminate D Style 14550 Pattern 73 (Non-Enhanced)	Laminate E Style 14550 Pattern 73 (Surface Enhanced) Version C	Laminate F Style 14550 Pattern 73 (Surface Enhanced) Version B
Laminate ID	I-551-12	I-551-11	**I-551-13**	I-554-28	I-554-34	**I-554-35**
Resin Type	Vinyl Ester	Vinyl Ester	**Vinyl Ester**	Vinyl Ester	Vinyl Ester	**Vinyl Ester**
Resin Content (%)	25.9	27.1	**27.4**	28.3	26.5	**28.5**
Thickness (in.)	0.152	0.156	**0.159**	0.150	0.146	**0.155**
Water Absorption (%)	0.146	0.174	**0.035**	0.066	0.043	**0.024**
Plies	10	10	**10**	7	7	**7**
Flex Strength (PSI - Dry)	73,820	94,980	**120,900**	109,700	104,000	**124,600**
Flex Strength (PSI - 2 hr. Boil)	68,230	78,960	**107,800**	99,310	88,360	**109,100**
Flex Modulus (PSI - Dry)	3.64	3.5	**3.68**	3.96	3.91	**3.87**
Flex Modulus (PSI - 2 Hr. Boil)	3.51	3.44	**3.74**	3.95	3.88	**3.64**
Compressive Strength (PSI - Dry)	42,294	60,529	**50,347**	51,130	41,760	**58,230**
Compressive Strength (PSI - 2 Hr. Boil)	35,105	47,570	**44,812**	35,150	31,370	**53,210**
Tensile Strength (PSI - Dry)	74,230	74,870	**95,950**	103,500	99,360	**110,100**
Tensile Strength (PSI - 2 Hr. Boil)	64,510	75,650	**93,960**	97,540	92,060	**103,200**
Fiber Volume (%)	48%-total; 35%-fill	48%-total; 35%-fill	**48%-total; 35%-fill**	51.6% - total; 44%-fill	51.6% - total; 44%-fill	**51.6% - total; 44%-fill**

NOTE: Laminates A, B & C were made using the same fabric construction
Laminates D, E &F were made using the same fabric construction

PULTRUDED LAMINATE ENHANCEMENT

Pultruded laminates have gained wide spread acceptance and are one of the leading processes in fueling the growth of composite materials. Pultruded laminates have traditionally been significantly stronger in the machine direction than in the crosswise direction. This is due to the ease of processing continuous roving reinforcements in the machine direction.

It has been difficult to achieve high laminate mechanical properties in the crosswise direction with pultruded laminates. Continuous strand glass mats are traditionally used to provide the current level of crosswise directional properties in pultruded laminates. Today, a considerable amount of development work is being done to incorporate uni-weft woven and non-woven fabrics to enhanced the crosswise direction of pultruded laminates.

"Surface enhanced fabrics" have demonstrated their ability to take the level of crosswise directional properties to new highs. It is now possible to achieve the same level of properties in the crosswise direction as it is in the machine direction for pultruded laminates.

Phenolics offer the pultrusion industry a tremendous opportunity for growth. Phenolic resins are excellent in enclosed spaces where there is a concern for flammability and the generation of smoke. Table II below entitled "Surface Enhanced Fabric Performance With Phenolic Resins" demonstrates what can be achieved when a surface of a reinforcement is enhanced. Remember that the base reinforcement does have a sizing applied to the glass roving by the glass producer and the "enhancement" consist of an additional finishing step in which a surface enhancing chemical has been applied to improve the interfacial bonding between the resin matrix and the surface of the glass fiber. This sizing applied by the glass manufacturer has not been removed.

TABLE II - SURFACE ENHANCED FABRIC PERFORMANCE WITH PHENOLIC RESIN
(Properties Measured in Crosswise Direction)

	Laminate G - Style 15000 (Non-Enhanced)	Laminate H - Style 15000 (Surface Enhanced)
Laminate ID	551-22F	551-24F
Resin Type	Phenolic - Clark-Schwebel SC-1008	Phenolic - Clark-Schwebel SC-1008
Resin Content	20.4	18.8
Thickness	.129"	.126"
Water Absorption (%)	0.073	0.33
Flex Strength (PSI - Dry)	76,740	117,100
Flex Strength (PSI - 2 Hr. Boil)	75,090	115,800
Flex Modulus (PSI - Dry)	4.69	5.47
Flex Modulus (PSI - 2 Hr. Boil)	4.96	5.36
Compressive Strength (PSI - Dry)	58,236	67,499
Compressive Strength (PSI - 2 Hr. Boil)	53,665	67,945
Tensile Strength (PSI - Dry)	94,307	99,212
Tensile Strength (PSI - 2 Hr. Boil)	89,712	96,845
Fiber Volume (%)	67.4% - total; 52.7% - fill	67.4% - total; 52.7% - fill

PULTRUSION CASE HISTORY

One way in which to measure the cost effectiveness of a product improvement is to generate laminate mechanical property data and to express that information in relationship to the cost of the reinforcement. In pultrusion it is appropriate to measure cost performance in terms of the cost in dollars ($) per 1000 psi of flexural strength, flexural modulus, compression strength and/or tensile strength. A designer/engineer may have other factors to consider, but one of the inputs to his or her decision making will be the cost performance of a particular composite system.

The Table III below entitled "Phenolic Pultrusion Laminate Cost/Performance Analysis" presents data from four different pultruded composites and expresses the cost as a ratio of cost to 1000 psi of tensile strength, tensile modulus, compression strength, flexural strength, and compression modulus.

TABLE III - PHENOLIC PULTRUSION LAMINATE COST/PERFORMANCE ANALYSIS

NOTE: 1) Laminate data is measured in crosswise direction

2) Cost/performance data based on considering only glass-fiber material cost in crosswise direction with cost/yd^2 for material contributing to crosswise performance.

Composite Description	Property	Cost/Unit (Material)	Total Material Cost/Yd^2	Cost Per Performance Criteria
4 Layers (1 osf) High Performance CSM + Rovings (A) .090" thick	28,930 psi Flex. Str. (Dry)	$1.50/lb	$3.38/yd^2	$.117/1000 psi
2 Layers, Uni-weft Woven Enhanced Fabric + Rovings (B) .090" thick	84,350 psi Flex. Str. (Dry)	$2.75/lb	$7.32/yd^2	**$.087/1000 psi**
2 Layers, Uni-weft Non-Woven Fabric (Non-Enhanced) + Rovings (C) .090" thick	40,000 psi Flex. Str. (Dry)	$2.20/lb	$5.68/yd^2	$.142/1000 psi
(A)	16,101 psi Tensile (Dry)	$1.50/lb	$3.38/yd^2	$.209/1000 psi
(B)	37,340 psi Tensile (Dry)	$2.75/lb	$7.32/yd^2	**$.196/1000 psi**
(A)	1.103 x 10^6 psi Flex. Mod. (Dry)	$1.50/lb	$3.38/yd^2	$3.06/10^6 psi
(B)	4.43 x 10^6 psi Flex. Mod. (Dry)	$2.75/lb	$7.32/yd^2	**$1.65/10^6 psi**

The data presented in Table III is taken from the pultruded laminates described in Table IV.

TABLE IV - PULTRUDED LAMINATE PROPERTIES

Materials Description	Laminate I - Four Layers of Coarse Fiber Continuous Strand Mat Plus Rovings	Laminate J - Two Layers Style 14550, Uni-Weft Enhanced Fabric Plus Rovings
Laminate ID	I-551-5F	I-551-44F
Resin Type	Phenolic	Phenolic
Glass Content	67.3%	84.2%
Thickness	.089"	.093"
Water Absorption	2.563%	0.31%
Flex Strength (PSI - Dry)	28,930	84,350
Flex Strength (PSI - 2 Hr. Boil)	21,649	65,330
Flex Modulus (PSI - Dry)	1.103	4.43
Flex Modulus (PSI - 2 Hr. Boil)	.905	4.11
Compressive Strength (PSI - Dry)	13,115	15,900
Compressive Strength (PSI - 2 Hr. Boil)	5,010	11,810
Tensile Strength (PSI - Dry)	16,101	37,340
Tensile Strength (PSI - 2 Hr. Boil)	14,889	35,050

NOTE: The number of machine direction rovings used in pultruding laminate I and J were equal.

TABLE V - "SURFACE ENHANCED" UNI-WEFT DEVELOPMENT PRODUCTS

	Construction	Total Weight (OSY)	Warp Weight (OSY)	Weft Weight (OSY)	Weave	Features
Style 14550 Pattern 72 (Dimensionally stable in weft)	44 x 14 Fill: 450 yield roving	21.4	3.4	18.0	Crowfoot Satin	▪ Flat pultruded surfaces ▪ Phenolic resin compatible
Style 14550 Pattern 73 (Dimensionally stable in weft)	44 x 14 Fill: 450 yield roving	21.4	3.4	18.0	Crowfoot Satin	▪ Flat pultruded surfaces ▪ Vinyl ester and polyester resin compatible
Style 14550 Pattern 73	44 x 14 Fill: 450 yield roving	21.4	3.4	18.0	Crowfoot Satin	▪ Excellent wetout ▪ HLU-SU, Scrimp, RTM, etc. ▪ Vinyl ester and polyester resin compatible

The pultruded laminates in Table IV were made using the same die but changing the form of the glass fiber reinforcements in the cross-machine direction. The cost analysis presented in Table III shows clearly the superior cost performance of using a uni-weft fabric in which the surface characteristics of the glass have been modified to produce the significantly increased laminate mechanical properties. Although not presented in the above table, it is worthy of note that the cross-directional laminate properties were approximately equal to the machine direction properties for the pultruded laminate made with the uni-weft "surface enhanced" fabric. Glass content in machine and cross-machine direction was approximately equal.

The processing of the uni-weft "surface enhanced fabric" was excellent. Wetout of the fibers was very good, and the cross-directional continuous fibers maintained a perpendicular relationship to the longitudinal rovings and did not distort which contributed to the increase in laminate mechanical properties. The transfer of one roll of fabric to the other was efficient with the tensile strength of the uni-weft fabric being sufficient to allow a supply roll to be stitched to the feed roll during processing without breaking out.

CONCLUSION

Initial technical development work for "surface enhanced" uni-weft fabrics for application with vinyl ester resins and phenolic resins has been completed. Table V gives a description of the fabric forms available. Changes to these constructions can be made to meet customer's specific requirements.

Development will continue to understand more fully the reasons behind why the additions of certain chemicals to the surface of glass fibers enhances laminate mechanical property performance. With phenolic resins, Clark-Schwebel observed a difference in the degree to which laminate mechanical properties are increased depending on the phenolic resin system used. As such, Clark-Schwebel will insist on working closely with potential users of this new fabric to make sure that they receive the expected benefits and if not, to identify changes that may be necessary to insure the expected performance.

All of the laminate data with vinyl esters was generated using Dow's 8084. Because there are differences between vinyl ester resin producers, each customer must verify the property levels that they are expecting in their laboratory prior to the production of finished fabricated parts. Comparable increases would be expected with isophthalic polyester resins.

As composites continue to make progress against the traditional materials such as wood, steel and aluminum, we all expect composite materials to improve. One of the keys to producing the maximum amount of strength and performance from a composite is to design the best interface between the resin matrix and the surface of the reinforcement product. Clark-Schwebel believes its "surface enhanced woven and non-woven fabrics" will contribute significantly to this goal of increasing the competitiveness of composite material systems by increasing their cost effectiveness.

Clark-Schwebel wants to thank Morrison Molded Fiber Glass Company for their cooperation in pultruding laminates using their phenolic resin system.

BIOGRAPHIES

Gordon L. Brown, Jr.
Gordon Brown received his undergraduate degree from the University of North Carolina in Charlotte and his graduate degree from Clemson University. He is Director of Commercial Development for Clark Schwebel Tech-Fab Company in Anderson, South Carolina.

Mansfield (Buddy) H. Creech, Jr.
Buddy Creech received his undergraduate degree in Chemistry from St. Andrews College. He holds the position of Commercial Development Chemist with Clark Schwebel Tech-Fab Company in Anderson, South Carolina.

State-of-the-Art Developments Yield Braided Reinforcement Designs that Meet Unique Composite Product Requirements

ANDREW A. HEAD AND JEFF MARTIN

ABSTRACT

Braiding is perhaps the oldest textile process known to man. Almost certainly, it was developed as a way of intertwining decreet lengths of plant or animal matter into longer lengths that could be used as rope. Braiding was also recognized as a work of art. Many ancient cultures used braiding to create ornate ceremonial belts and necklaces.

Until about 300 years ago, braid was produced by hand. Because the formation of a braid requires the constituent elements to move relative to one another, it was a very difficult task to keep track of where and when each element was to be moved. As a result, even complex decorative braids comprised relatively few elements and the applications for braid were quite limited. A typical example of a hand made product is the braided rug fabricated by America's frontiersmen. Because of the interlocking process, small scraps of fabric could be joined into long lengths and then stitched together into a flat coil.

With mechanization, more complex structures could be fabricated more cost effectively. Using essentially the same technology that was developed three centuries ago, modern braiding machines produce a broad range of products that are a part of our everyday life. Many applications go without notice. In addition to ordinary things like shoe laces, sash cord and clothesline, braid is broadly used in automobiles, appliances and computers for wire insulation and harnessing. The vast majority of candles and oil lamps use braided wick. Braid also finds a myriad of decorative applications such as shopping bag handles, gift cord and decorative trim on apparel.

While braided reinforcing fibers have begun to find application in a variety of composite products, their penetration has been slow for a number of reasons. But recently the technology in custom braiding equipment has permitted faster processing speeds, lower scrap and tighter processing specifications. This has led to lower costs and higher sale volume for braided reinforcements. Now, even more sophisticated braiding processes are yielding exciting new products that will likely redefine how the composites industry views braids. This paper will explore these new products made possible with this advanced braiding technology.

Andrew A. Head and Jeff Martin, A&P Technology Inc.

INTRODUCTION

For all it's influence on our everyday life, braid has never enjoyed the wide spread use that have knit, woven and non-woven materials. This is because it has been a relatively expensive process and even the largest machines were only capable of making very narrow fabrics.

Because of the size limitations, the use of braid as a reinforcement for composites has been traditionally limited to items with relatively small cross sections. Perhaps the oldest and most prolific use of braid in a structural composite dates back to the early 50's where a fiber glass braid was impregnated with a phenolic resin and hardened into a porous tube and incorporated into a lead-acid battery. More recently, braid has been incorporated in small cross section composites such as canoe paddles, baseball bats and snow skis.

The effectiveness of braid as a reinforcement had been constrained because of the nature of braiding equipment. Since braid has been a minute subset of the total textile industry, braiding machines were designed to utilize fibers that were being supplied to the weaving and knitting industries such as cotton, silk, rayon, polyester, nylon and other man-made fibers. Because these materials are relatively tough and flexible, braiding machines could handle the material quite vigorously without causing harm to the material.

THE COMPOSITES CHALLENGE

As braiding companies have sought to expand the use of braid as a reinforcement for composites, they have been confronted with materials that are neither tough nor flexible, namely glass and carbon. It became clear that the braiding process, while capable of making a textile structure of great value as a reinforcement material, was critically deficient in its ability to cost effectively do so.

In addressing the deficiencies, the braiding industry has advanced the state-of-the-art at a revolutionary rate. During this revolution, issues of size, cost, fiber architecture and fiber damage have all been addressed to create an entirely new industry within the world of braiding that is capable of meeting the most demanding advanced technology applications.

NEW BRAIDING TECHNOLOGIES YIELD NEW PRODUCTS
AIRCRAFT ENGINE CONTAINMENT

One of the first major advances in the state-of-the-art came in response to General Electric's desire to use a lighter weight, more efficient containment wrap around the primary fan of their commercial jet engines. In subscale testing, braid proved to be more efficient than woven, knit and non-woven structures. The problem was, however, that the production width of the braided tape was to be in excess of 26". In order to meet this need, a machine capable of incorporating the number of ends necessary to create the width had to be designed and built.

A & P Technology responded by building a 400 carrier machine which is almost 30 feet in diameter. At the time it was built, it was the largest braiding machine in the world. This remarkable machine not only consistently braids aramid yarn to the desired width, but also performs the braid to fit the precise contour of the engine case, allowing it to be wrapped without wrinkles or sags. This braid is capable of stopping a liberated 35 pound fan blade whose tip speed is traveling at mach two, which is equivalent to the force of an automobile falling off a five story building. Only braided fiber architecture and aramid materials could handle the containment of this force.

SIDE IMPACT AIR BAGS

Some of the technology that was developed in connection with the containment project became useful in meeting another high technology application for braided fabric: side impact airbags. In this application, the natural "Chinese handcuff" action of the braid is used in reverse. By inflating a biaxial braid internally it expands in diameter and shrinks in length. ASD Simula, Phoenix, Arizona, harnesses this phenomenon to create a device that is stowed in an elongated configuration. Upon inflation, it self deploys into a position to cushion a vehicle occupant's head in event of a side impact. While a braid made on any size machine would accommodate the requirement of shrinking in length while

Braided Side Impact Air Bag Deployed

expanding in diameter, only braid made on an extremely large braider would meet the rigid requirements of this market. The tube needs to have full coverage of the fabric at the six inch inflated diameter so that it can handle internal pressurization without damage to the bladder material. At the same time, the bag fabric needs to have a thin enough profile to be stowed in a small volume beneath the trim above the door. This led to the creation of the 600 carrier braiding machine, which is currently the largest in the world.

These large machines have inspired the use of braid in a variety of applications which in turn, have spawned the development of an entire family of large braiding machines ranging from 172 carriers to 600. These large machines allow for the braiding of light weight, full coverage sleevings for applications such as water skis, wakeboards and snowboards.

At the same time as providing fabrics for articles with large cross sections, the need has been very apparent to reduce the costs of braided fabrics. In creating this new line of equipment, A&P Technology has addressed all the cost drivers and yielded a process capable of competing with not only traditional textile processes like knitting and weaving, but also

An A&P Technology Mega Braider Processing Aramid Fiber Reinforcement

new composite reinforcement processes like those used to make stitched fabrics.

BROAD GOODS

The combination of these advances has allowed the braiding process to shatter a centuries old barrier. Braiding is defined in textile dictionaries as a narrow fabric process. A narrow fabric is distinguished from a broad good by having a width dimension of less than 18 inches. By taking these large dimension machines and cost effectively producing sleevings in excess of six inches in diameter, these bias fabrics can be slit and laid open to create a new family of cost competitive broad goods. A 16 inch diameter sleeve is used to create the standard 48 inch width of many broad goods.

These products, created by A&P Technology, are sold under the trade names of Bimax™ and Trimax™. The principle benefits of braided broad goods are that they provide the continuous length of biased reinforcement currently available in stitched fabrics with the handling and physical properties of woven fabric. Bimax™ and Trimax™ brand fabrics can be supplied in a variety of weights and fiber orientations, allowing a designer to cost effectively optimize a composite laminate.

CONFORMABLE TRIAXIAL BRAID

In addition to new technology involving increased size and reduced cost, new demands put on braid manufacturers by the composites market have inspired all new braid constructions.

For example, many composites are non-linear, having complex compound sections or both. Biaxial braid is great for these types of structures because of its excellent drape and conformability. As such, it is used extensively in RTM and SCRIMP molding applications. However, biaxial braid cannot always provide the desired axial, in-plane properties desired by the designer.

Triaxial Braid Architecture

Traditionally, the fix has been to cut out irregular shapes of unidirectional material and wrap it into the desired shape, employing a variety of creative ways of addressing the gaps, overlaps and constraint of the material until it gets consolidated.

To simplify this, A&P Technology developed a new braided construction trade named Unimax™. This patent pending product is a triaxial braid incorporating a very small percentage of the total weight in bias elastic yarns. This creates an axial reinforcing sleeve that is capable of stretching over tapered, curved or other irregular shapes and fitting

Biaxial Braid Architecture

snugly while evenly distributing the reinforcement around the perimeter of the part. Unimax™ is an ideal material for tapered structures like flag, light and utility poles, irregular shapes like snow boards and water skis and curved elements such as those in commercial furniture.

Another problem encountered in composites manufacturing is solved using elastic in another patent pending triaxial braid construction. In this approach, the reinforcement material is laid up in the bias direction. A small percentage of the total fabric weight is elastic yarn laid in the axial direction. This produces a biaxial reinforcement that orients itself at high angles while maintaining superior drapeability. This can be applied to sleevings as well as slit and braided flat fabrics. Flat braided tape utilizing this technology makes a cost effective, easy-to-use tabbing material for boat builders.

Unimax™ Braid on a Tapered Structure

TWO SIDED FABRIC

Another advance in the state-of-the-art braiding came in response to one product need, but may find more application in unrelated products.

The original need came from the non-lubricated composite bearing industry. They sought a reinforcement construction which would integrate a good bearing material, like PTFE, with a material that suitably reinforces a rugged composite, like fiberglass.

A traditional biaxial braid is not suitable because it presents equal areas of its constituent yarns on each side of the fabric. Glass fiber on the friction side of the fabric detracts from the tribological properties and PTFE on the composite side diminishes the strength properties. So, a "two sided" material was necessary. By a proprietary technique, A&P Technology tricked the braiding process into making a material that meets this need.

The patent pending sleeving is constructed with four sets of yarns where two sets are larger in cross section and/or stiffer than the other two sets. They are integrated so that one of the larger yarns is located on the inside surface, coiled in a helix and the other larger yarn is situated on the outside surface in a counter rotating helix. For the bearing reinforcement, the inside material is PTFE and the outside is glass. These two materials are tied together with the two smaller, more flexible yarns, one on the inside and the other outside.

Double-Sided Bearing Braid

Again, the inside material is PTFE and the outside is glass. The net result is an inexpensive, fully automated tubular perform that is tailored to have completely different materials on either side, while still having through the thickness yarns integrating both layers and the natural conformability of braided sleeving that easily forms flanges and irregular shapes.

CRIMPLESS BRAIDED REINFORCEMENT

What is even more remarkable about this construction is that it provides two sets of counter rotating sets of yarn that are uncrimped.

Hugo Kruesi and his team from U.S. Composites first pioneered this field of non-crimped braided reinforcement. Their effort led to a patented material that utilizes a heavy yarn in one direction and a lighter more flexible yarn in the other. Dubbed "Asymmetric Braiding" the lighter yarn does all the crimping, allowing the heavy yarn to maintain straightness and hence its in-plane properties.

This new material does the same thing, only symmetrically. The resulting structure is, in essence, like stitched fabric in a tubular form. If regular tubular braid could be considered a hybrid of filament winding and weaving, this new material would be a hybrid of filament winding and stitching or knitting.

This new structure features essentially straight fiber properties in-plane, but gives up some of the strengths that are unique to ordinary braided reinforcement. The reason seems to be that ordinary braid very efficiently distributes applied loads evenly throughout a structure. The results are good strength and toughness properties and remarkable fatigue strength. The new material, currently, does not appear to have the same broad combination of material properties.

SCREW THREAD REINFORCEMENT

From the oil field came another composites problem that created a new construction and a new opportunity for braided reinforcements.

For a number of reasons, fiberglass composite is a good material for oil well pipe. In service, lengths of pipe up to two miles are joined together. While the pipe itself handles the tensile load very reliably, the connections between the pipes were cumbersome, expensive, unreliable or all three.

Because braid orients yarns in a helix, it seems, intuitively, that an answer may lie in some adaptation of the technology. The answer finds its roots in the automotive industry.

The car makers had a problem passing tubes and wires by hot engine parts, like the exhaust manifold. In order to achieve a high degree of insulation cost effectively, they needed to go beyond typical fiberglass or texturized fiberglass braided sleeving. It was then discovered that, by incorporating several larger tow bundles into an ordinary sleeving, the thicker "ribs" would cause the main sleeve to stand off of a hot surface and efficiently insulate its contents with less glass fiber.

The ribs, because they are integrally woven into the structure, are great candidates for screw threads. BF Goodrich Company received a patent on the composite structure several years after the automotive work.

The problem with basic rib structure is, however, that the ribs, or screw threads, need to be oriented at a high angle relative to the axis of the part to be effective in a joint. While the braiding process can accommodate this, it comes at the expense of other difficulties in processing. More importantly, high angle screw thread ribs are reinforced, or intertwined,

Original Tubular Screw Thread

with yarns at an equally high angle in the other direction. So, the benefit to the tensile strength of the joint that is derived from increasing the angle of the screw thread rib is lost by increasing the angle of fiber holding it in place.

In a proprietary adaptation to this basic approach, A&P Technology has developed a patent pending reinforcement that solves this problem cost effectively. Essentially, the new construction's braid axis is rotated to an acute a ngle off the axis of the part.

In other words, an ordinary triaxial ribbed braid would have bias fibers at -35 degrees, ribs at +35 degrees and axial, laid in fibers at zero degrees, where the braid axis and part axis are identical. This new construction tilts the three elemental sets 35 degrees so that the ribs comprising the screw threads are now at 70 degrees, the laid in yarns are at 35 degrees and the bias fibers that reinforce the ribs are at 0 degrees, the tensile direction of the part.

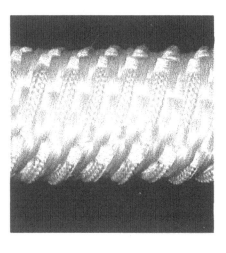

New Proprietary High-Angle Tubular Screw Thread Braid

This construction can be laminated into a filament wound or braided pipe and co-cured or cured separately and then incorporated into the pipe later.

The net result is a net shape thread structure that is user friendly, can handle enormous tensile loads, and can be formed and incorporated into the pipe cost effectively. In addition to fiberglass pipe, any structure that incorporates helical threads, such as composite bolts or impellers, can be improved with this invention.

CONCLUSION

While ordinary braid is a remarkable material for composites and other advanced technology applications, the depth and breadth of its opportunities are only recently coming to light. Without a doubt, the most important of recent advances in the state-of-the-art has been the reduction in the cost of braiding. Because of lower costs, the unique characteristics that braid contributes to a composite system are now more affordable. At the same time, the variety of useful forms that a braid can take are infinite and only now being discovered. So, while many in the market place are discovering the value of traditional braids and using them in new, innovative ways, braid manufacturers are pushing the envelope of cost and construction at a rate never before experienced in this ancient art. As more is understood about braid and what it can do, the pace will only quicken. As a result, advanced processing technologies promise to elevate braid from an art to a science - a cost effective choice for composite processors.

BIOGRAPHIES

Andrew A. Head, is the president of A & P Technology, Inc., originally part of Atkins & Pearce, Inc., the world's largest braiding company. Atkins & Pearce has been in business since 1817. A & P Technology was established in 1995 to address the industry's needs for braided reinforcements and braided materials for advanced technology applications.

Andy is a 1986 graduate of Dartmouth College with a degree in economics. He is a member of SPI and numerous other trade associations.

Jeff Martin, is a sales and marketing consultant with 30 years of experience in composite materials, processes and markets. His background includes nearly ten years with Owens Corning Fiberglas in reinforcement marketing. He assists A & P Technology, Inc. with their marketing and sales efforts to customers in all composite processes.

Jeff is a member of the Society of the Plastics Industry's (SPI) Composites Institute (CI) division, where he has served as a member of the Board of Directors for four years and Conference Chairman for their Annual Conference. He has managed the Pultrusion Sessions of that conference for the past 20 years. He is also a member of the European Pultrusion Technology Association (EPTA). Jeff has also lectured extensively on composite materials and processes, world-wide.

A Cost-Effective Approach to High-Strength Pultruded Composites

MICHAEL W. KLETT

ABSTRACT

This study describes the results from a collaborative effort between material suppliers to develop fiber glass reinforcement technology for pultrusion applications. The new technology utilizes an engineered (triaxially stitched) fabric constructed from a phenolic-specific fiber glass strand. The potential for engineered fabrics to compete economically with traditional nonwoven mats has been established. The purpose of this study is to characterize the physical, mechanical, processing, and cost advantages of engineered fabrics in pultruded composites. The results clearly show that stronger, higher modulus, thinner, lighter weight, and cost-effective pultruded (phenolic) composites can be obtained with engineered fabrics.

INTRODUCTION

Phenolic resins known for their low flame, low smoke, low smoke toxicity, and high service temperature performance have captivated the attention of pultruders and material suppliers in recent years, resulting in numerous publications (see references). Early efforts to pultrude phenolic fiber reinforced plastics resulted in various levels of success that depended on the material selection and processing conditions used. With the evolution of resin, reinforcement, and processing technology, the quality, structural properties, and cost of pultruded phenolic composites have become more competitive with traditional fire retardant styrenic systems. Driving the development efforts in phenolic fiber reinforced composites is the need to improve public safety in fire situations. The combination of low flame, low smoke, smoke toxicity, low weight, high strength, and corrosion resistance makes pultruded phenolic composites the material of choice for applications in offshore oil platforms, chemical facilities, mines, and tunnels. Current development efforts by various pultruders around the world are focusing on grating, cable tray, and conduit.

Dr. Michael W. Klett, PPG Industries, Inc., Fiber Glass Research Center, 201 Zeta Drive, Pittsburgh, PA 15238

OBJECTIVE

The primary purpose of this study is to demonstrate the mechanical, structural, processing, and cost advantages of a phenolic-specific engineered fabric over a "phenolic-compatible" continuous strand mat in pultruded phenolic fiber reinforced plastics.

EXPERIMENTAL

For the purpose of this study, both mat-reinforced and fabric-reinforced composites were pultruded using INDSPEC's Resorciphen[TM] resorcinol-modified phenolic resin and PPG 788 phenolic-specific roving (current designation is 2788). The engineered fabric designated as NPFMP 120 was produced by Advanced Textiles, Inc. The weight of the fabric was 1.4 oz/ft^2. The 1.0 oz/ft^2 phenolic-compatible continuous stand mat is described elsewhere (Gauchel, Lehman and Beckman, 1992).

Flat panel composites were fabricated at the PPG Fiber Glass Research Center in Pittsburgh, PA, using a Pulstar 804 pultruder and a 34-inch long flat panel die. Panel width was 4 inches, and thickness was adjusted through the use of shims. For composite property determinations, the heating platens were set at 350°F/375°F for Zones 1 and 2, respectively, and pull rate was held constant at 12 inches per minute (ipm). Standard fiber handling methods and a straight-through resin impregnation system were used. Coupons were cut from the flat sheets and randomized before testing. Flexural and shear properties were tested according to ASTM procedures D 790 and D 2344, respectively. Compressive strengths were measured by the straight 1.5-inch compression method described in ANSI A14.5-1982.

DISCUSSION

MATERIAL PROPERTIES

Traditional reinforcements used in pultrusions are combinations of roving and continuous strand mat. The reinforcing value of the roving is predominantly along the longitudinal axis relative to the pulling force or machine direction. The mat provides the transverse and torsional properties. It is generally accepted that the maximum benefit derived from a fiber reinforcement is along the fiber's length. Because of the random orientation of fibers in the mat, optimum transverse and diagonal properties are not possible. In contrast, engineered fabrics consist of precisely aligned fibers that can be directed such that the tensile properties of the fibers can be fully utilized. Fiber placement in the engineered fabrics is maintained

---

Resorciphen is a registered trademark of INDSPEC Chemical Corporation.

by stitching, thus eliminating the need for an insoluble binder typically used in mats. Incompatibility of the binder with the resin matrix and limitations due to binder-sizing interactions are avoided with stitching. The opportunities to engineer fiber orientation and sizing chemistries "optimized" for specific resin systems are greater for engineered fabrics.

A triaxially stitched fabric made from a phenolic-specific fiber glass strand was used in this program. The fabric was constructed of three consecutive layers of parallel laid fiber bundles with orientations of +45°, -45°, and 90° to the roll length. Glass contents of each of the layers were 36%, 36%, and 26%, respectively. The layers were held together by a polyester thread stitching.

PROCESSING

The basic tenet in manufacturing is to produce a saleable product in a cost-effective manner. Wide processing windows, fast production speeds, and a defect-free product are required for a profitable manufacturing operation. The processing window for pultrusion is complex and interdependent on such factors as resin type, resin additives, reinforcements, part complexity, and part thickness. The importance of these factors cannot be underestimated as they directly affect manufacturing costs and customer acceptance and satisfaction.

High glass loadings were necessary to achieve good surface and to minimize porosity and surface blisters. Typically, glass contents of 62-65% by weight gave the best results with the current filler package and filler level.

The effects of die temperature and pull speeds on product quality using each of the respective reinforcements are summarized in Table I. Pull speeds obtainable with the mat were limited to 12-18 ipm. Higher pull speeds resulted in severe delaminations originating at the center mat ply and extending across the width of

TABLE I--PULTRUSION PROCESSING WINDOW FOR MAT AND FABRIC REINFORCEMENTS

Platen Temperature (°F)					
Zone 1	375	375	350	320	350
Zone 2	400	390	390	390	375
Mat					
12 ipm	←--------Severe Delaminations--------→			←----------OK----------→	
18 ipm	←--------Severe Delaminations--------→			←----------OK----------→	
21 ipm	←----------------Severe Delaminations----------------→				Blisters
26 ipm	←--------------------------Severe Delamination--------------------------→				
Fabric					
12 ipm	←Scorched Surface→		←--------------------OK--------------------→		
18 ipm	←----------NA----------→		←--------------------OK--------------------→		
21 ipm	←--------------------NA--------------------→			←----------OK----------→	
26 ipm	←--------------------------NA--------------------------→				←-OK-→
30 ipm	←--------------------------NA--------------------------→				Small Blisters
36 ipm	←--------------------------NA--------------------------→				Small Blisters

the laminate. Surface blisters ranging from 0.5-2.0 inches in diameter also marred these panels. In contrast, the fabric was pulled at 26 ipm without incident of blisters or delaminations. Fabric pulled at speeds of 30-36 ipm resulted in 0.25-0.5 inch diameter blisters and no delaminations.

Production of blister-free mat-reinforced composites was a function of both pull speed and processing temperature, indicating a narrow processing window. High processing temperatures resulted in severe delamination, even at low pull speeds. Attempts to induce delamination of the fabric-reinforced composite by increasing the processing temperature was unsuccessful. The wider processing window afforded by the fabric is attributed to the phenolic-specific sizing chemistry. This is supported by the temperature-induced delaminations observed when a "general purpose" fabric was substituted for the phenolic-specific fabric. The "general purpose" designation used in the industry typically implies compatibility with epoxy and styrenic resins and does not guarantee optimum properties and processing in phenolics.

Pull loads of the fabric-reinforced and mat-reinforced panels at various glass contents are compared in Figure 1. The 30% lower pull forces required by the fabric are a consequence of the fabric thickness (0.02-0.03 inches vs. 0.03-0.05 inches for the mat) and the resin pickup. Resin pickup was indirectly

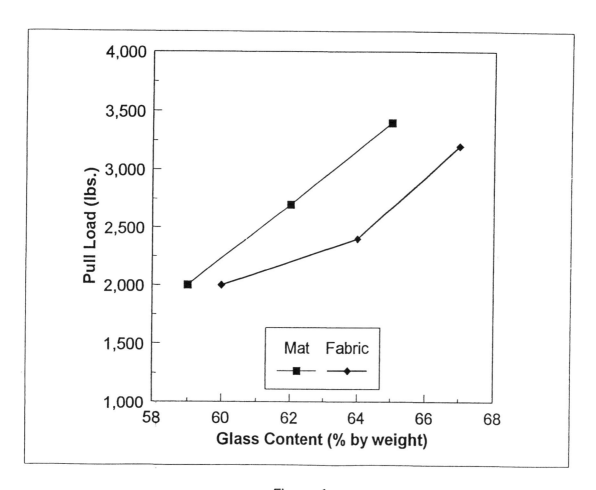

Figure 1.

determined by the "bulking" or "swelling" of the reinforcement at the die entrance and was significantly higher for the mat.

The implications of lower pull loads inherent in the fabric are thinner composite panels and potential improvements in line speed and cross sectional cure consistency. The ability to produce thinner panels was verified by pultruding a 0.10-inch thick 3-ply fabric-reinforced composite. Attempts to fabricate 3-ply mat composite at thicknesses less than 0.125 inches resulted in mat tears at the die entrance.

COMPOSITE PROPERTIES

Composite property comparisons for 0.10-inch thick 2-ply mat and fabric constructions are summarized in Table II. Weight differences between the mat and fabric (1.0 vs. 1.4 oz/ft^2) required adjustment to the number of rovings to maintain equal glass contents. Despite the lower (14%) roving content in the fabric-reinforced composite, flexural and shear properties along the machine direction were comparable to those obtained with the mat. Only compressive strength in the machine direction was off by 6%. The 25-50% improvement in transverse and diagonal properties observed for the fabric is undoubtedly due to the high degree of fiber orientation. Because of the inherent mechanical property advantages observed from the fabric, the effects of material reduction through roving content and panel thickness were evaluated for 3-ply constructions. Composite property comparisons between 0.115-inch and 0.125-inch thick 3-ply constructions of fabric and mat, respectively, are summarized in Table III. It should be noted that the 25% reduction in roving content associated with the fabric-reinforced composite negatively affected machine direction compressive strength only. Flexural strength along the machine direction was 39% higher for the fabric while shear strength and flexural modulus were comparable to the mat. In addition, transverse and diagonal properties for the fabric were 46-70% higher than those

TABLE II--MECHANICAL PROPERTY COMPARISONS FOR 2-PLY CONSTRUCTION

Properties[1]	Machine Direction		Transverse Direction		Diagonal Direction	
	Mat	Fabric	Mat	Fabric	Mat	Fabric
Glass Content (%)	65	65	65	65	65	65
No. of Rovings	56	48	56	48	56	48
Compressive Strength	50.0	47.2	11.9	15.4	15.2	17.7
Shear Strength	5.4	5.4	1.2	1.6	2.2	2.9
Dry Flexural Strength	76.8	77.3	17.0	25.5	29.1	39.4
Wet[2] Flexural Strength	60.8	71.5	13.0	25.3	21.4	37.1
Dry Flexural Modulus	3092	3066	1560	1954	2084	2740
Wet[2] Flexural Modulus	2664	2709	1477	1899	1904	2718

[1] Properties are in Kpsi (psi x 1000) unless otherwise noted.
[2] Coupons were subjected to boiling water for 48 hours.

TABLE III--MECHANICAL PROPERTY COMPARISONS FOR 3-PLY CONSTRUCTIONS

Properties[1]	Machine Direction		Transverse Direction		Diagonal Direction	
	Mat	Fabric	Mat	Fabric	Mat	Fabric
Glass Content as % by wt.	62	64	62	64	62	64
(no. of rovings)	(64)	(48)	(64)	(48)	(64)	(48)
Laminate Thickness (in.)	0.125	0.115	0.125	0.115	0.125	0.115
Compressive Strength	58.6	44.0	11.1	19.0	14.6	21.6
Shear Strength	4.2	4.2	1.7	2.5	2.4	3.5
Dry Flexural Strength	57.7	78.0	19.4	30.7	25.4	39.8
Wet[2] Flexural Strength	54.1	73.3	13.1	29.4	22.6	34.6
Dry Flexural Modulus	3088	3081	1525	1923	1543	2403
Wet[2] Flexural Modulus	2947	2861	1502	1857	1656	2539

[1] Properties are in Kpsi (psi x 1000) unless otherwise noted.
[2] Coupons were subjected to boiling water for 48 hours.

with the mat. Structural properties, such as load carrying capacity and deflection under load, for each of the 3-ply constructions were comparable in the transverse direction but correspondingly lower for the fabric in the machine direction. This structural deficiency along the machine direction was addressed by a 2-ply construction with a corresponding increase in the level of rovings. This included 60-roving and 68-roving inputs designated as Fabric-60 and Fabric-68, respectively. Composite properties are summarized in Table IV, and structural property comparisons are summarized in Table V. The 2-ply fabric-reinforced composites consistently showed higher strengths and modulus than those possible using the mat. Load carrying capacity (machine and transverse) and stiffness (transverse) for the fabric-reinforced composites were equal to or better than those

TABLE IV--MECHANICAL PROPERTIES OF 2-PLY FABRIC-REINFORCED COMPOSITES

Properties[1]	Machine Direction		Transverse Direction		Diagonal Direction	
	Fabric-60	Fabric-68	Fabric-60	Fabric-68	Fabric-60	Fabric-68
Glass Content as % by wt.	65	68	65	68	65	68
(no. of rovings)	(60)	(68)	(60)	(68)	(60)	(68)
Laminate Thickness (in.)	0.115	0.115	0.115	0.115	0.115	0.115
Compressive Strength	52.8	60.3	17.2	16.4	17.4	17.4
Shear Strength	5.8	6.2	1.9	2.1	3.6	3.5
Dry Flexural Strength	81.3	91.0	27.0	25.2	41.2	37.6
Wet[2] Flexural Strength	78.1	81.4	25.0	25.1	38.9	37.0
Dry Flexural Modulus	3037	3551	1927	1992	2456	2516
Wet[2] Flexural Modulus	2999	3345	1893	1940	2416	2599

[1] Properties are in Kpsi (psi x 1000) unless otherwise noted.
[2] Coupons were subjected to boiling water for 48 hours.

TABLE V--STRUCTURAL PROPERTY COMPARISONS

	Machine Direction		
	Mat	Fabric-60	Fabric-68
Maximum Load[1] (lbs.)	152.1	158.7	184.5
Deflection (in.)			
@ 50 lbs.	0.035	0.056	0.045
@ 100 lbs.	0.071	0.117	0.093
@ 140 lbs.	0.099[2]	0.172	0.134
@ 180 lbs.	NA	NA	0.142[3]
	Transverse Direction		
Maximum Load (lbs.)	53.0	53.4	52.7
Deflection (in.)			
@ 20 lbs.	0.031	0.033	0.030
@ 30 lbs.	0.052	0.052	0.045
@ 40 lbs.	0.076	0.073	0.066
@ 50 lbs.	0.102[4]	0.101	0.092[3]

[1] All testing used 2-inch span.
[2] Failure rate of 33%.
[3] Failure rate of 17%.
[4] Failure rate of 50%.

obtained from the mat. Only stiffness along the machine direction was off, but it could be tailored to meet or exceed that of the thicker 3-ply mat-reinforced composite by increasing the roving content.

Not to be overlooked is the consistently high wet strength property retention for the fabric-reinforced composites as shown in Tables II-V. Resistance of the composite to property degradation from boiling water is primarily a function of the sizing chemistry. This was previously demonstrated in phenolic composites pultruded with biaxially stitched fabrics containing a "general purpose" sizing chemistry (Dailey and Klett, 1993).

COST ANALYSIS

Figure 2 illustrates the relative cost components of mat-reinforced and fabric-reinforced composites having comparable structural properties (i.e., 3-ply mat and 2-ply [Fabric-68] construction). Costs were based on meters rather than weight due to differences in thickness and density of the two competing materials. For the current example, the comparison was based on 100,000 feet of composite using pull speeds of 18 ipm for the mat and 24 ipm for the fabric. These were the highest pull speeds achievable in our lab that provided blister-free parts.

Although the cost of the fabric is considerably higher than that of the mat, manufacturing costs are essentially identical for each of the composites. The higher fabric cost is offset by savings from material and labor reductions. The

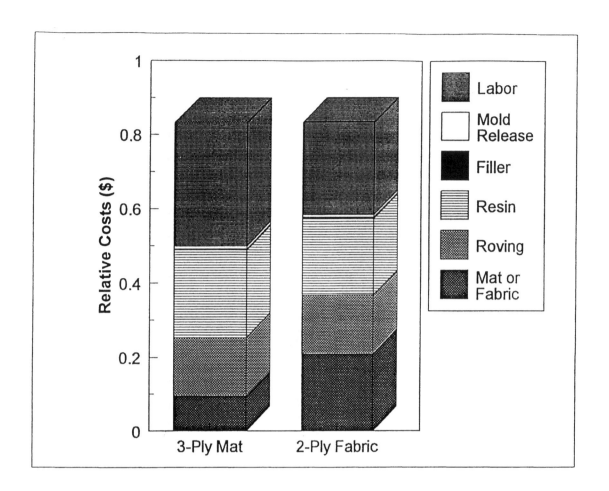

Figure 2.

lower labor component was a direct consequence of the higher pull speeds capable with the fabric. The fabric provides the opportunity to realize capacity improvements through more effective equipment utilization. The potential cost benefits derived from fabrics are specific to phenolic composites and need to be demonstrated in an actual application.

CONCLUSIONS

The ability to pultrude high-strength and high-modulus phenolic composites with good surface quality and low porosity is currently available using standard pultrusion equipment and materials. Previous limitations, such as low transverse strengths, slow pull speeds, and frequent blisters and/or delaminations associated with the "phenolic-compatible" continuous strand mat, have been overcome with a phenolic-specific triaxially stitched fabric. The potential to engineer thinner, lighter weight, less costly, and less complex constructions while maintaining comparable structural properties with a thicker, 3-ply mat reinforced composite was demonstrated. The opportunities for the fabric-reinforced pultruded phenolic composites are expected to increase as designers, specification writers, and regulatory agencies become more aware of this technology.

ACKNOWLEDGMENT

The author wishes to acknowledge several individuals for providing testing, raw materials, and typing of this manuscript. Included among these helpful people are Thomas Vorp and Sandy Nalesnik of PPG Industries, Inc.; Pete DeWalt of Advanced Textiles, Inc., and Harve Dailey of INDSPEC Chemical Corporation. I also wish to thank the management of PPG for authorization of this project.

REFERENCES

Dailey, T. H., M. W. Klett, and R. W. Allison. 1992. *An Investigation of Flame, Smoke, Toxicity, and Mechanical Properties of Pultruded Phenolic Composites,* 48th Annual Conference, Session 21-B, Composites Institute, SPI, Inc.

Dailey, T. H. and M. W. Klett. 1993. *Phenolic Pultrusion: Innovations for an Emerging Technology,* 49th Annual Conference, Session 9-D, Composites Institute, SPI, Inc.

Fanucci, J. P., S. C. Nolet, and C. Koppernaes. 1990. *Low Cost Manufacturing Technology for Fire-Resistant Phenolic Matrix Composites,* Proceedings of the International Offshore Mechanics and Arctic Symposium, 3(8):21-28.

Forsdyke, K. L. 1988. *Phenolic Matrix Resins - The Way to Safer Reinforced Plastics,* 43rd Annual Conference, Session 18-C, Composites Institute, SPI, Inc.

Gauchel, J. V., N. R. Lehman, and J. J. Beckman. 1992. *Pultrusion Processing of Phenolic Resins,* 48th Annual Conference, Session 2-B, Composites Institute, SPI, Inc.

Sheard, P. A. *The Potential for the Pultrusion of Phenolics,* Thermosets 1991, Session 3, Physical Performance and High Volume Manufacturing Conference.

BIOGRAPHY

Michael W. Klett is a Research Associate at the PPG Fiber Glass Research Center in Pittsburgh, PA. He received his B.S. degree in Chemistry from California University of Pennsylvania in 1979 and his Ph.D. degree in Physical Organic Chemistry at Iowa State University in 1985. Since joining PPG in 1985, Dr. Klett has worked in a variety of areas in fiber glass product development for RP applications and has authored numerous technical publications on pultrusion. He is currently project leader for the SMC program.

Six Major Testing Techniques for Composites: Choosing and Using the Right One

Nondestructive Testing Techniques Used to Evaluate the Progression in Damage of a Swirl Mat Composite

D. C. WORLEY II, P. K. LIAW, R.S. BENSON,*
W. A. SIMPSON JR. AND J. M. CORUM

Abstract

Confocal microscopy and Ultrasound were employed to evaluate progressive damage in continuous swirled mat composites (CSMC's). The samples consisted of an Isocyanurate-based matrix with a continuous glass fiber bundle reinforcement, which were produced using Structural Reaction Injection Molding (SRIM). These samples were subjected to 20%, 40%, 60%, and 80% of their ultimate tensile stress (UTS). The objective was to determine whether fiber/matrix debonding, fiber/fiber debonding, or matrix cracking could be observed and quantified by correlating the results from the two techniques. Ultrasound was utilized to evaluate progressive changes in the matrix as a function of incremental load. These changes were quantified based on the differences in attenuation of the initial C-scan and the incrementally loaded specimen's C-scan. Quantification was obtain by using image analysis to evaluate the digitized C-scan's median color value for the initial and subsequently loaded specimen's C-scan. The areas that showed an increase in attenuation were probed with Confocal microscopy. Confocal microscopy enabled the observation of some damage in samples loaded above 60% of the UTS. Presently, the damage observed has occurred at air voids; voids are present due to the processing technique. The presence of fiber/fiber and fiber/matrix debonding remains inconclusive.

INTRODUCTION

The use of continuous swirled glass mat composites, CSMC's, are growing in popularity in many industrial and domestic applications. An understanding of the micromechanisms of progressive failure in these materials are far from being understood completely. The most common composites employed are generally unaxially oriented fibers that are arranged in plies and rotated relative to one another which produces a unique material with properties which are different from the reinforcement and matrix.(1) CSMC's are produced using the structural reaction injection molding process, SRIM. The presence of continuous fibers aids in the distribution of energy. Therefore, by increasing the amount of a continuous fiber, there may be an increase in the amount of absorbable energy. Another characteristic of CSMC's was that the mechanical properties obtained were nearly isotropic.

D. C. Worley II, P. K. Liaw and R. S. Benson, Department of Material Science and Engineering, The University of Tennessee, Knoxville, TN 37996
W. A. Simpson Jr. and J. M. Corum, Lockheed-Martin Energy Research Corp., Oak Ridge National Laboratory, Oak Ridge, TN 37831-6285
*Contact for further correspondence.

The application of CSMC's was to reduce weight, and lower cost, while maintaining mechanical integrity.

From composite studies, it is a well accepted fact that the weak link is the fiber/matrix interface. The interface between the fiber and matrix is extremely important in the transport of energy from the matrix to the fiber. If the fiber and matrix are not in contact with each other, the fiber never sees the on-coming energy resulting in a material with low energy absorption. The problems associated with the use of CSMC's are in part based on the need to understand the mechanisms by which failure is introduced. Therefore, this investigation begins with initial observations of the glass mat and matrix condition. The implications associated with the inhomogeneity of the fiber surface, fiber wetting, and the intrinsic voids present due to processing, are discussed with the hope of providing some answers to the possible failures of the CSMC's.

EXPERIMENTAL

ULTRASONIC C-SCANS

The ultrasonic system used was manufactured by SONIX™. The transducer utilized was 15 MHz (focused) with a 1.27 cm (0.5 in) diameter element focused on the back surface. The scans were performed with 8 bit resolution using the pulse echo technique. Shown in the following schematic is the basic set-up for the pulse echo experiment.

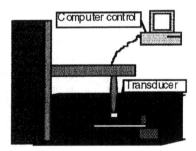

FIGURE 1. SCHEMATIC OF EXPERIMENTAL C-SCAN SET-UP.

The sample shown was placed in a water medium in order for the ultrasonic signal to be transmitted and received by the transducer. Note that these experiments were performed with samples having intrinsic voids, due to processing. This approach was used with the as received hope of understanding the combined deleterious effect of the voids, the glass mat, the matrix, and/or the fiber/matrix interface.

SCANNING ELECTRON MICROSCOPY

The investigation of the CSMC's was performed using an Environmental Scanning Electron Microscope, ESEM. The ESEM is an SEM that is designed to have different gases purge through the sample chamber. It is also used for the observation of specimens in their natural environment. The equipment used was the Hitachi S-3200N. This microscope is equipped with an x-ray detector, back scatter detector. The images that will be presented were produced using secondary electrons. Two sources for capturing the image were 1) Polaroid 55 film and 2) Computer connected with ESEM.

CONFOCAL AND LIGHT MICROSCOPY

Nikon's XeHg confocal microscopy was used with confocal and transmission modes. A 20x Plan M long working distance lens was mounted to evaluate damage in CSMC's as a function of depth. The confocal microscope is also equipped with a z-direction controller having a range of 0 - 50µm. Specimens were viewed through the first 50µm from the sample surface. The Confocal microscope provided a means to resolve defects as a function of depth.

SAMPLES

The swirled mat was composed of bundles containing ~300 monofilaments per bundle. The individual fibers were measured using SEM and found to be approximately 17µm in diameter. These bundles were swirled into a layer, which were then stacked with approximately 7 other individual layers. These layers were coated with a thermoplastic to act as a coupler between layers and an interface between the fiber and the matrix. The specimen dimensions were 20.3 cm (8.0 in) long and 2.5 cm (1.0 in) wide. Tabbing of these samples resulted in a 10.2 cm (4.0 in) observation length. The matrix material used in this study will be referred to as an isocyanurate-based polyurethane.

RESULTS AND DISCUSSION

An understanding of the composite's response to a load can be obtained from the observation of the fiber ends. From a fracture mechanics point of view, the presence of a flaw produces a stress riser. The shape of a flaw can alter the magnitude of the stress concentrations. A typical flaw is illustrated in Figure 2. The stresses present due to a flaw has been mathematically modeled and follows Equation 1.(2,3)

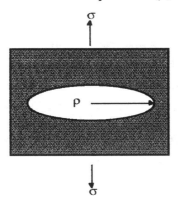

FIGURE 2. WITH RADIUS, ρ.

$$\sigma_A = \sigma\left(1 + 2\sqrt{\frac{a}{\rho}}\right) \qquad (1)$$

σ_A = Stress at the crack tip, ρ = crack tip radius of curvature, σ = applied stress.

The above expression, Equation 1, was first introduced by Inglis in the early 1900's.(2) This equation shows an increase in the crack tip stress as the radius of curvature decreases, which would essentially result in a sharp crack. The opposite would occur if the crack tip were blunted, that is, had a large radius of curvature. Figures 3 shows fiber ends that were present within the typical cross section of a CSMC matrix introducing blunted and sharp-edge cracks.

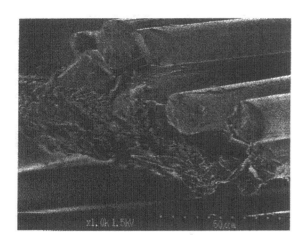

FIGURE 3. INTRINSIC FLAWS DUE TO CUTTING AND HANDLING.

These types of cracks would occur regardless of the degree of fiber wetting. These cracks are due to the existence of fiber ends being surrounded by the matrix. However, the fewer fiber ends terminating within the matrix, the fewer the number of stress risers- that could lead to catastrophic failure.

The challenge associated with the optical evaluation of debonding for a loaded specimen is exploited by the micrographs in Figure 4 and 5. There is evidence which indicates incomplete wetting of the fibers and fusion of fiber bundles along the bundle axis. This is accomplished through the fusing of sequential layers via the thermoplastic coating. The coating is thought to exist mainly at the junctions where two or more fiber bundle intersect. However, it is evident that the coating also exists along the bundle axis, which is an internal defect and leads to other potential defects upon production of the CSMC plaques. Two additional observations from these images are 1) The splayed fiber bundles are not wetted by the coupling material and thermoplastic coating, and 2) Lone fibers are shown penetrating perpendicular to the fiber bundle. These two points bring to light the difficulty in trying to conclusively discuss fiber/matrix debonding using optical techniques. Therefore, fiber/matrix debonding is extremely difficult to evaluate optically due to the initial state of the glass mat. The glass mat, as discussed previously, revealed the existence of gaps between fibers. These gaps had similar characteristics, under optical techniques, to fiber/matrix debonding. To work around this discrepancy, real time studies are being performed. In real-time, the loading of a specimen while simultaneously being observed optically will show regions around and within fiber bundles becoming dark in appearance due to the introduction of damage at points of fiber/matrix debonding.

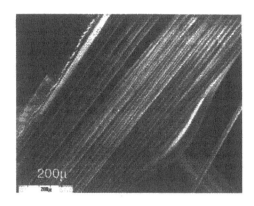

FIGURE 4 WELD POINTS, SINGLE
PENETRATION OF BUNDLE-

FIGURE 5. FIBERS SPLAYED FIBER
IRREGULAR SHAPED COATING.

CONFOCAL AND LIGHT MICROSCOPY

The value of the confocal microscopy is based on its capability to permit the observation of subsurface cracking. Initial confocal microscopy revealed no damage to samples loaded below 40% UTS. The specimen presented in Figure 6 a-d was loaded above 40% UTS. Figure 6a shows a surface which appears to be rough. The roughness is due to primarily to incomplete wetting of the fiber by the matrix at the surface as well as surface dimples due to processing. The area shown in Figure 6a-d is the same, but at incremental depths of 10 μm below the surface.

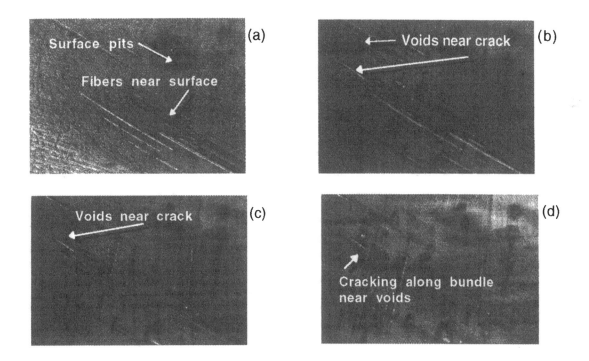

FIGURE 6a-d. CONFOCAL AND LIGHT OBSERVATIONS AT DIFFERENT DEPTHS.

Increments of 10 μm ensures that any fiber, diameter being ~17 μm, that appears at a chosen depth will disappear after penetrating 20 μm from the chosen depth. Figure 6 b through d shows a subsurface feature which persist after a penetration of 30 μm. Such a feature is undoubtedly a crack.

A void and a crack can be misinterpreted when using only confocal microscopy due to similarities in their optical signature. Therefore, transmitted light is used in conjunction with confocal microscopy to differentiate between a void and a crack. Simultaneous utilization of techniques showed the crack to initiate due to the presence of a void, see Figure 6c. The voids shown in Figure 6b-d are present due to trapped gases around fiber bundles from processing.

ULTRASONIC C-SCANS

Representative C-scan images are presented in Figure 7 and 8. This technique as presented only allows qualitative comparison between samples. The potential of C-scan as an NDE technique to study distributed damage hinges on quantification of the images. The quantification of C-scan images enables an investigation of distributed damage in a progressive manner. The use of digitized data in conjunction with image analysis produces a color range value that when related to damage shows a general decrease. A decrease in color range is due to an increase in attenuation. While an increase in attenuation is caused by the introduction of damage either in the form of matrix cracking, fiber/fiber debonding, or fiber/matrix debonding. This method is only effective if the same areas scan can be matched. Figure 7 shows tensile specimen P7-0-29's digitized C-scan results for the initial state, and incremental loads, in percentages of the ultimate tensile strength, of 20% UTS, 40% UTS, 60% UTS, and 80% UTS, respectively. As loading increased, there was a shift to darker gray scale. Shown in Figure 8 are the C-scan results for tensile specimen P7-0-30. Again, the progressive shift to a darker gray scale is observed.

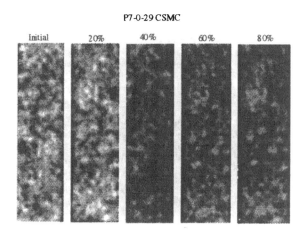

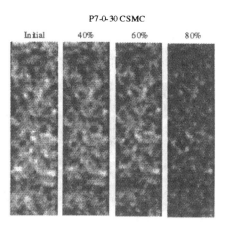

FIGURE 7. C-SCANS OF INCREMENTALLY LOADED SAMPLE P7-0-29.

FIGURE 8. C-SCAN OF INCREMENTALLY LOADED SAMPLE P7-0-30.

Because the same area for each subsequential loading can be obtained, an image analysis was performed. This analysis measured the average color range in gray scale for matched areas of incrementally loaded specimens. This value was then plotted against percent loading. The results as shown in Figure 9 exhibited a general trend towards a decreasing color range value, which corresponded to an increase in damage and a decrease in modulus, not presented here.(4) Shown also in Figure 9 was sample P7-0-31 which showed an increase at its highest loading, while sample P7-0-1 showed essentially no differences. The difference in attenuation from sample to sample are real changes. However, these tests revealed the importance of focusing the ultrasonic signal being reflected off the back surface of the sample. To optimize the ultrasonic signal the transducer must be focused by increasing or decreasing the sample to transducer distance. However, the position of the transducer, once optimized, should be maintain for each subsequential scan of the same specimen after loading. The mostly likely cause for samples P7-0-31 and P7-0-1 not maintaining the decreasing trend was due to focusing.

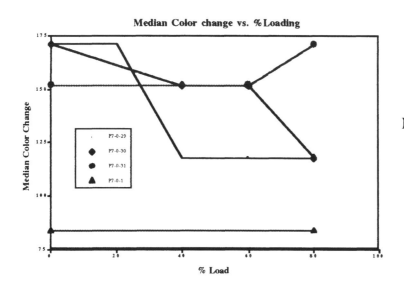

FIGURE 9. COLOR VALUE CHANGE VS. PERCENT LOADING.

CONCLUSION

The utilization of scanning electron microscopy and Confocal microscopy proved to be valuable tools in the evaluation of the continuous swirl mat composites, CSMC's. SEM showed conclusively that the initial condition of the mat is flawed which introduces an intrinsic flaw, around fibers and within the matrix having the potential of reducing the mechanical properties. The SEM also showed incomplete wetting of the coupling agent, thermoplastic, along the fiber surface. This was manifested through globs which acted as spot welds along the bundle axis. Ultrasound along with confocal and light microscopy together proved the existence of damage in the matrix. Ultrasound showed the progression in damage and gave hint to the locations to probe with microscopy. Confocal and light microscopy techniques showed the existence of damage occurring at void spaces beneath the surface. However, fiber/matrix debonding was inconclusive. Quantification of the ultrasound data using image analysis showed a decreasing trend in the median color value where the decreasing color value corresponded to an increase in attenuation, which has been associated with an increase in damage.

ACKNOWLEDGMENT

The research is supported by U.S. Department of Energy, under subcontract 11B-99732C, T21 to the University of Tennessee with Lockheed Martin Energy Systems, Inc.

REFERENCES

1.   R.T. Schwartz and H.S. Schwartz,"Fundamental Aspects of Fiber Reinforced Plastic Composites", John Wiley and Sons, 1968.
2.   Inglis, C.E., "Stresses n a Plate Due to the Presence of Cracks and Sharp Corners", Transactions of the Institute of naval Architects, Vol. 55, p. 219-241, (1913)
3.   T. L. Anderson, "Fracture Mechanics", CRC Press, p. 39-42, (1991).
4.   J.M. Corum, "Durability of Lightweight Composite Structures for Automotive Applications: Progress Report Sept. 30, 1995" p1-15, ORNL.

Comparison of Three-Point and Four-Point Flexural Bending Tests

DENISE THEOBALD, JACK McCLURG AND JAMES G. VAUGHAN

ABSTRACT

ASTM D790-92 provides a standard for both three-point and four-point flexural bending tests. It has long been observed that these tests result in different reported values of flexural strength. Use of flexural strengths as determined by D790-92 requires a proper understanding of the limitations of three-point and four-point testing. The objective of this research is to examine these tests and investigate the variation in reported properties in an effort to better define the flexural strength and modulus of composite materials.

Epoxy and polyester reinforced with either E-glass or graphite were pultruded in a 2.54 cm x 0.318 cm (1 in. x 0.125 in) profile. Flexural bend testing was performed on an MTS 25KN (5 kip) testing machine at the University of Mississippi. The samples were cut to length according to ASTM D790-92 for the various loading conditions. For glass composites, the typical span-to-depth ratio is 16:1, and for graphite reinforced composites the typical span-to-depth ratio is 32:1 or 40:1. Five samples were tested for each product under various loading conditions including quarter point and third point loading for four-point tests and standard three-point tests. Flexural strengths were recorded and plotted for each sample tested at each load span. Results indicate flexural strength is directly dependent on load span and decreases as the span is changed from the standard three-point span toward the four-point test configurations. Flexural modulus is independent of load and support span.

INTRODUCTION

The flexural bending test method is a popular and simple test method for determining a number of mechanical properties of composites; however, the test does not provide basic material property information useful for design. Three-point and four-point bending test fixtures are the standard configurations for testing flexural strength. Past experience has shown that

Denise Theobald, University of Mississippi, 124 Carrier Hall, University, MS 38677
Jack McClurg, University of Mississippi, 124 Carrier Hall, University, MS 38677
James G. Vaughan, University of Mississippi, 201F Carrier Hall, University, MS 38677

three-point and four-point flexural bending tests result in different values for flexural strength and that flexural strength is also affected by load span and support span (Theobald, 1994; McClurg, 1994; Blount, 1994). Due to these differences, when flexural strength is reported, it should be accompanied by the test method and by the load span and support span used in testing the material. It was the objective of this study to show how flexural strength and modulus change with test method, load span and support span.

Flexural strength testing methods were developed as a means for quick and simple tests for quality control and for specification purposes. Even though flexural strength cannot be used as a design criterion, standardized methods of reporting flexural strength are important. Different methods of testing produce different values for flexural strength, and depending on the relative values of the tensile, compressive, and shear strengths of the composite, any one of the three strengths may be the limiting value that causes failure. This can be a problem if flexural strength is reported without specifying the conditions under which the material was tested. A lack of standardization exists in the testing of composite materials and must be overcome if the true nature of mechanical properties of composites is to be fully understood (Tech Spotlight, 1991).

TEST PROCEDURES

Three unidirectional pultruded composite systems were tested for this study, including Shell EPON® Resin 9420/ EPON CURING AGENT® 9470/ EPON CURING AGENT ® 537 (EPON 9420) epoxy resin system reinforced with Hercules AS4-12K graphite fibers, EPON 9420 reinforced with E-glass and polyester reinforced with E-glass. All products were produced at the University of Mississippi Composite Materials Research Laboratory on a PTI Pulstar 804 pultruder. These composites were produced in a 2.54 cm x 0.32 cm (1 in x 0.125 in) profile and tested according to ASTM D-790-92 using span-to-depth ratios of 40:1, 32:1, and 16:1 (ASTM D-790-92, 1992). A span-to-depth ratio of 32:1 is recommended for graphite composites, and 16:1 is recommended for glass composites. ASTM D-790-92 includes standard methods for testing by applying the load at the center of the beam (three-point) and at quarter points and one-third points (four-point) on the beam. Included in the standard are the equations for calculating stress and modulus, span-to-depth ratios based on sample dimensions, and suggested load rates for set span-to-depth ratios. If span-to-depth ratios greater than 16:1 are used and the maximum deflection at the center of the beam exceeds 10% of the support span, correction factors must be applied to the standard equations to obtain the flexural strength.

An MTS universal testing machine, equipped with standard three-point and four-point test fixtures, was used for the testing. A deflection gage with a maximum deflection of 1.27 cm (0.5 in) was used during testing by the four-point method in order to determine maximum deflection.

In order to determine the effect of load span on flexural strength, the load span was varied by small increments from the maximum allowed by the four-point test fixture to the minimum allowed by the test fixture. This resulted in a minimum load span of 0.127 cm (0.5 in). The maximum load span varied according to span-to-depth ratio.

MODIFICATION OF ASTM STANDARD EQUATIONS

The standard equations given in ASTM D-790-92 assume small deflection and are applicable only to three-point loading, four-point quarter point loading, and four-point one-third point loading. These equations were modified using simple mechanics theory to calculate flexural strength for other load configurations. The general equation is given by

$$S = \frac{3P(L-l)}{2bd^2} \tag{1}$$

where S is the stress in the outer fiber throughout the load span, P is the load at a given point on the load deflection curve, L is the support span, l is the load span, b is the width of the beam and d is the depth of the beam.

Maximum deflection of the glass samples exceeded 10% of the support span for the 32:1 and 40:1 span-to-depth ratios. This necessitated the incorporation of correction factors to account for nonlinear responses in calculating the flexural strength. The equation for large deflection given in ASTM D-790-92 for three-point loading is

$$S = \frac{3PL}{2bd^2}[1 + 6(D/L)^2 - 4(d/L)(D/L)] \tag{2}$$

Four-point one-third point loading is

$$S = \frac{PL}{bd^2}[1 + (4.70D^2/L^2) - (7.04Dd/L^2)] \tag{3}$$

Four-point quarter-point loading is

$$S = \frac{3PL}{4bd^2}[1 - (10.91Dd/L^2)] \tag{4}$$

where D is the maximum deflection of the center of the beam and S, P, L, b and d are the same as for Equation (1). The correction factors provided in ASTM D790-92 are approximate correction factors for the case where deflections in the test sample are large enough that significant end forces develop at the supports. They appear to be based on simple corrections for rotation of the beam at the support such that a horizontal force is developed that must be included in the moment term for the flexural stress. Large deflection beam theory does not appear to have been applied. It is questioned as to whether the correction factors provided in D790-92 are correct/appropriate for the present test conditions, but this will be addressed in a later paper. Due to the inconsistencies related to these equations, regression techniques were used to develop correction factors for an equation to calculate flexural strength at large deflections based on the equations provided in the D790-92 standard.

FLEXURAL STRENGTH

Figure 1 shows a typical load versus deflection curve for the EPON 9420/glass 16:1 span-to-depth ratio and large load span. The linearity of the load-deflection curve is typical for other 16:1 span-to-depth tests for all composites tested. Figure 2 shows the nonlinearity in the load versus deflection curve for the polyester/glass 40:1 span-to-depth ratio and large load span. This is also a typical load-deflection curve for samples tested at large span-to-depth and large load spans. It is evident that the nonlinearity will affect the mechanical properties determined at large load and support span settings.

Maximum, minimum and average flexural strengths for the EPON 9420/E-glass product tested at the recommended support span for glass can be seen in Figure 3. Maximum, minimum and average flexural strengths for the EPON 9420/graphite product tested at the recommended support span for graphite can be seen in Figure 4. Three-point and standard four-point load spans are indicated on the figures. The horizontal bars represent maximum and minimum values of flexural strength and the line represents the average flexural strength. There is very little variation in the flexural strengths within a given support span for the glass product, but flexural strength decreases with increasing load span. The same trend of decreasing flexural strength with increasing load span was seen for the polyester/glass product. The flexural strength of the graphite product varies more than the glass product but still shows a slight downward trend with increasing load span for four-point tests.

Correction factors, as suggested by D790, were used to determine flexural strength at deflections greater than 10% of the support span as was the case for both glass products tested at 32:1 and 40:1 span-to-depth ratios. Maximum deflection for four-point flexural tests resulting in deflections greater than the maximum capability of the deflection gage were found by linear extrapolation. Corrected versus uncorrected flexural strength for the EPON 9420/glass tested at a span-to-depth ratio of 32:1 are plotted as a function of load span in Figure 5. The uncorrected flexural strength increases dramatically with increasing load span while the corrected flexural strengths are constant over a mid-range of load spans and decrease with increasing load span. This decrease in flexural strength with increasing load span is in agreement with trends seen for flexural bending tests performed at standard span-to-depths ratios. The average flexural strengths for all span-to-depth ratios for the polyester/glass, EPON 9420/glass and EPON 9420/graphite composites are shown in Figures 6-8, respectively. Flexural strength typically showed a decreasing trend with increasing load span. In most cases the three-point flexural strength was very close to the four-point flexural strength at the 0.5 inch load span, which is expected. For the 16:1 span-to-depth ratio, a rapid decrease in flexural strength with increasing load span was observed. When load spans reached the maximum value for the 40:1 span-to-depth ratio, it is apparent mechanisms such as excessive bending and slippage came into play resulting in a significant reduction in flexural strength. It is evident from these plots that flexural strength is not a material property.

MODULUS

ASTM D-790-92 provides equations for determining modulus when testing at standard three-point and four-point load spans. Regression techniques were used to determine a correction factor for calculating modulus at nonstandard load spans. Maximum deflection as recorded by a linear velocity displacement transducer (LVDT) was recorded for calculation of modulus for the three-point tests. A deflection gage located at the center of the sample support span with a maximum deflection of 1.27 cm (0.5 in) was used during testing by the four-point method in order to determine maximum deflection. Maximum deflection for four-point flexural tests resulting in deflections greater than the maximum capacity of the deflection gage were found by linear extrapolation to maximum load. Moduli were calculated for each three-point and four-point test performed and the maximum, minimum, and average moduli for the polyester/glass tested at a 16:1 span-to-depth ratio are presented in Figure 9. The variation seen in the figure is typical for composites tested at other span-to-depth ratios. Average flexural strengths for the polyester/glass, EPON 9420/glass and the EPON 9420/graphite tested at all load spans are plotted in Figures 10 through 12, respectively. For the 40:1 span-to-depth ratio, moduli calculated at the maximum load span varied slightly from values calculated at other load spans. Again, this is most likely due to the sample experiencing excessive deformation and slipping on the support rollers. For all test methods the variation of the modulus with load span and support span is not significant and values obtained at different span-to-depth ratios provide a reasonable estimate of the material parameter of modulus.

CONCLUSION

Flexural strength varies with test method, load span and support span. Flexural strength typically decreases with increasing load span and decreases more rapidly at small span-to-depth ratios with increasing load span. When reporting flexural strength, specification of test method, load span, and support span used in testing is important for consistency in material characterization. Modulus does not vary significantly with test method, load span or support span.

ACKNOWLEDGEMENTS
The authors would like to acknowledge the support of NSF/EPSCoR grant number EPS-9452857, the State of Mississippi, and the University of Mississippi.

REFERENCES

Theobald, D. 1994. "The Effects of Processing Parameters on the Mechanical Properties of an EPON 9420 Epoxy/Graphite Composite," Masters Thesis, University of Mississippi, Oxford, MS.

McClurg, J. 1994. "The Effects of Pultrusion Processing Parameters on the EPON 862/W Epoxy/Fiberglass System," Masters Thesis, University of Mississippi, Oxford, MS.

Blount, K. R. 1995. "The Effects of Processing Parameters on the Mechanical Properties of Various Graphite and Fiberglass Epoxy Systems," Masters Thesis, University of Mississippi, Oxford, MS.

Tech Spotlight 1991. "Standardized Testing of Advanced Composites," *Advanced Materials and Processes.*

ASTM D-790-92 1992. "Standard Test Methods for Flexural Properties of Unreinforced and Reinforced Plastics and Electrical Insulating Materials D790-92," *Annual Book of ASTM Standards.*

BIOGRAPHIES

Denise Theobald received her B.S. in Mechanical Engineering in 1991 and her M.S. in materials science and engineering in 1994 from the University of Mississippi. She will be completing the requirements for her Ph.D. in materials science and engineering in 1997 and plans to pursue her research and teaching interests. Ms. Theobald has been involved with the Composite Materials Research Group at the University of Mississippi since 1990. She is currently involved in the application of mechanical testing to pultruded structural shapes.

Jack McClurg received his B.S. in mechanical engineering from Colorado State University in 1991 and his M.S. in materials science and engineering from the University of Mississippi in 1994. He will be completing the requirements for his Ph.D. in materials science and engineering in 1997 and plans to continue his research while involved in academia. Mr. McClurg joined the Composite Materials Research Group at the University of Mississippi in 1991. He is currently involved in the research of environmental effects on the mechanical properties of pultruded composites.

Dr. Vaughan received his B.S. degree in Electrical Engineering in 1971, and his M.S. and Ph.D. in Materials Science and Metallurgical Engineering (1974, 1976), from Vanderbilt University. He worked with the Bendix Corporation before joining the University of Mississippi in 1980. Dr. Vaughan serves as Professor of Mechanical Engineering and Associate Dean of the School of Engineering at the University of Mississippi. He has worked extensively with pultrusion over the past ten years and has developed a pultrusion laboratory at the University where he conducts various experimental and analytical studies of the pultrusion process.

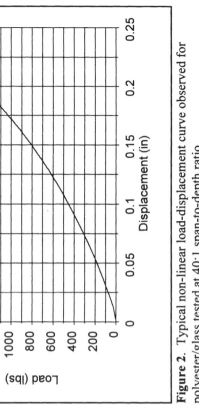

Figure 1. Typical linear load-displacement curve observed for 9420/glass tested at 16:1 span-to-depth ratio.

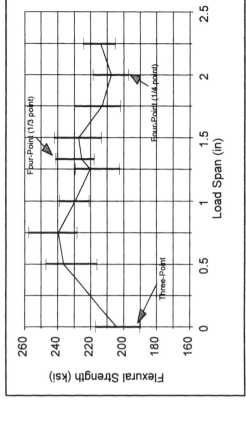

Figure 2. Typical non-linear load-displacement curve observed for polyester/glass tested at 40:1 span-to-depth ratio.

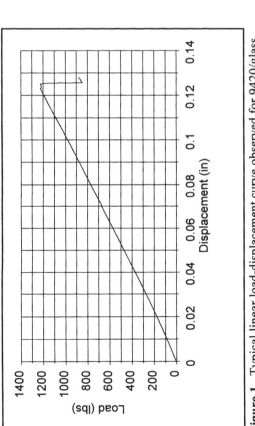

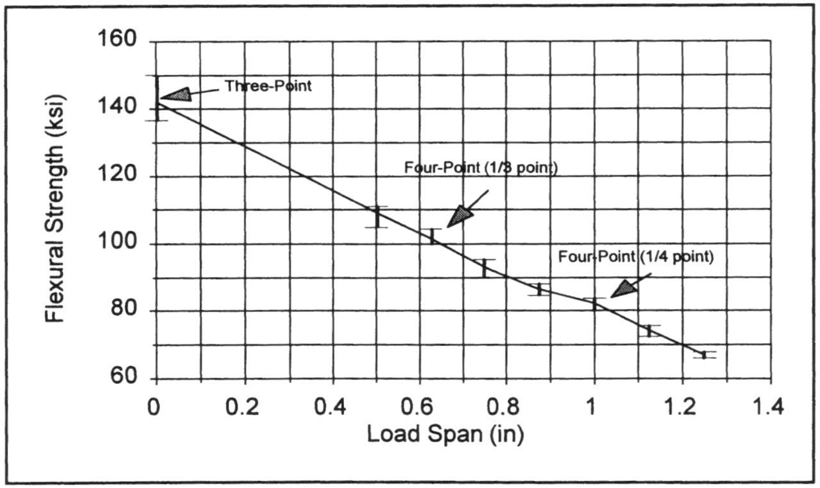

Figure 3. High, low, and average flexural strength data for 9420/glass tested at 16:1 span-to-depth ratio.

Figure 4. High, low, and average flexural strength data for 9420/graphite tested at 32:1 span-to-depth ratio.

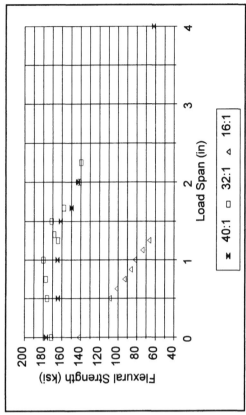

Figure 5. Degree of correction for 9420/glass tested at 32:1 span-to-depth ratio.

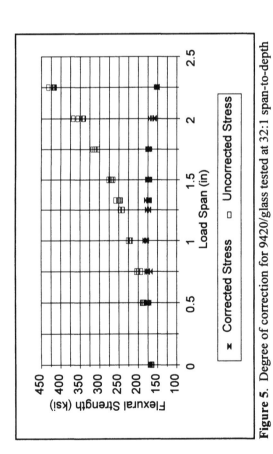

Figure 6. Average flexural strength data for polyester/glass.

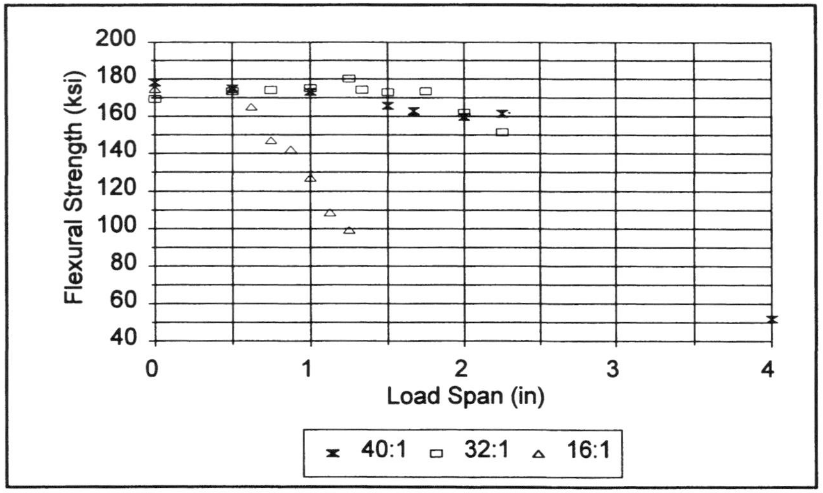

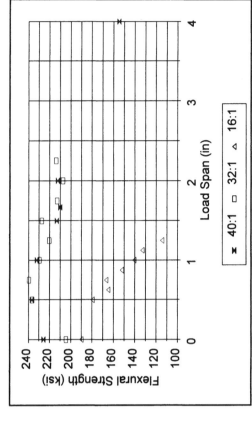

Figure 7. Average flexural strength data for 9420/glass.

Figure 8. Average flexural strength data for 9420/graphite.

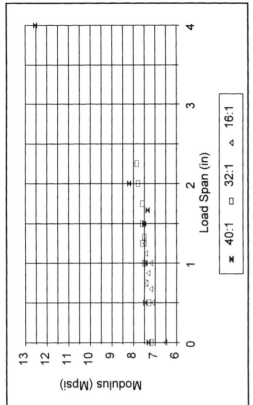

Figure 9. High, low, and average modulus data for polyester/glass tested at 16:1 span-to-depth ratio.

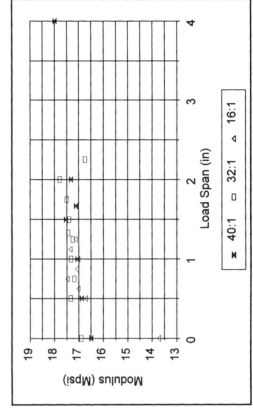

Figure 10. Average modulus data for polyester/glass.

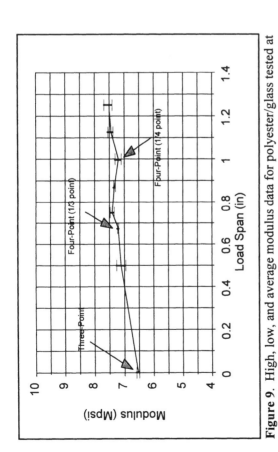

Figure 11. Average modulus data for 9420/glass.

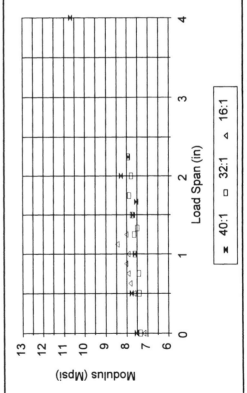

Figure 12. Average modulus data for 9420/graphite.

Progressive Fracture and Damage Tolerance of Composite Pressure Vessels

CHRISTOS C. CHAMIS, PASCAL K. GOTSIS AND LEVON MINNETYAN

ABSTRACT

Structural performance (integrity, durability and damage tolerance) of fiber reinforced composite pressure vessels, designed for pressured shelters for planetary exploration, is investigated via computational simulation. An integrated computer code is utilized for the simulation of damage initiation, growth, and propagation under pressure. Aramid fibers are considered in a rubbery polymer matrix for the composite system. Effects of fiber orientation and fabrication defect/accidental damages are investigated with regard to the safety and durability of the shelter. Results show the viability of fiber reinforced pressure vessels as damage tolerant shelters for planetary colonization.

INTRODUCTION

Future Lunar and planetary exploration and colonization attempts require the planning, deign, and construction of shelters to accommodate expeditionary communities for extended periods of time. As a first priority, a Lunar shelter must provide a breathable atmosphere with sufficient interior pressure for pulmonary function. The most efficient shelter configurations will more than likely be a pressure vessel type which is self contained and will also provide structural support.

Candidate ply layups for the fiber composite for a half dome-type pressure vessel are investigated with regard to progressive damage and fracture of the shelter due to pressurization. The performance of aramid fibers in a rubbery polymer matrix is evaluated. For a standard thickness and geometry, hemispheres with different fiber orientations are investigated to examine the influence of ply fiber layup on the burst pressure and damage tolerance. Results indicate that structural fracture (burst) pressure is sensitive to ply fiber orientations.

In addition to defect-free hemispheres, the behavior of hemispheres with fabrication defects at the surface and at the mid thickness is examined. An additional case with local through-the-thickness damage is evaluated for damage tolerance by the use of structural progressive fracture. In general, the design of composite structures requires an evaluation of safety and durability under all loading conditions. The objective of this paper is to describe an integrated computational code that has been developed to quantify the durability of fiber reinforced composites via the simulation of damage initiation, growth,

Christos C. Chamis and Pascal K. Gotsis, NASA Lewis Research Center, Cleveland, OH 44135
Levon Minnetyan, Clarkson University, Potsdam, NY 13699

accumulation, progression, and propagation to structural fracture under loading. Quantification of the structural fracture resistance is also fundamental for evaluating the durability/life of composites structures. The most effective way to obtain this quantification is through integrated computer codes which couple composite mechanics with structural analysis and with fracture mechanics concepts. The COmposite Durability STRuctural Analysis (CODSTRAN) computer code (Chamis, 1978) has been developed for this purpose. Composite mechanics, (Murthy and Chamis, 1986) finite element analysis, (Nakazawa, Dias and Spiegel, 1987) and damage progression modules (Chamis, 1978) are incorporated to implement the durability analysis. The CODSTRAN code is used as a virtual computational laboratory to simulate composite damage progression mechanisms in substantial detail. The simulation of progressive fracture by CODSTRAN has been validated to be in reasonable agreement with experimental data from tensile tests (Irvine and Ginty, 1986). Recent additions to CODSTRAN have enabled investigation of the effects of composite degradation on structural response, (Minnetyan, Chamis and Murthy, 1992) composite damage induced by dynamic loading, (Minnetyan, Murthy and Chamis, 1990) composite structures global fracture toughness, (Minnetyan, Murthy and Chamis, 1990) effect of the hygrothermal environment on durability, (Minnetyan, Murthy and Chamis, 1990 (Minnetyan, Murthy and Chamis, 1992) structural damage/fracture simulation in composite thin shells subject to internal pressure, (Minnetyan, Chamis and Murthy, 1991) composite pressure vessel durability assessment, (Minnetyan, Chamis and Murthy, 1992) an overall evaluation of damage progression in composites, (Chamis , Murthy and Minnetyan, 1992) damage progression in stiffened shell panels, (Minnetyan, et al., 1992) general design concerns for progressive fracture in composite shell structures, (Minnetyan and Murthy, 1992) and damage progression in a stiffened composite panel subjected to a displacement controlled loading (Minnetyan, et al., 1993). The objective of this paper is to demonstrate an investigation of the viability of a fiber reinforced as a pressurized Lunar shelter. The terms shelter and pressure vessel are used interchangeably.

CODSTRAN METHODOLOGY AND ARCHITECTURE

CODSTRAN is an open-ended computer code integrating select modules on composite mechanics, damage progression modeling, and finite element analysis. The damage progression module (Chamis and Smith, 1978) keep a detailed account of composite degradation for the entire structure and also acts as the master executive module that directs the composite mechanics module (Murthy and Chamis, 1986) to perform micromechanics, macromechanics, laminate analysis and synthesis functions. It also calls the finite element analysis module (Nakazawa, Dias and Spiegel, 1987) with anisotropic thick shell analysis capability to model laminated composites for global structural responses. A convenient feature of the utilized finite element module is that structural properties are input and generalized stress resultants are output at the nodes rather than for the elements. The anisotropic generalized stress-strain relationships for each node are revised according to the composite damage evaluated after each finite element analysis. Subsequent to damage and composite degradation, the model is automatically updated with a new finite element mesh and properties and the structure is reanalyzed for further deformation and damage.

Figure 1 shows a schematic of the computational simulation cycle in CODSTRAN. The ICAN composite mechanics module is called before and after each finite element analysis. Prior to each finite element analysis, the ICAN module computes the composite properties from the fiber and matrix constituent characteristics and the composite layup. The laminate properties may be different at each node. The finite element analysis module accepts the composite properties that are computed by the ICAN module and performs the analysis at each load increment. After an incremental finite element analysis, the computed generalized nodal force resultants and deformations are supplied to the ICAN module that evaluates the nature and amount of local damage, if any, in the plies of the composite laminate. Individual ply failure modes monitored by CODSTRAN include the failure criteria associated with the negative and positive limits of the six ply-stress components (σ_{l11}, σ_{l22}, σ_{l33}, σ_{l12}, σ_{l23}, σ_{l13}) and a combined stress or modified distortion energy (MDE) failure criterion, and interply delamination due to relative rotation (RR) of the plies (Murthy and Chamis, 1986).

The design example for this paper consists of a composite hemisphere shelter subjected to internal pressure. Loading is applied by imposing a gradually increasing uniform lateral pressure from the underside of the membrane. Large displacements are taken into account before and during damage initiation and progression.

The pressurized hemisphere has a circular foundation of 20.32 m (66.67 ft) diameter. The mid height is raised to 6.1 m (20 ft) relative to the edges. The computational model of the hemisphere vessel consists of 220 rectangular shell elements with 237 nodes, as shown in Figure 2. The quadrilateral finite element properties are determined from the laminate configuration of each node.

The composite system is made from Kevlar aramid fibers in a rubbery polymer matrix. The fiber volume ratio is 50 percent. The fiber and matrix properties used are listed in Tables 1 and 2, respectively. The hemisphere vessel is manufactured from twelve 0.127 mm (0.005 in) plies, resulting in a total thickness of 1.524 mm (0.060 in).

RESULTS AND DISCUSSIONS

Three different ply layups are considered to investigate the influence of fiber orientation on load carrying capability and durability. The three layups are [0/90]₆, [0/±45/90]₃, and [0/±60]₄. Each layup is independently investigated due to gradual pressurization of the shelter. Results are summarized in Table 3. Also, the damage initiation and growth stages are depicted in Figure 3. The ordinate in Figure 3 shows the percent of damage based on the volume of the membrane that is affected by the various damage mechanisms. The percent damage parameter is used as an overall indicator of damage progression.

The results in Figure 3 show the [0/±60]₄ layup to be the most effective for this shelter. Damage initiation and global fracture stages are consistently related to all three cases. Damage initiation is typically by local fiber failures near the apex of the pressurized hemisphere. After damage initiation, damage usually grows through the thickness before global fracture. Global fracture is initiated by tearing near a through-the-thickness damage with coalescence of multiple damage zones.

After selecting the design layup as [0/±60]₄, damage tolerance of the shelter is investigated by prescribing local defects near the apex of the shelter. Three cases are examined as follows: 1) Surface defect prescribed by cutting the first three surface plies ; 2) interior defect where three interior plies (plies 5-7) are cut; and 3) a through-the-thickness defect where all plies are cut. Ply cuts in all cases are 700 mm long and made perpendicular to the fiber directions. The defective shelters are simulated by subjecting them to gradually increasing pressure. A summary of the pressures for further damage growth from the existing defects is given in Table 4. Comparisons of the initial damage growth stages for these cases are shown graphically in Figure 4.

Results indicate that the $[0/\pm60]_4$ shelter has excellent damage tolerance. Even though the initial damage pressure is considerably lower for the defective shelters, the global fracture pressure is about the same for the defect-free, the surface-defective, and the internally defective shelters. The worst case is that with through-the-thickness damage which has 18 percent lower structural fracture pressure compared to the defect-free case.

CONCLUDING REMARKS

The significant results from this investigation in which CODSTRAN (Composite Durability STRuctual Analysis) is used to evaluate damage growth and propagation to fracture of a pressurized composite hemisphere vessels for planetary shelters are as follows:

1. Computational simulation, with the use of established composite mechanics and finite element modules, can be used to predict the influence of existing defects as well as loading, on the safety and durability of fiber composite pressure vessels.

2. CODSTRAN adequately tracks the damage growth and subsequently propagation to fracture for initial defects located at the surface, or in the mid-thickness of composite vessel, as well as through-the-thickness defects.

3. Initially defective vessels begin damage growth at a lower pressure compared to a defect free vessel. However, the ultimate pressure is not significantly reduced for partial-thickness defects. For the vessel with a through-the-thickness defect the ultimate pressure is reduced by 18 percent.

4. The CODSTRAN methodology and architecture is flexible and applicable to all types of constituent materials, structural geometry, and loading. Hybrid composite shells, composite containing homogenous materials such as metallic foils, as well as binary composites can be readily simulated.

5. Fracture toughness parameters such as the structural fracture load/pressure are identifiable for any structure with any defect shape.

6. Computational simulation by CODSTRAN represents a new global approach to structural integrity durability and damage tolerance assessments for design investigations.

REFERENCES

Chamis, C.C. and G.T. Smith, "Composite Durability Structural Analysis, " NASA TM-79070, 1978.

Chamis, C.C., P.L.N. Murthy and L. Minnetyan, "Progressive Fracture of Polymer Matrix Composite Structures: A New Approach," NASA TM-105574, January 1992.

Irvine, T.B. and C. A. Ginty, "Progressive Fracture of Fiber Composites," *Journal of Composite Materials*, Vol. 20, March 1986, pp. 166-184.

Minnetyan, L., P.L.N. Murthy and C.C. Chamis, "Progression of Damage and Fracture in Composites under Dynamic Loading," NASA TM-103118, April 1990.

Minnetyan, L., P.L.N. Murthy and C.C. Chamis, "Composite Structure Global Fracture Toughness via Computational Simulation," *Computers & Structures,* Vol. 37, No. 2, pp. 175-180, 1990.

Minnetyan, L., C.C. Chamis and P.L.N. Murthy, "Damage and Fracture in Composite Thin Shells," NASA TM-105289, November 1991.

Minnetyan, L., P.L.N. Murthy and C.C. Chamis, "Progressive Fracture in Composites Subjected to Hygrothermal Environment," *International Journal of Damage Mechanics,* Vol. 1, No. 1, January 1992, pp. 60-79.

Minnetyan, L., J. M. Rivers, P.L.N. Murthy and C.C. Chamis, "Structural Durability of Stiffened Composite Shells," Proceedings of the 33rd SDM Conference, Dallas, Texas, April 13-15, 1992, Vol. 5, pp. 2879-2886.

Minnetyan, L., C.C. Chamis and P.L.N. Murthy, "Structural Durability of a Composite Pressure Vessel," *Journal of Reinforced Plastic and Composites,* Vol. 11, No.11, November 1992, pp. 1251-1269.

Minnetyan, L., C.C. Chamis and P.L.N. Murthy, "Structural Behavior of Composites with Progressive Fracture," *Journal of Reinforced Plastics and Composites,* Vol. 11, No. 4, April 1992, pp. 413-442.

Minnetyan, L. and P.L.N. Murthy, "Design for Progressive Fracture in Composite Shell Structures," Proceedings of the 24th International SAMPE Technical Conference, Toronto, Canada, Ocober 20-22, 1992, pp. T227-T240.

Minnetyan, L., J. M. Rivers, C.C. Chamis and P.L.N. Murthy, "Damage Progression in Stiffened Composite Panels," Proceedings of the 34th SDM Conference, LaJolla, California, April 19-22, 1993, Vol. 1, pp. 436-444.

Murthy, P.L.N and C.C. Chamis, *Integrated Composite Analyzer* (ICAN): *Users and Programmers Manual,* NASA Technical Paper 2515, March 1986.

Nakazawa, S., J.B. Dias and M.S. Spiegel, MHOST *Users' Manual,* Prepared for NASA Lewis Research Center by MARC Analysis Research Corp., April 1987.

BIOGRAPHY

Dr. Chamis is a Senior Aerospace Scientist in the Structures and Acoustics Division of NASA Lewis Research Center. His major research has focused on the development of computational simulation methods for composite mechanics, composite structures progressive fracture, probabilistic structural analysis and probabilistic composite mechanics. His current research is in the development of computational simulation methods for coupled multidiscipline problems and for concurrent engineering.

TABLE I - KEVLAR ARAMID FIBER PROPERTIES

Number of fibers per end = 580
Fiber diameter = 0.0117 mm (0.460E-3 in)
Fiber Density = 3.94E-7 Kg/m^3 (0.053 lb/in^3)
Longitudinal normal modulus = 152 GPa (22.00E+6 psi)
Transverse normal modulus = 4.14 GPa (0.60E+6 psi)
Poisson's ratio (ν_{12}) = 0.35
Poisson's ratio (ν_{23}) = 0.35
Shear modulus (G_{12}) = 2.90 GPa (0.42E+6 psi)
Shear modulus (G_{23}) = 1.52 GPa (0.22E+6 psi)
Longitudinal thermal expansion coefficient = -0.40E-5/°C (-0.22E-5 /°F)
Transverse thermal expansion coefficient = 0.54E-4/°C (0.30E-4 /°F)
Tensile strength = 2,758 MPa (400 ksi)
Compressive strength = 517 MPa (75 ksi)

TABLE II - GV6S RUBBERY POLYMER MATRIX PROPERTIES

Matrix density = 3.42E-7 Kg/m^3 (0.0460 lb/in^3)
Normal modulus = 68.9 MPa (10 ksi)
Poisson's ratio = 0.41
Coefficient of thermal expansion = 10.3E-3/°C (0.57E-4/°F)
Tensile strength = 48.3 MPa (7.0 ksi)
Compressive strength = 145 MPa (21.0 ksi)
Shear strength = 48.3 MPa (7.0 ksi)
Allowable tensile strain = 0.014
Allowable compressive strain = 0.042
Allowable shear strain = 0.032
Allowable torsional strain = 0.038

TABLE III - EFFECT OF PLY LAYUP ON DURABILITY

	PRESSURE (KPa)		
PLY LAYUP	DAMAGE INITIATION	DAMAGE THROUGH THE THICKNESS	GLOBAL FRACTURE
[0/90]$_6$	95.95	99.22	115.42
[0/±45/90]$_3$	124.01	125.99	129.35
[0/±60]$_4$	132.72	134.36	140.65

TABLE IV - EFFECT OF INITIAL DEFECT ON DURABILITY

INITIAL DEFECT	PRESSURE (KPa)		
	DAMAGE INITIATION	DAMAGE THROUGH THE THICKNESS	GLOBAL FRACTURE
NONE	132.72	134.36	140.65
SURFACE	122.92	130.91	140.61
INTERIOR	122.92	129.46	131.45
THROUGH	95.95	99.22	115.38

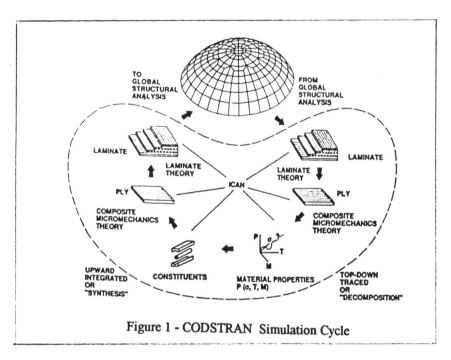

Figure 1 - CODSTRAN Simulation Cycle

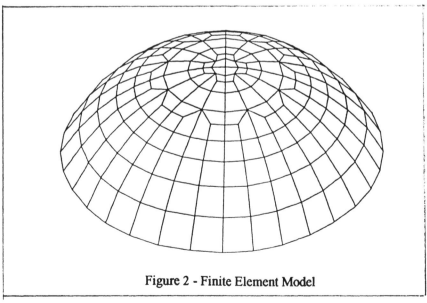

Figure 2 - Finite Element Model

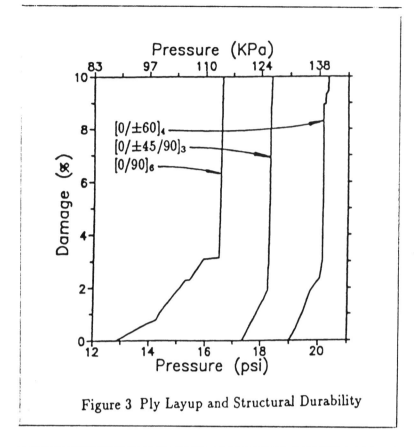

Figure 3 Ply Layup and Structural Durability

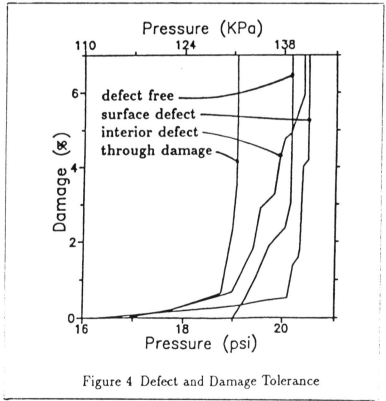

Figure 4 Defect and Damage Tolerance

Development of Accelerated Test Methods to Determine the Durability of Composites Subject to Environmental Loading

T. RUSSELL GENTRY, LAWRENCE C. BANK, AARON BARKATT,
LUCA PRIAN AND FENG WANG

ABSTRACT

Research at The Catholic University of America has explored the effects of aqueous environments and elevated temperatures on the mechanical and physico-chemical properties of various composite materials. The composite materials tested were produced using the pultrusion process. Round rods and rectangular specimens cut from larger pultruded plates were conditioned and tested in tension, flexure, short beam shear, and diametral compression. Resins tested include vinylester and polyester. Fibers included in the study were E-glass, carbon, and aramid. Environments included were air, deionized water, acetic acid at two concentrations, and ammonia at two concentrations. Conditioning temperatures were room temperature, 50°C, and 80°C.

Discussion centers on the behavior of different fiber and resin systems under the aqueous environments. Predictive models relating the degradation of composites conditioned in extreme environments to those conditioned in milder "service" environments are discussed. The linear rate model, outlined in ASTM E632 is presented. Relationships between mechanical and physico-chemical tests are also discussed. Physico-chemical tests performed as part of this study include dry weight, thermogravimetric analysis (TGA), and differential scanning calorimetry (DSC).

INTRODUCTION

Fiber reinforced plastic composite materials are currently being developed for use in highways, commercial buildings, and in residential construction. These organic matrix composites may take the form of rods for concrete reinforcements, thin plates for repair of existing concrete and steel structures, and thin wall beams for new construction. Many of these composites will be manufactured using the pultrusion process (Tarricone, 1993; Erki and Rizkalla, 1993) In bridge deck, marine fender, and wall applications, molded composites with sandwich cores are being developed. These composites are manufactured using open mold or closed mold processes.

Matrix materials for most infrastructure applications will be thermoset resins, primarily vinylester and polyester, due to their reasonable costs. Epoxies may be used in high performance applications like prestressing tendons for reinforced concrete. Most composites under development for infrastructure application use E-glass fiber systems, though some use of S-glass, carbon, and aramid fibers is anticipated. Once again, the use of high modulus, high cost fibers like aramid and carbon have been concentrated in the development of prestressing tendons for concrete.

T.R. Gentry, L.C. Bank and F. Wang, The Catholic University of America,
Department of Civil Engineering, Washington, DC 20064.

A. Barkatt and L. Prian, The Catholic University of America, Department of Chemistry,
Washington, DC 20064.

ACCELERATED TEST METHODS

An accelerated test method uses an accelerating mechanism to increase the rate of degradation of a given material. Accelerating mechanisms may include temperature, frequency of loading, concentration of a given environment, or mechanical stress level, for example. The rate of degradation under the accelerated condition is then related to the degradation rate under everyday or "service" conditions by a degradation relationship or factor. It is hoped that these degradation factors can then be used to extrapolate to the desired lifespan of the material–to demonstrate that the materials are suitable for long term use under defined service conditions. A broad framework for the development of tests is described in ASTM E632-82, Standard Practice for Developing Accelerated Tests to Aid in the Prediction of the Service Life of Building Components and Materials (1982).

Under everyday service conditions, infrastructure facilities are subject to a wide range of temperature and moisture. The presence of deicing salts and petro-chemical solvents must also be considered. An environment of special interest is concrete pore water, which is known to be highly alkaline. The degradation of glass fibers in alkaline environments is well known and is thus of concern to users of glass-based composite reinforcements in concrete (Bank and Puterman, 1995).

Finally, all environmental factors must be considered along with the mechanical loading to which the composite components will be subjected. Long-term or creep-type loading as well as periodic or fatigue-type loading must be considered. The combination of environmental (i.e., non-mechanical) and mechanical loading complicate the development of an accelerated test method.

This paper describes the overall test method and presents selected results from the first experimental task completed in the overall testing program. Emphasis is placed on comparisons of different fiber and matrix systems. A previous paper describes the testing program in greater detail and emphasizes results from a wider range of environmental exposures on vinylester-glass and polyester-glass systems (Gentry et al., 1996).

ENVIRONMENTAL EFFECTS ON COMPOSITE MATERIALS

Composites change over time due to degradation of the resin matrix or the fiber reinforcements, or by breakdown at the fiber/matrix interface. Degradation of organic polymer matrices often manifests itself as molecular rearrangements, with accompanying loss of compounds of low molecular weight (Prian et al., 1995). This degradation is accelerated by elevated temperatures and the presence of moisture. Degradation of fiber reinforcements vary with fiber type. Unprotected glass fibers dissolve in highly basic environments (Litherland et al., 1981). Aramid fibers are themselves organic polymers and are subject to depolymerization and UV degradation (Katawaki et al., 1992). Carbon fibers are subject to galvanic corrosion as they are electrical conductors (Alias and Brown, 1992). Little work has been done on degradation of fiber/matrix interfaces. Chateauminois et al. inferred the presence of fiber/matrix debonding in glass-epoxy composites subjected to high temperatures and saturated conditions (1994).

OUTLINE OF TESTING PROGRAM

As part of the overall development of an accelerated test method, an experimental program including glass, carbon, and aramid fibers, and both polyester and vinylester resins, was implemented. The following materials were selected to represent the range of materials to be used in infrastructure applications: polyester/glass rod, 6.35 mm diameter (PGR); vinylester/glass plate, 6.35 mm thick (VGP); vinylester/aramid tendon, 6.35 mm diameter (VAT); vinylester/carbon tendon, 6.35 mm diameter (VCT); and vinylester/glass tendon, 6.35 mm diameter (VGT). Information on the constituents and manufacture of the various composites is listed in Table 1.

Table 1–Materials Used in Accelerated Testing Program

Material	Description, Size and Manufacturer	Fiber [2]	Resin and Filler [1]
PGR	• Polyester/E-glass pultruded rod • 6.35 mm diameter • Creative Pultrusions Inc.	56% ± 3% E-glass Owens Corning 113 yds/lb	44% ± 3% Aropol 2036C Ashland Chemical $T_g = 90.5$ °C
VGP	• Vinylester/E-glass plate (Extren 625) • 6.35 mm thick • MMFG Inc.	22% ± 3% E-glass (supplier unknown)	75% ± 3% vinylester ≈3% clay filler (supplier unknown) $T_g = 101.8$ °C
VAT	• Vinylester/Aramid pultruded tendon • 6.35 mm diameter • DFI Pultruded Composites Inc.	57% aramid fiber Akzo-Twaron 309 yds/lb	43% DFV-1004-D Ashland Chemical
VCT	• Vinylester/Carbon pultruded tendon • 6.35 mm diameter • DFI Pultruded Composites Inc.	59% carbon fiber Akzo-Fortafil 127 yds/lb (50k)	41% DFV-1004-D Ashland Chemical
VGT	• Vinylester/Glass pultruded tendon • 6.35 mm diameter • DFI Pultruded Composites Inc.	61% glass E-fiber Akzo-Fortafil 127 yds/lb (50k)	39% DFV-1004-D Ashland Chemical

(1) None of the rod or tendon material contained any measurable filler. Some slight amount of kaolin clay filler, less than 0.5% by volume, may have been added to aid in processing in the polyester/glass (PGR) and vinylester/glass rods (VGT).

(2) Vinylester glass plate was laid up as follows (CSM = continuous strand mat): surface veil / CSM / 0° / CSM / 0° / CSM / surface veil. All rod and tendon materials contained unidirectional fibers only.

(3) Volume fractions for glass-reinforced composites taken from burnout tests performed using procedures outlined in ASTM D-2584 (1996). Volume fractions for aramid and carbon composites calculated directly from component specific gravities.

(4) Glass transition temperature, T_g, taken from measurements made on Perkin Elmer Model 7 calorimeter.

TEST METHODS SELECTED

Any accelerated test method is based on a set of measurements made periodically during the duration of the test to note changes in the mechanical, physical, thermal or chemical properties of the materials. These changes are used as indicators of the amount of degradation that has taken place. Researchers have used a wide variety of tests to determine changes in composite materials. An extensive review of test methods available for composites was completed as an earlier part of this research (Bank et al., 1995). The test methods selected for use in this research are discussed below.

Mechanical Testing

Of prime importance are changes in the mechanical properties of composite materials with time. Both changes in modulus and strength are important design parameters. Additionally, the anisotropic nature of composites dictates that mechanical tests be carried out in the fiber and off-axis directions where possible. Finally, in this research, it was decided to select test methods likely to indicate whether changes in mechanical properties were due to changes in the matrix or in the fiber.

To meet these requirements, the following mechanical test methods were selected: tension testing, flexural testing, short beam shear testing, and diametral compression testing. All mechanical tests were carried out on an MTS 810 Materials Test System with continuous acquisition of load, cross-head displacement, and strain data. Discussion of mechanical test methods and results are limited to tension testing and flexural testing in what follows.

Tension testing was carried out on unidirectional rod/tendon material and on coupons of plate material cut from larger pultruded plates. Both tensile strength and modulus were recorded. Plate specimens were straight sided, 6.35 mm thick x 12.7 mm wide, and were tested without tabbed ends. Rod and tendon specimens were tested without potted ends. Strain in the tension specimens was measured by a clip-on extensometer over a 25 mm gage length. The recorded modulus was taken as the linear regression of the first one-third of the data points between zero load and the peak loading. From results of preliminary testing, it was determined that tension testing provided the most reliable measure of modulus and strength of plate specimens.

Flexural testing was carried out on a span of 125 mm in three-point bending. This span equates to a span-to-depth ratio of 20 for the rod, tendon, and plate material. Flexural strain was calculated from loading head displacement readings. For baseline or unconditioned material, the flexural modulus measured on an 125 mm span was within 5% of the tension modulus measured with a clip-on extensometer. For flexural testing of rod and tendon material, the loading head and supports were rounded to minimize local bearing deformation at the load points. From results of preliminary testing, it was concluded that flexural tests provided the most reliable strength and modulus measurements for rods and tendons.

Thermal Testing

Thermogravimetric analysis (TGA) was performed on all composite samples. TGA measurements were completed in an ATI/Cahn high-mass TGA. In the TGA, samples were heated from room temperature to 335 degrees at a rate of 2 degrees per minute. The output from the TGA is a continuous plot of the weight of the specimen as a function of the temperature. This output is most commonly presented as percent change from initial weight. As recommended by Prian et al., weight loss between 150°C and 300°C was taken as a measure of the resin degradation (1995).

ENVIRONMENTS

Composites in infrastructure can be expected to see a wide range of environments. From an extensive review of the literature, environmental exposures were selected that were considered to be most likely to be present and most likely to cause degradation of the composite materials. The environments considered in the study were temperature, moisture, and pH (both acidic and basic). Each environment was present in at least one "service" concentration and one "accelerated" concentration and for at least three durations. Environments considered in the study are listed along with the results of mechanical testing in Fig. 1.

Inclusion of a basic environment was considered critical due to the attack of basic solutions on silicate glass fibers. Concrete pore water is highly basic. Ammonia in two concentrations was selected as the basic environment because it volatilizes completely on heating, leaving no precipitates. Other basic environments, e.g., sodium hydroxide and potassium hydroxide, leave precipitates as they are heated, and thus hinder the interpretation of gravimetric and thermogravimetric measurements. Acetic acid in two concentrations was selected as the acidic environment. Acidic environments are known to increase the rate of depolymerization as compared to neutral solutions. At the milder concentration, the acetic acid solution selected models EPA-specified groundwater conditions.

DISCUSSION OF TEST RESULTS

Discussion of mechanical test results is limited to two material types, polyester glass rod (PGR) and vinylester glass plate (VGP). Results of physical/thermal testing is presented for the full range of materials included in the study. Examples of mechanical test data and TGA results are presented to illustrate the effect of various environments on the matrix and fiber phases of the composites.

MECHANICAL TEST RESULTS

Overall, mechanical testing showed that environmental conditioning effected the strengths of the various composites to a much larger degree than they effected the moduli. All mechanical test results reported are mean values from 5 identically conditioned and tested coupons.

Flexural test results from polyester glass rods (PGR) and tensile test results from vinylester glass plate (VGP) are presented in Fig. 1. Baseline or un-conditioned strengths are depicted as the first bar in the graphs. The next three data points reflect the change in material strengths in ambient conditions (denoted AIR, RT or air at room temperature) over the course of the experiment. A slight increase in tensile strength and flexural strength with time was noted in all unconditioned materials. This is attributed to post-curing and/or physical aging of the materials under ambient conditions.

The data show that conditioning temperature has a profound effect on the strength loss in PGR. Change in strength of PGR conditioned in de-ionized (DI) water at room temperature is just detectable for durations up to 224 days. Rods conditioned in (DI) water at 50°C have a moderate loss in flexural and short beam shear strength. Rods conditioned in DI water at 80°C have an severe loss of both shear and flexural strengths.

Ammonia at room temperature at a concentration of 0.3% has a mild effect on the strength of PGR. Increasing the concentration of ammonia to 3%, while holding the temperature constant, appears to increase the degradation rate slightly. Increasing the temperature of the 3% ammonia solution from room temperature to 80°C increases the rate of degradation dramatically. Two factors contribute to this increased loss in mechanical properties: (1) the rate of solution ingress into the polyester matrix is accelerated by the increase in temperature due to the Fickian nature of the diffusion and thus more of the glass fiber is subject to ammonia attack and (2) the rate of hydrolytic depolymerization of the matrix material increases with temperature. In this case, because two different degradation mechanisms are accelerated by the increase in temperature, it is difficult to predict the effect of each contribution separately. Gravimetric and thermogravimetric measurements can aid in this separation.

In all environments, the loss of mechanical properties in VGP is milder than in PGR. Vinylester resin is known to be less susceptible to hydrolytic depolymerization as vinylester has fewer ester groups available for hydrolysis (Harper and Naeem, 1989). Furthermore, because the volume fraction of glass

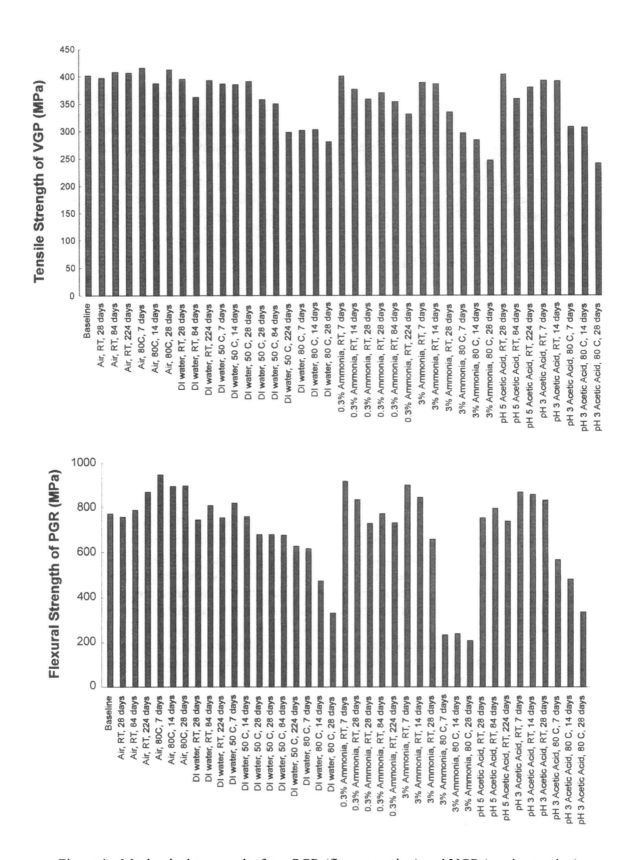

Figure 1 - Mechanical test results from PGR (flexure testing) and VGP (tension testing).

fiber is significantly lower in the plate, as compared to the uni-directional rod (see Table 1), the effect of fiber dissolution on overall mechanical properties may be attenuated.

THERMOGRAVIMETRIC TEST RESULTS

Results from thermogravimetric analyses of PGR, VAT, VCT, and VGT are shown in Fig. 2. Three TGA curves are depicted for each material: a baseline curve for un-conditioned specimens, a curve for specimens conditioned for 28 days in DI water at 80°C, and a curve for specimens conditioned for 28 days in 3% ammonia at 80°C. The DI water at 80°C accelerates the degradation of the resin matrix. The ammonia at 80°C accelerates the degradation of the matrix and is thought to accelerate the degradation of silicate glass fibers.

In the polyester glass rod (PGR), the dominant feature in the basic environment is the substantial weight loss between 23°C and 140°C for PGR conditioned in 3% ammonia at 80°C. It is believed that this weight loss represents a large volume of moisture that was trapped in the specimen. It is further believed that this large volume of moisture could not have been present unless an equally large volume of glass had eroded from the specimen due to conditioning in the basic environment. This thesis is supported by the fact that the TGA curve for DI water at 80°C does not display as dramatic a weight loss in the 23°C to 150°C range.

In the vinylester aramid tendon (VAT), all TGA curves show a significant loss in weight on heating, even the baseline material. Two potential reasons are offered for this loss: (1) the cure of the vinylester resin in the VAT was incomplete, leaving a large amount of material with a relatively low molecular weight–material which could be easily evolved from the VAT during heating, and (2) poor interfacial bond between the vinylester matrix and aramid fiber permitted easy evolution of low molecular weight material along the interface during heating. The transverse mechanical properties of the VAT, i.e., short beam shear and diametral compression, were quite low compared to the other vinylester tendons which used the same resin system, indicating a potential problem at the fiber/matrix interface. The weight loss for specimens conditioned in DI water and ammonia was significantly accelerated compared to the baseline.

Results from TGA of the vinylester carbon tendon (VCT) and vinylester glass tendon (VGT) are nearly identical. The overall weight loss during heating to 300°C was much less in these two materials (note the change in vertical scale for these two graphs). The fact that the VGT did not lose more weight due to conditioning in the ammonia than the VCT leads to a conclusion that the glass fiber in the VGT was not attacked. The significant drop in weight at low temperatures, seen in PGR and attributed to fiber dissolution, did not occur in the VGT.

In all materials, the ammonia environment causes a greater initial weight loss in the 30°C to 175°C range than the DI water at the same temperature. This may be due to the high volatility of the ammonium hydroxide solution and its increased ability to penetrate into and evolve from the specimen. It is thought that the degradation of the resin is best described by the weight loss between 150°C and 300°C. Consequently, the weight loss from 30°C to 175°C is a measure of moisture in the specimen–not a measure of resin degradation.

ACCELERATION FACTORS

The goal of an accelerated testing program is to relate the degradation of a material in an accelerated environment to the degradation of the material in its service environment. In Fig. 3, accelerating factors based on conditioning temperature, concentration of ammonia, and concentration of acetic acid are presented for PGR. The graphs on the left-hand side of the page represent the change in mechanical property as a function of time. The graphs on the right-hand side represent the relative degradation rate R as a function of the severity of the accelerating environment. In their simplest form,

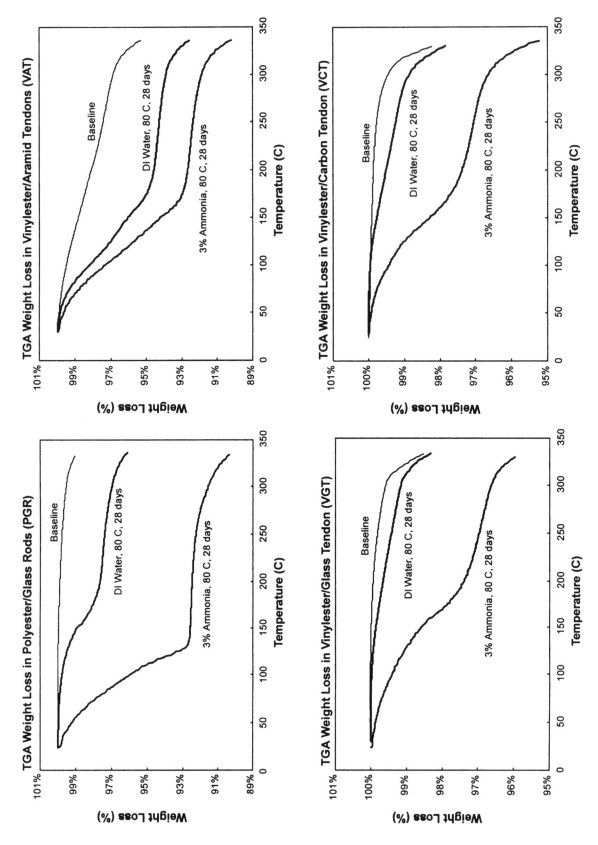

Figure 2 - Thermogravimetric test results for PGR, VAT, VCT, and VGT conditioned in DI water and ammonia at 80 C.

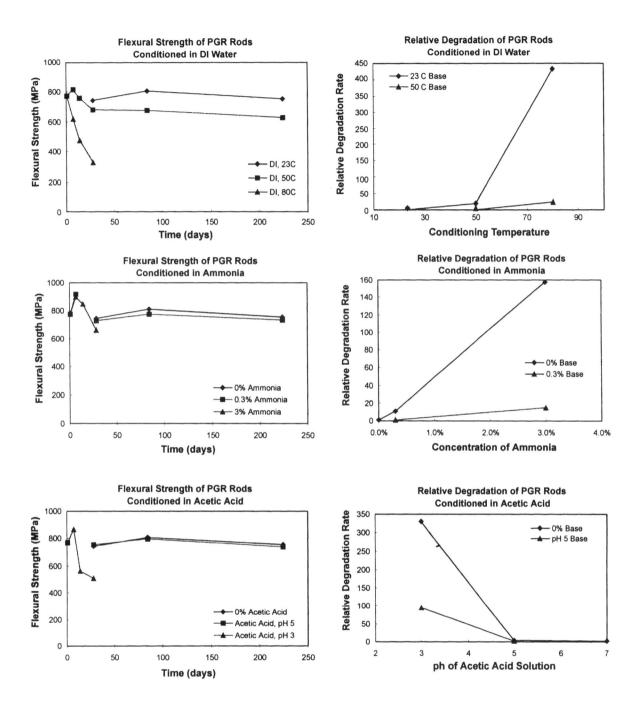

Figure 3 - Strength degradation and relative degradation rates (acceleration factors) for PGR in DI water, acidic, and basic environments.

accelerating factors are calculated as the degradation rate in the accelerated environment over the degradation rate in the service environment:

$$R = \frac{dA/dt}{dS/dt} \qquad (1)$$

For the service and accelerated environments, the degradation rates (dA/dt and dS/dt) are calculated as the linear regression of the strength versus time plots.

Due to its broad scope, this study had relatively few time periods during which the materials were sampled. Consequently, it cannot be determined whether the degradation relationships are linear in time or whether some other regression might be more appropriate. A follow on study has been designed to overcome this limitation. It is felt that the accelerating factors presented below are illustrative of the methodology of accelerated testing, but may need the refinement provided by more durations and levels of accelerating environment.

For temperature as the accelerating mechanism, a relative accelerating rate (R) of 420 is achieved if a temperature of 23°C is considered the service temperature and 80°C is considered the accelerating temperature. The accelerating factor R drops to 25 if 50°C is considered the service temperature and 80°C is considered the accelerating temperature. The importance of a having a well-defined service environment and a well-understood degradation curve for the service environment is clear. For concentration of ammonia as the accelerating mechanism, an acceleration factor of 160 is achieved if a zero concentration of ammonia (i.e., DI water) is considered to be the service environment and a factor of 14 is achieved if 0.3% ammonia is considered to be the service environment. Finally, for concentration of acetic acid as the accelerating mechanism, an accelerating factor of 320 is achieved if a neutral pH is considered to be the service environment, and an accelerating factor of 95 is achieved if pH of 5 is considered to be the service environment.

SUMMARY AND CONCLUSIONS

Degradation mechanisms in fiber reinforced plastic composites work in both the fiber and matrix phases. Accelerated testing promises to quantify the rate of this degradation in various environments and thus allow for the prediction of service life of composite materials. Accelerating mechanisms proven successful by this research include temperature and concentration of a given environment. Thermogravimetric analyses are useful indicators of the changes taking place in the fiber and matrix phases. Judicious use of these weight measurements will indicate whether degradation is taking place in the fiber or matrix phases, or in both.

The degradation of mechanical properties in polyester/glass and vinylester/glass composites impacts material strengths to a larger degree than moduli. Strength-loss acceleration factors of up to 400 can be achieved using temperature as the accelerating mechanism. Acceleration factors using acidic or basic environments at ambient temperature are somewhat lower. Determination of accurate acceleration factors requires a test plan that provides many durations in time and severity of accelerating mechanism. The rate of degradation under service conditions must be known with a high degree of confidence if accurate acceleration factors are to be determined.

ACKNOWLEDGMENTS

The authors gratefully acknowledge the support of the Federal Highway Administration under Contract No. DTFH61-93-C-00012. Any opinions, findings, and conclusions or recommendations expressed in this publication are those of the Authors and do not necessarily reflect the view of the

Federal Highway Administration.

REFERENCES

Alias, M. N. and Brown, R. 1992. "Damage to Composites from Electrochemical Processes", <u>Corrosion</u>, Vol. 48, pp. 373-378.

ASTM E-632. 1982. "Standard Practice for Developing Accelerated Tests to Aid in the Prediction of the Service Life of Building Components and Materials," American Society for Testing and Materials, Philadelphia, PA.

ASTM D 2584-68. 1996. "Standard Test Method for Ignition Loss of Cured Reinforced Resins," Annual Book of ASTM Standards, Vol. 8.01, American Society for Testing and Materials, Philadelphia, PA.

Bank, L.C., Gentry, T. R. and Barkatt, A. 1995. "Accelerated Test Methods to Determine the Long-Term Behavior of FRP Composite Structures: Environmental Effects," *Journal of Reinforced Plastics and Composites,* Vol. 14, pp. 559-587.

Bank, L. C., and Puterman, M. 1995. "Microscopic Study of Surface Degradation of Glass Fiber Reinforced Polymer Rods Embedded in Concrete Castings Subjected to Environmental Conditioning," *2nd Symposium on High Temperature and Environmental Effects on Polymer Composites,* ASTM, Philadelphia, PA.

Chateauminois, A., Vincent, L., Chabert, B., and Soulier, J.P. 1994. "Study of the Interfacial Degradation of a Glass-Epoxy Composite during Hygrothermal Ageing using Water Diffusion Measurements and Dynamic Mechanical Thermal Analysis," *Polymer,* Vol. 35, No. 22, pp. 4766-4774.

Erki, M.A., and Rizkalla, S.H. 1993. "FRP Reinforcement for Concrete Structures," *Concrete International,* V. 15, No. 6, June, 48-53.

Gentry, T. R., Bank, L.C., Barkatt, A., and Prian, L. 1996. "Accelerated Test Methods to Determine the Long-Term Behavior of Composite Highway Structures Subject to Environmental Loading," ASTM Symposium on Composite Materials in Non-Aerospace Applications, ASTM STP-1323, to appear.

Harper J.F. and Naeem, M. 1989. "The Moisture Absorption of Glass Fibre Reinforced Vinylester and Polyester Composites," *Materials and Design,* Vol. 10, No. 6, 1989, pp. 297-300.

Katawaki, K., Nishizaki, I., and Sasaki, I. 1992. "Evaluation of Durability of Advanced Composites for Applications to Prestressed Concrete Bridges," in *Advanced Composite Materials in Bridges and Structures,* (eds, K.W. Neale and P. Labossiere), CSCE, Montreal, pp. 119-128.

Litherland, K. L., Oakley, D. R., and Proctor, B. A. 1981. "The Use of Accelerated Aging Procedures to Predict the Long Term Strength of GRC Composites," *Cement and Concrete Research,* Vol. 11, pp. 455-466.

Prian, L., Pollard, R., Shan, R., Mastropietro, W., Gentry, T. R., Bank, L. C., and Barkatt, A. 1995. "Use of Thermogravimetric Analysis to Develop Accelerated Test Methods to Investigate Long-Term Environmental Effects on Polymer Composites," *2nd Symposium on High Temperature and Environmental Effects on Polymer Composites,* ASTM, Philadelphia, PA.

Tarricone, P. 1993. "Plastic Potential," *Civil Engineering,* ASCE, NY, August, 62-63.

Residual Structural Integrity of a Polymer Matrix Composite Structural Component

A. H. CARDON, M. BRUGGEMAN, Y. QIN AND S. HAROU-KOUKA

ABSTRACT

Polymer matrix composites exhibit time dependent behaviour induced by the viscoelastic nature of the polymer matrix. As consequence the material properties of such composite systems, stiffness and strength, are changing with time, function of the stress history and the environmental variations. The design of a polymer matrix composite structural component today with an imposed life time needs not only the knowledge of the instantaneous material properties but also the prediction of the variations of those properties and a certain number of real time control measurements on test specimen at different scales.

Three principal methods for the prediction of the residual structural integrity, or residual life, are proposed by different groups and applied to some specific composite systems. We discuss briefly those three methods and we develop the prediction method based on the generalised time-temperature-stress superposition principle, as proposed by H. Brinson (1978) on the basis of our results obtained over the last decade.

An accepted, validated, methodology for the prediction of the long term residual material properties, is the key for larger industrial applications of polymer matrix composites in transport, off shore and infrastructure. If the final answers will be obtained by the understanding of the mechanisms on microlevel, atomic or molecular scale, and the formulation of reliable micro-macrorelations, we are forced today to develop test methods on macrolevel in order to obtain the necessary inputs for our design programs. This has to be worked out within a viscoelastic-viscoplastic modelling, in combination with damage analysis and models for the influence of ageing effects.

MECHANICAL BEHAVIOUR OF POLYMERS AND POLYMER MATRIX COMPOSITES

The stress-strain relations of polymers are time-dependent, viscoelastic-viscoplastic, presenting creep and relaxation behaviour, strongly influenced by temperature variations and other environmental factors. Internal changes of properties, without any loading are also possible and such ageing effects has to be analysed for long term structural applications.

As consequence of this general hygrothermomechanical behaviour it is necessary at the moment of the design to have complete information on the residual material properties after the imposed life time following a complex loading history.

Albert H. Cardon, Michel Bruggeman, Yang Qin and Soumaila Harou-Kouka, Composite Systems and Adhesion Research Group of the University Brussels (COSARGUB), VUB-TW (KB), Pleinlaan 2, B1050 Brussels, Belgium.

For primary structural applications long fibre reinforced polymer matrix composites are the interesting candidates under laminate form or obtained by filament winding, pultrusion or robotised tape placement.

Time dependent and environmental related effects are not so important in polymer matrix composites as in bulk polymers but those effects have an anisotropic character and stress transfers occur internally between matrix and fibres, over the interface-interphase, even under external constant loading.

BASIC ELEMENTS FOR THE LONG TERM ANALYSIS AND POSSIBLE METHODS

For laminates the basic element is the unidirectional reinforced lamina. The results obtained on this transverse isotropic element has to be introduced in a laminate program with some assumptions on the interlaminar properties or in a finite element program with some interface elements. Control tests on laminates with different stacking sequences are necessary for a reliable structural application.

For filament wounded components the results of unidirectional reinforced lamina can be used after the analysis of the limit behaviour or failure mode.

For pultruded elements full scale results are needed for the long term predictions and for components obtained by robotised manufacturing processes a prototype testing seems necessary in order to overcome scaling effects.

Three methods are actually available. One of them is the critical element method proposed by K. Reifsnider and the Materials Response Group at Virginia Tech, (1986, 1992, 1996). This method is based on the failure mode analysis of a basic material element, the definition of a failure function and the computation of the remaining strength and life. The complete analysis can be performed by the MRLife code.

The second method is based on the extension of the classical Time-Temperature superposition principle (WLF-principle ; master curve), strictly speaking only valid for simple thermorheological materials, to a generalised Time-Temperature-Stress superposition principle validated for some polymer matrix composite systems. This method was proposed by H. Brinson (1978, 1985, 1991) and further developed by C. Hiel (1983), M. Tuttle (1986, 1995, 1996) and other research groups, see A.H. Cardon et al. (eds., 1996). This approach is based on nonlinear viscoelastic constitutive equations formulated by R. Schapery (1966, 1969, 1996). A viscoplastic modified Schapery model was proposed by A. Pasricha et al. (1995) and by X. Xiao et al. (1994). In this method stiffness maps are obtained by the simple or modified Schapery equations. Strength maps are obtained by the time and temperature modelling of the transverse strength and shear strength characteristics in one of the possible failure criteria.

A third method is based on the direct application of damage mechanics to composite systems as proposed by R. Talreja (1985, 1991, 1994, 1996) and by the group of P. Ladevèze (1989, 1993, 1996). This method is simple but needs a good experimental information not only of the damage initiation and development, but also of the damage level where the residual strength, or residual life time, has reached the limit acceptable for a safe residual structural integrity.

VISCOPLASTIC MODIFIED NONLINEAR VISCOELASTIC MODEL

The general Schapery equations are

$$n_a = \left(-\frac{\partial G_R}{\partial F_a} \right) + \frac{\partial \hat{F}_q}{\partial F_a} \int_0^\psi \Delta S_{qr}(\psi-\psi') \frac{d[\hat{F}_r/a_G]}{d\psi'} \, d\psi' \tag{1}$$

n_a being the generalised coordinates (e.g. strains) ;

F_a the associated generalised forces (e.g. stresses) ;

$\hat{F}_q$ the interior forces ;

G the Gibb energy, subscript R is for the reference state ;

ΔS_{qr} is the transition compliance $\displaystyle\sum_p \frac{d^r_{ps}\, d^r_{pq}}{d_p} \left(1 - e^{-\frac{\psi - \psi'}{r_p}} \right)$,

$d\psi = \dfrac{dt}{a_\sigma}$ is the reduced time, to be smaller than 1 for the prediction potential.

For one dimension (1) can be reduced to :

$$\varepsilon(t) = g_0(\sigma^*, T...)\, S_0\, \sigma(t) + g_1(\sigma^*, T,...) \int_0^\psi \Delta S_{qr}(\psi - \psi') \frac{d[g_2(\sigma^*, T,...)\, \sigma(\psi')]}{d\psi'}\, d\psi' \quad (2)$$

When the nonlinearizing functions g_0, g_1, g_2 and a_σ are equal to 1, equation (2) is the constitutive equation for non ageing linear viscoelastic materials.

The modified viscoplastic form of equation (2) is given by :

$$\varepsilon(t) = \varepsilon_{NVE}(t) + (C\, \sigma_0^{Nm})\, t^m \,. \quad (3)$$

If
$$S(t) = S_0 + (S_\infty - S_0)\, F(t) \quad , \quad (4)$$

different kernel functions F(t) can be used in order to obtain a good curve fitting of the material behaviour in the linear domain.

For a complete prediction of the long term behaviour of the unidirectional reinforced lamina, tests are necessary in the transverse direction and in one off axes direction θ. Depending on the type of application it is also important to obtain the result of the interaction between mechanical loading modes : quasi static loading, low frequency dynamical loading, fatigue loading ; between different environmental variations : temperature, moisture diffusion; and between the mechanical loading modes and the environmental variations.

A reduction of the number of tests is possible function of the specific expected loading modes and environmental variations. Taking into account the fact that in a laminate the lamina with fibre directions in the loading direction are constraining the time dependent evolution of the lamina where the loading direction corresponds to the transverse direction a limited number of tests on a (0,90) symmetric laminate can give an important insight in the real time dependency of the laminate properties.

For the strength analysis it is possible to use directly the damage mechanics approach or some limit criterium with time dependent reference strengths in the transverse direction and for the shear behaviour if we start the analysis on the UD-lamina level.

For the lamina level strength criterium characteristics it is possible to introduce a time to rupture rule as was proposed for bulk polymers :

$$\log t_r = A + B\sigma \quad , \quad (5)$$

where A and B are material characteristics, function of temperature and other environmental factors.

For a composite lamina the longitudinal strength, X_{11}, is assumed to be time independent but is function of the environmental factors. The transverse strength, X_{22} and the shear strength, X_{12}, are time dependent and function of the environmental variations as proposed by D. Dillard et al. (1981) and further developed by M. Tuttle, (1996).

Different types of expressions for the time to rupture can be used as proposed by H. Brinson (1985).

Finally we obtain the lamina characteristics of stiffness and strength as function of time, stress and environmental variations. From the lamina level the design of the laminate and the resulting structural component can be performed with the properties predicted for an imposed life time.

INTERNAL STRESS TRANSFERS

Within a UD-lamina under loading stress transfers occur when matrix cracking forces the fibres to take up the small stress initially supported by the matrix. The same type of stress transfers occur in a laminate after failure of some of the lamina.

Even without any cracks stress transfers occur in a lamina between matrix and fibres and in a laminate between the lamina, function of the stacking sequence, due to the relaxation in the viscoelastic component as was mentioned by A.H. Cardon (1991). Those internal stress transfers occur even under constant external mechanical and environmental loading. They depend on the stiffness properties of the interaction region between elastic and viscoelastic components and are maximum for the perfect interface conditions and disappear when debonding occurs. Those internal stress transfers under loading, unloading and environmental variations can result in local stress concentrations, crack forming and crack propagation. It seems interesting to quantify those internal stress transfers for different polymer matrix composites in order to be able to take this effect into account in a design procedure or to disregard it.

A CASE STUDY (FIBREDUX 920-C-TS-5-42)

In an interuniversity European research program between Brussels (VUB), the University of Patras and the University of Porto, we have considered as composite the CIBA EIGY Fibredux 920-C-TS-5-42 with Toraya 300B fibres. The instantaneous stiffness and strength properties of the UD-lamina with 60 % V_f were measured by direct methods and for the stiffness also by a mixed numerical-experimental method.

Following values were obtained :

E_1=120 GPa	E_2 = 8.5 GPa	v_{12} = 0.34	v_{21} = 0.02	G_{12} = 4.5 GPa	X_{22} = 65 MPa

The transient creep compliance was approximated by a general power law of the form

$$\Delta S(\psi) = \frac{S_\infty - S_0}{1 + (\frac{\tau_0}{\psi})n} \quad . \tag{6}$$

We obtained Y. Qin (1996),

$S_\infty = 2.29 \ 10^{-3} (\text{MPa})^{-1}$	$S_0 = 2.29 \ 10^{-5} (\text{MPa})^{-1}$	$\tau_0 = 5.41 \ 10^{15} \text{s}$	$n = 0.5072$

and from the creep and creep recovery testing :

$g_0 = 1 - 5.10^{-4}(\sigma^* - \sigma_c^*)$	$g_1 = 1 - 9.8 \ 10^{-4}(\sigma^* - \sigma_c^*)$
$g_2 = 1 - 51.64 \ 10^{-4}(\sigma^* - \sigma_c^*)$	$g_1 = 1 - 1.9 \ 10^{-4}(\sigma^* - \sigma_c^*)$

Those bilinear approximations are obtained with $\sigma_c = 19.5$ and $\sigma^* = \dfrac{\sigma}{\text{MPa}}$.

For the viscoplastic component (3) we obtained following values :

$C = 2.48 \ 10\text{-}17$	$N = 1.45$	$m = 2.36$

The predictions by the viscoplastic corrected model are in better agreement than those obtained by the nonlinear viscoelastic model. Using a Neural Network Fuzzy Inference System, Q. Yan (1996) confirmed these results in a much simpler and direct way avoiding, by the parallel processing, comulative errors occurring in the classical methods.

INTERACTION EFFECTS

Interaction effects between quasi static loading and a small number of low frequency oscillations were studied. The small number of oscillations has an important accelerating effect on the creep evolution and results in higher permanent strains after recovery.

CONCLUSIONS

Function of the estimated loading program, including the environmental variations, a test program for stiffness and strength properties of polymer matrix composite candidates can be worked out. The results obtained on UD-laminates have to be corrected by some (0,90) symmetric laminates testing. The design for a safe residual structural behaviour after an imposed life time can be performed with a simple classical design code where the predicted material properties after the considered life time are used as input. The consequences of this design on the short term behaviour may result in the introduction of smart components able to change the material properties during the life time and the loading program. Non destructive follow up will be necessary and also testing on real time basis of representative specimens of the real structure.

Ageing effects have to be analysed in order to predict their influence on the long term material properties.

ACKNOWLEDGEMENTS

We wish to thank the Science Council of the University (OZR-VUB) and the Belgian National Science Foundation (NFWO) for their financial support. Many thanks also to all staff members of COSARGUB, to the Composite Group of the University of Patras (G. Papanicolaou, S. Zaoutsos and Y. Mitropoulos) and the Composite Group of the University of Porto (A. Torres-Marques and R. Miranda-Guedes).

REFERENCES

Brinson J.R., Morris D.H. and Yeow Y.T. 1978. "A new experimental method for the accelerated characterisation and prediction of the failure of polymer based composite laminates", Proceedings of the 6th International Conference for Experimental Stress Analysis, pp. 395-400.

Brinson H.F. 1985. "Viscoelastic behaviour and lifetime (durability) predictions", Mechanical Characterisation of Load Bearing Fibre Composite Laminates (eds. A.H. Cardon, G. Verchery), Elsevier Appl. Sc., pp. 3-20.

Brinson H.F. 1991. "A nonlinear viscoelastic approach to durability predictions for polymer based composite structures", Durability of Polymer Based Composite Structures (eds. A.H. Cardon, G. Verchery), Elsevier Appl. Sc., pp. 46-64.

Cardon A.H. 1991. "Global and internal time dependent behaviour of polymer matrix composites". Inealstic Deformation of Composite Materials (ed. G.J. Dvorak), Springer, pp. 489-499.

Cardon A.H., Fukuda H. and Reifsnider K. (eds.). 1996. Progress in Durability Analysis of Composite Systems, A.A. Balkema.

Dillard D.H. 1981. "Creep and creep rupture of laminated graphite epoxy composites", PhD Dissertation Virginia Polytechnic Institute and State University.

Hiel C., Cardon A.H. and Brinson H.F. 1983. "The nonlinear viscoelastic response of resin matrix composites", Composite Structures 2 (ed. I. Marshall), Appl. Science Publ., pp. 271-281.

Ladevèze P. 1989. "About a damage mechanics approach", Mechanics and Mechanisms of Damage in Composite and Multimaterials (ed. D. Baptiste) MEP, pp. 119-142.

Ladevèze P. 1993. "On an anisotropic damage theory", Failure Criteria for Structured Media (ed. J.P. Boehler), A.A. Balkema, pp. 355-363.

Ladevèze P. 1996. "A damage computational approach for composites : Basic aspects and micromechanisms strength/stiffness indicatiors", Progress in Durability Analysis of Composite Materials" (eds. A.H. Cardon, H. Fukuda, K. Reifsnider), A.A. Balkema, pp. 131-138.

Pasricha A., Tuttle M.E. and Emery A.F. 1994. "Time dependent response of IM7-5260 composites subject to cyclic thermo-mechanical loading", Proceedings of the SEM Spring Conference, pp. 471-484.

Qin Y. 1996. "Nonlinear viscoelastic-viscoplastic characterization of a polymer matrix composite", PhD Dissertation, Free University Brussels (VUB).

Reifsnider K.L. and Stinchcomb W.W. 1986. "A critical element model of the residual strength and life of fatigue loaded composite coupons", Composite Materials : Fatigue and Fracture (ed. H.T. Hahn), ASTM-STP 907, ASTM, pp. 298-313.

Reifsnider K.L. 1991. "Performance simulation of polymer based composite systems", Durability of Polymer Based Composite Systems (eds. A.H. Cardon, G. Verchery), Elsevier Appl. Sc., pp. 3-26.

Reifsnider K.L., Case S. adn Xu Y.L. 1996. "A micro-kinetic approach to durability analysis : the cirtical element method", Progress in Duraiblity Analysis of Composite Systems (eds. A.H. Cardon, H. Fukuda, K.L. Reifsnider), A.A. Balkema, pp. 3-11.

Schapery R.A. 1966. "A theory of nonlinear thermoviscoelasticity based on irreversible thermodynamics". Proceedings 5th US Nat. Congress of Appl. Mechanics., ASME, pp. 511-530.

Schapery R.A. 1969. "On the characterization of non-linear viscoelastic materials", *Polymer Eng. Science*, vol 9 (N°4), pp. 295-310.

Schapery R.A. 1996. "Characterization of nonlinear, time-dependent polymers and polymeric composites for durability analysis", Progress in Duraiblity Analysis of Composite Systems (eds. A.H. Cardon, H. Fukuda, K.L. Reifsnider), A.A. Balkema, pp. 21-38.

Talreja R. 1985. "A continuum mechanics characterization of damage in composite materials", Proceedings Royal Society London, A.399, pp. 195-216.

Talreja R. 1991. "Damage mechanics of composite materials based on thermodynamics with internal variables", Durability of Polymer Based Composite Systems (eds. A.H. Cardon, G. Verchery), Elsevier Appl. Sc., pp. 65-79.

Talreja R. 1994. "Damage characterisation by internal variables", Damage Mechanics of Composite Materials (ed. R. Taljera), Elsevier, pp. 53-78.

Talreja R. 1996. "A synergistic damage mechanics approach to durability of composite material systems", Progress in Duraiblity Analysis of Composite Systems (eds. A.H. Cardon, H. Fukuda, K.L. Reifsnider), A.A. Balkema, pp. 117-129.

Tuttle M.E. and Brinson H.F. 1986. "Prediction of the long term creep compliance of general composite laminates", *Experimental Mechanics*, pp. 89-102.

Tuttle M.E., Pasricha A. and Emery A. 1995. "The nonlinear viscoelastic-viscoplastic behaviour of IM7/5260 composites subjected to cyclic loading", *J. Comp. Materials*.

Tuttle M.E. 1996. "A framework of long-term durability predictions of polymeric composites", Progress in Duraiblity Analysis of Composite Systems (eds. A.H. Cardon, H. Fukuda, K.L. Reifsnider), A.A. Balkema, pp. 169-176.

Xiao X.R., Hiel C. and Cardon A.H. 1994. "Characterization and modeling of nonlinear viscoelastic response of peek resin and peek composites", *Comp. Eng.*, vol. 4, n°7, pp. 681-702.

BIOGRAPHIES

Albert H. Cardon
PhD - University Brussels (ULB), 1971
Full Professor, University Brussels (VUB)
Head Department - Coordinator COSARGUB

Michel Bruggeman
Master in Engineering Sciences, University Brussels (VUB), 1992
Assistant professor, University Brussels (VUB)

Yang Qin
PhD University Brussels (VUB), 1996
Research Assistant, University Brussels (VUB)

SESSION 6

Enhancing the Properties of Concrete Composites:
An Effective R_X for a Traditional Material

An FRP Grid for Bridge Deck Reinforcement

HABIB RAHMAN, CHARLES KINGSLEY AND JOHN CRIMI

ABSTRACT

A fibre-reinforced plastic (FRP) grid was investigated to determine its suitability as reinforcement for concrete bridge decks. Various mechanical and durability properties relevant to the application in Canadian climate were determined. It was found that the short-term mechanical properties of the material are acceptable. However, the glass-fibre reinforced grid appears to be vulnerable to alkaline environment under stress. The carbon-fibre reinforced grid also shows some signs of deterioration and therefore needs to be studied further to predict its service life and thus determine its cost-effectiveness.

INTRODUCTION

Corrosion of steel causes billions of dollars of damage to constructed facilities in North America. Along with protective measures to prevent or reduce corrosion of steel, researchers have been looking for alternative materials for infrastructure construction. A more durable alternative to steel would mostly benefit bridges as they constitute the largest segment of the infrastructure worst affected by corrosion of steel.

FRPs have drawn researchers' attention all over the world as a potential alternative to steel reinforcement for concrete structures. Several companies have developed FRP products in the past ten years specifically for use as concrete reinforcement. One such product has been investigated over the past few years by an extensive research programme developed by the authors to determine its suitability as reinforcement for concrete bridge decks and barrier walls. This paper summarizes the research programme and a part of its outcome focusing on the suitability of the product as non-prestressed reinforcement for bridge decks.

RESEARCH PROGRAMME

A 3-phase research programme was developed to study the suitability of the proprietary FRP for bridge decks. The first phase was planned to deal with laboratory testing of the

Habib Rahman and Charles Kingsley, Institute for Research in Construction, National Research Council of Canada, M-20 Montreal Road, Ottawa, Ontario, CANADA K1A 0R6

John Crimi, Autocon Composites Inc., 203 Toryork Drive, Weston, Ontario, CANADA M9L 1Y2

product to evaluate its basic mechanical and durability properties. The tests were selected and designed to evaluate the promise of the product quickly, inexpensively, and reliably prior to carrying out more in-depth tests. Thus tests were to be conducted directly on FRP specimens, rather than on concrete specimens reinforced with it, to facilitate unambigous interpretation of the test results and to keep costs low. Preliminary results from the exposure tests indicated the need for further durability tests. Therefore, a systematic durability study with the ultimate goal of predicting the service life of bridge decks reinforced with the FRP was later proposed to be included in Phase 3.

Barring serious shortcomings of the product discovered during the Phase-1 tests, finite element (FE) analysis and load testing of a FRP-reinforced bridge deck were planned to be carried out in Phase 2. The FE analysis was to evaluate various parameters governing the performance of the bridge deck relatively inexpensively. The load test, to be carried out under cyclic service loads and monotonic load to failure, was expected to provide data on the deck's performance over a reasonable length of service life.

However, the ultimate test for the product would be a field test on an actual structure, to be carried out under Phase 3 of the investigation. According to the original plan, the performance of the product in an actual bridge would be monitored over a period of at least five years. Design criteria and guidelines were also to be developed in this phase. After preliminary results from the Phase-1 durability tests were available, tests to study the mechanism of deterioration, fatigue tests of concrete specimens reinforced with the FRP, and transverse thermal expansion tests were added to the Phase-3 programme.

The first two phases and a part of the third have been completed so far. Details of several tests have been published before. A summary of the outcome of all testing completed so far and its implications on the suitability of the product are presented in this paper.

DESCRIPTION OF THE FRP REINFORCEMENT

The FRP reinforcement investigated, commercially known as NEFMAC, is made of carbon and/or glass fibres, and vinyl-ester resin. It is available as 2-dimensional grids for reinforcing concrete slabs and 3-dimensional cages for beams. Two types of the 2-dimensional grid shown in Fig. 1 were investigated: Type C made of carbon fibres only, and Type H of 88% glass and 12% carbon. The volume fraction of fibres is about 34% in Type C and 42% in Type H.

Both grid spacing and bar size can be varied to satisfy a wide range of demand of reinforcement area in either direction. The grid spacing for both types used in this study was 100 mm and the sizes were C10 for Type C and H12 for Type H with a bar cross sectional of 46 mm^2 and 109 mm^2, respectively. Characteristically, this product develops bond with concrete almost solely through mechanical anchorage at the grid intersections. As evident from the grid construction shown in Fig. 1, this anchorage is dependent on the shear strength of the resin at the interfaces of the laminae.

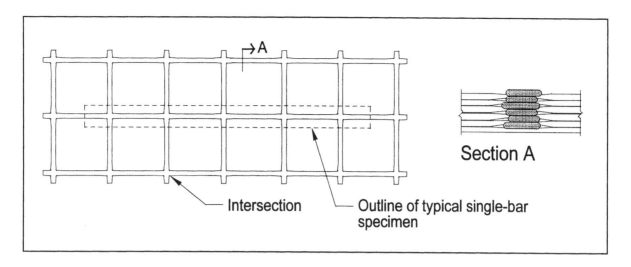

Figure 1. NEFMAC Grid and bar construction

TESTS FOR MECHANICAL PROPERTIES

The following properties of the product were experimentally derived:

1. Ultimate tensile strength (UTS), modulus of elasticity and elongation at failure, and the effect of temperature on these properties
2. Fatigue life under several ranges of tensile stresses and the effect of temperature
3. Creep and residual UTS under loads up to 75% of the original UTS sustained for 10 000 hours and the effect of temperature
4. Axial and transverse coefficients of thermal expansion
5. Shear strength and junction strength
6. Durability indicators such as residual UTS and junction strength after exposure to salt, alkali, ultraviolet radiation, or freeze-thaw action, for a full year under both sustained stress and no stress

These properties are described below and listed in Table I.

TENSILE PROPERTIES:

The average UTS obtained from testing five specimens of each type was 797 MPa for Type H and 1476 MPa for Type C with a coefficient of variation (COV) of 5.7% and 1.8%, respectively. The modulus of elasticity obtained from the same tests is 41.5 GPa and 85 GPa, respectively, with a COV of 1.9% and 2.2% (Rahman, Taylor and Kingsley, 1993). These properties remain practically unchanged in the temperature range of −30 °C to 50 °C.

FATIGUE PROPERTIES:

In Phase 1, fatigue tests were carried out on bare specimens, i.e. without any encasement in concrete, under several ranges of loading (Rahman and Kingsley, 1996). After 4 million cycles of tension-tension loading, more than 75% of the UTS was retained by Type C when loaded between 30% and 60% of its UTS and by Type H between 10% and 30%. Also, fatigue performance of Type H was found to deteriorate with increasing temperature although the decrease in fatigue life from room temperature to 50 °C is small.

TABLE I PROPERTIES OF NEFMAC

Property	Type C	Type H
Ultimate tensile strength, UTS (MPa)	1476	797
Modulus of elasticity (GPa)	85	41.5
Percent elongation at failure	1.7	1.9
Loss of UTS due to rise in temp. (MPa/°C)	Negligible	2
Cyclic stress range from bare-specimen tests corresponding to 75% retention of UTS (% of UTS)	30 - 60	10 - 30
Under tension of 75% UTS for 10 000 hrs		
Creep strain (%)	< 2	< 2
Residual UTS (% of original UTS)	101	102
Coeff. of thermal expansion:		
Axial (microstrains/°C)	1.9	8.4
Transverse (microstrains/°C)	50	30
Shear strength (MPa)	169	101
Junction strength (kN)	6.5	18.9
Residual UTS (% of original) after 120, 240 and 360 days of exposure to		
No stress + salt (% of original UTS)	96, 97, 97	101, 103, 104
No stress + alkali (% of original UTS)	97, 98, 94	97, 102, 104
No stress + salt & alkali	97, 97, 99	103, 105, 104
No stress + freeze-thaw	100, 99, 98	102, 103, 105
No stress + UV + water spray	99, 99, 102	106, 105, 105
Residual junction strength (% of original) after 120, 240, and 360 days of exposure to		
No stress + salt	77, 92, 97	80, 72, 77
No stress + alkali	94, 95, 97	74, 79, 78
No stress + salt & alkali	80, 95, 100	70, 67, 87
No stress + freeze-thaw	99, 89, 105	99, 74, 86
Residual UTS (% of original) after 360-day exposure to		
stress equal to 60% of UTS + salt & alkali	88	Failed
stress equal to 25% of UTS + salt & alkali	91	97

From more realistic fatigue tests carried out recently in which the FRP specimens were encased in concrete, fatigue life appears to be substantially lower than that from testing bare specimens (Rahman, Adimi and Benmokrane, 1996).

CREEP PROPERTIES:

For both types of the FRP bars, creep is negligibly small as found from sustained-tension tests with stresses varying from 10% to 75% of the UTS (Rahman, Kingsley and Crimi, 1995). The residual UTS of the specimens were about the same as the original UTS for both types. The elevated temperature of 50°C did not show any detrimental effect on either creep strains or the residual UTS.

THERMAL EXPANSION

The coefficient of axial thermal expansion of Type H bars is about $8.2 \times 10^{-6}/°C$, somewhat lower than that of concrete; that of Type C is much smaller, about $1.9 \times 10^{-6}/°C$ (Rahman, Taylor and Kingsley, 1993). However, the transverse thermal expansion of the bars is much larger, about $30 \times 10^{-6}/°C$, for Type H and $50 \times 10^{-6}/°C$ for Type C (Rahman, Kingsley and Taylor, 1995). These values are for a temperature range close to room temperature. In the temperature range of 20°C to 50°C, the axial coefficient varies little with temperature whereas the transverse coefficient varies significantly. At 50°C, the transverse coefficient increases to $55 \times 10^{-6}/°C$ for Type H and $80 \times 10^{-6}/°C$ for Type C.

SHEAR STRENGTH

The shear strength is about 101 MPa for Type H and 167 MPa for Type C, about 12.7% and 11.3% of their respective UTS. These are much smaller fractions of tensile strength compared to that of steel. However, this has little significance as the shear strength of reinforcement is ordinarily not relied upon in concrete structures.

JUNCTION STRENGTH

Junction strength is the maximum force resisted by a bar of the grid with the reaction provided by a cross bar (Rahman, Kingsley and Crimi, 1996). Failure typically occurs by shearing of the resin at the laminar interfaces (Fig. 1). The strength, a measure of the anchorage strength of a bar, is about 19 kN for Type H and 6.5 kN for Type C. The higher junction strength of Type H is explained by its larger size, i.e. larger interlaminar resin area resisting shear. The fibres have no contribution to this strength. It should be noted that the junction strength is likely to be higher under confinement of concrete. Pull out tests in concrete cylinders indicated that two cross bars embedded in the concrete were sufficient to develop the full tensile strength of the bar.

DURABILITY

The results of the Phase-1 durability tests indicated some potential limitations (Rahman, Kingsley and Crimi, 1996). All Type-H specimens exposed to a saline-alkaline solution failed within 8 to 115 days when subjected to a sustained stress of 60% of UTS, but survived a full year under 25% of UTS. Under the same exposure for a full year, Type C indicated a loss in UTS of about 9% and 12% under sustained stresses of 25% and 60% of UTS, respectively. On the other hand, both types exposed separately to freeze-thaw action, saline, alkaline, and saline-alkaline solutions for one year but not subjected to any stress, retained their full UTS.

Both types showed a loss in junction strength when exposed to salt, alkali or freeze-thaw action — a loss in the range of 20% was observed within 120 to 240 days of exposure; however, an unexpected recovery was observed with longer exposures.

More durability tests were later conducted on smaller model specimens made from the same type of fibres and resin as NEFMAC's. The weakness of the glass-fibre FRP was confirmed as all the specimens stressed to 50% of UTS and exposed to an alkaline solution of a lower concentration (5.8% by wt.) failed within a few days. The carbon-fibre specimens showed signs of physical damage in the form of etching, cracking and spalling of the resin after a full year's exposure to saline and alkaline solutions.

CONCLUDING REMARKS

The tensile strengths of the FRP grids are high compared to that of ordinary reinforcing steel, but the moduli of elasticity are lower. In addition, both types of the grid fail without undergoing any plastic deformation like steel. Therefore, stresses in the FRP under service loads may have to be limited to keep deflections and cracking within acceptable limits and to reduce the risk of brittle failure.

It appears that the usable strengths of the grids for bridge deck applications will be limited by fatigue considerations. More fatigue testing of specimens encased in concrete is necessary to realistically determine this limit of the strength.

Creep strain and loss of strength under sustained loads appear to be negligible for both types of the grid.

The effect of temperature on the tensile, fatigue and creep properties within the range of the outdoor temperature in the Canadian climate, i.e. between -30°C and 50°C, are noticeable albeit small. The coefficient of axial thermal expansion of Type H is close to the coefficient of expansion of concrete, and although that of Type C is much smaller, no appreciable stress will develop either in concrete or in the FRP bars due to this difference. However, the coefficient of transverse thermal expansion of both types are much higher than the coefficient of expansion of concrete. The stress induced in concrete due to the transverse expansion of the FRP bars is large enough to cause cracking in concrete (Rahman, Kingsley and Taylor, 1995). These cracks have already been experimentally observed and their effect on the bond between FRP and concrete is now being investigated.

Observations from durability tests indicate the need for a careful assessment of durability of the product. The short life of Type H, made predominantly of glass fibres, in an alkaline environment and under stress renders it unsuitable for bridge deck application. However, as the stress level appears to be a major determinant of its durability, use of Type H may still be possible in some structures by limiting the stress developed in it under service loads. This stress limit can be determined by further investigation. Type C, made of carbon fibres, shows superior durability but, nevertheless, sure signs of deterioration. It is therefore particularly important to develop the capability of predicting the product's service life so that its cost-effectiveness as a substitute for steel can be assessed.

ACKNOWLEDGEMENT

This study was supported by Autocon Composites Inc. and by the Industrial Research Assistance Programme of the National Research Council of Canada through its financial support to Autocon.

REFERENCES

Rahman, A.H., C.Y. Kingsley and J. Crimi. 1996. *Durability of a FRP Grid Reinforcement*, 2nd International Conference on Advanced Composite Materials in Bridges and Structures, Montreal, Quebec, Canada pp.681-90

Rahman, A.H., M.R. Adimi and B. Benmokrane. 1996. *Fatigue Behaviour of FRP Reinforcements Encased in Concrete*, 2nd International Conference on Advanced Composite Materials in Bridges and Structures, Montreal, Quebec, Canada, pp.691-98.

Rahman, A.H. and C.Y. Kingsley. 1996. *Fatigue Behaviour of a Fibre-Reinforced-Plastic Grid as Reinforcement for Concrete*, First International Conference on Composites in Infrastructure, Tucson, Arizona, USA, pp.427-39.

Rahman, A.H., C.Y. Kingsley and J. Crimi. 1995. *Behaviour of FRP Grid Reinforcement for Concrete under Sustained Load*, Non-Metallic (FRP) Reinforcement for Concrete Structures, Second International RILEM Symposium (FRPRCS-2), Ghent, Belgium, pp.90-99.

Rahman, A.H., C.Y. Kingsley and D.A. Taylor. 1995. *Thermal Stress in FRP-Reinforced Concrete*, CSCE Annual Conference, Ottawa, Ontario, Canada, Vol. II, pp.605-14.

Rahman, A.H., D.A. Taylor and C.Y. Kingsley. 1993. *Evaluation of FRP as Reinforcement for Concrete Bridges*, Fibre-Reinforced Plastic Reinforcement for Concrete Structures -- International Symposium, Vancouver, Canada, ACI SP-138, pp.71-86.

BIOGRAPHY

Habib Rahman is a Research Officer at the Institute of Research in Construction, National Research Council Canada. His areas of research include FRP as reinforcement for concrete and the evaluation and repair of concrete structures.

Charles Kingsley is a Technical Officer at the Institute of Research in Construction, National Research Council Canada. He has assisted in research on FRP as a reinforcement for concrete and on the structural performance of light-gauge steel wall panels.

John Crimi is the President of Autocon Composites Inc. His company manufactures FRP grids for use as reinforcement for concrete.

Constructability Assessment of FRP Rebars

DAVID HOLMAN DIETZ AND ISSAM E. HARIK

ABSTRACT

One potential solution to the corrosive deterioration of reinforced concrete bridge decks in harsh environments is to use FRP rebars rather than conventional epoxy coated steel rebars. A concern with the use of FRP rebars to reinforce concrete bridge decks is the behavior of the material during construction. Changes in construction techniques from conventional methods using steel rebars might be required due to the lightweight and flexibility of the FRP rebar. To assess the characteristics of FRP rebars during construction two bridge decks were built as slabs on grade. One of the slabs was reinforced with FRP rebars and constructed using plastic ties and chairs. The other deck was built using conventional epoxy coated steel rebars and epoxy coated steel construction materials. Comparisons between construction procedures and material behavior during the construction of the two decks were made.

Results of the study showed that decks reinforced with the different material types can be constructed using similar procedures. Observations showed that some of the problems encountered during the construction were due to the flexibility and strength of the plastic chairs and plastic ties used. These problems can be overcome by using epoxy coated steel ties and epoxy coated steel chairs during construction.

INTRODUCTION

The Kentucky Transportation Center in cooperation with the Federal Highway Administration and with the Kentucky Transportation Cabinet are conducting a research project involving the use of FRP rebars in concrete bridge decks. Use of FRP rebars instead of conventional epoxy coated steel rebars could provide a solution to the corrosive deterioration of reinforced concrete bridge decks in the commonwealth.

David Holman Deitz, Ph.D. Candidate, University of Kentucky, Kentucky Transportation Center, 371 CE/KTC Building, Lexington, Kentucky 40506-0281.

I.E. Harik, Professor of Civil Engineering, Head, Structures Section, Kentucky Transportation Center, University of Kentucky, 371 CE/KTC Building, Lexington, Kentucky 40506-0281

Before laboratory testing, engineers at the transportation department wanted to insure that FRP rebars could withstand the construction sequence of a bridge deck. The constructability assessment would address possible changes in construction specifications and techniques involved with the use of FRP rebars instead of conventional epoxy coated steel rebars.

To identify these changes, two 10.25 ft x 15.25 ft decks were built as slabs on grade. One of the decks was reinforced using epoxy coated steel rebars; the other was reinforced with FRP rebars. Both decks were constructed using the bridge deck reinforcement layout shown in Figure 1. The FRP reinforced deck was constructed with only plastic materials, including plastic ties and plastic rebar chairs. The epoxy coated steel reinforced deck was constructed using epoxy coated steel chairs and plastic coated steel ties that are typically used in bridge deck construction.

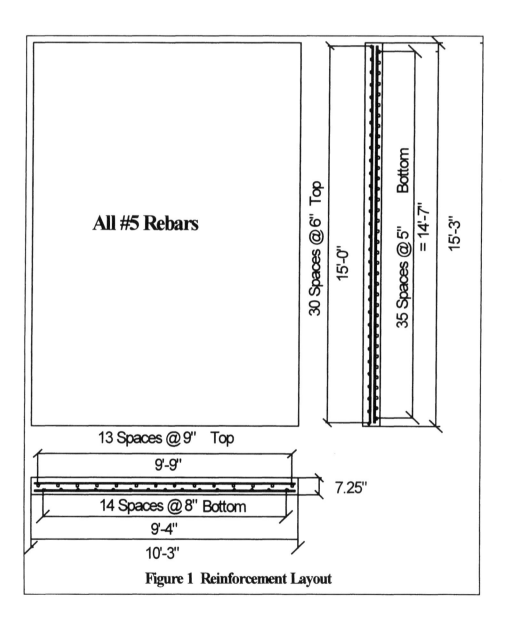

Figure 1 Reinforcement Layout

The constructability assessment addressed the following problems during the construction of the decks:

- adequacy of plastic ties,
- potential for bar displacements,
- ability of rebars to support construction loads,
- rebar support (i.e., chair requirements), and
- potential for bar damage from concrete placement tools and equipment.

Comparisons were made between the decks built with the different reinforcing materials. The epoxy coated steel reinforced deck was considered a benchmark for the test since it typifies current bridge construction.

ADEQUACY OF TIES

Two types of ties were used during construction. Plastic zip ties available at most hardware stores were used in the FRP reinforced deck. Plastic coated steel ties were used for the epoxy coated steel reinforced deck.

Observations during construction of the deck showed that the plastic zip ties were more difficult to tie. Gloves were worn while handling the FRP reinforcement making it hard to grip the end of the zip ties while zipping them. Tools used to tie coated steel ties made it quick and simple to tie the reinforcement in place.

Next, the plastic zip ties were not strong enough to support construction workers on the mat. Often the ties would unzip or break. To avoid the problem on the experimental decks, two ties had to be used at each tie point. This increased the amount of construction time greatly. No problems with the coated steel tie strength were found in the epoxy coated steel deck.

Finally, the plastic ties could not provide an adequate force between rebars at the tie points to prevent transverse rebars in the mat from slipping. A similar problem was encountered with movement of the plastic chairs supporting the mat. Steel ties in the epoxy coated steel reinforced deck kept the rebars and chairs in their correct positions.

POTENTIAL FOR BAR DISPLACEMENTS

Before testing, concerns arose that the lightweight and flexibility of the rebars could cause trouble during construction. Problems with FRP rebar movement were a major focus. Another concern involved the floating of the FRP rebars. Since the specific gravity of the FRP rebar was less than that of concrete, a specific gravity of 1.92 and 2.4 respectively, floating could possibly occur.

During the test, the FRP rebars would deflect greatly in the vertical direction under the weight of construction workers. However, the FRP rebars would return to their original position when the weight was removed even after the concrete was in place around the rebars. Though the mat wobbled under the weight of construction workers, the continuity of the reinforcing mat prevented any permanent movements. Neither of these problems were observed with the epoxy coated steel reinforcing mat.

Floating of the reinforcing mat was not observed in the construction of the decks. The slump of the concrete was too low to allow free movement of the rebar in this case. However, this could be more of a problem if the slump of the concrete is higher, as in situations where a concrete pump is used. Obviously, floating was not a problem in the epoxy coated steel reinforced deck.

These test results could change significantly based on the size of the bridge deck. This test consisted of a small slab section, much smaller than typical bridge decks. An increased deck size contributes greatly to the stiffness and weight of the mat which could reduce these problems significantly.

ABILITY TO SUPPORT CONSTRUCTION LOADS

In most cases, reinforcing mats are required to support construction workers and their equipment during the placement of reinforcement and concrete. During the construction of the FRP deck, there were no problems with supporting construction loads from a strength standpoint. However, the flexibility of the FRP rebars made the mats difficult for workers to walk on. Workers sometimes had difficulty in keeping their balance. The problem could be reduced by providing more chair supports. These problems were not seen in the stiffer epoxy coated steel reinforced deck.

REBAR SUPPORT, CHAIR REQUIREMENTS

Individual plastic chairs were used in FRP reinforced deck. Conventional individual epoxy coated steel chairs were used in the epoxy coated steel reinforced deck. Approximately three times as many chairs were used in the FRP reinforced deck than in the epoxy coated steel reinforced deck.

The lightweight of the FRP mats caused problems with the plastic chairs. The low weight allowed the plastic chairs to move around under the mat. Chairs that were tied to the mat twisted and slid around the rebar, providing no support. Chairs that were not tied moved from underneath the reinforcing mats. The flexibility of the plastic chairs compounded the problem associated with the flexibility of the mat. This made walking on the reinforcing mats more difficult.

Strength problems were found with the plastic chairs. Many chairs broke and could not support the weight of a construction worker. Again the use of more chairs might alleviate this problem somewhat. However, plastic chairs would break if a worker stepped directly on one. This problem cannot be solved by additional chairs.

POTENTIAL DAMAGE DUE TO CONSTRUCTION EQUIPMENT

Engineers involved in the project were concerned that the dropping of shovels or mishaps with other equipment during construction might damage the FRP rebars. Observations showed that the construction tools could scratch the outer surface of the rebars, but no major physical damage such as the loss of rebar surface deformations was encountered. Epoxy coated rebars suffered similar damage from the construction tools. Shovels or trowels could scratch the protective epoxy coating from the bars.

Though there was no physical damage, the small scratches on the surface of the FRP rebar could lead to problems later with deterioration. Scratches could remove the protective coating on the surface of the rebar applied to prevent deterioration of the FRP rebar in the harsh environment of the concrete.

CONCLUSIONS & RECOMMENDATIONS

The constructability assessment showed that the FRP rebars could withstand normal bridge deck construction. Results showed that many of the problems encountered during construction were due to the plastic chairs and plastic zip ties, rather than the rebars themselves.

The problems with the plastic zip ties and plastic chairs could be overcome by using epoxy coated steel chairs and epoxy coated steel ties rather than plastic materials with FRP rebars. The strength and stiffness of the steel materials would defeat the flexibility and strength problems of the plastic materials. Plastic coated steel ties are also much quicker and easier to place than the plastic zip ties, saving valuable construction time. Additional chairs would still be required to decrease the flexibility of the mat making it easier to walk on. However, there would no longer be a problem with the chairs breaking.

The coatings on the steel chairs and ties help to prevent their deterioration. The epoxy coating would also prevent direct contact between the steel and the FRP reinforcement. Preventing the direct contact lowers the chance of the steel scratching the protective coating of the FRP rebars.

There would still be a problem with the deterioration of the steel ties and chairs due to the application of deicing salts. However, the protective epoxy coating in combination with the small amount of steel in the deck would greatly reduce this problem.

BIOGRAPHIES

David Holman Deitz

David Holman Deitz is a Dwight D. Eisenhower fellow pursuing his Ph.D. in Civil Engineering at the University of Kentucky. Mr. Deitz obtained his Bachelor of Science in Civil Engineering at the University of Kentucky, and his Master of Science in Civil Engineering at the University of Cincinnati. He has experience in bridge design at the Kentucky Transportation Cabinet.

Issam E. Harik

Issam E. Harik joined the University of Kentucky in 1982. He is currently a Professor of Civil Engineering, and Head of the Structures Section at the Kentucky Transportation Center.

Dr. Harik has been involved in numerous research projects, and most recently, the research deals with seismic analysis and retrofit of bridges, vessel impact on bridges, and fiber reinforced plastic components and structures.

Dr. Harik currently serves as a member of eleven societies, and holds offices in four of them. He is also a member of a Federal Highway Administration Council, and a member of the earthquake advisory panel for the Governor of Kentucky. He has authored and co-authored over 100 technical publications.

He is currently directing the research of 3 visiting professors, 8 Ph.D. dissertations, and 3 M.S. theses and reports. Dr. Harik is very active with student organizations and outreach programs dealing with science and engineering in middle schools. He has received numerous teaching awards.

Yield Line Study of Concrete Slabs with FRP Rebars

ZIA RAZZAQ, MOHAMED ADNAN, E. DABBAGH,
MIKE M. SIRJANI AND RAM PRABHAKARAN

ABSTRACT

The ultimate load-carrying capacity of concrete slabs with steel rebars is often determined by the so-called yield line theory. In the present investigation, experiments are first conducted on simply-supported square concrete slabs with FRP rebars and subjected to a gradually increasing concentrated central static load. Similar experiments on slabs are then repeated, with steel rebars. The experimental load-carrying capacities are compared to those obtained using the yield line analysis. To conduct a dimensionless comparison of the strength of a slab with FRP rebars to that with steel rebars, a 'relative strength factor (Ω)' is also defined. The Ω values for the slabs tested show that FRP rebars result in a 17% to 24% better performance than the steel rebars. Lastly, practical examples are presented for square and rectangular concrete slabs with FRP rebars and subjected to uniformly distributed load, utilizing existing analysis and design expressions which are modified herein to account for the use of FRP rebars. The study demonstrates that FRP rebars, with all of their known advantages, may be used quite effectively in concrete slabs in practical structures.

INTRODUCTION

In civil engineering practice, yield line theory is used to predict the ultimate or maximum load-carrying capacity of uniformly reinforced concrete slabs. For a rigorous study of the known theories and experimental results for a wide variety of slabs with steel rebars, the reader may refer to the treatise by Park and Gamble, 1980. The yield line theory was initiated by Ingerslev, 1923. In this theory, the ultimate load of a given slab with specified boundary conditions is estimated by postulating various types of collapse mechanisms. The collapse mechanism resulting into the lowest ultimate load is considered the correct one. In the present paper, experimental results are first presented for simply supported square concrete slabs with

Zia Razzaq, Dept. Of Civil & Env. Engineering, Old Dominion Univ., Norfolk, VA 23529
Mohamed Adnan E. Dabbagh, Faculty of Civil Eng., Univ. of Aleppo, Syrian Arab Republic
Mike M. Sirjani, Norfolk State University, Norfolk, VA 23504
Ram Prabhakaran, Dept.of Mechanical Engineering, Old Dominion Univ., Norfolk, VA 23529

Fiber Reinforced Plastic (FRP) rebars. The rebars are manufactured by International Grating, Inc., 7625 Parkhurst, Houston, Texas 77028. Each slab is subjected to a gradually increasing concentrated central static load up to the maximum load-carrying capacity. Similar experiments are then repeated on slabs with steel rebars. The experimental ultimate loads are compared to loads predicted by the yield line theory. For slabs with FRP rebars, the use of a limiting rebar stress is proposed in this paper and its use demonstrated with practical examples of uniformly loaded slabs.

PROPOSED LIMITING FRP REBAR STRESS

Figure 1 shows a general FRP rebar stress-strain curve AB in which f_{uf} is the ultimate strength. It is proposed herein that a limiting rebar stress, f_f, for use in the yield line theory of slabs may tentatively be defined as the smaller of the following two values:

$$f_f = \text{Smaller of} \begin{cases} 0.66f_{uf} \\ \\ f_{os} \end{cases} \qquad (1)$$

where f_{os} is obtained using a 0.2 percent strain off-set. As seen from Figure 1, f_{os} is obtained by drawing the line CD at a strain value of 0.002 in./in. and parallel to the tangent at A to the curve AB. The intersection of the line CD with the curve AB that is, point D defines the f_{os} value. The off-set approach has been used in the past for certain types of materials with nonlinear stress-strain relations.

EXPERIMENTAL STUDY AND RESULTS

Table 1 summarizes experimentally obtained material properties for concrete, steel rebars, and FRP rebars used in this study. In this table, E is the Young's modulus; f'_c is the 28-day ultimate compressive strength of concrete based on tests on 6-in. diameter cylinders with 12-in. height; f_y and f_u are the yield and ultimate stresses for #3 steel rebars (3/8 in. dimeter); f_{uf} and f_f are the ultimate and 0.2 percent off-set stresses for #3 FRP rebars. The normal stress-strain relation for #3 FRP rebars was found to be initially linear but somewhat nonlinear at higher strains. The value of f_f may conservatively be used in lieu of the yield stress f_y of steel rebars to assess the ultimate load of a slab with FRP rebars.

A total of four square slabs are tested each of overall dimensions 20 in. x 20 in. and a total thickness of 3 in. The effective slab depth, d, is 2.5 in. Slab 1 and 2 are reinforced with #3 FRP rebars with 3 in. and 6.5 in. center to center spacing, respectively, in each of the slab principal directions. Figure 2 shows the details of Slab 1. The boundaries of each slab are simply-supported resulting in a clear span of 18 in. in each principal direction. Figure 3 shows a typical slab and the support structure. The slab is loaded in a Tinius-Olsen machine through a solid steel cylinder of 3 in. diameter placed at the top center of the slab. Slab 3 and 4 are reinforced with #3 steel rebars with the spacing used in Slab 1 and 2, respectively. The concentrated load is increased gradually and the central deflections and strains in the bottom rebars measured at each load level until the ultimate load is reached. Figure 4 shows the experimental load versus central deflection relations, and Figure 5 shows the load versus central bottom rebar strain relations, respectively, for each of Slab 1 through 4. The yield lines observed on the bottom (tensile) face of the four slabs are shown in Figures 6 through 9. No substantial cracking was observed on the top face of the slabs. The 3 in. diameter steel loading cylinder left an indentation on the top face of each slab. The two possible theoretical yield line patterns for the square slabs are shown in Figures 10(a) and 10(b) with the corresponding ultimate load given by the smaller of the following two estimates (Park and Gamble, 1980):

$$P_{ua} = 8m_u \tag{2}$$

$$P_{ub} = 2\pi m_u \tag{3}$$

in which m_u is formed from:

$$\frac{m_u}{\phi bd^2} = \rho\frac{f_r}{b}(1 - 0.59\rho\frac{f_r}{f_c'}) \tag{4}$$

where:

ϕ = resistance factor = 0.9,
b = unit width of slab (one foot),
d = effective depth to the the reinforcement,

$$\rho = \frac{A_s}{bd} \tag{5}$$

A_s = area of tension reinforcement per unit width of slab,

f_r = limiting reinforcement stress, and

f_c' = 28-day concrete ultimate strength.

Equation 4 is a known expression used in the presence of steel rebars for which $f_r = f_y$ is used, however, in this paper it is adapted for FRP rebars with $f_r = f_f$ as proposed via Equation 1.

Table 2 summarizes the theoretical ultimate loads, P_{ut}, and the experimental ultimate loads, P_{ue}, for Slab 1 through 4. In each of these cases, Equation 3 corresponding to the yield line pattern of Figure 10(b) governed the theoretical load. This is further confirmed by observing the yield line patterns in Figures 6 through 9 which resemble Figure 10(b) rather than Figure 10(a). For Slab 1, P_{ut} is in good agreement with P_{ue}. For Slab 2, P_{ut} is a very conservative estimate of the observed P_{ue} value. For Slab 3, P_{ut} is in excellent agreement with P_{ue}, while for Slab 4, P_{ut} is smaller than P_{ue}. It should be noted that the P_{ut} estimates for Slab 1 and 2 utilized f_f value for the FRP rebars, unlike those for Slabs 3 and 4 which utilized f_y. One reason for testing Slab 3 and 4 with steel rebars was to see if the observed behavior resembles that predicted by the yield line theory before it is applied to Slab 1 and 2. Another reason was to conduct a dimensionless comparison of the strength of the slabs with FRP rebars to those with steel rebars. To this end, Equations 3 and 4 were first applied to a slab with FRP rebars, and then to a slab with steel rebars. Next, using a ratio between the resulting expressions led to a term herein called the 'relative strength factor,' and represented by Ω:

$$\Omega = \frac{1}{\lambda} \frac{\dfrac{P_{uf}}{P_{us}}}{\dfrac{f_f}{f_y}} \tag{6}$$

in which P_{uf} and P_{us} are the ultimate loads of slabs with FRP and steel rebars, respectively, and λ is given by:

$$\lambda = \frac{1-0.59\rho f_f /f_c'}{1-0.59\rho f_y /f_c'} \tag{7}$$

In Equations 6 and 7, the same percentage (ρ) is used for both FRP and steel rebars. Table 3 presents the relative strength factors, Ω, wherein the experimental ultimate loads (P_{ue}) from Table 2 are utilized for P_{uf} and P_{us}, while comparing Slab 1 to 3, and then Slab 2 to 4. The respective Ω values are found to be 1.244 and 1.166 indicating that FRP rebars resulted in an approximately 17% to 24% better performance than the steel rebars.

Although additional research needs to be conducted in order to develop practical yield line analysis and design procedures for slabs with FRP rebars, the results for Slab 1 and 2 indicate that the use of the f_f value may perhaps be a possible approach to a conservative estimate of the ultimate loads. If such an approach is acceptable, the ACI expressions for the balanced, maximum, and minimum reinforcement percentages, namely, ρ_b, ρ_{max}, and ρ_{min} may tentatively be modified to take the following form:

$$\rho_b = 0.85\beta_1 \frac{0.003}{\dfrac{f_f}{E_f}+0.003} \frac{f_c'}{f_f} \tag{8}$$

$$\rho_{max} = 0.75\rho_b \tag{9}$$

$$\rho_{min} = \frac{200}{f_f} \tag{10}$$

In Equations 8, β_1 may be obtained from the following existing expressions:

$$\beta_1 = 0.85 \qquad \text{for} \qquad f_c' \leq 4,000 \text{ psi.} \tag{11}$$

$$\beta_1 = 0.85 - 0.05 \; (\frac{f_c'-4,000}{1,000}) \geq 0.65 \qquad \text{for} \quad f_c' > 4,000 \text{ psi.} \tag{12}$$

PRACTICAL EXAMPLES OF CONCRETE SLABS WITH FRP REBARS

The following examples demonstrate the use of the above expressions, in combination with known ultimate load expressions based on yield line theory for slabs subjected to a uniformly distributed load.

Example 1

Design a square slab 20 ft x 20 ft simply supported on all four edges and carrying dead and live loads of 50 psf and 250 psf, respectively. Assume dead and live load factors of 1.4 and 1.7, respectively. Given f_f = 40 ksi ; E_f = 7,210 ksi; f_c' = 4 ksi.

Solution:

Try a slab depth of 6 in. Thus, the slab self-weight is 6/12 x 150 psf = 75 psf, and the total ultimate load is:

$$w_u = 1.4(50+75) + 1.7(250) = 600 \text{ psf}$$

The yield line pattern of an isotropically reinforced square slab has the form shown in Figure 10(a), and its ultimate load is given by (see Park and Gamble, 1980):

$$w_u = \frac{24m_u}{l^2} \tag{13}$$

from which m_u = 10,000 lb-ft/ft. Also, Equations 8-10 give:

$\rho_b = 0.05357$

$\rho_{max} = 0.019$

$\rho_{min} = 0.005$

If ρ = 0.01 is adopted, then Equation 4 provides:

d = 5.43 in. ≈ 5.5 in.

A total slab depth of 7.0 in. may be used, including a 1.5 in. effective cover. Thus, Equation 5 gives:

A_s = 0.01 bd = 0.01(12)(5.5) = 0.66 in² per foot width of slab.

Final design: Use #6 FRP rebars at 8" c/c in each direction. Figure 11 shows a typical section of the slab.

Example 2

A rectangular slab 15 ft x 20 ft simply supported on all edges is orthotropically reinforced by # 6 FRP rebars at 7 in. center to center placed parallel to the x direction, and by #6 FRP rebars at 9 in. center to center placed parallel to the y direction. The overall and effective depths of the slab are 8.0 in., and 6.5 in., respectively. The f_c' and f_f values are 4000 psi and 60,000 psi, respectively. Determine the ultimate load intensity of the slab and the corresponding yield line pattern.

Solution:

The x and y FRP reinforcement percentages are:

$$\rho_x = \frac{0.76}{12(6.5)} = 0.00974$$

$$\rho_y = \frac{0.59}{12(6.5)} = 0.00756$$

These ρ values are found to be within the limits defined by Equations 9 and 10. The ultimate moment capacities in the two directions can be readily found by substituting the given f_c' and f_f values into Equation 4 with ρ_x and ρ_y, respectively, to obtain:

$$m_{ux} = 20,306 \text{ lb-ft/ft}$$

$$m_{uy} = 16,094 \text{ lb-ft/ft}$$

The ultimate uniformly distributed load for an orthotropically reinforced slab with simple supports is given by (see Park and Gamble, 1980):

$$w_u = \frac{24 m_{uy}}{l_y^2 \left\{ \left[3 + \frac{m_{ux}}{m_{uy}} \left(\frac{l_y}{l_x} \right)^2 \right]^{1/2} - \frac{l_y}{l_x} \left(\frac{m_{ux}}{m_{uy}} \right)^{1/2} \right\}^2} \tag{14}$$

in which l_x and l_y are the overall slab dimensions. With $l_x = 20$ ft, $l_y = 15$ ft, and the m_{ux} and m_{uy} values computed above, Equation 14 gives:

$$w_u = 1,649 \text{ psf}$$

The yield line pattern corresponding to this load is shown in Figure 12 in which l_1 is found from (see Park and Gamble, 1980):

$$\frac{l_1}{l_y} = \frac{1}{2} \left\{ \left[\left(\frac{l_y}{l_x} \frac{m_{ux}}{m_{uy}} \right)^2 + 3 \frac{m_{ux}}{m_{uy}} \right]^{1/2} - \frac{l_y}{l_x} \frac{m_{ux}}{m_{uy}} \right\} \tag{15}$$

which gives a value of $l_1 = 9.12$ ft.

CONCLUSION

Yield line theory may be a possible approach for the analysis and design of concrete slabs with FRP rebars. Further experimental and theoretical research is needed for the development of slab yield line analysis and design approaches of practical use.

REFERENCES

Park, R., and Gamble, W.L., 1980, Reinforced Concrete Slabs, John Wiley & Sons, New York.

Ingerslev, A., 1923, "The Strength of Rectangular Slabs," *J. Inst. Eng.*, Vol.1, No.1, pp. 3-14.

BIOGRAPHIES

Dr. Zia Razzaq is Professor of Civil Engineering at Old Dominion University, Norfolk, Virginia. Dr. Mohamed Adnan E. Dabbagh is Associate Professor at the University of Aleppo, Syrian Arab Republic. Mr. Mike Sirjani is Assistant Professor at Norfolk State University, Norfolk, Virginia. Dr. Ram Prabhakaran is a Professor of Mechanical Engineering at the Old Dominion University, Norfolk, Virginia.

TABLE 1. MATERIAL PROPERTIES

Material	Test	E (ksi)	Other properties (psi)
Concrete	Compression	3,537	$f_c' = 5,305$
Steel	Tension	30,965	$f_y = 61,930$; $f_u = 111,230$
FRP	Tension	6,790	$f_{uf} = 49,991$; $f_f = 32,000$

TABLE 2. THEORETICAL AND EXPERIMENTAL ULTIMATE LOADS

Slab	Reinforcement	P_{ut} (lb.)	P_{ue} (lb.)
1	FRP, #3 @ 3" c/c	15,722	18,680
2	FRP, #3 @ 6.5" c/c	7,361	12,860
3	Steel, #3 @ 3" c/c	28,859	27,560
4	Steel, #3 @ 6.5" c/c	13,922	20,850

TABLE 3. SLAB RELATIVE STRENGTH FACTORS

Comparison of	Relative Strength Factor Ω
Slab 1 (FRP) to Slab 3 (Steel)	1.244
Slab 2 (FRP) to Slab 4 (Steel)	1.166

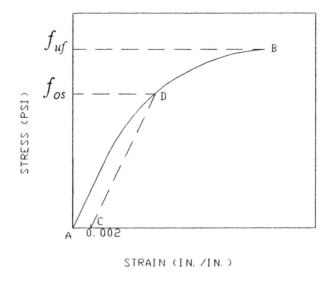

Figure 1. FRP rebar tensile stress- strain curve

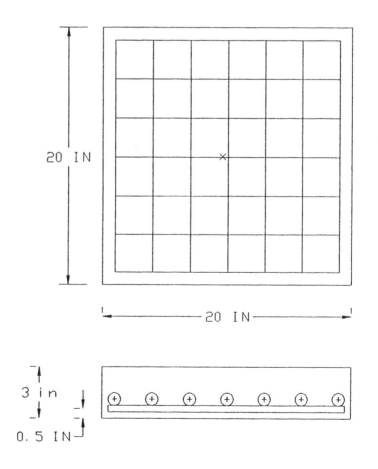

Figure 2. Slab 1 dimensions and reinforcement details

Figure 3.   Slab and support structure

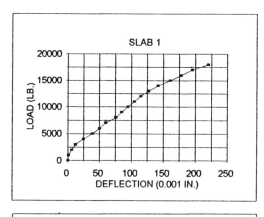

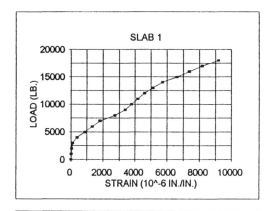

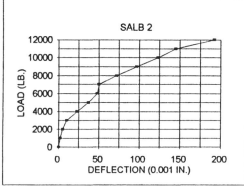

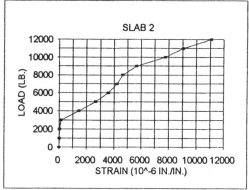

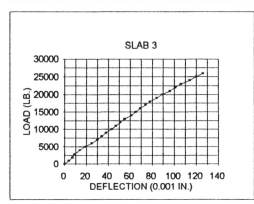

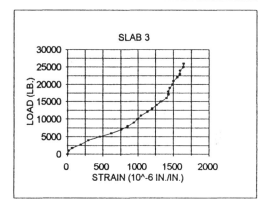

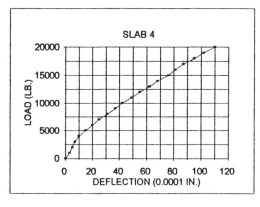

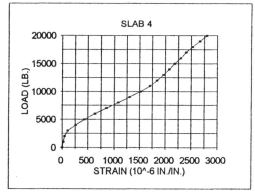

Figure 4. Experimental load versus central deflection relations

Figure 5. Experimental load versus central bottom rebar strain relations

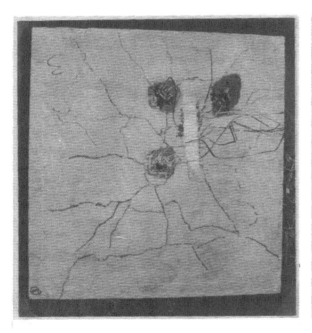

Figure 6. Yield lines for Slab 1
with #3 FRP rebars at 3" c/c

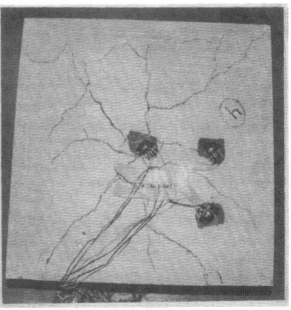

Figure 7. Yield lines for Slab 2
with #3 FRP rebars at 6.5" c/c

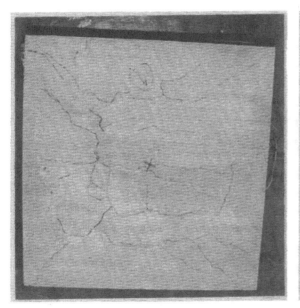

Figure 8. Yield lines for Slab 3
with #3 Steel rebars at 3" c/c

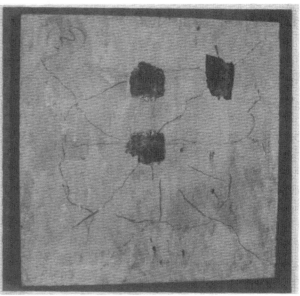

Figure 9. Yield lines for Slab 4
with #3 Steel rebars at 6.5" c/c

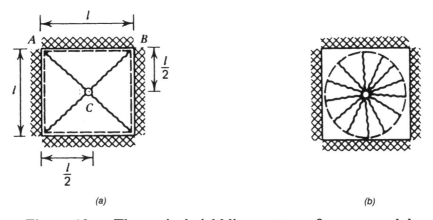

(a)   (b)

Figure 10.   Theoretical yield line patterns for square slab

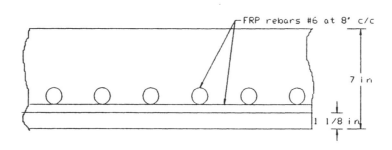

Figure 11.   Square slab section, Example 1

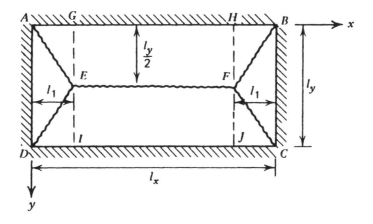

Figure 12.   Yield line pattern for uniformly loaded rectangular slab

Aging of Glass Fiber Composite Reinforcement

MAX L. PORTER, JACOB MEHUS, KURT A. YOUNG,
ED O'NEIL AND BRUCE A. BARNES

ABSTRACT

Over the last decade fiber-reinforced-plastic (FRP) reinforcement such as glass, carbon, or aramid fibers embedded in a resin such as vinyl ester, epoxy, or polyester has emerged as one of the most exciting and promising solutions to the deterioration problem caused by the corrosion of steel reinforcement in structural concrete. However, before FRP reinforcement can become a commonly used construction material, more knowledge about the short term load-carrying capabilities, physico-mechanical characteristics, and long term weathering performance must be determined. The high pH of the concrete porewater creates a potentially damaging environment for the FRP reinforcement, and there is a strong need for long term durability studies of different fibers and resin systems that may be used in this application.

This short term need to long term weathering data has necessitated the development of analytical techniques such as accelerated aging as a supplement to real-time weathering testing. The research conducted in this investigation involved a study of accelerated aging effects on commercially available (off-the-shelf) FRP reinforcement for regular and prestressed concrete, simulating approximately 50 years of real-time aging. The specimens tested included carbon and glass fiber composite tendons for prestressed applications, and glass fiber composite reinforcement for non-prestressed applications. The specimens were exposed to an aging solution with a high pH (simulating a concrete environment) and an elevated temperature to accelerate the aging process. The specimens were tested both unloaded and under constant load (stress rupture and stress corrosion) and results were compared to control specimen properties. This paper will present some of the results from this research.

Max L. Porter, Dept. of Civil & Constr. Engrg., Iowa State University, Ames, IA 50011
Jacob Mehus, REINERTSEN Engrg. ANS, Oslo, Norway
Kurt A. Young, Walter P. Moore & Assocs., Inc., 3131 Eastside, Second Floor, Houston, TX 77098-1919
Ed O'Neil, Army Corps of Engrs., ATTN: CEWES-SC-A, Waterways Experiment Station, 3909 Halls Ferry Rd., Vicksburg, MS 39180-6199
Bruce A. Barnes, Black & Veatch, 11900 E. Cornell Ave., Suite 300, Aurora, CO 80014

AGING OF GLASS FIBER COMPOSITE REINFORCEMENT

Introduction

Corrosion of steel reinforcement in structural concrete is causing severe deterioration of the infrastructure around the world. Studies show that large investments are needed to keep the infrastructure at an economically efficient and structurally safe operating level (Reinforced Plastics, 1994). This has caused government and private owners in Europe, Japan, and the US to look for better and more cost efficient materials. Over the last decade, fiber-reinforced-plastic (FRP) composites have emerged as one of the most exciting and promising solutions to the deterioration problem caused by the corrosion of steel.

Civil Engineering structures are required to be structurally safe and long lasting with reasonable maintenance costs, even when exposed to sustained stress and constantly changing environments. Thus, the success of FRP materials as a substitute to steel is highly dependent on the long term performance of the FRP materials, and consequently, on the ability to perform accurate long term aging predictions. Due to the lack of real-time aging data from structures utilizing FRP reinforcement, other means must be used to predict the long term response of FRP to sustained load and adverse exposures. Although FRP composites exhibit excellent virgin properties and appear to be a viable alternative to regular steel reinforcement, concern has surfaced with respect to the response of these materials to the high pH of concrete (pH of 12.5-13.0) (Anderson, 1994, Proctor, 1982). The hydration process which creates the high pH is expected to continue for months and years, as long as moisture is present (Knofel, 1978). Hence, the continuous hydration process is expected to keep the structural reinforcement in concrete exposed to a highly alkaline environment for an extended period of time. This exposure could possibly last throughout the entire lifespan of the structure.

This paper will present some of the results from the investigation 'Aging Degradation of Fiber Composite Reinforcement for Structural Concrete' (Porter, 1995). The investigation addressed several significant issues related to possible effects of corrosion and/or aging of commercially available FRP reinforcement in a highly alkaline environment. The research was performed by researchers at Iowa State University (ISU), supported by the joint sponsorship of the U.S. Army Corps of Engineers Waterways Experiment Station and the U.S. Navy's Naval Facilities Service Center. This paper will present some of the results obtained from the testing of glass FRP rebars.

OBJECTIVE

The objective of the research project focused on accelerated long term durability testing of commercially available FRP products for structural concrete. More specifically the objective of the part of the study presented in this paper was to:

- Evaluate the structural behavior and tensile strength of unaged FRP rebar test specimens.
- Evaluate the structural behavior and tensile strength of aged FRP rebar test specimens after direct exposure to an accelerated aging solution.

SCOPE

The scope of the research project presented in this paper was to:
- Develop a gripping technique suitable for tension testing of FRP rebar specimens investigated in this study.
- Develop and perform an accelerated aging procedure, expanding on the experimental technique developed by the Pilkington Bros. Ltd. (Aindow, 1983) and also based on previous research at ISU (Porter, 1992).
- Perform tensile testing on aged and unaged FRP specimens, and study the effects of simulated aging (49 test specimens).
- Perform flexural testing on concrete beams cast after the aging period with aged and unaged FRP rebar reinforcement (12 specimens).

The overall scope of the investigation included a total of 120 specimens plus several SEM (scanning electron microscope) and light microscope specimens (Porter, 1995).

MATERIALS

Three different rebar types from two U.S. companies were obtained for this study. These bars were designated A, B, and C for confidentiality. All the FRP rebar specimens were #3 bars, manufactured using E-glass and a helical fiberglass wrap. This wrap gave the bars a deformed shape for bond purposes. Each rebar type was manufactured with a different resin system. The first company, which supplied Rebar A specimens, used an isopthalic polyester resin. The second company, which furnished both Rebar B and Rebar C, used an isopthalic polyester resin for Rebar B and a bisphenol vinyl ester resin for Rebar C. In addition, Rebars B and C specimen were manufactured using a polyester veil on the outer surface, giving a rougher surface texture. The veil was also believed to give a more resin rich outer surface, supposedly providing better protection of the FRP rebars in a alkaline environment.

Table 1 lists the effective cross sectional area, along with the weight and volume fractions for each specimen group. The effective area was determined through submersion testing, while the ASTM D2584-68 'burnout' test was used to determine weight fractions of each group.

TABLE 1. EFFECTIVE AREA, WEIGHT AND VOLUME FRACTIONS

Specimen Group	Effective Cross Sectional Area (in²)	Average Weight Fractions of Resin (%)	Average Weight Fractions of E-glass (%)	Average Volume Fractions of Resin (%)	Average Volume Fractions of E-glass (%)
Rebar A	0.1317	19.856	80.144	33.44	66.56
Rebar B	0.1214	29.755	70.245	49.50	50.50
Rebar C	0.1123	34.302	65.698	56.50	43.51

ACCELERATED AGING

The accelerated aging procedure developed for this investigation was based on studies performed by Pilkington Bros. Ltd. (Aindow, 1983) and expanded further by Porter (1992).

Researchers at Pilkington Bros. showed that long term aging predictions made over a very short period of time correlate well with real weather aging. These predictions were made based on testing performed on specimens subjected to elevated temperatures to speed the chemical reactions within the concrete. Based on these findings, researchers at ISU developed two equations for accelerated aging in a local Iowa environment (Porter, 1992). The two equations provide a theoretical relationship between the number of days aged per days of exposure to an accelerated aging solution at a specified temperature. The accelerated aging procedure briefly described above was used in the investigation with the following criteria:

- A highly alkaline aqueous solution was used ($Ca(OH)_2$ and $NaOH$) to simulate the high pH of the porewater solution. The target pH was set to 12.5-13.0 at room temperature.
- The specimens were exposed to the aging solution for 2-3 months (depending on the bath temperature), simulating approximately 50 years of real weather aging.
- The aging tank was operated at a target temperature of 140°F.

The rebar specimens were submerged in the aging bath after the chemicals were added. The pH remained steady at a level of 12.5-12.9, measured at room temperature, throughout the aging period. Based on daily temperature recordings, the aging bath was terminated after 81 days of accelerated aging when the specimens had been aged to an equivalent of approximately 50 years of real time aging.

RESULTS

The FRP rebar tensile tests were performed using a universal testing machine at a low load rate. During testing, data for load, strain, and load table movement (total elongation) were recorded. The gripping method developed for this investigation, using internally threaded steel pipes epoxied on to the ends of the tensile test specimens, was a modification of a gripping technique previously developed by researchers at ISU (Porter, 1991). As in the case of the 200+ tensile tests previously performed at ISU, the modified gripping technique (Porter, 1995) yielded consistent results with no apparent influence from the grips. The average tensile test results from the unaged and aged FRP specimens are presented in Tables 2-4.

TABLE 2. AVERAGE PROPERTIES FROM TENSILE TESTING OF UNAGED FRP REBAR

Specimen Group	Ultimate tensile stress σ_U (ksi)	Maximum strain ϵ_U (in/in)	Modulus of elasticity, E (ksi)
Rebar A	50.2	0.0146	5.28×10^3
Rebar B	52.2	0.0217	3.78×10^3
Rebar C	53.3	0.0188	4.33×10^3

Specimens from all three rebar groups exhibited consistent, ultimate tensile strengths, but strengths were lower than expected based on the manufacturer's data. Both manufacturers specified ultimate tensile strengths for their # 3 bars to be greater than 100 ksi. The consistently low tensile strengths exhibited by all unaged specimens created a need for verification of the experimental test procedure. Two Rebar A and two Rebar B specimens were gripped and tested independently by Dr. Charles W. Dolan at University of Wyoming (UW), using his own gripping technique (Dolan, 1994). The differences between the tests performed at ISU (Table 2) and UW (Table 3) are not of a magnitude that would indicate any procedural problems. Therefore, the verification testing confirmed that the low tensile test results were due to the characteristics of the FRP rebars, and not the gripping technique and/or testing procedure.

TABLE 3. TENSILE TESTING OF UNAGED FRP REBARS AT UW

Specimen Group	Area (in^2)	Ultimate Tensile Stress, σ_U (ksi)
Rebar A	0.121	54.5
Rebar B	0.117	59.4

Two different series of aged FRP rebars were tested in tension. The first test series included a total of six additional rebar specimens that were removed from the aging bath after only 19 days of accelerated aging. Based on the aging equations, the aging effect after 19 days was determined to be 5.4 years of real time aging. Tensile testing of these specimens (Table 4) included recording of the breaking load only. The results showed that the FRP rebar specimens had experienced a significant strength loss after a relatively short period of exposure. The second series of tensile tests were performed on specimens that had been exposed to the aging solution for the full 81 days, which is equivalent to 50 years of real time aging. These tensile tests (Table 5) included recordings of strain and load table movement in addition to breaking load. The aging effects presented in Tables 4 and 5 are based on a comparison to test results of unaged specimens in Table 2. The results presented in Tables 4 and 5 clearly showed that the strength loss was not linearly related to the number of days aged. This indicated that the specimens experienced an initial region of rapid strength loss followed by a second more stable region. Similar behavior has previously been reported by (Aindow, 1983).

TABLE 4. RESULTS FROM TESTING OF AGED FRP REBARS (5.4 YEARS)

Specimen Group	Ultimate Tensile Strength, σ_U (ksi)	Aging Effect (%) (Decrease)
Rebar A	35.6	29.0
Rebar B	20.5	60.7
Rebar C	28.9	45.8

TABLE 5. RESULTS FROM TESTING OF AGED FRP REBARS (50 YEARS)

Specimen Group	Ultimate Tensile Strength, σ_U (ksi)	Maximum Strain ϵ_U (in/in)	Modulus of Elasticity, E (ksi)
Rebar A	26.3	0.0069	5.21×10^3
Aging Effect (%)	47.6	50.2	1.3
Rebar B	17.9	0.0070	3.88×10^3
Aging Effect (%)	65.8	67.7	2.5
Rebar C	18.6	0.0060	4.15×10^3
Aging Effect (%)	65.1	68.1	4.2

After termination of the aging bath, a total of twelve concrete beams were constructed with either unaged or aged FRP reinforcement. The beams were 6 feet long with an 8 by 8 inch cross section, and each beam was reinforced using a single FRP rebar. Flexural testing of the beams was performed in a universal testing machine using center-point loading. Load, deflection, and slip data was recorded. All the specimens failed in flexural tension without any slip between the concrete and the reinforcement. Based on this data, the ultimate tensile stress in the reinforcement was calculated. Though flexural testing gave somewhat higher tensile strength results (Table 6) than the regular tensile tests (Tables 4 and 5), the results were in the same range and served as additional verification of the normal axial tensile testing. The aging effect listed in Table 6 is a comparison between unaged and aged ultimate tensile strength properties from the flexural testing.

TABLE 6. RESULTS FROM CONCRETE BEAMS WITH FRP REINFORCEMENT

Specimen Group	Unaged Ultimate Tensile Stress, σ_U (ksi)	Aged Ultimate Tensile Stress, σ_U (ksi)	Aging Effect (%) (Decrease)
Rebar A	55.3	33.0	40.3
Rebar B	62.3	31.6	49.7
Rebar C	63.7	40.1	40.1

SUMMARY AND CONCLUSIONS

The average unaged tensile strength of the FRP rebars was found to be 50-55 ksi, approximately 50% lower than that specified by the manufacturers. The large discrepancy between these results and the manufacturer specifications appeared to be due to the characteristics of the FRP rebars. The low ultimate tensile strengths exhibited by the unaged FRP rebars could be caused by poor fiber alignment and/or orientation, the formation of stress concentrations in the fibers, and imperfections from production. The ultimate tensile capacity of any fiber is sensitive to the angle between the fiber and the axis of loading, but only in the center portion of the cross section will the fibers likely be parallel to the axis of loading.

The aged specimens exhibited a severe reduction of ultimate tensile strength and maximum strain capacity. After accelerated aging equivalent to approximately 50 years of real time aging, the reduction in both ultimate tensile strength and maximum strain was as high as 50-70%. Significant reductions were also seen after less than six years of aging. The effects of only limited exposure (approximately 5.4 years of real time aging) showed that the FRP reinforcement could experience significant strength loss with only short periods of high pH in the concrete. The magnitude of strength loss and reduction of maximum strain capacity after an equivalent of approximately 50 years of real time aging did not appear to be affected by the type of resin used. However, the rate of strength loss differed among the three specimen groups. The indications of an initial region with rapid strength loss, followed by a slower aging rate, appeared to correlate well with previous research performed by researchers at Pilkington Bros. (Aindow, 1983).

The test results presented in this paper do not necessarily indicate that the resins behaved poorly in the highly alkaline environment. Since the resin is expected to protect the E-glass (known to be highly susceptible to alkaline attacks), the efficiency of the protective seal is probably more important. If the seal protecting the E-glass is not perfect, the raw glass fiber strands are subject to possible degradation, regardless of the corrosive resistance of the resin in a highly alkaline environment. The relatively complex geometric shape of the specimens tested in this investigation make a perfect resin seal even more difficult to achieve. However, the results presented in this paper clearly show the importance of a continued research effort to develop better FRP bar designs, which yield high tensile strengths and have resin seals which effectively protect the FRP rebars. Furthermore, the results demonstrate the need for future real time aging research to provide a larger data base for combinations involving new materials.

REFERENCES

"Building an infrastructure for the future". 1994. Reinforced Plastics, Volume 38, Number 1, pp:24-28.

Anderson, G.R., Bank, L.C., and Munley, E. 1994. "Durability of Concrete Reinforced with Pultruded Fiber Reinforced Plastics", Proceedings-49th Annual Conference, Composite Institute, The Society of the Plastics Industry, Inc., Cincinnati Convention Center, Cincinnati, OH, Session 2-B.

Proctor, B.A., Oakley, D.R., and Litherland, K.L. 1982. "Developments in the Assessment and Performance of GRC Over 10 Years", Composites, Volume 13, Number 2, pp: 173-179.

Knofel, D. 1978. Corrosion of Building Materials, Van Nostrand Reinhold, New York, NY, pp:94-95.

Porter, M.L., Mehus, J., and Young, K.A. 1995. Aging Degradation of Fiber Composite Reinforcements for Structural Concrete", Final Report, Iowa State University, Engineering Research Institute.

Aindow, A.J., Oakley, D.R., and Proctor, B.A. 1983. "Comparison of the Weathering Behavior of GRC with Predictions Made from Accelerated Aging Tests", Cement and Concrete Research, Volume 14, Number 2, pp: 271-274.

Porter, M.L., Lorenz, E.A., Barnes, B.A., and Viswanath, K.P. 1992. Thermoset Composite Concrete Reinforcement, Department of Civil and Construction Engineering, Engineering Research Institute, Iowa State University, Ames, IA.

Porter, M. L., B. A. Barnes, "Tensile Testing of Glass Fiber Composite Rod". 1991. Proceedings of the Speciality Conference - Advanced Composite Materials in Concrete Structures, S. L. Iyer, Ed. American Society of Civil Engineers, Flamingo Hilton, Las Vegas, NV, pp: 123-131.

Dolan, C.W. 1994. "Non-Metallic Prestressing Tendons", U.S. Department of Transportation, Fiber Reinforced Plastics Workshop, Washington, D.C.

BIOGRAPHIES

Max L. Porter, P.E., Ph.D., Parl., is a Professor of Civil Engineering at Iowa State University, where he has taught over 30 different courses, conducted over 70 research projects principally on various kinds of composite structures, reinforced concrete, and masonry, and published over 70 papers and 200 reports. He is a member of the Board of Governors of the new Structural Engineering Institute (of ASCE), Chair of the ASCE Codes and Standards Activities, past Chair of the Executive Committee of the Building Standards Division, Chair of the Structural Composites and Plastics Standards Committee, Chair of the Steering Committee for the ASCE/PIC-LRFD Standards, Chair of the Masonry Standards Joint Committee, and past Chair of the Composite Deck Standards Committee, as examples of over 50 such leadership or officer positions in several professional organizations.

Jacob Mehus has a three year engineering degree from Bergen College of Engineering (BCE) in Bergen, Norway. He received a B.S. from Iowa State University in the fall of 1993 and M.S. from ISU the fall of 1995. He is currently employed doing structural design work related to highway and railroad structures at REINERTSEN Engineering in Oslo, Norway.

Kurtis A. Young is a structural design engineer with Walter P. Moore and Associates, Houston, Texas. He has worked on a variety of building structures in the U.S. and overseas. Mr. Young holds a Bachelor of Science degree in Aerospace Engineering and a Master of Science degree in Civil (Structural) Engineering from Iowa State University. He was previously employed as a structural analyst on the Space Shuttle program for Thiokol Corporation, Ogden, Utah.

Ed O'Neil is Chief of Engineering of the Mechanics Branch at U.S. Army Corps of Engineers Waterways Experiment Station at Vicksburg, MS. Mr. O'Neil has been with the U.S. Army Corps of Engineers for 22 years. He has conducted numerous research projects in the ares of Concrete materials and structural elements. He is an acting member of Society of American Military Engineers, American Concrete Institute, Post Tensioning Institute, and other professional organizations.

Bruce A. Barnes received his B.S. and M.S. degrees from Iowa State University in 1987 and 1990 respectively. Following his master's degree, Mr. Barnes worked as a research project manager at Iowa State University. During his tenure in this position, Mr. Barnes managed projects encompassing bond and corrosion of composite elements, flexural and axial load capacities of masonry veneer anchors and composite sandwich wall connectors, and seismic resistance of unreinforced masonry walls. Mr. Barnes is currently employed by Black & Veatch in Denver, Colorado as a structural engineer.

FRP Grids for Reinforced Concrete: An Investigation of Fiber Architecture

RENATA S. ENGEL, CHARLES E. BAKIS,
ANTONIO NANNI AND MICHAEL CROYLE

ABSTRACT

The objective of this paper is to illustrate the manufacture and performance of FRP (fiber reinforced plastic) grids used in reinforced concrete. Composite manufacturing methods, such as fiber placement and closed molding, easily lend themselves to the production of structures that have intricate shapes and contours such as the grid or lattice structures used in reinforced concrete panels. In this investigation, continuous carbon fibers are impregnated with vinylester resin and placed via filament winding into a mold to form a two-dimensional grid. Once filled, the mold is closed and the part is cured under low temperature and pressure. During filament winding, fiber placement is controlled such that two general types of layer architectures at the grid joints are obtained: (a) cross ply, i.e., alternating unidirectional tapes without fiber connectivity between layers; and (b) weaves with fiber connectivity between adjacent layers. The influence of fiber architecture on the grid stiffness and the ability of the grid to transmit loads was studied using stand-alone tensile tests and pull out tests, wherein the reinforcement was pulled from the concrete. While the magnitudes of the joint strength and post damage stiffness vary little for the two different architectures, the damage progression, characterized by permanent damage via shear displacement of a transverse grid element for a given load, is lower for the cross-ply joint.

INTRODUCTION

The construction industry can take advantage of some recent advances in automated composites manufacturing and apply them to grids and cages that are used as concrete reinforcement. Currently, steel grids are put in place at the construction site by tieing steel rods into the desired shape. A natural extension to incorporate FRP has been the replacement of steel with pultruded rods that are joined with plastic ties (GangaRao, et al., 1995). Regardless of the material, multi-component structures, such as bent and tied rods, result in labor intensive, on-site construction; therefore, more efficient alternate designs that involve FRP

Composites Manufacturing Technology Center, 227 Hammond Building, The Pennsylvania State University, University Park, PA 16802

reinforcement have developed. For example single structures such as bolted pultruded grates (Bank, et al., 1992; Larralde, 1992) and stacked grids that are pressed together (Sugita, 1993; Schmeckpeper and Goodspeed, 1992, 1994) have been developed and investigated in concrete slabs and beams. More complex geometries, like 3-D lattices (Yonezawa, 1993) and 3-D fiber configurations (Nakagawa, 1993) have been used in partition walls and slabs. These single structures take advantage of the contribution of the transverse members to transfer the load to the concrete. This research extends the investigations of 2-D FRP grid structures by examining the fiber architecture at the joints in an effort to maximize the benefit of the transverse members.

EXPERIMENTAL

A closed molding method was developed to manufacture two-dimensional FRP grids for use in reinforced concrete panels or flat work. The manufacturing method has two attributes that allow for flexibility in design. First, a range of grid spacings, with no limitation on the minimum grid spacing, is possible. For steel, the grid development from individual steel rods is labor intensive, therefore maximum spacing is used. By reducing grid spacing, the reinforced concrete crack spacing, crack width and overall structural deflections are reduced. For molding, effective use of grid spacing is particularly critical because the fiber volume fraction in individual grid lines is likely to be less than the pultruded grid lines. However, since grids can be close, the fiber reinforcement per unit area can be comparable. The second attribute of the method is that fiber placement can be controlled, thereby developing a prescribed fiber continuity and optimum fiber arrangement at the joint. To illustrate the effect of the latter attribute, two fiber arrangements with one grid spacing will be investigated. The manufacturing method for making grids with different fiber architectures at the joints, and the pull out experimental tests for evaluating the stiffness and strength of the joints in concrete are discussed in detail.

GRID MANUFACTURING

Grids were constructed in a two-stage process: fiber placement in a mold and consolidation/cure in a smart press. The two part mold, made of silicone flexible molding compound, makes a single lattice (61 cm x 61 cm) with 10.2 cm grid spacing. The number of fiber tows in each leg was held constant, so the thickness of the lattice varied depending on the winding pattern, but nominally was 0.5 cm. The fiber was placed by wet winding vinylester (Ashland 922) impregnated carbon fiber (Zoltek Panex 33, 48K) tows into the female half of the mold. The mold was secured to a 60 cm diameter mandrel while the payout eye of the filament winder traced the grid pattern of the mold, placing fiber tows along the grid lines. The flexible mold permitted the use of an automated method, i.e., filament winding, to construct the grids. The male half of the mold and edge supports were attached before placing the mold in the smart press for consolidation and cure under low pressure (20 psi, based on grid area) and temperature (80°C).

Since this research focuses on understanding and improving the fiber architecture at the grid joints, two fiber patterns shown in Figure 1 were studied. The first is a unidirectional cross ply such that the fibers are placed in alternating directions for each fiber layer (Figure 1b). The second pattern consists of a weave structure such that alternating fiber tows overlap at each joint; thereby providing an integrated fiber structure (Figure 1c). For the former pattern, no fiber connectivity exists at the joints, whereas in the latter one, the weave structure connects adjacent fiber layers.

TESTING

Tensile tests were performed on stand-alone grids to determine the stiffness of a repeating unit (one grid and joint). *In situ* performance was evaluated using a pull out test. The test requires that a portion of the FRP grid be embedded in concrete with the concrete resting on the crosshead (Figure 2); the other end of the specimen is loaded axially. The tensile load is resisted by one 5.2 cm long transverse segment of the grid in concrete. Without transverse members the embedded grid legs would be supported by frictional forces and chemical adhesion; however, the transverse members distribute the load to the concrete thus providing the load carrying capability. Grid performance was evaluated by recording load versus displacement at the load end and free end using linear variable-differential transformers (LVDT's) with a sensitivity of 15 μm. The free end LVDT measures the displacement from the free end of the grid to the concrete surface and the load end LVDT was positioned 38 cm below the embedded grid segment.

RESULTS

The results are tabulated for all stand-alone and pull out specimens (Table I). The data collected from the tensile tests performed on the grid without concrete (stand-alone specimens) indicate that the ultimate strength, P_U, and the stiffness of the repeating unit (leg and joint) k_{L+J}, for both architectures are the same within experimental error. These values were used to interpret the results from the *in situ* pull out tests.

Typical load versus displacement plots for the cross ply and the weave specimens are shown in Figures 3 and 4, respectively. The two curves on each plot show the load-displacement at the free end and load end. Key values from the pull out tests, shown in Table

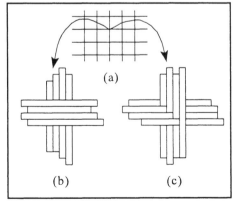

Figure 1. (a) Grid structure with exploded views of two joint architectures: (b) cross-ply and (c) weave.

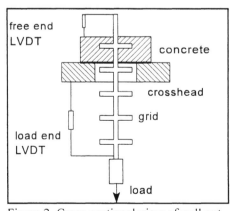

Figure 2. Cross-sectional view of pull out test specimen.

I, are taken from the free end load-displacement curve. They are: the ultimate strength, P_U, the critical load at the onset of free end displacement, P_I, and the stiffnesses, k_I and k_{II}, for the two distinct slopes after the critical load.

In order to determine whether the region after load P_I corresponds with linear elastic deformation, nonlinear elastic deformation or damage (rigid body motion), cyclic tests were conducted. An example of the data (Figure 5) shows that the displacement at the free end was nonrecoverable--suggesting the formation of damage. The load results support this and show that the displacement at the load end is due to a combination of elastic deformation and rigid body motion.

Visual inspection of the cyclic test specimens after cutting them open showed no visible damage to the concrete and that longitudinal tows debonded and slipped relative to the transverse tows. The observation that the fibers slip past each other was confirmed by comparing P_I to the predicted interlaminar shear strength of a carbon/vinylester composite.

Finally, the fiber architecture does influence the performance of the grid in concrete, but only to the extent that the damage progression is affected. The stiffness immediately after P_I for the cross ply is significantly higher than that of the weave. But that trend is reversed as damage progresses: the stiffness of the weave joint becomes greater than the cross ply joint stiffness. For the cross ply, the fibers may act as a knife that progressively cut through the resin between layers; whereas the woven fiber tows may act as entanglements that impede the sliding of transverse tows.

TABLE I. Averages and the number of specimens tested for stand-alone and pull out tests

Fiber architecture	Stand-alone - S		Pull out - P				
	P_U kN	k_{L+J} kN/mm	P_U kN	P_I kN	k_I kN/mm	k_{II} kN/mm	$P_{U\text{-}P}/P_{U\text{-}S}$
Cross ply (# of specimens)	38.7 (3)	44.9 (3)	32.3 (2)	14.0 (3)	40.1 (2)	3.10 (2)	0.83
Weave (# of specimens)	40.5 (3)	47.1 (3)	32.9 (2)	13.6 (3)	32.6 (2)	5.25 (2)	0.81

CONCLUSIONS

Several conclusions can be drawn from the stand-alone and pull out tests for the cross ply and weave FRP grids. First, the onset of slip is not significantly affected by these particular fiber architectures. However, the damage progression is dependent on the architecture. In both cases, pseudo-ductile failure was observed, and the amount of permanent deformation is significant. Even so, over 80% of the ultimate strength of the stand-alone grid is retained before the grid fails when embedded in concrete. The ability of the grid to carry load even after interlaminar shear damage, which starts at 35% P_U, is valuable in determining the allowable stress in the reinforcement at service conditions. No conclusion can be made about the response of the joint prior to P_U, due to the limited sensitivity of the LVDT's.

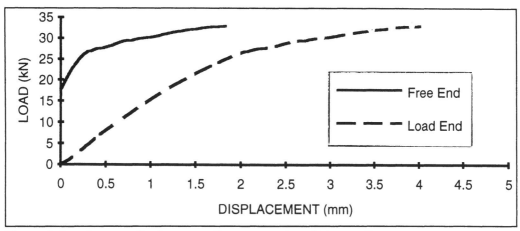

Figure 3. Load versus displacement for a cross ply grid in a pull out test.

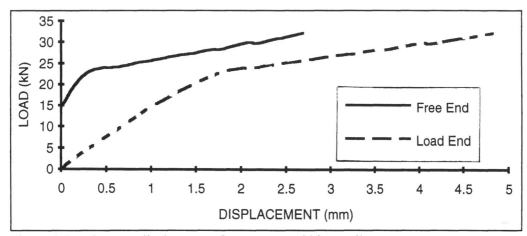

Figure 4. Load versus displacement for a weave grid in a pull out test.

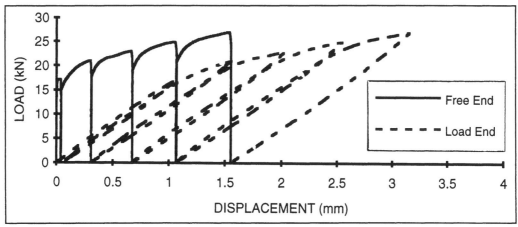

Figure 5. Load versus displacement for a weave grid in a cyclic pull out test.

ACKNOWLEDGMENTS

This work is supported through the National Science Foundation (CMS-9531998). The contributions by Ashland Chemical Co. and Zoltek Corp. are greatly appreciated.

REFERENCES

Bank, L. C., Z. Xi, and E. Munley. 1992. *Proc. of the Materials Engineering Congress*, ASCE, pp. 618-631.

GangaRao, H. V. S., S. S. Faza, S. V. Kumar, and R. W. Allison. 1995. *J. Reinf. Plast. and Comp.*, 14:910-922.

Larralde, J. 1992. *Proc. of the Materials Engineering Congress*, ASCE, pp. 645-654.

Nakagawa, H. and M. Kobayashi, T. Suenaga, T. Ouchi, S. Watanabe and K. Satoyama. 1993. A. Nanni, Ed., pp. 387-404.

Schmeckpeper, E.R. and C. H. Goodspeed. 1992. *Proc. of the Materials Engineering Congress*, ASCE, pp. 632-644.

Schmeckpeper, E.R. and C. H. Goodspeed. 1994. *J. Composite Materials*, 28(14):1288-1304.

Sugita, M. 1993. *Fiber-Reinforced-Plastic (FRP) Reinforcement in Concrete Structures*, A. Nanni, Ed., pp. 355-385.

Yonezawa, T., S. Ohno, T. Kakizawa, K. Inoue, T. Fukata, and R. Okamoto. 1993. *Fiber-Reinforced-Plastic (FRP) Reinforcement in Concrete Structures*, A. Nanni, Ed., pp. 405-419.

BIOGRAPHIES

R. Engel is an Associate Professor in the College of Engineering. She is interested in manufacturing composites and modeling the transport processes of composite materials manufacturing.

C. Bakis is an Associate Professor of Engineering Science and Mechanics. His professional interests are in the areas of designing, manufacturing and testing composite materials.

A. Nanni is interested in construction materials, their structural performance, and field application. Dr. Nanni is an active member in the technical committees of ACI, ASCE, ASTM and TMS. He is founding Chairman of ACI Committee 440 - FRP Reinforcement.

M. Croyle has a BS in Civil Engineering and is pursuing a MS in Engineering Mechanics. He is interested in composites as they relate to civil structures.

McKinleyville Bridge: Construction of the Concrete Deck Reinforced with FRP Rebars

SANJEEV V. KUMAR, HEMANTH K. THIPPESWAMY AND HOTA V. S. GANGARAO

ABSTRACT

McKinleyville Bridge, located in Brooke County in the northern panhandle of West Virginia, is the first vehicular bridge in this country whose concrete deck is reinforced with FRP rebars. The construction of this bridge was recently completed and opened to traffic beginning October 1996. Some of the first hand experience in handling and placement of FRP bars in concrete deck have been presented in this paper. The lightweight of FRP rebars is attractive for construction crew in terms of easy handling and placement. It is recommended to use leather gloves to avoid bruises and itching caused by fibers. Other issues like spacing of chairs for FRP mesh and flotation of FRP rebars while vibrating wet concrete have been addressed. Improper construction practices and adverse weather conditions while casting concrete bridge deck can cause excessive shrinkage and thermal cracks in the deck. Moreover in FRP reinforced concrete, unlike steel, once these cracks are formed, they could greatly enhance in their size due to lower modulus of elasticity of FRP bars. Hence, it was necessary to re-consider some of the precautionary measures to minimize cracking in concrete deck.

INTRODUCTION

A new vehicular bridge constructed across Buffalo Creek in McKinleyville, West Virginia, is the first bridge in the U.S.A to use Fiber Reinforced Plastic (FRP) rebars in the concrete deck. The owner of the bridge, the West Virginia Department of Transportation-Division of Highways (WVDOT-DOH) joined its research partners (the U.S. Army Corps of Engineers, the Federal Highway Administration, the Constructed Facilities Center-West Virginia University, and structural plastics producers) to design and construct this integral abutment bridge with FRP rebar reinforced concrete deck and steel stringers (Figures 1 and 2).

Sanjeev V. Kumar, Constructed Facilities Center, West Virginia University, Box 6103, Morgantown, WV 26505-6103
Hemanth K. Thippeswamy, P.O. Box 6103, Morgantown, WV 26505-6103
Hota V.S. GangaRao, P.O. Box 6103, Morgantown, WV 26505-6103

The McKinleyville Bridge is WVDOT-DOH, Project No. S305-27/4-0.03 and is located in Brooke County in the northern panhandle of West Virginia. The new bridge is a 177 feet long, 3-span continuous structure accommodating two lanes of traffic and located approximately 200 feet downstream of the existing bridge. The alignment of the new bridge allows a straight entrance to the bridge for traffic from County Route 27/4 traveling to WV Route 67. The existing alignment requires the motorist to maneuver two sharp corners before getting to the bridge. By improving the alignment, sight lines of the bridge were also improved. All this and more make the McKinleyville bridge an ideal location for featuring an FRP reinforced bridge deck and integral abutments.

DESIGN OF THE MCKINLEYVILLE BRIDGE

The design of the McKinleyville bridge was accomplished jointly by the WVDOT-DOH and the CFC. The CFC was responsible for the design of FRP reinforced concrete deck; all other design aspects were the responsibility of the WVDOT-DOH.

The design of the FRP reinforced concrete deck was based on a design method developed at the CFC (GangaRao, 1996). The design method is similar to the procedure described in American Association of State Highway Transportation Officials' *Standard Specifications for Highway Bridges* (AASHTO, 1992) working stress design of transversely reinforced concrete decks. The design requires a deck thickness of 9 inch and #4 (0.5") FRP rebars as the main transverse reinforcement at 6 inch spacing. The main reinforcement is tied to #3 (0.375") FRP bars for distribution reinforcement, also at 6 inch spacing, with 1 ½ inch clear cover on top and 1 inch clear cover on bottom surface. The distribution reinforcements consisted of straight bars and long loops. The purpose of using long loops was to confine the deck concrete and create an integrity between top and bottom reinforcement.

To support the dead and live loads M270 grade 50 steel W33X130 rolled beams spaced at 5 feet centers were chosen. The integral abutment design used a single row of piling at the bearing centerline. The steel pile webs were parallel to the bearing centerline to offer minimum resistance to longitudinal movement. The piles were driven in a pre-bored hole filled with loose sand for the first 20 feet, and then driven to refusal.

FRP REBARS FOR CONCRETE DECK

Two types of FRP bars, C-bars and Sand Coated bars, were supplied by two manufacturers for the concrete deck as shown in figure 3. The variety of materials and processes used in just these two types of bars illustrates both the complexity and the opportunities available in the design of FRP products. The former types have an E-glass fiber, polyester resin core combined with a "compression-molded, chopped and continuous E-glass fiber reinforced, urethane-modified vinyl ester sheet molding compound (SMC). The latter types were manufactured of E-glass fibers and isopthalic unsaturated polyester. After the pultruded bar was formed, the rod was wrapped with two additional fiber chords in a helical pattern and then coated with another layer of epoxy resin and rolled in sand. The wrapped chords and sand coating provide a mechanical bond with the surrounding concrete.

Specifications prepared by the CFC and the WVDOH require certain materials and processes be used to assure the expected quality product. Many of the manufacturing process variables are not specified, however there was not significant differences in their mechanical and thermal properties. In general, both types of FRP bars were to have 75% fiber by weight and 25 % resin. The E-glass fibers with isopthalic unsaturated polyester or urethane modified bisphenol vinylester were used in manufacturing these bars. The recommendation, for the type of resins to be used in FRP rebars, was based on cooperative work with Reichhold Inc., in screening resins for FRP rebars under different environmental conditions (Altizer et al., 1995).

The minimum mechanical properties required were an ultimate tensile strength of 80,000 psi, tensile modulus of 6,000,000 psi, and bond strength of 1500 psi in concrete.

HANDLING AND PLACEMENT OF FRP REBARS

From the standpoint of the installer, the FRP bars have the advantage of somewhat reduced manpower requirements because more bars can be carried into position at one time due to lightweight. The bars also stayed in position well, once they were tied. In general, the construction crew handled the FRP bars in a manner similar to steel bars. The FRP bars withstood typical on-site handling problems during concreting. However, to avoid any bar distress careful handling of bars in the field is recommended.

Other factors to consider include the fact that some construction crew reacted negatively to FRP rebars and also the presence of razor sharp edges (while fabrication the reinforcement mesh) could create a lot of misery and bloodshed if the rodbusters get careless. However, these problems can be solved as the use of the bars becomes more common. Moreover the use of leather gloves should be recommended while handling these bars (Martin, 1996).

Spacing of Chairs

FRP bars were found to be more flexible when walked over. The spacing of chairs was based on limiting bar deflections under construction loads like concreting equipment and construction crew on FRP mesh while concreting. Epoxy coated steel chairs with 1 inch height, for bottom cover, were spaced at 4 feet apart to support main reinforcement in the bridge at the bottom. In addition, 7 inches high chairs were placed between 1 inch high chairs to support the top bars as shown in figure 4. Further, the long loops in the longitudinal direction also provided the support for top bars.

Flotation of FRP Bars in Wet Concrete

Density of FRP bars with 75 % glass fiber by weight is approximately 100 lb./cft, while the density of normal wet concrete is approximately 150 lb./cft (Lydon 1982). Hence, there is a possibility of flotation of FRP bars while vibrating wet concrete which can alter the specified clear covers for FRP bars. Therefore, the FRP mesh was tied down to the form-work, against the chairs, at four foot intervals in both the transverse and longitudinal directions in order to prevent their floating and maintain adequate cover.

CASTING AND CURING OF CONCRETE BRIDGE DECK

Cracking of concrete due to improper construction practices and adverse conditions has been well probed in Concrete Technology. The improper construction practices are in terms of concrete pour size and sequence, admixtures, vibration, surface finishing, and curing.

Adverse conditions include high winds, extreme low and high temperatures, and low humidity. Moreover in FRP reinforced concrete, unlike steel, once these cracks are formed, they could greatly enhance in their size due to lower modulus of elasticity of FRP bars. Hence, it was necessary to consider precautionary measures to minimize cracking in concrete bridge deck due to the above said improper practices and conditions.

The concrete bridge deck was cast in two pours. The first pour included casting of the portion of the bridge deck between points of inflection in the end spans, i.e., portion of the bridge deck excluding 10 feet length on either end of the bridge. The first pour was in the evening hours (7:30 P.M. to 2:30 A.M.) on August 1, 1996. The second pour included casting of the remaining portion of bridge deck, abutment above bridge seat level, and approach slab, started in the morning at 7.00AM and was concluded at 10.00 AM. The evening and early morning pours were preferred because of lower air temperatures (75-80° F existed while concreting), lower solar radiation, and higher air relative humidity. The above conditions ensured minimum loss of moisture from wet concrete, which in turn should minimize shrinkage and temperature associated cracking. The concrete was vibrated for about 10 seconds at each location spaced 20 inches apart. This optimum vibration was necessary to prevent honeycombing, settlement cracking, and segregation and bleeding of concrete. Figures 5, 6 and 7 show the pouring, screeding and finishing sequences for the concrete deck, respectively.

The concrete deck including abutments and approach slabs was cured for 14 days with wet burlap and plastic cover on them. Constant wetting of concrete surfaces were ensured by continuous flow of water over the surfaces. The above curing method was adopted to minimize plastic shrinkage cracking in concrete which occurs when evaporation rate exceeds the bleed rate of plastic concrete.

CONCLUSION

The construction of McKinleyville bridge with the use of FRP rebars in the concrete deck is an attempt to increase the longevity of deck by eliminating corrosion which is very common with steel rebars. The lightweight of FRP rebars is attractive for construction crew in terms of easy handling and placement. It is recommended to use leather gloves to avoid bruises and itching caused by fibers. FRP rebar mesh was found to be more flexible at 4 feet chair spacing than steel, when walked over. However, no major distress or disturbance in spacing of FRP rebars was reported. There is a general concern on the flotation of FRP bars while vibrating wet concrete, even though it has not been quantified yet. Hence, anchoring of these bars to concrete forms should be a good practice. Shrinkage and plastic cracks in concrete should be minimized, especially for concrete decks reinforced with FRP rebars, by stringently adhering to the specifications provided by the American Concrete Institute or American Standard Testing of Materials on admixtures (ASTM C 494-92, C 1017-92, C 94-94, C 260-94), vibration (ACI 309R-87), surface finishing, and curing (ACI 308-92).

ACKNOWLEDGMENT

The construction work described in this paper was done by the Orders Construction Company, Incorporated. The "C-bars" and "Sand-Coated FRP bars" were donated by Marshall Industries Inc., and International Grating Inc. respectively.

REFERENCES

Altizer S.D., Vijay P.V., and GangaRao H.V.S., 1995., "Performance Evaluation of Conditioned GFRP Bars with Different Resins Systems," *Report Submitted to Reichhold Chemical, Inc.*

AASHTO, 1992., Standard Specification for Highway Bridges, *15th Edition, Washington, D.C.*

GangaRao H.V.S., 1996., "Concrete Beams Reinforced with FRP Rebars Under Static and Fatigue Loads," *Constructed Facilities Center Report, CFC RP # 230-96.*

Horn M.W., Stewart, C.F., and Boulware R.L., 1995., "Factors Affecting the Durability of Concrete Bridge Decks: Construction Practices", *Interim Report No. 3., Bridge Department, California Division of Highways, CA-DOT-ST-4104-4-75-3.*

Lydon F.D., 1982., "Concrete Mix Design", II Edition, Applied Science Publisher, N.Y.

Martin D., 1996., Field Officer for the McKinleyville Bridge, *WVDOT-DOH, Based on written communication.*

BIOGRAPHIES

Sanjeev V. Kumar

Sanjeev Kumar is a Research Engineer at the Constructed Facilities Center, College of Engineering and Mineral Resources, West Virginia University. Holds a B.S. in Civil Engineering from University of Madras, India and M.S. in Civil Engineering with Structural Engineering Major from West Virginia University.

Hemanth K. Thippeswamy

Hemanth K. Thippeswamy is a Research Assistant Professor at the Constructed Facilities Center, College of Engineering and Mineral Resources, West Virginia University. Thippeswamy earned his B.S. in Civil Engineering and M.S. in Structural Engineering from University of Mysore, India. He earned his Ph.D. from West Virginia University in Civil Engineering with Structural Engineering Major. Thippeswamy's areas of research interest include application of composites in civil infrastructures, field testing and monitoring of bridges, and analytical modeling of bridges.

Hota V. S. GangaRao

Hota V. S. GangaRao is a Professor of Civil Engineering and Director of the Constructed Facilities Center, College of Engineering and Mineral Resources, West Virginia University. Professor GangaRao earned his B.S. from IIT Madras; M.S. and Ph.D. from North Carolina State University, Raleigh, NC. He has published more than 150 research papers in structural analysis and design of buildings, bridges, transmission towers, railroad ties and water tanks. He has set up precast, prestressed concrete plants in the USA and in India. Dr. GangaRao has earned many citations for his technical activities. His recent experience and publications have focused on stressed timber bridge system development and fiber reinforced plastic structural shapes and rebars.

Figure 1   New McKinleyville Bridge (Side Elevation)

Figure 2   Bridge from South Abutment

Figure 3   FRP Deck Reinforcement (Sand Coated bars in the left, C-bar in the right)

Figure 4   Chairs for Top and Bottom FRP Rebars

Figure 5   Concrete Pouring and Vibration

Figure 6   Surface Screeding of Concrete

Figure 7   Finished Surface of the Concrete Deck

SESSION 7

Understanding the Regulatory Grab-Bag

NOTE: No formal papers were submitted for this session.

SESSION 8

Composites in Action

The format of this session was a hands-on demonstration; therefore, there were no formal papers.

SESSION 9

Rebars, Reinforcements and Repairs: Composites to the Rescue

Strengthening of Concrete Structures—State of the Art and Future Needs

P. H. EMMONS, A. M. VAYSBURD, J. THOMAS AND M. VADOVIC

ABSTRACT

Deterioration, increased load-carrying requirements, inadequate design or construction errors are some of the main reasons for repair and strengthening of existing concrete structures.

Historically, conventional strengthening techniques such as section enlargement, post-tensioning and steel plate bonding were used to address these problems. But in some cases, these methods are not viable due to constraints imposed by durability, constructibility and aesthetics limitations. New needs require new technical solutions.

This paper focuses on strengthening using externally applied carbon fiber plastics (FRP). Along with being lightweight and having a very high strength-to-weight ratio, FRP have the benefits of being non-corrosive and highly resistant to chemicals.

To date a few testing methods for bond strength between FRP and existing concrete have been suggested, but these are tests which can be carried out under controlled laboratory conditions. Their applicability to in-situ conditions is very questionable. This paper describes the results of in-situ bond testing program used to determine the applicability of proposed tensile and shear bond strength testing techniques. The program was performed at the Hollidaysburg Waste Water Treatment Plant in Pennsylvania.

INTRODUCTION

Concrete repair is an art which has been practiced for many centuries. Today it is evolving into a complex science as the demand for repair and maintenance of our crumbling infrastructure increases. The art and science of concrete repair engineering includes both "old" and, unfortunately, "new" structures, including conventional cement-based materials and so-called high performance materials.

A large number of different repair and strengthening materials and techniques are being proposed and used to provide increased strength capacity and/or ductility to concrete structures in an effort to extend their useful service life.

There has never been an era in which the evolution of materials was faster and the range of their properties more varied. Today, with more materials than ever before, the

P. H. Emmons, A. M. Vaysburd, J. Thomas and M. Vadovic, Structural Preservation Systems, Inc.

opportunities for innovation are unlimited. An understanding of materials enables advance in engineering design. The number of materials available to the engineer today is vast: somewhere between 40,000 and 80,000 different materials are at his disposal (Ashby, 1993). The menu of materials for construction is shown in Figure 1. Some of the new materials do better and are cheaper than the older ones; others have combinations of properties that entirely enable new products to be made or quite new affects to be achieved.

Advanced composites combine the attractive properties of the other classes of materials while avoiding some of their drawbacks. They are light, stiff and strong, and they can be tough. Most of the composites which at present are available to the engineer have a polymer matrix - epoxy or polyester, usually - reinforced by fibers of glass, carbon or Kevlar.

Advanced composites, fiber reinforced plastics (FRP), are more expensive than conventional materials. But numerous non-commercial materials started this way. Aluminum, in Napoleon's time, was a scientific wonder. He commissioned a set of aluminum spoons for which he paid more than those of solid silver. Aluminum was not, at the time, a commercial success as it is now.

Corrosion of steel reinforcement is now recognized as the major factor affecting concrete structures, new and repaired. In severe environments, material degradation and distress from steel corrosion dramatically affects the strength, stability and safety, and may lead to premature failure. Advanced composite materials in the form of fiber reinforced plastics (FRP) offer the potential to eliminate many problems associated with adverse environmental effects resulting in corrosion problems with metals.

This paper discusses the history, present, and future needs in the area of repair and strengthening of concrete structures. It considers and comments on the present and future opportunities offered by advanced composite materials.

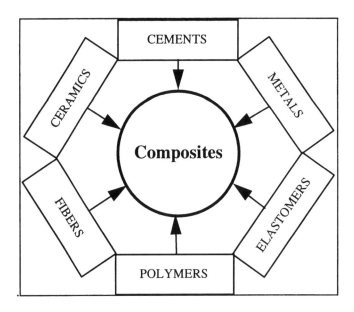

Figure 1. The menu of materials for construction.

A GLIMPSE OF CONVENTIONAL STRENGTHENING METHODS

Concrete repairs may be broadly classified as structural (stress-carrying) and protective. It should be indicated, however, that structural repair also must protect itself and underlying concrete and reinforcing steel from premature deterioration and corrosion. Increase in live load on the structure, deterioration of concrete and corrosion of embedded reinforcement, increase in fatigue capacity, reduction of crack widths, reduction of excessive deflection are just some of the reasons why a structure might become structurally deficient. Once this has happened, two alternatives of fixing the problem are to replace the structure or to strengthen it. Economically, repair and strengthening are often the only viable solution.

Section enlargement method is as old as concrete construction. Enlargement is the placement of additional reinforced concrete on an existing structural member. Columns, beams, slabs, bridge decks and walls can be enlarged to add load-carrying capacity. The enlargement should be bonded to the existing concrete to produce a monolithic member - a composite system. Concrete and mortar have been used for these enlargements.

Although relatively easy and economically effective, disadvantages include corrosion of embedded reinforcing steel and concrete deterioration. The way to make this strengthening technique effective in the future is to use materials with higher tensile strain capacity, with low shrinkage properties. Some of the materials available today are capable of satisfying these requirements.

Post-tensioning techniques have been employed with great success to correct excessive and undesirable deflections in existing structures. They have also been used to strengthen existing concrete structures to carry additional loads. Post-tensioning may be used on the inside of box girders or the outside of I-girders to increase the capacity of existing bridges and to provide improved resistance to fatigue and cracking.

The advantages include good constructibility, possibility of inspection during the lifetime of the structure, replaceability of strands and tendons. The disadvantages include vulnerability to corrosion, fire, and acts of vandalism.

The principle of strengthening with bonded steel plates is quite simple: steel plates or other steel elements are glued to the concrete surface by a two-component epoxy adhesive creating a three phase concrete-glue-steel composite system.

The wide acceptance and attractiveness of using epoxy-bonded external plates can be attributed to the development of high quality epoxy adhesives, simple and rapid methods of construction, together with negligible changes to overall dimensions of the structure and minimum disruption to its use. It was demonstrated that steel plates bonded to the tension face of concrete beams can lead to increase in flexural capacity, along with increases in flexural stiffness and associated decreases in deflection and cracking.

When a structure is strengthened by bonding of any type of external reinforcing, the most important aspect of its behavior is the composite action in the system – capability of the adhesive to transfer stresses. This in turn depends on the adhesive-concrete and adhesive-plate bonding and interface shear stresses, as well as the stiffness, flexibility and the viscosity of the adhesive. "Overstrengthening" and excessive deflection shall be avoided.

Problems with steel plates include weight restriction for handling, difficulties to shape for complex profiles, and corrosion.

STRENGTHENING WITH FIBER REINFORCED PLASTICS

The FRP can provide additional reinforcement for flexure, shear, confinement, tension, and energy absorption. Composites may be tailored to suit the required shape. Being light in weight, they are easy to handle, transport, manipulate, and bond to the structure. The high tensile strength of composite materials allows thinner plates to be used which are more flexible than steel plates. The advantages of strengthening with prefabricated FRP laminate plates include high chemical resistance to acids and bases, corrosion resistance, no increase of dead weight of the structure, easier handling, minimal interruption in the use of the structure.

The carbon fiber reinforced plastic (CFRP) laminates consist of uni-directional continuous carbon fibers impregnated with an epoxy resin. The laminate strength in direction of fibers is proportional to fiber strength, and amounts to about 3,000 MPa (depending on fibers volume fraction

The pioneering application of composite plate bonding was started by Professor Meier in Switzerland in 1982. However, in spite of many research works and satisfactory practical applications, this method still has a very limited use for strengthening of concrete structures, and it is still not well known and used in structural engineering practice. Therefore, we would like to present an overview of important features of strengthening with CFRP.

CFRP have high strength, lightweight, resistance to chemicals, good fatigue strength and non-magnetic and non-conductive properties, and is an ideal material for repair and strengthening of existing concrete structures.

As with any other composite system, bond of the strengthening composite to existing concrete is very critical, therefore, the surface preparation of both phases of the system, concrete and CFRP plates, is very important. The plates should be abraded on the bonding side and, immediately before the bonding, the surface should be cleaned with acetone.

After mixing the epoxy glue components, it should be placed on the plate without delay. After assembling the plate in the designed position, a slight pressure shall be applied to squeeze out excessive adhesive.

The modification to the CFRP prefabricated plates is the use of "cast-in-place" carbon fiber sheets which was developed about nine years ago in Japan. The acceptance of this approach is reflected in the fact that over 1,000 strengthening projects using carbon fiber sheets have been performed (Thomas, and Kline, 1996). In 1996, the largest manufacturer of carbon fiber sheet in Japan installed approximately 4,000,000 sq. ft. of this material (Emmons, 1996).

Epoxy based primer is rolled or brushed onto the properly prepared existing concrete surface. After the primer has dried yet is still tacky, the first epoxy coat is applied by a roller. The carbon fiber sheets are applied to a fresh epoxy using a ribbed roller to remove air bubbles. The sheets may be spliced, if necessary (Figure 2).

After the sheet application, about thirty minutes shall be allowed for the application of the first epoxy coat to saturate the carbon fibers. Then the second and final epoxy coat is applied, and allowed to cure completely. In areas where special protection is required, for instance, fire protection or prevention against UV degradation of epoxy resin, a special topcoat may be applied.

Figure 2. Beam strengthened with carbon fiber. Reserve Square Parking Garage, Cleveland, Ohio.

The disadvantages of the method include lack of codes of practice and design standards, and necessity in fire and UV protection. Fire resistance is an important problem for many applications which must be examined and addressed. But it should not be a brake on the development of these techniques. The problem of fire is not new and exists in other exterior strengthening techniques, and has caused many less accidents than corrosion of steel.

FRP materials are being used in most of the cases to repair and strengthen concrete beams and columns. But lately it is being used in the repair of concrete slabs. The concrete deck slab of a bridge of the Tokyo Metropolitan Expressway was strengthened using two-way strips of CFRP to reduce the stresses in the existing steel reinforcement caused by excessive wheel loads (Ichimasu et al, 1993). The concrete slab of the City Hall of Gossau, Switzerland, was reinforced around the rectangular area of the slab that was to be cut for a new elevator shaft (McKenna, 1993).

CFRP is superior to steel properties when used for concrete confinement (Figure 3). Concrete confined with composite material demonstrates different behavior than concrete confined with steel. Uni-directional composites exhibit linear stress/strain behavior to failure and fracture at strains much higher than typical yield strains of steel. With sufficient

Figure 3. Column strengthening (confinement) to correct construction error. Municipal Parking Garage, Charleston, South Carolina.

confinement, composite confined concrete reaches its highest axial stress and strain simultaneously while steel confined concrete slowly loses strength with increased axial strain after the steel yields. Also, radial strains at peak axial stress are much higher in composite confined concrete than steel confined (Harmon, Slattery, and Ramakrishnan, 1995).

FIELD TESTING PROGRAM

When a concrete structure is strengthened with an externally bonded FRP, the most critical aspect of its behavior is that the composite action in the system must be preserved during the designed service life of the structure. This behavior is governed primarily by the ability of the bond to transfer stresses, and this in turn, depends on the bond between two entirely different components of the system: existing concrete substrate, and FRP strength and durability. Deficiencies and weaknesses in the bond when exposed to long term severe environment can be detrimental to overall performance of the composite system. There is limited information concerning the durability of the described strengthening systems in severe environments. More data is necessary before confidence and reliability can be assured.

The overall objective of the testing program performed by Structural Preservation Systems, Inc. and Gannet Fleming, Inc. was to perform durability testing of CFRP strengthening system in severe environment. More specifically, the objective was first to study the bond behavior between the concrete and CFRP under different environmental exposures and, second, to evaluate several unconventional bond testing techniques which have been developed and applied to meet the actual demand for improved field evaluation and quality control.

In 1994, a supplemental strengthening of reinforced concrete basin walls was performed at the Hollidaysburg Sewage Treatment Plant in Pennsylvania. Carbon fiber sheets have been applied on two walls to control potential overstress and existing cracking of the underreinforced concrete walls. The durability testing program performed in 1996, after more than 2.5 years in service, included visual examination and natural exposure bond testing in three zones: (1) dry, exposed to ultraviolet (UV) rays and frost action; (2) splash, exposed to freezing and thawing action in water saturated condition, wetting and drying, UV, and chemical attack by waste water elements; (3) submerged, exposed to chemical attack by waste water elements (Figures 4, 5).

Figure 4. Testing operation.

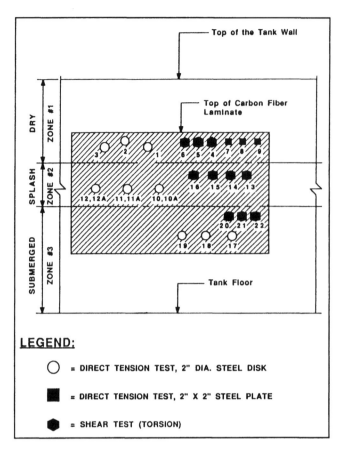

Figure 5. Test locations.

Although a variety of different test methods were analyzed, two test methods were selected by the authors for use in site testing – the pull-off tensile bond test and the torsion shear bond test. In the modified ACI 503 pull-off test, a 1/8-in. partial depth core is cut to below the interface between the concrete and the CFRP and a steel 2-in. diameter steel disk is applied with epoxy (Figure 6).

Since the use of a core drill is not always easy in on-site quality assurance, especially where access is difficult, a pull-off test utilizing 2-in. square steel plate was also performed. The CFRP laminate was cut along the perimeter of the plate.

The transfer of force from the concrete substrate into the carbon fiber laminate occurs by shear stress. Therefore, it is important to be able to evaluate this property of the composite system for design and quality control. In torsion shear test procedure, a cylinder of CFRP the thickness of the strengthening phase is failed in torsion. Test probes 1-in. and 2-in. diameter were glued to the CFRP surface with epoxy adhesive. Torsion was applied using a calibrated torque wrench with a series of hinges to eliminate any possible bonding moment due to the eccentricity (Figure 7). The torque was gradually increased manually until failure. Maximum value at failure and the failure mode were recorded.

Figure 6. Pull-off tensile testing.

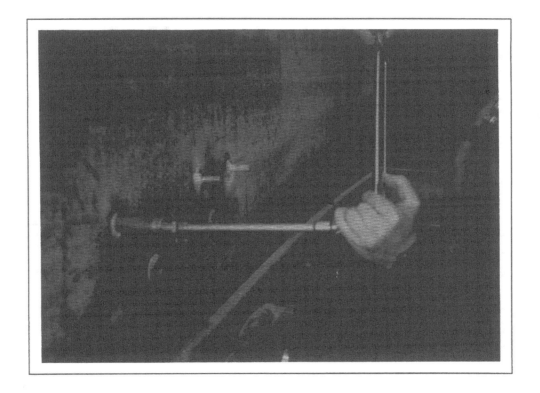

Figure 7. Shear bond testing.

TABLE I - TENSILE AND SHEAR BOND BETWEEN CFRP AND CONCRETE

Exposure Zone	Test No.	Pull-off Tensile Bond, psi		Shear bond (torsional test), psi	Failure Mode
		2-in. dia. disk	2-in. sq. plate		
Dry Zone #1	1	216			Existing concrete
	2	261			Between existing concrete and composite
	3	153			Between composite and steel disk
	4			1,830	Between probe and composite
	5			2,140	Between probe and composite
	6			1,220	Between probe and composite
	7		235		Between composite and steel plate
	8		225		Between composite and steel plate
	9		275		Existing concrete, composite
Splash Zone #2	10	191			Between composite and steel disk
	10A*	541			Composite and composite-steel disk
	11	172			Between composite and steel disk
	11A*	242			Between composite and steel disk
	12	153			Between composite and steel disk
	12A*	509			Composite-concrete
	13			1,830	Between probe and composite
	14			0	No bond between composite and probe
	15			1,220	Between probe and composite
	16			1,220	Between probe and composite
Submerged Zone #3	17	407			Composite-concrete, composite
	18	274			Composite-concrete, composite
	19	306			Composite-concrete, composite
	20			915	Composite
	21			1,220	Between probe and composite
	22			1,830	Between probe and composite

* Tests were repeated with more accurate composite surface cleaning procedures

It is worth mentioning that the mere magnitude of bond strength obtained in tension should not be compared with that of the shear, due to the fact that the stress mechanisms that cause failure are different in both methods. However, the comparison between the shear and the tensile bond strength can be very useful for the prediction of the real performance of a strengthening system in practice.

Test results of the tensile and shear bond strengths using different testing systems are presented in Table I.

In direct tension the predominant failure mode was a combination of partial failure in concrete and between concrete and CFRP. The average tensile stress at failure was 329 psi which is more than adequate tensile bond between parent concrete and CFRP. The results of the pull-off tests in Zone #1 when square probes were used, demonstrate a close numerical correlation with the test results with circular probes.

In shear tests the predominant mode of failure was in the bond between the steel probe and CFRP. The average shear stress at failure was 1,492 psi, which exceeds more than three times the required value of 370 psi.

The overall averages were calculated by averaging all the readings irrespective of the mode of failures.

The following conclusions can be drawn from this study:

- The bond between the existing concrete and CFRP strengthening system in severe environment demonstrates good performance after 2.5 years in service.
- The pull-off test using square plates and torsion shear bond test demonstrate that both are practical test methods that can be performed in the field. But both methods need more development and precision.
- No numerical correlation of pull-off tensile strength and torsion shear strength was possible.
- The look at the described test methods shows an urgent need for further research and adaptation of the standard test methods, materials and procedures to on-site demands including surface preparation of FRP, surface preparation of steel probes, quality of the epoxy glue, curing conditions and duration.
- Further extensive tests ought to be carried out to confirm the above correlation.

CONCLUDING REMARKS

Reinforced concrete structures are often required to resist loads over a long time under potentially corrosive environments. The use of FRP materials may provide a corrosion resistant alternative to the use of steel as tension reinforcement in structural concrete. Composites appear to exhibit excellent resistance to the aggressive environments that normally cause corrosion of typical steel reinforcement.

Modern materials, including fiber reinforced plastics, are available for our use. Now it is up to us to "engineer" our repair projects to be able to intelligently and without cost penalty to exploit the outstanding properties of these materials. To achieve this successfully, a significant input by educators, scientists, owners, engineers, material manufacturers and contractors is needed. We have to change the paradigm.

One must reach a new level of understanding that the future not only requires, but demands that the concrete repair industry look to innovative materials and technologies to achieve excellence in the 21st Century, exploiting the full potential of advance materials and methods. One has to recognize the fact that in spite of many advantages of the FRP materials, we have very limited experience of the material and the structural implications of the use of these materials for repair and strengthening of concrete structures. The use of FRP for repairs is still hampered by the lack of standards, and the lack of training documents to educate engineers, contractors, and laborers. Standard test methods are needed for the fibers, epoxy, and combined FRP including the FRP bonded to the substrate. To make innovation possible by the use of new materials, the reliable data on their properties must be available.

The applied research programs are necessary to study the durability of fiber reinforced plastic strengthening systems, to enhance the understanding of their performance characteristics under different environmental conditions, to develop reliable standard test methods, and to develop design guidelines that will ensure technically and economically successful use.

REFERENCES

Ashby, M.F., 1993. Material Selection in Mechanical Design," Pergamon Press, 249 pp.

Emmons, P.H., 1996. Personal communications with Tonen Corporation.

Harmon, T., Slattery, K., and Ramakrishnan, S., 1995, "The Effect of Confinement Stiffness on Confined Concrete." Proceedings of the Second International RILEM Symposium (FRPRCS-2), Non-metallic (FRP) Reinforcement for Concrete Structures, E&FN Spon:584-591.

Ichimasu, H. et al, 1993. "RC Slabs Strengthened by Bonded Carbon FRP Plates: Part 2 - Application," Proceedings of the International Symposium Fiber-reinforced Plastic Reinforcement for Concrete Structures, ACI SP-138:957-970.

McKenna, J.K., 1993. "Post Strengthening of Reinforced Concrete Members Using Fibre Composite Materials," Material Engineering Thesis, Royal Military College of Canada.

Thomas, J. and Kline, T., 1996. "Strengthening Concrete With Carbon-Fiber Reinforcement," Concrete Repair Digest:88-92.

Experimental and Analytical Studies of Adhesive Property for Rehabilitation of Concrete Structures by Using Composite Materials

HIROYUKI HAMADA, ASAMI NAKAI,
SHIGEO URAI AND ATSUSHI YOKOYAMA

ABSTRACT

Rehabilitation or retrofitting old and/or damaged infrastructure is quite emergency problems.

In this paper, adhesive properties of concrete structure with adhering FRP were investigated. A new three point bending type test for evaluating adhesive properties was proposed. The test was performed by using several reinforcing fibers, matrix resin and adhesive resin. As a result, adhesive properties between concrete and FRP increased with increase of elastic modulus of FRP and decrease of elastic modulus of adhesive resin. One of effective material selections for rehabilitation of concrete structure also could be obtained from the above new type bending test.

INTRODUCTION

Unsafe old infrastructures such as bridges, highways, railroad, canals and so on are increasing continuously because most of that have a design life of 50~75 years(V.M.Karbhari. 1995., HUGH C. BOYLE. 1994.). Therefore, old infrastructure have to be retrofitted the condition. Also rehabilitation of infrastructure is quite emergency problems in the country where earthquake often occurs, particularly in Japan(M.R.Ehsani and H. Saadatmanesh. 1996.). There are many materials and many operation system for rehabilitation or retrofitting. Basically, high modulus and strength materials are suitable to maintain infrastructure safety. However, operation system is also very important issue, because difficult operation system and/or long operation time can not be accepted in architecture field and in busy city.

According to these circumstance, rehabilitation of infrastructure have been

Hiroyuki Hamada, Faculty of Textile Science, Kyoto Institute of Technology, Matsugasaki, Sakyo-ku, Kyoto 606, JAPAN
Asami Nakai, Faculty of Textile Science, Kyoto Institute of Technology, Matsugasaki, Sakyo-ku, Kyoto 606, JAPAN
Shigeo Urai, Graduate School, Kyoto Institute of Technology, Matsugasaki, Sakyo-ku, Kyoto 606, JAPAN
Atsushi Yokoyama, Faculty of Education, Mie university, Kamihama-cho, Tsu, Mie 514, JAPAN

constructed by using fiber reinforced plastic(FRP)(M.Xie et al. 1995.). FRP can bring easy operation because of their light weight and also bring short operation time. FRP, of course, achieves high modulus and strength properties. Therefore FRP is the most attractive material for rehabilitation and retrofitting. The most important factor in rehabilitation of damaged concrete structure by FRP is adhesive properties between them. The selection of materials includes reinforcing fibers, matrix resin and adhesive agents. Usually high corrosive materials are often selected and used. However, rehabilitation can be regarded to construction of adhesive structure. From the mechanics view point the adhesive properties becomes to be important, and understanding these properties leads to predict a long-life of rehabilitated structure.

There are many research works for mechanical properties rehabilitated concrete structure(Kazuro Kageyama. et al. 1995., Masahiko Uemura et al. 1994.). However, very few works attempted to understand adhesive properties between composite materials and concrete. Authors made co-research work between academic and company in Society of Interfacial Materials Science which is a new society founded in Japan.

As results of co-research work, we proposed tensile-peel test for evaluating adhesive properties and also developed finite element model(Atsushi Yokoyama et al. 1996.). The most attractive result was adhesive properties increased with increase of elastic modulus of composite materials and with decrease of elastic modulus adhesive. The purpose of this study was to confirm above mentioned preliminary result by using three-point bending type experimental method. The proposed tensile-peel test contained some difficulties. For example, specimen costs are expensive because two steel rods are set at the center of specimen and special tensile machine is needed. Also unsymmetric fracture behavior often occurs, so that the evaluation would be complex. Instead of tensile-peel test three-point bending type testing method was proposed in this paper. Moreover, flexible resin was introduced as adhesive.

EXPERIMENTAL METHOD

Fig.1 shows schematic diagram of bending test specimen. Two concrete columns were prepared. Dimension of the column was 100mm × 100mm and 400mm long. FRP was patched on one side surface of concrete columns. Adhesive length was 400mm. Along the butting surface some lubricant grease was applied.

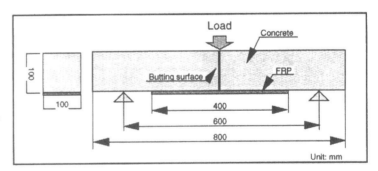

Fig.1: Schematic diagram of bending test specimen.

Load was applied at the butting point. Span length was 600mm. According three point bending type test method was conducted. The butting line would become a notch and it was expected that delamination between the concrete and FRP initiated from this notch. For reference, schematic diagram of tensile-peel test specimen is shown in Fig.2. Apparently, proposed bending type method is easier than tensile-peel test.

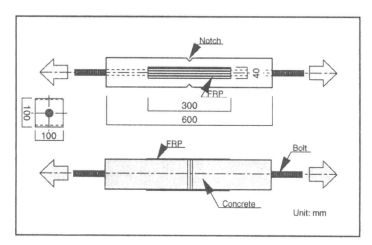

Fig.2: Schematic diagram of tensile peel test specimen.

Table 1: Specification of adhesive resin.

	Elastic Modulus(GPa)	Tensile strength(GPa)	Failure strain
E2300(Rigid)	3.15	27.1	0.008
E2400(Flexible)	1.00	13.0	> 0.03

Table 2: Specification of FRP(matrix: E2300(Rigid)).

Reinforcemen	Elastic Modulus(GPa)
Unidirectional fiber(Carbon)	53.5
Unidirectional fiber (Glass)	19.3
Cloth(Glass)	12.1
Braid with middle-end-fiber(Glass)	2.0

MATERIALS

Materials used in this paper were listed in Table1 and 2. Two different adhesive resins were used, one is rigid resin and the other is flexible resin. The flexible resin has low elastic modulus and very high failure strain. Four different reinforcements were used as listed in Table2. The matrix in all of FRP was the rigid resin as shown in Table1. The elastic modulus of used FRP was varied from 54 to 2 GPa. According to these materials the effects of elastic modulus on adhesive properties could be discussed.

RESULTS AND DISCUSSION

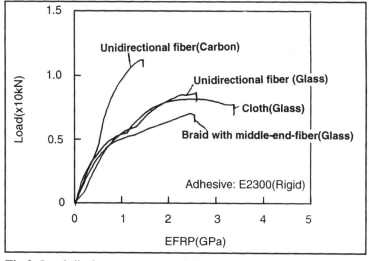

Fig.3: Load-displacement curves of three point type bending test with varying reinforcing fibers and adhesive resin.

Fig.3,4 show load-displacement curves for rigid and flexible adhesive specimen respectively. Most of the curves indicated knee point where inclination was changed. That point was expected as the initiation of delamination. After the knee point the inclination became lower than before. Fig.5 shows photographs of fractured specimen. Delamination fracture appeared in all specimen with rigid adhesive. According to observation by eyes, the delamination initiated from butting part which was located at the center. The delamination region became large and propagate to the end of the FRP with increase of bending load. This phenomenon was very stable. When the delamination reached the end, finally the specimen was separated completely. On the other hand, some of the specimen with the flexible adhesive resin showed

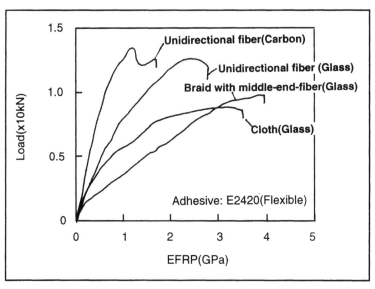

Fig.4: Load-displacement curves of three point type bending test with varying reinforcing fibers and adhesive resin.

concrete fracture. The fracture initiated from the end of adhesive region in the case of unidirectional carbon fibers composite and in the case of unidirectional glass fibers composites the crack in the concrete initiated from the half of the FRP in longitudinal

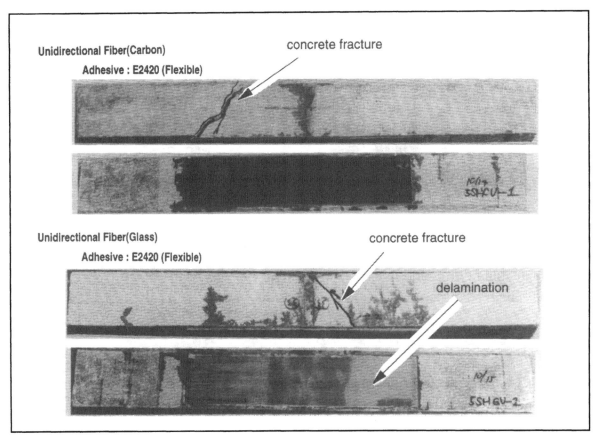

Fig.5: Photographs of fractured specimen.

direction. After the occurrence of crack, the delamination started. Finally white part in the photograph was delamination part. Therefore in both cases crack in concrete was former phenomenon than delamination between FRP and concrete. The occurrence of concrete fracture means that the delamination suppressed and adhered FRP were still in good health condition From the fracture aspects view point, the materials system with concrete fracture could be regarded asgood adhesive properties. This is the most important and interesting point. The delamination did not occur in the case of high modulus composites and flexible adhesive resin. Fig.6 shows the relation between the maximum load and elastic modulus of FRP. In the case of rigid adhesive resin the maximum load increased with increase of elastic modulus of FRP. This tendency was similar to our former results which were obtained from tensile peel tests. On the other hand the specimen with flexible adhesive

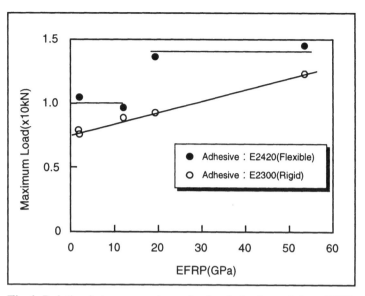

Fig.6: Relation between maximum load and elastic modulus of FRP.

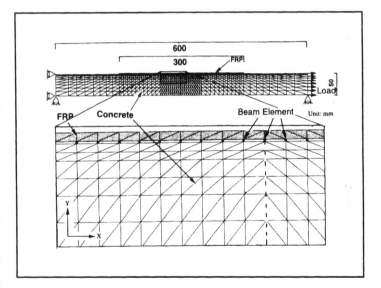

Fig.7: Finite element division for tensile peel test specimen.

layer showed different tendencies. The specimen with concrete fracture showed almost same maximum load and the maximum load in other two specimens was lower. Fig.7 shows finite element division for tensile peel test. Two dimensional plane strain element was used in the concrete and FRP parts and beam element was used in adhesive layers. This modeling succeed to predict fracture mode and adhesive properties in tensile-peel test. In further study this modeling would be used for three point bending type test as described in this paper. The result in which low elastic modulus of adhesive layer can give higher adhesive properties was obtained from this finite element calculation.

CONCLUSION

A new evaluation method for adhesive properties between concrete and FRP was proposed. That was three point bending type specimen, which was easy to operate. The most important result was that higher modulus FRP and lower modulus adhesive resin gave higher adhesive properties.

REFERENCES

Atsushi Yokoyama, Akihiro Fujita, Shigeo Urai and Hiroyuki Hamada. 1996. "Estimation of initial debonding stress in adhesive structure between concrete and composites". Advanced Composites Letters. Vol.5. No.3. pp.91-94.

HUGHC.BOYLE and VISTASP M.KARBHARI. 1994. "INVESTIGATION OF BOND BEHAVIOR BETWEEN GLASS FIBER COMPOSITE REINFORCEMENTS AND CONCRETE". POLYM.-PLAST. TECHNOL. ENG. 33(6). pp.733-753.

Kazuro Kageyama , Isao Kimpara and Koji Esaki. 1995. "FRACUTRE MECHANICS STUDY ON REHABILITATION OF DAMAGED INFRASTRUCTURES BY USING COMPOSITE WRAPS". Proceedings of ICCM-10. Whistler. B.C. Canada. August. pp.Ⅲ-597

Masahiko Uemura, Kazuo Aoyama and Howard S. Kliger. 1994. "NEW REINFORCING FIBER SHEET MATERIAL AND ITS APPLICATIONS IN THE REPAIR OF CONCRETE STRUCTURES". Proceedings of 39th International SAMPE Symposium, April 11-14 pp.347-354.

M.R.Ehsani and H. Saadatmanesh. 1996. "REPAIR AND STRENGTHENING OF EARTHQUAKE-DAMAGED CONCRETE AND MASONRY WALLS WITH COMPOSITE FABRICS". Proceedings of the First International Conference on Composites in Infrastructure. ICCM'96. pp.1156-1167.

M.Xie , S.V.Hoa AND X.R.Xiao. 1995. "Bonding Steel Reinforced Concrete with Composites". Journal of REINFORCED PLASTICS AND CONPOSITES. vol. 14 - September. pp.949-964.

V.M.Karbhari. 1995. "CHARACTERISTIC OF ADHESION BETWEEN COMPOSITES AND CONCRETE AS RELATED TO INFRASTRUCTURE REHABILITATION". 27th International SAMPE Technical Conference October 9-12. pp.1083-1094.

BIOGRAPHIES

Hiroyuki Hamada
Dr. Hiroyuki Hamada is presently an associate professor of Faculty of Textile Science at Kyoto Institute of Technology.   He received his B. Eng. degree from Doshisha University in 1978, M. Eng. degree from Doshisha University in 1980 and Dr. Eng. degree from Doshisha University in 1985.

Asami Nakai
Miss. Asami Nakai is presently a graduate student at The University of Tokyo.   She received his B. Eng. degree from Kyoto Institute of Technology in 1994 and M. Eng. degree from Kyoto Institute of Technology in 1996.

Shigeo Urai
Mr. Shigeo Urai is presently a graduate student at Kyoto Institute of Technology.   He received his B. Eng. degree from Kyoto Institute of Technology in 1996.

Atsushi Yokoyama
Dr. Atsushi Yokoyama is presently an associate professor of Faculty of Education Mie University.   He received his B. Eng. degree from Doshisha University in 1981, M. Eng. degree from Doshisha University in 1983 and Dr. Eng. from Osaka City University in 1988.

Tensile Reinforcement by FRP Sheets Applied to RC

ANTONIO NANNI, CHARLES E. BAKIS, THOMAS E. BOOTHBY,
ELIZABETH M. FRIGO AND YOUNG-JOO LEE

ABSTRACT

Unidirectional FRP sheet materials, applicable for repair and strengthening of concrete structures, were studied to characterize the change in stiffness in reinforced concrete. In this work, the FRP sheets were attached with epoxy to the two opposite surfaces of a concrete prism. The variables characterized included five FRP material systems, ply number, and a ±45° mixed orientation. The specimens were subjected to tensile loading directly applied to a deformed steel bar embedded along the central axis of the prism. Each prism was pre-loaded before the application of the FRP sheets. Notches at three locations along the specimen controlled the position of the primary cracks due to pre-load. Direct and continuous measurements included load, elongation, crack opening, and strain.

MATERIALS AND SPECIMENS

Three materials were used in the construction of the test specimens. Concrete and steel made up the rectangular reinforced concrete prism, and FRP was used to repair the cracked prism. The concrete had an average compressive strength of 21.8 MPa. The concrete prism was 700 mm long, 100 mm in width, and 50 mm in height. The cross-sectional dimensions of the prism were chosen because similar specimens were to be used in corrosion studies on the FRP sheets. The length of the concrete element was chosen to ensure an adequate evaluation of cracking. The concrete was notched at three locations (200 mm apart).

The steel reinforcement used was No. 3 Grade 60 (yield strength of 450 MPa). The reinforcement consisted of a center bar measuring 1200 mm long, with two 375 mm long bars welded to each end of the longer bar. This was done to assure that the steel reinforcement would not yield within the grips. A welded wire mesh with 12.7 mm openings was used to limit cracking in the concrete at the specimen ends.

Antonio Nanni, Prof., Penn State University, University Park, PA 16802
Charles E. Bakis, Assoc. Prof., Penn State University, University Park, PA 16802
Thomas E. Boothby, Asst. Prof., Penn State University, University Park, PA 16802
Elizabeth M. Frigo, Res. Asst., Penn State University, University Park, PA 16802
Young-Joo Lee, Res. Asst., Penn State University, University Park, PA 16802

Two manufacturers provided the five FRP material systems that were tested: high modulus carbon (C5), high-strength carbon (C1), E-glass (GE), carbon prepreg (R-2), and carbon prepreg (R-3). The C1, C5, and GE fiber sheets were dry unidirectional fiber fabrics (Tonen 1994). Properties of the C1, C5, and GE fiber sheets are shown in *Table 1*. The R-type sheets were a pre-preg material form (Mitsubishi Chemical 1994). The R-2 and R-3 differed in number of fibers per unit area (see *Table 1*).

ASTM D3039 was used to determine the in-plane tensile properties of the FRP sheets. The ultimate strength of the material was determined from the maximum load carried prior to failure. Furthermore, the stress-strain response of the material could be determined since two strain gauges were aligned opposite to one another on the specimen. These experimental results were compared with the theoretical values that were given in the manufacturer's literature (see *Table 1*). In addition to the tensile tests performed on the FRP composites, the fiber volume fractions for three of the composite types were also experimentally determined according to ASTM D3171 (see *Table 1*).

Table 1: **Properties of FRP sheets**

Material Form	Dry		Dry	Pre-preg		Dry	
Type	Carbon		Glass	Carbon		Carbon	
Fiber	C1	C5	GE	R-2	R-3	C5T	C5A
Fiber density (g/cm^3)	1.82	1.82	2.55	1.07	1.07	1.82	1.82
Fiber Areal Weight (g/m^2)	300	300	300	200	300	300	300
FRP sheet width (mm)	500	500	500	250	250	500	500
Theor. f_u (MPa)	2950	2500	900	2500	2500	5000	N/A
Exper. f_u (MPa)	2074	2087	886	2185	1464	4311	1038
Theor. E (GPa)	38.4	62.6	8.1	48.4	48.4	62.6	N/A
Exper. E (GPa) [SD]	42.7 [0.90]	67.8 [3.27]	15.6 [0.82]	53.4 [11.3]	39.6 [4.01]	64.1 [7.22]	6.0 [0.96]
Theor. E (fibers) (GPa)	233	371	72.7	235	235	371	371
Thickness (mm/ply)	0.165	0.165	0.118	0.111	0.167	0.165	0.165
Exper. ε_u (µmm/mm) [SD]	8764 [913]	6599 [859]	8983 [91]	8870 [1008]	8013 [245]	6960 [1413]	N/A N/A
Exper. V_f [SD]	0.175 [0.01]	N/A N/A	0.235 [0.04]	N/A N/A	0.476 [0.03]	N/A N/A	N/A N/A

N/A = not available

The FRP sheet dimensions for the repair of the specimens in the study were dependent upon the number of plies applied. The one layer system used a 650 mm by 60 mm FRP sheet. The two layer system consisted of two 650 mm by 40 mm tow sheets, offset by 25 mm at either end. To prevent failure at the specimen end (i.e., delamination or splitting), FRP wraps were added (60 mm wide and 325 mm long). The length was chosen so that the wrap would fully contain the end of the specimen, while also having the 25 mm overlap recommended by the manufacturer.

After fabrication and curing specimens were pre-loaded, repaired with FRP, and finally, tested in tension. A summary of the tensile test specimen types is shown in *Table 2*.

Table 2: Summary of Tensile Test Specimen Types

FRP designation	Specimen No.	No. of Layers	Epoxy system
Control	1, 2, 3, and 4	None	None
C1	1, 2, and 3	1	A1/B1
C5	1, 2, and 3	1	A1/B1
GE	1, 2, and 3	1	A1/B1
R-2	1, 2, and 3	1	A2/B2
R-3	1, 2, and 3	1	A2/B2
C5T (two plies)	1, 2, and 3	2	A1/B1
C5U (un-cracked)	1, 2, and 3	1	A1/B1
C5A (±45°)	1, 2, and 3	2 at ±45°	A1/B1

During pre-loading, load was applied at a rate of 0.1 mm/minute. The load was taken to 18 kN, based on $\sqrt{f_c'}A$, where f_c' was taken to be 21 MPa and A was the notched cross-sectional area of the specimen.

Immediately following pre-loading, the surface of the specimen was cleaned, and primer was applied using a small paint brush. Two different epoxy systems were used, depending on the type of FRP sheet to be applied. Both used a 2:1 mixing ratio of resin to hardener. The application of the FRP was conducted in accordance to manufacturers' specifications (Tonen 1994, Mitsubishi Chemical 1994).

The instrumentation mounted on the specimen consisted of two linear variable differential transducers (LVDT's), four strain gauges, and two clip gauges (*Figure 1*). Total elongation of the specimen was measured using the two LVDT's located on the two non-repaired opposing sides of the rectangular prism. Opening of the central precrack was measured using two clip gauges, also fixed to the non-repaired opposing sides of the specimen. The four strain gauges were used to examine the strains at two locations on the FRP: the central precrack, and an intermediate location where a crack was anticipated to form (i.e., equidistant from the center and outer precracks).

Three specimens (C1, C5, and GE) were also examined using the photoelastic method of stress analysis. This method provided full-field data across the specimen, enabling visualization of the complete distribution of surface strains. The testing procedure for the photoelastic full-field stress analysis was very similar to the tensile testing of the specimens. The tests were performed under controlled loading conditions, at a rate of 0.1 mm/minute. A reflective polariscope using white light was set up such that the fringe patterns on the specimen could be easily documented. The loading was paused every 10 kN to take photographs of the newly developing cracks, as well as the existing crack's propagation patterns throughout the specimen.

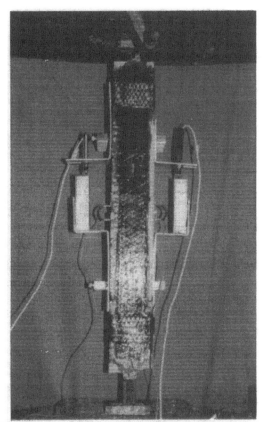

Figure 1: Instrumentation on tensile test specimen

EXPERIMENTAL RESULTS

Load-elongation curves, as well as load-strain curves for each of the 28 specimens examined were recorded. The typical load-elongation summary plot (Fig. 2) is made up of 6 curves, two per specimen. The total elongation is measured by LVDT's across a base length of 350 mm, whereas opening of the center pre-crack is measured by the clip gages across a gauge length of 50 mm. The typical load-strain summary plot (Fig. 3) contains seven curves: two per specimen plus a theoretical one. Two strain gauges bridging the center pre-crack, and the two strain gauges bridging the anticipated intermediate crack location were used for the experimental curves. The theoretical curve for load-strain was derived as follows. The steel yield load was calculated using the relationship:

$$P_y = \varepsilon_s (E_s A_s + E_f A_f) \qquad (1)$$

where ε_s is the strain in the steel, E_s is the elastic modulus of the steel, A_s is the cross-sectional area of the steel, E_f is the elastic modulus of the FRP, and A_f is the cross-sectional area of the FRP. This relationship assumes that there is a crack at the location of measurement. After steel yielding, the slope of the load-strain curve is equal to. The yield load was then used to calculate the theoretical elongation at a crack based on the expression:

$$\delta = \frac{P_y L}{AE} \qquad (2)$$

This expression was calculated for a unit length and then multiplied by the number of cracks observed before theoretical yielding to give an approximate theoretical value for the total elongation of the specimen when the steel begins to yield. Likewise, this estimate does not take into account the fact that precracks exist in the specimen.

For the specimens with FRP, first intermediate crack load was calculated based on the expression:

$$P_{cr} = \varepsilon_{cr} E_c A_c \qquad (3)$$

where A_c is the area of concrete by transformed area:

$$A_c = A_{concrete} + \frac{E_f}{E_c} A_f + \frac{E_s}{E_c} A_s \qquad (4)$$

and the cracking strain, ε_{cr}, is based on:

$$\frac{0.62\sqrt{f_c'}}{E_c} = \varepsilon_{cr} \qquad (5)$$

This is only an estimate because it neglects to account the effect of pre-crack spacing on crack development. Crack pattern diagrams corresponding to each of the specimens were also recorded. For brevity, only the results of the E-glass fiber sheet (GE) are shown in this paper. All of the specimens that were repaired with the E-glass fiber sheet failed in virtually the same manner; discoloration in the epoxy indicated that the FRP sheet debonded at locations bridging cracks and eventually the fibers failed across the full width of the specimen. Summary diagrams for load-elongation and load-strain curves are shown in *Figures 2* and *3*. Figure 3 shows the theoretical load-strain curve.

Specimen GE-3 can be used to describe the typical performance. The epoxy debonding phenomenon, *Figure 4,* began in at 23 kN. The total elongation increased greatly after a load slightly above 40 kN was reached. This is the point at which the steel began to yield. The total elongation of the specimen at failure was measured to be 3.5 mm. The first sign of intermediate cracking in GE-3 was at 12.8 kN, when a crack centered between the center and lower precracks was observed propagating from the left to the right. At 48 kN, a crack just above the upper precrack formed. None of these cracks fully crossed the specimen before the maximum load was reached, at 50.6 kN. The failure mode, *Figure 5,* was due to fiber failure across the intermediate crack that had fully developed between the center and lower precracks..

Several observations were made in examining the three specimens with photoelastic coating. Fringe orders indicate intensity of the principal stress or strain difference on the surface of an object (Dally et al. 1991). The photoelastic method of stress analysis was used to determine how cracks develop and propagate beneath the FRP sheets as well as to develop full-field stress data in the FRP. It was found that secondary cracks develop on one side of the specimen and propagate across its width. The photoelastic results further showed that the elastic properties of the FRP sheets have a significant impact on the strain field. The photoelastic program is more fully described in Frigo (1996).

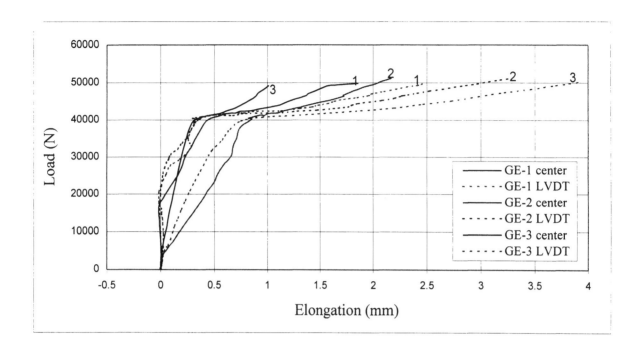

Figure 2: Summary of Load-Elongation for GE

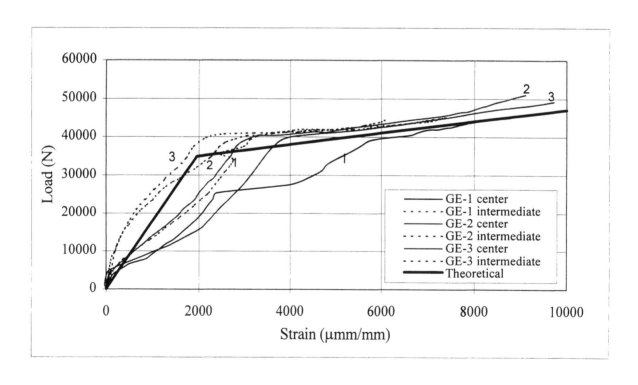

Figure 3: Summary of Load-Strain for GE

Figure 4: Debonding on E-glass FRP

Figure 5: E-glass fiber failure

CONCLUSIONS

The experimental results have shown that the glass FRP sheet is the most ductile of the material systems, allowing several secondary cracks to develop after repair, with a total elongation of 3 mm over an initial length of 200 mm. Both the dry carbon and dry high-modulus carbon exhibited a maximum total elongation of approximately 1 mm, with the high modulus system reaching a greater ultimate load. The pre-preg sheets were similar in behavior, each reaching a maximum total elongation of about 1 mm as well as similar ultimate loads. Secondary crack formation was nearly non-existent in the two-ply application. The ±45° orientation produced a ductile behavior. Ultimate strain for this specimen type reached nearly 0.025 mm/mm.

REFERENCES

Dally, J.W. and Riley, W.F., Experimental Stress Analysis, 3rd edition. McGraw-Hill, Inc., 1991.

Frigo, E., The Mechanics of Externally Bonded FRP for the Repair of Reinforced Concrete, , M.S. Thesis, The Pennsylvania State University, 1996.

Mitsubishi Chemical, "REPLARK: Carbon Fiber Prepreg for Retrofitting and Repair Method", Manufacturer Publication, Tokyo, Japan, 1994, 18 pp.

Tonen Corporation, "FORCA Tow Sheet Technical Notes", Manufacturer Publication, Rev. 1.0, Tokyo, Japan, 55 pp.

BIOGRAPHIES

A. Nanni. Dr. Nanni is interested in construction materials, their structural performance, and field application. He is an active member in the technical committees of ACI, ASCE, ASTM and TMS. He is founding Chairman of ACI Committee 440 - FRP Reinforcement.

T. Boothby. Dr. Boothby conducts a research program in the assessment, maintenance, repair, and rehabilitation of old and deteriorated structures. He is an active member of the ASCE and TMS.

C. Bakis. Dr. Bakis is an Associate Professor of Engineering Science & Mechanics. His professional interests are in the design, manufacture, and testing of composite materials.

E. Frigo. Elizabeth Frigo is a Research Assistant of Architectural Engineering. Her interests are the mechanics and repair of FRP Reinforcement.

Y. Lee. Young-Joo Lee is a Graduate Student of Architectural Engineering. She is interested in the analysis of tensile behavior in composite material and new resins for fire resistance.

Evaluation of the Hybridization of Wood Crossties with Glass Fiber-Reinforced Composite (GFRC)

H. V. S. GANGARAO,[1] N. RAGHUPATHI,[2] T. H. DAILEY,[3]
E. A. MARTINE,[2] S. S. SONTI[1] AND M. C. SUPERFESKY[1]

Abstract

The following paper discusses the results of an investigation focusing on the hybridization of the timber crosstie with glass fiber reinforced composite (GFRC). The paper addresses some of the service problems which hinder the performance of wood crossties, and discusses the potential benefits that may be achieved by encasing the wood crosstie with GFRC. The applicability of filament winding as the manufacturing process is discussed. Results from bending tests indicate significant increases in stiffness, strength and ductility of the GFRC encased ties over conventional wood ties.

Motivation

The demands on the world's timber supply has made it necessary to find new methods for improving timber's structural and long-term performance. One possible technique is hybridizing timber with fiber-reinforced composites. The combination of high performance composites with lower cost timber may benefit many applications. One potential application is wood crossties. In

[1]Respectively: Director and Professor of Civil Engineering, Research Engineer and Graduate Reserach Assisstant, Constructed Facilities Center, West Virginia University, Morgantown, WV 26505-6103

[2]Respectively: Manager, Senior Research Engineer, Rienforcement Products Research, PPG Industries, Fiber Glass Research Center, P.O. Box 2844, Pittsburgh, PA 15238

[3]Group Leader, High Performance Resins Department, Indspec Chemical Corporation, 1010 William Pitt Way, Pittsburgh, PA 15238

1994, over 10 million new wood crossties were installed on class I tracks alone. The major reasons for replacement were degradation due to repetitive mechanical laoding, biological attack, and horizontal splitting.

Analysis and consideration of the problems and issues which currently dictate the technology level and serviceability limits of crossties indicate that encasing the wood crosstie with a jacket system which has the ability to strengthen as well as protect would result in elevated service performance. Using the qualities of fiber reinforced composites to wrap or encase wood crossties would offer the following advantages; 1) the enhancement of the environmental resistance of the wood tie due to the ability of the GFRC to form a protective boundary between the wood and the environmental agents of destruction: 2) the reduction or elimination of wood splitting by the ability of the GFRC jacket to provide confinement; 3) the creation of a tie with high performance mechanical properties due to the high specific strength and modulus of GFRC.

Objective and Scope

The primary goals of this study were to: 1) establish the feasibility of manufacturing GFRC-wood crossties using filament winding process, and 2) to demonstrate that significant improvements in strength, stiffness, and ductility can be achieved by encasing wood crossties with GFRC. The scope of this study involved the manufacturing, testing, and evaluation of twenty-seven filament wound specimens. Thirteen of the specimens were wound using epoxy as a matrix material and the other 14 specimens utilized resorcinol formaldehyde based resins. The evaluation of the mechanical properties of all specimens was accomplished using static bend tests.

Materials

Timber Crossties

Northern red oak timber was used to manufacture test specimens because of its frequent use as a crosstie material. Each wood test specimen had nominal dimensions 4" x 4" x 60". The test specimen dimensions are approximately half that of typical crossties used in track. Prior to encasing the specimens, the four surfaces of each sample was planed using a wood molder. After planing, the corners of each sample was rounded using a quarter-inch router bit. The rounded edges help reduce stand separation during filament winding.

Fiber System

Fiber glass was selected as the composite reinforcement because of its balance of material properties and cost. Single-end 250-yield roving packages were supplied by PPG Industries. The product was chosen based on compatibility with the matrix resin and process considerations.

Resin Systems

The selection of adhesives used in this investigation were based upon two main issues. First, it was desirable to use an resin system which had the ability to perform in harsh environments. Secondly, resins were sought that could be utilized for automated manufacturing processes such as filament winding and tape winding. Based upon this, epoxy and Resorcinol Formaldehyde were selected as the resins for the GFRC-wood crosstie manufacturing process. Resorcinol Formaldehyde primers were used with epoxy matrix resins because previous coupon testing showed this combination provided enhanced resin durability in accelerated aging tests (Dailey et al. 1996). Additional resins were considered and the details of the resin selection process can be found in Sonti et al. (1996).

GFRC-wood Crosstie Manufacturing (Filament winding)

Filament winding was used in this study because of the control it brings to the manufacturing of GFRC-reinforced crossties. The filament winding process has been used for decades in the manufacture of composite tanks, pipes and pressure vessels. The process offers flexibility of wind angles, a high degree of fiber alignment, and a high fiber volume fraction. In this application, the wood crosstie served as the filament winding mandrel. The filamanet winding of test specimens was performed at the PPG Industries in Pittsburg, PA and Industrial Fiberglass Specialities in Dayton, OH. Figure 1 is an illustration of the crosstie configuration and winding pattern.

It should be noted that other manufacturing processes such as banding and tape winding were also considered for GFRC-wood crossties. But it was found that filament winding process provided better product in terms of proper resin distribution, wood-composite bond integrity and adherence to desired winding angles. Details of banding and tape winding can be found in GangaRao et al. (1996).

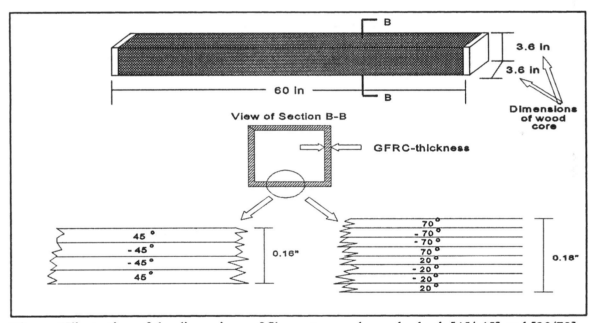

Figure 1 Illustration of the dimensions of filament wound samples both [45/-45] and [20/70].

Performance Evaluation (Bending tests)

To quantify the mechanical perform and assess the effect that the filament winding had on increasing the stiffness, strength, and ductility of each sample, static bend tests were performed following the procedures of ASTM standard D198 (1992) using a Baldwin Universal Testing Machine. The Baldwin machine allowed the loading to be applied automatically, using pressurized air to control the rate of hydraulic fluid in and out of its two reservoirs. The samples were loaded using two loading points applied at third points of the span. The loading rate used to achieve failure within six to ten minutes of the start of loading was 0.2 in/min of midspan deflection. During testing, a computerized data acquisition system was used to collect load, strain and deflection data. Midspan deflections, which were needed to quantify enhancements in stiffness and ductility were measured using a linear variable displacement transducer (LVDT).

Evaluation of the increase in stiffness to a wood specimen due to filament winding with GFRC required that stiffness testing be conducted on all the samples prior to filament winding them with GFRC and then after GFRC encasement. The deflections measured for stiffness evaluation were obtained from load ranges which varied from zero to 25 % of ultimate load. Linear regression analysis, using the least-square method, was used to calculate the apparent modulus of elasticity, E_f, for each specimen. For third point loading, the modulus of elasticity is calculated using equation 1.

$$E_f = (\frac{P}{\delta})(\frac{a}{4bh^2})(3L^2 - 4a^2) \tag{1}$$

where,
P/δ = slope of the load-deflection curve of test sample
a = distance from support reaction to the nearest load point
L = support span
b = cross sectional width of test sample
h = cross-sectional height of test sample

The quantification of enhancement to specimen strength due to encasement with GFRC had to be approached differently than the procedure used for evaluation of the increase in stiffness due to encasement with GFRC. Because of the destructive nature of flexural strength testing, the enhancement to crosstie strength has to be performed on a comparative basis. The comparison was accomplished by testing a group of control samples which were not filament wound and establishing their average strength. In conjunction with testing the non-filament wound samples, strength evaluation testing was also performed on the filament wound specimens. From the individual bending strength values obtained for each filament wound specimen, and the average value of the non-filament wound control groups, an average strength increase could be made on the enhancement to specimens flexural strength due to filament winding with GFRC. The numerical quantification of flexural strength of a beam is based upon the maximum load achieved during bend testing, the cross-sectional dimensions of the beam, and the load and support spans. For a rectangular beam loaded at its third point, the flexural strength and modulus of rupture, S_R can be calculated using equation 2.

$$S_R = \frac{3aP_u}{bh^2} \tag{2}$$

where
S_R = flexural strength or modulus of rupture (MOR)
Pu = maximum load obtained during testing
h = height of the test sample or cross-sectional dimension perpendicular to the axis of bending
b = width of the sample or cross-sectional dimension parallel ro the axis of bending
a = distance from the support reaction to the nearest point of loading

Results and Discussion

From the experimental work, it was observed that significant increases in stiffness, strength and ductility were achieved by encasing the wood crossties in composite fabrics. Also, there was a change in the mode of failure from "catastrophic" to "progressive plastic" in nature. Figure 2 shows a typical load-deflection plot comparing the performance of a GFRC-wood crosstie and a wood crosstie. Additional details regarding the experimental program can be found in GangaRao et al. (1996).

The experimental data indicates that the samples wound with the epoxy performed slightly better than the samples which were wound using resorcinol formaldehyde (RF) based resins in terms of stiffness. In both cases, stiffness increases were significant ranging from 15% to 40%. Notable was the reduction in the variability of material properties due to the encasement (wood is highly variable in terms of material properties).

The addition of the GFRC-Jacket system by filament winding was believed to increase the strength of a beam by two mechanisms. First, by placing the GFRC on the top and bottom faces of the crosstie, the GFRC would carry the larger bending stresses. This would reduce the stresses in the outer surfaces of the wood core and result in an increased failure load above that of the wood core itself. The second mechanism of strength enhancement deals with the issue of confinement. It is possible that the GFRC can act to reduce the Poisson effect that a beam exhibits during bending by preventing deformation in the lateral outward direction in the compression region of the beam. If the GFRC is able to reduce the Poisson effect, an increase in bending strength should result. From the experiments, strength increases of 35-70% were observed across all sample types. Also, ductility enhancements of 2-3 times were observed.

Conclusions

The results of the experimental data indicate that the composite fabric wrapping did increase the bending stiffness (15-40%) and strength (35-70%). In addition to stiffness and strength increases, the GFRC wrap assisted in increasing the ductile behavior (2-3 times) which is desirable in most structural members. The increases in stiffness and strength were higher for the filament-wound samples than other processes that were used in this study (banding and tape winding), due to the consistency and control that can be achieved in this automated process.

It is recommended that filament-winding process be used for the manufacturing of new GFRC-wood ties due to better product characteristics and banding/tape winding processes as a cost-effective alternative for rehabilitation of degraded ties.

References

ASTM. 1992. Standard methods of static tests of timbers in structural sizes. ASTM D198-84. Philadelphia, PA: American Society for Testing and Materials.

GangaRao, H.V.S. et al., "Development and evaluation of glass fiber reinforced composite-wood railroad crossties- Phase I," Quarterly Report submitted to Federal Railroad Administration, May 1996.

Dailey, T.H., Sonti, S.S., GangaRao, H.V.S., Talakanti, D.R. "Using Phenolics foe Wood Composite "Hybrid" Members" SPI Composites Institute, 1996.

Sonti, S.S., GangaRao, H.V.S., Talakanti, D.R. "Accelerated Aging of Wood Composite Members," 41st International SAMPE Symposium and Exhibition, Anaheim, California. 1996.

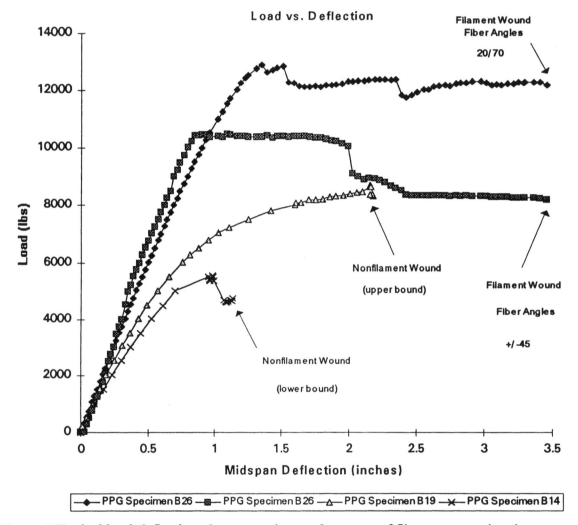

Figure 2 Typical load-deflection plot comparing performance of filament wound and non-filament wound samples

The Tom's Creek Bridge Rehabilitation and Field Durability Study

JOHN J. LESKO,* RICHARD E. WEYERS, JOHN C. DUKE, MICHAEL D. HAYES,
JOE N. HOWARD, DANIEL E. WITCHER, GLENN BAREFOOT, RANDY FORMICA,
JOSE GOMEZ, ERNESTO VILLALBA AND JULIUS F. J. VÖLGYI

ABSTRACT

Questions concerning the long-term durability of composites in the infrastructure prohibit their routine use in present designs. A central obstacle to answering these questions lies in the complex interaction of stress, time, temperature and environmental exposure. Furthermore, the sheer number of possible combinations of fiber and matrix systems disallows a universal generalization of durability for composites in all environments and applications. Thus, the team formed by Virginia Tech, the Town of Blacksburg, Morrison Molded Fiber Glass, the Virginia Transportation Research Council, and the Virginia Department of Transportation have approached the durability problem by examining the specific performance of a composite structural box I beam in the Tom's Creek Bridge. This experimental program aims to properly implement the structural shape in this small bridge and to develop a field site monitoring program which will assist in advancing a predictive methodology for composite performance in the infrastructure.

INTRODUCTION

Many groups have made the claim that fiber reinforced plastic (FRP) composites will offer improved durability and corrosion resistance over metallic materials for infrastructure and commercial applications. However, these claims have not been substantiated in the open literature. The group on International Research on Advanced Composites for Construction (IRACC) stated in their last meeting in Montreal, August 1996 that the understanding and mechanistic description of composite durability is the primary obstacle to routine use of these material systems in civil infrastructure (IRACC, 1996). There, however, exists a great body of enviro-mechanical data on primarily glass and carbon / epoxy and polyester composites (Agarwal, et al., 1990, Springer, 1988a, Seymour, 1988, Delre, et al., 1988, Adist, 1987 ASTM STP's 602 & 768). Yet, no comprehensive approach exists that describes what fraction of the as-processed stiffness, strength, and life of the composite material system will be retained after exposure to moisture, temperature, and UV. Schutte points out in an extensive review

John J. Lesko, Richard E. Weyers, John C. Duke, Michael D. Hayes and Joe N. Howard, Virginia Tech
Daniel E. Witcher and Glenn Barefoot, Morrison Molded Fiber Glass
Randy Formica, Town of Blacksburg
Jose Gomez, Virginia Transportation Research Council
Ernesto Villalba and Julius F. J. Völgyi, Virginia Department of Transportation
*To whom all correspondence should be addressed, at Engineering Science & Mechanics Dept., Virgina Tech, Blacksburg, VA 24061-0219.

article on the environmental durability to glass composites, that *"the state of the art in predictive methodologies for life-time behavior of composites lacks assimilation of the large base of information available (on glass/ epoxy and polyester composites). Furthermore, some phenomenological approaches accompanied by experimental work exist for this purpose; however, it is most desirable that this methodology be based on a more fundamental understanding of processes that are responsible for the mechanisms of failure."* (Schutte, 1994).

Furthermore, there exist no comprehensive data sets on the synergistic effects of load and environment generated in the field. These realities can not be completely studied in the laboratory. The design lives for these civil engineering structures range anywhere from 50-75 years which is well beyond the design lives for traditional composite structures with which the community has experience. It is difficult to find experimental programs that examine the aging of composites for more than a few years (Bank, 1995).

Although, much of the work in this area has been experimental, some efforts to identify critical mechanisms and model environmentally induced stresses have been undertaken. The work has focused on either the laminate level or the individual constituents and their micromechanics, i.e. fiber matrix and interface (Bank, 1995, Schutte, 1994). The latter approach has yielded comprehensive descriptions of moisture diffusion and the moisture induced stresses (Shen et al., 1976, Lee et al., 1993). In many cases, moisture causes separation of the fiber and matrix, leading to irreversible changes in composite properties. The effects of the interface/phase are appreciated by this community, and this will obviously play a strong role in our understanding of long term performance. Yet, Springer suggests that "at the present time (1988), no [complete] theory exists which would predict the change in properties caused by environmental exposure (Springer, 1988b)." This is still true today, and the effects of exposure are still determined by experimental means.

Complicating efforts to develop a comprehensive assessment of composite durability is the myriad of combinations of resins (epoxies, vinyl esters, polyesters, etc.) and fibers (glass, carbon, aramid, thermoplastic) to form specific composite material systems. Each material system has its own strengths and weaknesses, including cost of material and fabrication. The sheer number of damage and degradation mechanisms, their combination, and their severity in combination prevents the community from making comprehensive generalizations about "composite material performance" in these conditions.

DISCUSSION

THE TOM'S CREEK BRIDGE REHABILITATION WITH COMPOSITE BEAMS

The present state of our understanding requires that we consider more realistic and long term investigations of composite material systems' response to infrastructure environments. To accomplish this goal civil engineers and composite material mechanicians at Virginia Tech (VT), the Virginia Transportation Research Council (VTRC), and the Virginia Department of Transportation (VDoT) are collaborating to examine composite durability in actual field conditions. The group has identified a number of potential field sites where specific composites could be integrated and monitored for long term degradation. One field site is close to home. Initial work with the Town of Blacksburg (ToB) on the rehabilitation of the Tom's Creek Bridge (see Figure 1) presents a unique opportunity to examine the performance of composites in actual field conditions. Recent inspections of the structure reveals that a number of the steel stringers in the understructure have lost significant section due to corrosion. The ToB has decided to rehabilitate the bridge with a design that includes composite structural members developed and manufactured by Morrison Molded Fiber Glass (MMFG) of Bristol, Virginia and Georgia Tech (NIST ATP, 1994). The box I beam is an 8 inch high sub-scale prototype for a potential beam designed to meet AASHTO standards for highway bridge structures (see Figure 2). The beam is a pultruded section constructed with both glass and carbon in a vinyl ester resin. The carbon is strategically positioned in the flanges to increase the effective bending stiffness of the section.

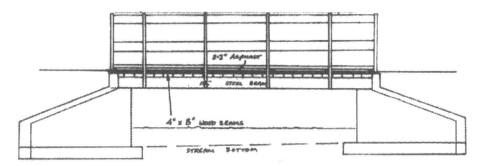

Figure 1. The Tom's Creek bridge crosses Tom's Creek in the town of Blacksburg, Virginia; two miles from the Virginia Tech Campus. Presently posted at 10 tons due to steel stinger corrosion. Built: 1932, Reconstructed: 1964, Geometry: 24' wide and a 17.5' span, 15° skewed, Construction: Twelve, 20' W 10"x21# steel stringers, 4"x8" wood decking, 2"-3" asphalt.

Figure 2. Composite beam cross section

The proposed design will replace each of the steel wide flange beams with two composite box I beams, as shown in Figure 3. These beams are oriented in the direction of traffic and lay on the ledge made by the concrete abutments. A glue-lam deck will replace the existing 4 inch x 8 inch wood planking that is laid perpendicular to the direction of traffic. The present and new designs do not enforce or rely on composite action for strength and stiffness. Thus, connections between the deck and beams are only made to ensure that each component stays in place and is therefore not structural. The guard rail system is under design by VDoT and will not attach to the beams as in the present bridge.

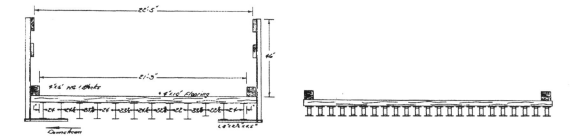

Figure 3. The present design (left), and the new design (right); two composite box I beams replace each of the steel W10x21 beams.

VERIFICATION OF THE DESIGN

A simple mechanics of material analysis of the proposed design has been conceived. The model assumes that only the composite beams carry the loads, and the loads are distributed by wood planking traversing the composites. Thus, composite action between the deck and the beams is not enforced and disregards any potential contribution of the deck to the bending stiffness of the bridge. The analysis may be characterized by a wood beam which is supported by axial springs representing the span deflection stiffness of the composite beams. A statically indeterminate problem, to the n-2 order, is constructed and solved for deflection as a function of position for the wood plank and the point loads on the composite beams. (In our case we consider 24 composite beams, i.e. n=24.) Multiple wheel loads can be placed anywhere within the skewed bridge.

Characterization of the section properties for the beam have yielded an effective bending modulus of 6.8 Msi from a four point bend configuration. A non-shear deformable description for bending of the composite beams accurately describes the strain and deflection characteristics at loads well above those encountered in this analysis. The composite beams are assumed to be simply supported at their ends. A 4 inch x 8 inch wood planking was modeled to be consistent with the present design. In the new design the deck may be changed to accommodate a glue-lam system. The dimensions for the bridge were used as described in Figure 1; a composite beam spacing of 12 inches (implied in Figure 3) was observed for this analysis. Upon loading for the maximum bending case (the worst case) under AASHTO HS-20 loading, a maximum beam load of 3,500 lb. is observed: i.e. loading by a 32,000 lb. axial located at center span, 2' from the curb. The maximum deflection observed was 0.75 inch. (A frame of the loading simulation for HS-20 is shown in Figure 4.) This maximum load is well below the 29,000 lb. measured capacity. In this strength test the beam was constrained to prevent failure by lateral buckling, inducing a bending failure at the mid-span.

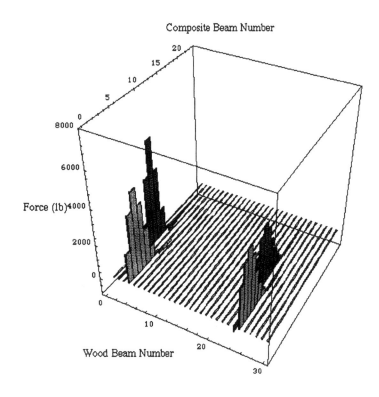

Figure 4. A full scale skewed model for the Tom's Creek Bridge under simulated HS-20 loading. The frame shown simulated two 32,000 lb. axles (16,000 lb. wheel loads separated by 6 ft.) separated by 14 ft. The bars represent the vertical force supported by each composite beam.

Thus, the analysis presented above suggests that a safe implementation of the composite structural shapes can be made. The implementation of composite materials for this structure is also appropriate for the following reasons. There are plans to widen the road in 10 years, so the repair will be in-place for a limited period. This makes the use of composites attractive as a "short-term, low-risk" experimental effort. Second, continued exposure to moisture from Tom's Creek, is a primary cause for the extensive corrosion to the understructure. The use of these hybrid glass/carbon/vinyl ester composites structural shapes may possibly prove more durable than the steel. (Vinyl esters have good resistance to corrosion and absorb 50-80% less moisture than epoxies.) And third, the traffic loads on this bridge are considerably light. Approximately, 1000 vehicles a day traverse this structure, 95% of which are light 2-axle vehicles. Thus, the bridge will experience low but reasonable loading.

PRESENT PLANS FOR THE TOM'S CREEK BRIDGE DURABILITY FIELD SITE

The Tom's Creek Bridge (TCB) effort is broken down into two separate phases: implementation of the composite beams and long term monitoring to assess composite durability in the field and lab. As there are presently no standards by which to guide the design, special attention is being paid to the beams and the manner in which they are to be implemented. Thus, a number of full scale tests on the beams and the structure are being conducted to provide the appropriate data needed for the design and certification by the engineer of record. This work continues and will be completed by Spring of 1997. Construction of the bridge is scheduled for mid June 1997.

The final phase of this effort deals with the assessment of the material durability as exposed to actual service conditions. From the present plan, the beams will remain in the bridge for 10 years allowing for an extended period over which to assess composite durability. At the conclusion of service, the beams will be reclaimed and examined to understand the effects of service life on their remaining strength and life. An extensive field site monitoring system will be put into place that allows the university and VDoT to remotely monitor all facets of the composite response in the context of the hygrothermal mechanical load history. (In addition, an early warning system will be installed to monitor critical strains and deflections for signs of distress.) This monitoring program will allow us to define the specific service related mechanisms of degradation and property change based on recorded hygrothermal mechanical history. The Tom's Creek Bridge rehabilitation and durability field site is a unique opportunity for the composites community to study the durability in an actual service environment with minimal risk. The team is actively working to include other groups that see value in conducting their work at this site.

REMARKS

The Tom's Creek Bridge rehabilitation with composite beams is a unique opportunity to implement composite structural shapes in a load carrying capacity. The present analysis of the structures demonstrates that the commercially available components can be implemented in the Tom's Creek Bridge safely with only minor adaptations. The limited time that the bridge will be in place provides an ideal demonstration effort for composites in the infrastructure. The design issues raised by the team to implement composites within this design demonstrate that it is possible to rehabilitate with composites. These discussions have also reinforced the need for further study of composites durability; this has been the number one concern raised by VDoT. Thus, efforts are underway to estimate and predict long-term performance with the data at hand and additional experimental efforts: i.e. full scale fatigue tests of the beams.

However, the bridge site also offers an opportunity to examine the durability of commercial composite components in an actual field environment. Careful co-monitoring of the hygrothermal mechanical environment and material representative of the beam material will allow for a comprehensive description of material durability. However, this is only accomplished for this particular material system, in this specific environment. This study will not provide a universal approach to the description of composite performance in the infrastructure, but will suggest a perspective on how to generalize a predictive methodology for other systems and applications.

ACKNOWLEDGMENTS

The authors would like to thank Jim Stewart of Virginia's Center for Innovative Technology for the final support for the Tom's Creek Bridge rehabilitation effort. Financial support for the initiation of this effort was made possible by Virginia Tech's NSF Science & Technology Center. Special thanks goes to the Town of Blacksburg for their willingness to support the experimental and research aspects of the implementation. The assistance of Malcom Kerley, Richmond office of VDoT, is especially appreciated. Mr. Kerley has provided the engineer of record, Ernesto Villalba, for this effort which has allowed us to move forward on the safe implementation of this design.

REFERENCES

Adist, N. R., 1987, "Mechanical Properties and Environmental Exposure Tests," in Engineered Materials Handbook: Vol. 1 - Composites. Metals Park, OH: ASM International, pp. 295-301.

Agarwal, B. C. and Broutman, L. J., 1990, Analysis and Performance of Fiber Composites. (2nd Edition). New York: John Wiley.

ASTM, 1975, Environmental Effects on Advanced Composite Materials, ASTM STP 602. Philadelphia, PA: American Society for Testing and Materials.

ASTM, 1982, Composites for Extreme Environments, ASTM STP 768. Philadelphia, PA: American Society for Testing and Materials.

Bank, L.C, Gentry, T.R., Barkatt, A., 1995, "Accelerated Test Methods to Determine the Long-Term Behavior of FRP Composite Structures: Environmental Effects," Journal of Reinforced Plastics and Composites, Vol. 14, pp. 459-587.

Delre, L. C. and Miller, R. W., 1988, "Characterization and Weather Aging and Radiation Susceptibility," in Engineered Materials Handbook: Vol. 2 - Engineering Plastics. Metals Park, OH: ASM International, pp. 576-580.

International Research on Advanced Composites for Construction (IRACC), 1996, Final Report: www.iper.net/co-force/iracc.htm,.

NIST ATP, 1994, "Composites Structural Shapes for Infrastructure," MMFG Primary,.

Lee, M.C., and Peppas, N.P., 1993, "Models of Moisture Transport and Moisture-Induced Stresses in Epoxy Composites", Journal of Composite Materials, Vol. 27, No. 12, pp. 1146-1171.

Schutte, C. L., 1994, "Environmental Durability of Glass-Fiber Composites," Materials Science and Engineering, R13 265-324.

Seymour, R. B. 1988, "Influence of Long-Term Environmental Factors on Properties," in Engineered Materials Handbook: Vol. 2 - Engineering Plastics. Metals Park, OH: ASM International, pp. 424-432.

Shen, C.H., and Springer, G.S., 1976, "Moisture Absorption and Desorption of Composite Materials," J. Composite Materials, Vol. 10, pp. 2-20.

Springer, G. S. ed. 1981, 1984, 1988a. Environmental Effects on Composite Materials - Vol. 1, Vol. 2, Vol. 3, Lancaster, PA: Technomic Publishing Co, Inc.

Springer, G. S. 1988b, "Environmental Effects," in Environmental Effects on Composite Materials - Vol. 3, G. S. Springer, ed., Lancaster, PA: Technomic Pub. Co. Inc., pp. 1-34.

SESSION 10

An Overview of Regulations Governing Emissions from Composites Manufacturing

NOTE: No formal papers were submitted for this session.

SESSION 11

Dealing with the Characteristics of SMC/BMC: Flow, Shear, Shrinkage, Sink Marks, and Painting

Predicting Sink-Mark Formation in Ribbed SMC Parts

SARI K. CHRISTENSEN, ESTHER M. SUN AND TIM A. OSSWALD

ABSTRACT

A wide variety of reinforced thermoset composites can provide weight reduction, heat resistance and higher toughness. Most applications have a combination of both structural and cosmetic requirements. The presence of integrated ribs in sheet molding compound (SMC) panels usually causes some degree of deformation on the cosmetic side. These deformations, known as sink-marks, result from material inhomogeneties caused by flow induced fiber orientation, fiber-matrix separation, curing, polymerization shrinkage and processing conditions. Within this research, a coupled heat transfer and stress-strain finite element simulation was developed to model residual stress build-up and sink-mark formation during the curing process. The simulation was used to study the effect of geometry and processing conditions on the depth of the sink-mark. It was found that resin-rich areas near the rib entrance significantly increased the depth of the sink-mark . On the other hand, both, small lead-in radii and fiber reinforcement on the cosmetic side of the ribbed part kept the sink-mark depth to a minimum. This type of analysis can be used to reduce or eliminate the sink marks in a finished product. The curing heat transfer simulation is able to predict the cure time, maximum temperature during cure, and the thermal history during cure for each location inside the part.

INTRODUCTION

Some major problems that occur when molding polymeric parts are the control and prediction of the shape and dimensions of the finished product. Hence, modeling and simulation of shrinkage and warpage are important when trying to understand and predict the complex thermomechanical behavior the material undergoes during processing. Shrinkage and warpage are directly related to residual stresses. Transient thermal and curing or solidification behavior and material anisotropies can lead to the build-up of residual stresses during a manufacturing process. Such process-induced residual stresses can have a significant effect on the mechanical performance of the component by inducing warpage or initiating cracks and delamination in composite parts. Utilizing simulation programs to analyze all of the parameters we are able to minimize sink-mark formation in SMC parts with ribs.

THERMOMECHANICAL ANALYSIS

Since geometries with ribs and bosses have thick sections, the traditional shell element simulation is inaccurate. A new plane-strain model and simulation had to be used to accurately predict the evolution of residual stresses during the process and the corresponding shrinkage, warpage and other defects of the finished parts.

University of Wisconsin-Madison, Department of Mechanical Engineering, Polymer Processing Research Group, 1513 University Avenue, Madison, WI, 53706

In the two-dimensional simulation, the stress analysis during solidification and part removal is derived by the principle of virtual work. The principle states that for a body in static equilibrium, the virtual internal work created by the stresses and the virtual strains, equals the virtual external work done by the external forces and the virtual displacements (Cook,1988). Equating the internal and external virtual work we have

$$\int_V \{\delta\varepsilon\}^T \{\sigma\} dV = \{\delta u\}^T \{f\} \tag{1}$$

where $\{\delta\varepsilon\}$ are the virtual strains, $\{\sigma\}$ the stresses, $\{\delta u\}$ contains the virtual displacements and $\{f\}$ the externally applied forces. The stresses are represented as a function of local strain and the residual stress $\{\sigma_0\}$

$$\{\sigma\} = [E]\{\varepsilon\} - [E]\{\varepsilon_0^{thermal}\} - [E]\{\varepsilon_0^{cure}\} + \{\sigma_0\} \tag{2}$$

In Eq. (2), the material tensor $[E]$ is anisotropic and depends on the local degree of cure. $\{\varepsilon_0^{thermal}\}$ represents the thermal strain caused by a temperature change and $\{\varepsilon_0^{cure}\}$ is the shrinkage due to cure which occurs in the time step under consideration. The detailed formulation of the model is given by (Sun, 1996) and (Tseng and Osswald, 1994). To determine the behavior during cure of complex geometries the above equation can only be solved numerically.

To compute the thermal strains, curing shrinkage, and modulus as a function of cure, we must first solve the energy equation, described by

$$\rho c_p \frac{\partial T}{\partial t} = k\nabla^2 T + \rho \dot{Q} \tag{3}$$

and $\dot{Q}$ represents the rate of internal heat generation due to cure. The model proposed by (Kumal and Sourour, 1973) was used and is given by

$$\dot{Q} = Q_T(a_1 e^{-\frac{b_1}{RT}} + a_2 e^{-\frac{b_2}{RT}} \cdot c^m)(1-c)^n \tag{4}$$

where Q_T, a_1, a_2, b_1, b_2, m and n are constants that depend on the matrix material.

MATERIAL PROPERTIES OF FIBER REINFORCED COMPOSITES

The mechanical properties of the fiber reinforced SMC composite parts were calculated using the Halpin and Tsai model (Halpin, 1969), where the orthotropic mechanical properties of the composites are determined in terms of the properties of the fiber, the matrix, and the relative volume fraction of the fibers and the matrix.

The thermomechanical properties were approximated using Schneider's model (Schneider, 1971). Here, we calculate the thermal expansion coefficient of the parts as a function of the properties of the fiber and matrix. A detailed discussion of the equations used to model mechanical and thermomechanical behavior of SMC parts is given by (Sun, 1996).

As the thermoset resin cures, the effective mechanical properties of the composite change from the behavior of a viscous liquid in its uncured state, toward an elastic solid in its fully cured state. The mechanical properties of the resin phase are governed by two competing mechanisms, chemical hardening and viscoelastic relaxation. This makes the elastic modulus and Poisson's ratio of the thermoset composites vary as the part cures. (Kim and Hahn, 1989) have developed a model that supplies a linear approximation between the material properties and degree of cure. The material is considered a fluid at the beginning of gelation (c=5%) and assumed to have reached its final strength at 80% cure.

During the process the polymer material tends to expand or contract, due to the temperature differences and polymerization effects. As the thermoset resin cures, its chemical shrinkage can reduce its volume anywhere between 2% to 6% for a typical thermoset material. In fiber reinforced parts, or parts with substructures such as a reinforcing rib, this change in volume will add to poor surface appearance. Also, the volumetric chemical shrinkage of the resin will induce significant macroscopic strains in the composites. These strains contribute to the internal loading in thick section composites, in addition to the traditional thermal loading. To account for the chemical shrinkage effect on the evolution of residual stresses in the thick section parts, a curing strain model is presented for chemical shrinkage due to cure.

In the model the unit volume of pure thermoset resin is assumed to shrink freely to its final volume when experiencing a uniform volumetric change caused by the cross linking polymerization.

With homogenous and isotropic conditions assumed, the contraction strain in all directions can be represented as ε. Where ε is defined as

$$\varepsilon = \frac{-1 + \sqrt{1 + \frac{4}{3}\Delta V}}{2} \tag{5}$$

By modeling the relationship between the polymerization shrinkage and conversion to be linear we were able to produce results that compare well with the experimental dilatometer results (Hill, 1996). Mathematically, the model can be represented as

$$\Delta V = \Delta c V^{tot} \tag{6}$$

where ΔV is the change in volume, and Δc is the incremental change of the degree of cure and V^{tot} is the total volume change of the resin during cure.

The effective longitudinal and transverse curing strain of the thermoset composites can be computed over each time increment. These strains can be calculated using the micromechanics model. To include the shrinkage due to cure we extended Schneider's model used to model the thermal expansion coefficients of unidirectional fiber reinforced composites. Since the volumetric shrinkage of the resin due to cure is treated as an equivalent curing strain, an analogous model is used here to incorporate the fiber effect on the curing strain of unidirectional, oriented and random fiber reinforced thermosets.

RESULTS

This section presents the results of residual stress and sink-mark prediction of thermoset components using the two-dimensional finite element model. First, a ribbed section shown in Fig. 1 is used to numerically simulate the sink-mark and residual stress field. Figure 1 shows the dimensions of the rib used in the simulation as well as the boundary conditions used in the stress analysis. Due to symmetry, only half of the geometry is used in the simulation. The center of the top surface was selected to have no displacement to prevent the rigid body motion. The volume fraction of the fiber is assumed to be 21% throughout the part. The simplified fiber orientation within the model is shown in Fig. 1.

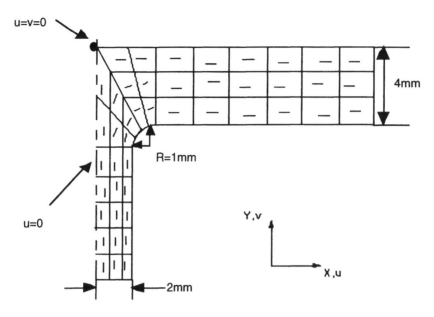

Figure 1. Assumed fiber orientation in the ribbed section.

The simulated displacement field is shown magnified in Fig. 2. Since the center point is assumed to have no displacement, the magnitude of the displacement is zero, and the top profile indicates a 4.5 μm sink-mark.

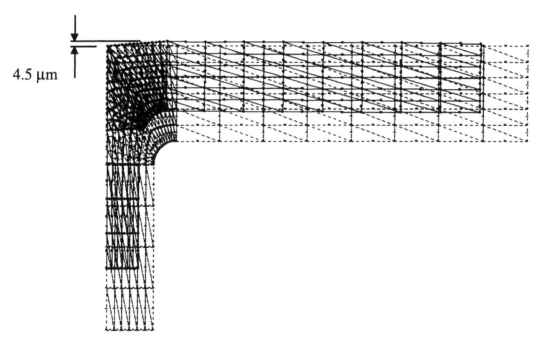

Figure 2. Simulated displacement and top surface profile of the ribbed section.

Figures 3 and 4 present the temperature and curing fields at the point when the entire rib has reached 80% cure. The initial charge temperature was 293K (20ºC) and the mold temperature was 423K (150ºC).

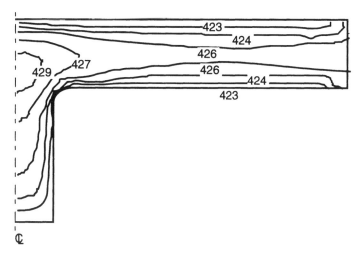

Figure 3. Simulated temperature contour at the end of curing (K).

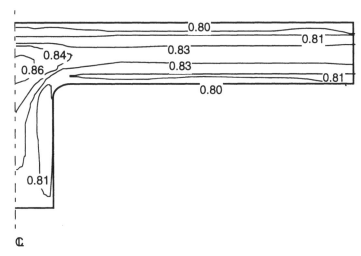

Figure 4. Degree of cure contour at the end of curing.

The normal stresses have a similar pattern, with compressive stresses in all locations due to the volumetric shrinkage taken place during the curing and solidification stage. The maximum compressive stresses occur in the rib area due to the higher temperature and degree of cure. The maximum shear stress occurs at the junction of the rib and plate.

To see how different geometric conditions affect the sink-mark depth, the rib thickness (L), the plate thickness (H), and the lead-in radius (R) were varied. Figure 5 shows the top surface profile with different plate thickness. The rib thickness was 3 mm and the lead-in radius was 1 mm. The plate thickness was varied from 1 mm to 5 mm. Generally, with increasing plate thickness, the depth of sink-mark increases, due to the greater amount of polymerization shrinkage near the rib-plate junction. This is in agreement with the experimental work by (Jutte, 1973) where he concluded that the thick section directly above a rib should be avoided for minimum sink. When the plate is 1mm thick a long wave develops at the top surface.

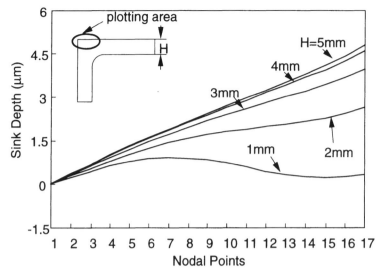

Figure 5. Predicted sink depth for the ribbed section with different plate thickness

Figure 6 shows the top surface profile for different rib thickness. For all cases, the plate thickness was 2 mm and the lead-in radius was 1 mm. The rib thickness ranged from a very thin case (L=2mm) to a very thick case (L=8mm). From Fig. 6 we see that the sink depth increases with rib thickness due to the larger thermal gradient encountered with increasing rib thickness. On the other hand, the single sink-mark is changed to a long wave when increasing the rib thickness from 6 mm to 8 mm.

The effect of lead-in radius is shown in Fig. 7. For all cases, the plate and rib thickness were both 3mm. With the increase in lead-in radius, the corresponding sink depth is increasing. This is due to the larger amount of polymerization shrinkage encountered at the rib entrance as well as the higher thermal gradient. This also matches the work by several researchers (Jutte, 1973) and (Smith,1979) where experimental studies found that for minimum sink, ribs should have a small lead-in radius, usually between 0.127mm and 0.254mm.

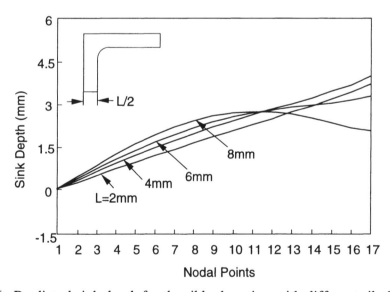

Figure 6. Predicted sink depth for the ribbed section with different rib thickness

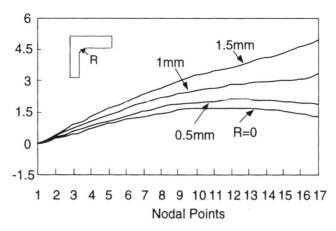

Figure 7. Predicted sink depth for the ribbed section with different rib lead-in radii

The above calculation assumed a constant fiber density distribution, and fibers were oriented along the body panel direction. However, it has been experimentally shown that there is a wedge of resin rich material above the ribs. The formation of this resin-rich region is a consequence of separation as the rib is filled. As a result, the thermal expansion coefficient is higher in the resin-rich area, which in turn contributes to the formation of sink-mark on the surface of the molded panels. Resin and fiber rich areas are assigned to the respective region in the calculation. The magnitude of the sink depth was predicted to be around 12 μm, which is much greater than the magnitude with no fiber-matrix separation. These resin rich areas at the rib entrance caused the corresponding variations in material properties, especially the thermal expansion coefficient and polymerization shrinkage, leading to significant sink-mark formation at the surface of the plate. The fiber-matrix separation is controlled by the flow behavior when the polymer composites fill the rib, which in turn depends on the rib design, as discussed in (Sun, 1996).

There have been several techniques proposed in the literature to minimize the sink-mark formation (Jutte, 1973; Kia, 1989, 1990; Smith, 1979; Boyd, 1976; Ampthor, 1978), such as material selection, molding conditions and mold design.

Within this research we simulated some cases which attempted to reduce sink-mark depth. The effect of glass content was considered first. Glass free samples can result in no sink, which is proved both experimentally and numerically. By adding fibers in the system, the overall mechanical performance of the material is improved. However, significant sink may occur due to the heterogeneity encountered in the ribbed section. When fiber content is increased to 31% and compared to the case with fiber content at 21% the sink depth is reduced by a third, due to the added stiffness. A similar experimental study (Jutte, 1973) where the fiber content was increased from 20% to 30% proved the same effects as predicted by our simulation.

Insignificant changes in sink-mark depth were predicted by varying the mold temperatures. Here, we compared results from mold temperatures of 130°C, 150°C, and 170°C. A slight increase in sink-mark depth was associated with an increase in temperature.

In order to reduce sink-mark depth we introduced a layer of continuous SMC on the composite side of the ribbed part. The continuous layer acts as a stiffener that makes the cosmetic side of the SMC part less sensitive to the rib or boss. The ribbed section with two different continuous layer thicknesses were simulated. One case with a thickness of 1mm and the other case with a thickness of 1.5mm. Both cases significantly reduced sink-mark formation.

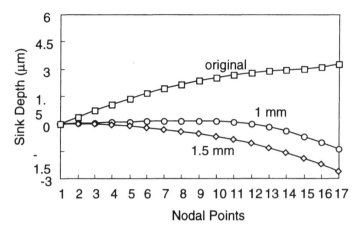

Figure 8. Predicted rib top surface profile by using composites.

It has been proposed that the use of parallel thin ribs in place of a single thicker rib would result in a reduction in the sink-mark depth. For this a plate with two 1.524 mm thick x 25.4 mm in deep ribs was compared to a plate with one 2.54 mm x 31.75 mm rib. Both plates have the same stiffness. The dimension of the case study is chosen to match the experimental study (Jutte, 1973) performed. Similar symmetries were chosen as in the previous analysis. Due to symmetry in the two rib case, only one rib was included in the simulation.

For the case of using two thin parallel ribs the maximum deflection is around 2.8μm, compared to the single thick rib which had a maximum deflection around 1.4μm. This is in accordance with the experimental findings of (Jutte, 1973) that the parallel ribs had significantly more sink than the thicker ribs.

REFERENCES

Ampthor, F. J., 33rd Annual Technical Conference, Reinforced Plastics/Composites Institute, The Society of the Plastics Industry, Inc., (1978).

Boyd, H. J., 31st Annual Technical Conference, Reinforced Plastics/Composites Institute, The Society of the Plastics Industry, Inc., Section 2-C, (1976).

Cook, R. D., D. S. Malkus and M. E. Plesha, Third Edition, John Wiley and Sons, NewYork,(1988).

Halpin, J. C., *J. Comp. Mater.*, 3:732 (1969).

Hill, R. R., S. V. Muzumdar, and L. J. Lee, *Poly. Eng. Sci.* (1996).

Jutte, R. B., International Automotive Engineering Conference, ASME 730171.

Kamal, M. R. and S. Sourour, *Polym. Eng. Sci.*, 13(1):59 (1973).

Kia, H. G., *Polym.-Plast. Technol. Eng.*, 28(1), 43-53, (1989).

Kia, H. G., 45th Annual Conference, Composite Institute, The Society of the Plastics Industry, Inc., Session 1-B, (1990).

Kim, K. S. and H. T. Hahn, *Comp. Sci. Tech.*, 36:121 (1989).

Schneider, W., Kunststoffe, 61:273, (1971).

Smith, K. L., and N. P. Suh, *Polym. Eng. Sci.*, 19(12):829-834, (1979).

Sun, E.M., Ph.D. Thesis, University of Wisconsin-Madison (1996).

Tseng S.C. and T.A. Osswald, *Polymer Composites*, (15),4, 270-277,1994

BIOGRAPHIES

Sari K. Christensen

Sari Christensen received her B.S. in Mechanical Engineering from the University of Wisconsin-Madison in 1996. She currently is earning her M.S. degree at the University of Wisconsin-Madison. She currently works as a research assistant in the Polymer Processing Research Group, in addition to teaching a polymer lab for the undergraduate program at the University. Her current area of research is in composites, specializing in fiber matrix separation. Ms. Christensen interned several semesters at Briggs and Stratton and Strattec Security corporation in Milwaukee, Detroit, and Mexico. She worked as a Quality Engineer, Manufacturing Engineer, Product designer, and Application Engineer.

Esther M. Sun

Dr. Sun, a native of the People's Republic of China, finished her M.S. in Mechanical Engineering in the area of polymer processing in 1993 at the University of Wisconsin-Madison. She received her Ph.D in Mechanical Engineering in the field of polymer processing from the University of Wisconsin-Madison in 1996. Dr. Sun is currently a Research Scientist at Union Carbide. Dr. Sun has published several papers and a book chapter in the area of fiber reinforced thermoset composites.

Tim A. Osswald

Tim Osswald is an Associate Professor in Mechanical Engineering and the Director of the Polymer Processing Research Group at the University of Wisconsin-Madison. Originally from Cúcuta, Colombia, he received his B.S. and M.S. in Mechanical Engineering from the South Dakota School of Mines and Technology and his Ph.D. in Mechanical Engineering at the University of Illinois at Urbana-Champaign in the field of Polymer Processing. He spent two and one half years at the Institute for Plastics Processing (IKV) in Aachen, Germany, as an Alexander von Humboldt Fellow. His current research projects include: mixing, extrusion, compression molding, thermomechanical behavior of fiber reinforced compression molded parts, boundary element simulation of the non-isothermal, non-Newtonian flow of polymer melts. Professor Osswald has published over 100 papers, one book in materials science of polymers and has contributed 8 book chapters.

Analysis of Cure and Flow Behavior of SMC During Compression Molding

HIROYUKI HAMADA, KEIGO FUTAMATA AND HAJIME NAITO

ABSTRACT

Cure and flow behavior of SMC during compression molding are rather complicated due to heat generation, anisotropy of fiber orientation and so on. This paper describes numerical analysis method for both cure and flow behavior of SMC in order to establish total CAE system for SMC compression molding. The differences between experimental and analytical without heat generation temperature change were caused by heat generation due to a chemical reaction. The heat generation was determined by both data and consequently was defined as function of temperature inside of SMC. The most important flow behavior which was clarified by the experiments was slippage flow in the initial stage of compression. The slippage flow occurs in both interlamina and intralamina , and this flow behavior influence on distribution of fiber content fraction in products, that leads to scattering for the mechanical properties of products. In order to understand slippage flow behavior three-dimensional large deformation elastic-plastic analysis was used. The calculated deformation states were in good agreement with the experimental results. Thermal deformation analysis was also performed using results of heat conduction analysis. These analyses proposed were included into total CAE system for SMC compression molding.

INTRODUCTION

SMC(Sheet Molding Compound) is typical fiber reinforced plastics for high volume processing and has been used in structural parts of transportation vehicles, channel shaped fabrication, bathtub and so on. However, material flow pattern during compression molding is very complicated because reinforcing fibers move together with matrix. Also the analysis of curing process is difficult due to heterogeneous materials containing many different components. Fiber orientation state and curing profile inside of SMC moldings greatly affect on not only mechanical properties but also post-deformation. On the other

Hiroyuki Hamada, Faculty of Textile Science, Kyoto Institute of Technology, Matsugasaki, Sakyo-ku, Kyoto 606, JAPAN
Keigo Futamata, Graduate School, Kyoto Institute of Technology, Matsugasaki, Sakyo-ku, Kyoto 606, JAPAN
Hajime Naito, Sekisui Chemical Co., Ltd., 2-2, Kamichoshi-cho, Kamitoba, Minami-ku, Kyoto 601, JAPAN

hand a set of mold would be expensive, so that changing mold design according to try and error method is inconvenient. Therefore CAE system is required for mold design. Several research studies have been conducted in numerical analysis for SMC molding. Some analyses were of use, however, geometry of analytical model was restricted to specific geometry(Lee et al, 1984 ; Barone and Caulk, 1986 ; Hirai et al., 1982).

In this paper we propose total CAE system as shown flow chart in the Figure 1. The CAE system was constructed by several parts. In each part the numerical analysis methods were developed by using commercial numerical software and/or original program which were mainly finite element method.

This paper describes usefulness of the total CAE system. Particularly, a new original method was introduced to analysis of curing process, which enable to understand curing profile during flow stage and curing stage.

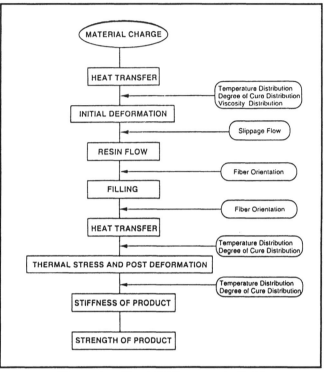

Figure 1. Total CAE system in SMC compression molding.

ANALYTICAL METHOD

HEAT CONDUCTION ANALYSIS

Heat generation curve was obtained by measuring temperature inside of SMC. The dimension of single ply SMC was $40 \times 40 \times 4$, $80 \times 80 \times 4$ mm. Three-plies materials were piled up. Next, three-dimensional heat conduction analysis was applied to calculate in the specimen for the case that heat generation is neglected. Figure 2 shows analytical model, which was a quarter of the specimen due to symmetry. As boundary conditions, the plane-ADM and the plane-ABK were adiabatic boundary because of neutral planes. Mold temperature, 140℃, was prescribed at nodes on the plane-KLM. After 30sec. interval when closing upper mold come into contact with specimen, mold temperature of 140℃ was also prescribed at nodes on the plane-ABCD. Initial nodal temperature are assumed to be 20℃. A comparison between experimental and numerical temperature change was made and it could be considered that the differences between them was caused by heat generation due to a chemical reaction(Hirai et al., 1984). The heat generation was determined by both data using following equation.

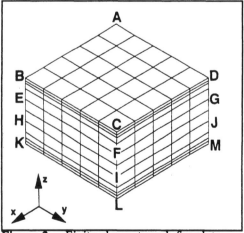

Figure 2. Finite element mesh for plate shaped model.

$$Q = \rho c \frac{\Delta T_1 - \Delta T_2}{\Delta t} \qquad (1)$$

where ρ is density, c is specific heat. ΔT_1 and ΔT_2 are increase of temperature at system for heat generation and without heat generation respectively, Δt is time change. In this study, heat generation was calculated every 5℃ increase and consequently could be defined as function of temperature inside of SMC. The degree of cure α was defined as the ratio of sum of heat generation $\sum Q$ to total heat generation $\left[\sum Q\right]_{Total}$.

Furthermore, curing was completed when α was 0.8 in the analytical region(Barone and Caulk, 1979 ; Sun et al., 1995). The curing profile could be predict by using this assumption.

ELASTIC-PLASTIC ANALYSIS

In order to express slippage flow behavior using numerical analysis, the material in the initial process stage was modeled as an elastic-perfect plastic material. Figure 2 shows analytical model. Nodes on the plane-KLM were fixed for x-, y- and z-direction, nodes on the plane-ADM and the plane-ABK of neutral planes were fixed for x- and y-direction respectively. Nodes on the plane-ABCD were fixed for x- and y-direction and were given prescribed displacement -0.1 mm every one step for z-direction. Higher elastic modulus was used in completed curing element. The plane-EFG and the plane-HIJ express intralamina, so that double nodes at the same coordinates were set. The slippage flow occurred when shear strain calculated between two adjacent layer exceed a certain critical shear strain. After occurrence of slippage flow two nodes were separated.

THERAMAL DEFOMATION ANALYSIS

A product would be taken out from a mold and put on the stand. At that time, the product cool and thermal deformation occurs due to thermal stress. Here thermal deformation analysis was performed using results of heat conduction analysis. Figure 3 shows analytical model. This model is a part of channel shaped fabrication with thickness of 4 mm.

Firstly, above-mentioned heat conduction analysis was performed for plate shaped model. This analysis was undertaken by adding the elements step by step. The temperature distribution was determined using that of previous step. Next, the product with temperature distribution after 100sec. would cool until room temperature. The boundary conditions for this model were as follows. Nodes on the plane-ABE of

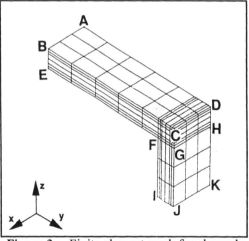

Figure 3. Finite element mesh for channel shaped fabrication.

neutral plane were fixed for y-direction. Nodes on the plane-BCGE and the plane-ADH of neutral plane were fixed for x-direction. Nodes on the plane-IJK were fixed for z-direction because of contact with the stand.

EXPERIMENTAL METHOD

Compression molding was performed for above-mentioned specimen. Mold temperature was 140℃. The upper and the lower surfaces of center layer is used colored material in order to understand flow behavior. After specimen was charged on the cavity, the upper mold was closed and contacted with specimen 30 sec. later. This state was kept until start of compression, and this time was defined as holding time. Curing profile at the start of compression could be changed by change of holding time. The mold closing speed was 10 mm/min.

In order to obtain heat generation curve, temperature inside of specimen was measured using thermocouple during holding time.

RESULTS AND DISCUSSION

Figure 4 shows temperature-time curves for experiment and analysis of without heat generation in the case of $40 \times 40 \times 4$ mm. The heat generation occurred from about 130 ℃. Relation between heat generation and temperature inside of specimen was conducted from these results as shown in Figure 5. In order to confirm validity of the method the heat generation function obtained was introduced into heat conduction analysis. Figure 4 also shows result of this analysis. Temperature-time curve obtained by numerical analysis was good agreement with the experimental result.

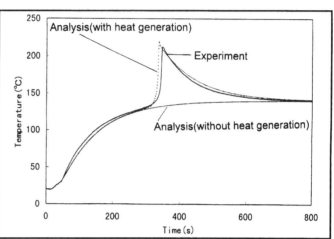

Figure 4. Temperature-time curve for plate shaped model.

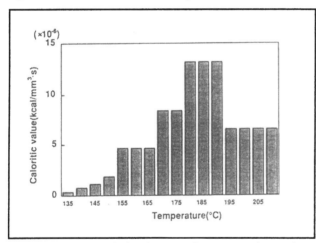

Figure 5. Relation between heat generation and temperature inside of SMC.

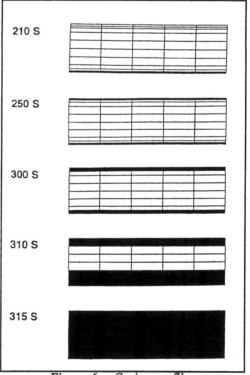

Figure 6. Curing profiles.

Figure 6 shows curing profiles of cross section obtained by numerical analysis. At holding time 210 sec., curing of the bottom layer was completed firstly. At holding time 250 sec., curing of the top layer was also done. Finally, At holding time 315 sec. curing of all region was completed. Cure behavior in the initial process stage was asymmetry through thickness direction due to considering with the upper mold closing time. Consequently, it seemed that practical curing profile could be obtained by this numerical analysis method.

Figure 7 shows schematic drawing of cross section of specimen by the experiments. At holding time 0 sec. outer layer flow, whereas at holding time 100, 250 sec. middle layer flow remarkably, because curing profile until starts of compression was different. Further more, flow length at each layer was different because of slippage flow in intralamina.

Figure 8 shows result of elastic-plastic analysis at the case of holding time 250 sec. Top and bottom layer were not deformed because of curing. However adjacent elements for these layers through thickness direction were deformed larger. Double nodes at interlamina were separated, so that middle layer jutted out, which means occurrence of slippage flow. It seemed that practical flow behavior in the initial process stage was expressed by this analysis.

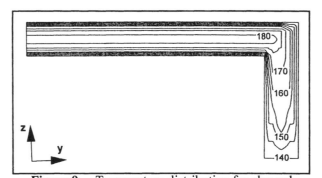

Figure 7. Schematic drawing of cross section of specimen.

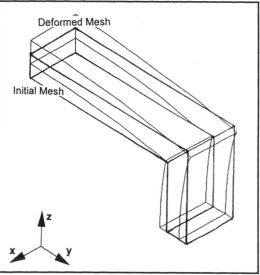

Figure 8. Deformed mesh obtained from elastic-plastic analysis.

Temperature distribution at 100 sec. for channel shaped fabrication is shown in Figure 9. The surface temperature was 140 ℃, on the contrary, maximum internal temperature was about 180 ℃ Figure 10 shows the original and the deformed mesh. The state of thermal deformation could be predicted in this study.

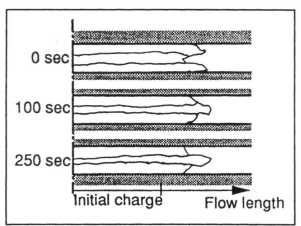

Figure 9. Temperature distribution for channel shaped fabrication.

Figure 10. Deformed mesh for channel shaped fabrication due to thermal stress.

CONCLUSION

In this study, numerical analysis method was proposed for cure and flow behavior of SMC during compression molding. The two analyses proposed were introduced in order to understand curing profile and slippage flow at the initial process stage. It was also possible to develop thermal deformation analysis using result of this heat conduction analysis. These analyses were useful in order to establish total CAE system for SMC molding.

REFERENCES

Barone, M. R., D. A. Caulk. 1986. "A Model for the Flow of a Chopped Fiber Reinforced Polymer Compound in Compression Molding," J. Appl. Mech., 53:361~371.

Barone, M. R., D. A. Caulk. 1979. "Effect of Deformation and Thermoset Cure on Heat Conduction in a Chopped Fiber Reinforced Polyester During Compression Molding," Int. J. Heat and Mass Transfer, 22:1021-1032.

Hirai, T., T. Katayama, H. Hamada. 1982. "FLOW STATE OF COMPOSITE MATERIALS IN THE FORGING DIE DURING THE MOLDING PROCESS CONTINUED : EFFECTS OF MOLDING TEMPERATURE," ADVANCES IN COMPOSITE MATERIALS, : 1606~1619

Hirai, T., H. Hamada, H. Saka. 1984. "Basic Study on Chemical Diffusion in Curing Process of Composite Material under Compression Molding," J. THE SOCIETY OF MATERIAL SCIENCE, 33(364):85~90.

Lee, C. C., F. Folgar, C. L. Tucker. 1984. "Simulation of Compression Molding for Fiber-Reinforced Thermosetting Polymers," ASME Trans., J. Eng. Ind., 106:114~125.

Sun, Esther M., Bruce A. Davis, Tim A. Osswald,1995. "Modeling and Simulation of Thick Compression Molded Parts," Proc. of the 50th Annual Conference, Session 15-B, Composites Institute, The Society of the Plastics Industry.

BIOGRAPHIES

Hiroyuki Hamada
Dr. Hiroyuki Hamada is presently an associate professor of Faculty of Textile Science at Kyoto Institute of Technology. He received his B. Eng. degree from Doshisha University in 1978, M. Eng. degree from Doshisha University in 1980 and Dr. Eng. from Doshisha University in 1985.

Keigo Futamata
Mr. Keigo Futamata is presently a graduate student at Kyoto Institute of Technology. He received his B. Eng. degree from Kyoto Institute of Technology in 1995.

Hajime Naito
Mr. Hajime Naito is presently an assistant manager at Sekisui Chemical Co. Ltd. He received his B. Eng. degree from Kyoto Institute of Technology in 1986.

Study of In-Mold Flow Behavior of SMC

KENICHI HAMADA, TETSUYA HARADA,
TAKASHI TOMIYAMA AND HIROKAZU YAMADA

ABSTRACT

One common defect in SMC moldings is pinholes. Previous work suggests that pinholes form while the SMC is flowing in a mold.[1] In this paper, SMC flow behavior was analyzed using an apparent viscosity and a shear rate as suggested by Menges, et al.[2] The number of pinholes was correlated with the apparent viscosity and the shear rate of flowing SMC in the mold. The shear rate depends on the molding conditions. The number of pinholes decreases as the shear rate increases. SMC flow is dominated by elongational deformation at high shear rates, but at low shear rates it is dominated by a softening of the SMC caused by the heat from the mold surfaces.

The relationship between the SMC composition and the number of pinholes was evaluated as well. A relationship was found between the apparent viscosity and the shear rate which can be represented by a power equation; therefore, in-mold SMC flow is non-Newtonian. A structural viscosity index, which is calculated from this equation, varies widely with the SMC composition. Pinholes are not formed when the structural viscosity index is low even under relatively low shear rate.

INTRODUCTION

Some studies have indicated that pinholes form during SMC flow in a mold. Many studies have been conducted for SMC flow analysis[3]-[9]; however, few of these studies have mentioned a quantitative relationship between the flow behavior and pinholes. In order to reduce pinhole defects, molding conditions such as molding pressure have been modified based on experience.

In this paper, the experiential procedures needed to decrease the pinholes are explained quantitatively, and the SMC in-mold flow behavior is analyzed with the apparent viscosity and the shear rate as suggested by Menges, et al.

EXPERIMENTAL

MOLDING CONDITIONS

An experiment with 21 trials was designed with three levels of molding force, closing rate, SMC charge coverage, and molding temperature as shown in Table 1. A low profile SMC for residential use was molded using a 400 mm diameter disk mold under these

---

Kenichi Hamada, Tetsuya Harada, Takashi Tomiyama and Hirokazu Yamada, Dainippon Ink and Chemicals, Inc., Osaka, Japan.

conditions (Figure 1). Round SMC charges were placed on the center of the mold. The number of pinholes on the molded part surface was counted.

Table 1. Levels of Molding Condition

Molding Force, tons	30	70	120
Closing Rate, mm/s	1	3	5
SMC Charge Coverage, %	30	45	60
Molding Temperature, °C	130	140	150

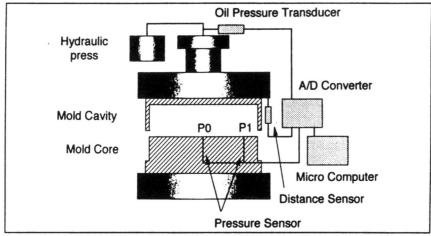

Figure 1. Equipment for Experiment

SMC COMPOSITION

SMC composition was also studied. SMC formulations with three levels of filler, thickener, and glass content as shown in Table 2, were molded under the molding conditions shown in Table 3. It was assumed that the number of pinholes varied with the molding conditions in Table 3.

Table 2. SMC Compositions

SMC Sample	A	B	C	D	E	F	G
UPE, phr	80	80	80	80	80	80	80
LPA, phr	20	20	20	20	20	20	20
Additives, phr	30	30	30	30	30	30	30
Catalyst, phr	1.5	1.5	1.5	1.5	1.5	1.5	1.5
Filler, phr	150	150	150	100	200	150	150
Thickener, phr	1.3	0.8	1.8	1.3	1.3	1.3	1.3
Glass Fiber, %	25	25	25	25	25	20	30

Table 3. Molding Conditions

Molding Condition	#1	#2	#3	#4	#5	#6	#7
Molding Force, ton	120	120	70	70	30	30	120
SMC Charge Coverage, %	30	60	30	45	30	60	30
Closing Rate, mm/s	5	5	5	1	5	5	5
Molding Temperature, °C	140	140	140	140	140	140	150

APPARENT VISCOSITY AND SHEAR RATE

A shear rate (γ) and an apparent viscosity (η) during the SMC molding were calculated with the Equation (1) and (2)[2] respectively. Figure 2 shows the mold cross-section.

$$\gamma = 3h'(t)r / h(t)^2 \qquad (1)$$

$$\eta = h(t)^3 / (6h'(t)\, r) \times (P_0 - P_1) / r \qquad (2)$$

where,
t: molding time
h(t): mold distance at t
h'(t): closing rate
r: distance from the mold center
P_0: pressure at the mold center
P_1: pressure at r
γ: shear rate at r
η: apparent viscosity at r

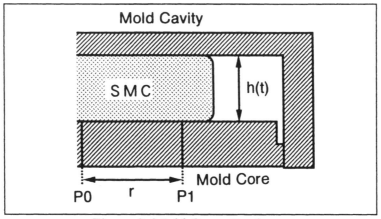

Figure 2. Mold Cross-Section

RESULTS AND DISCUSSIONS

PINHOLES AND MOLDING CONDITION

The designed experiments were analyzed with a statistical regression technique. The results of this analysis - relating the number of pinholes on the product surface as a function of molding conditions - are plotted in Figure 3 and Figure 4. The correlation coefficient of the regression equation is 0.88; this indicates that the pinholes depend on the molding condition significantly.

In Figure 3, as the molding force increases and the SMC coverage decreases, the number of pinholes decreases. In Figure 4, as the closing rate increases, pinholes decrease. However, the effect of the closing rate on the pinholes is not as significant as are the molding force and the SMC coverage. The molding temperature does not significantly affect the number of pinholes. Most pinholes form on the product outer edge. This result suggests that the pinholes are formed during the SMC flow in the mold.

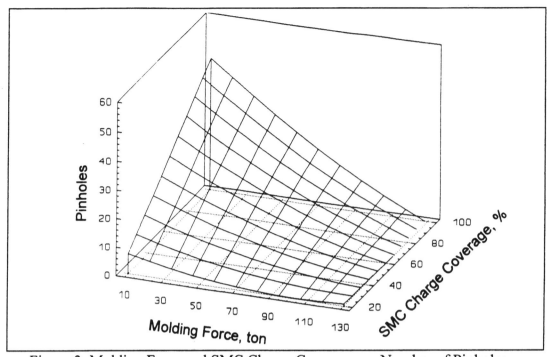

Figure 3. Molding Force and SMC Charge Coverage vs. Number of Pinholes
(Mold Temperature: 140°C, Closing Rate: 3 mm/s)

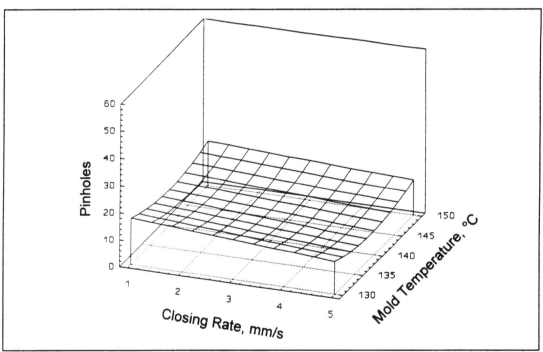

Figure 4. Closing Rate and Upper Mold Temperature vs. Number of Pinholes
(Molding Force: 70 ton, SMC Charge Coverage: 60%)

SMC flow distance decreases as the SMC coverage increases; however, the number of pinholes increases. It seems that the pinholes are not formed inevitably during the SMC flow, but the formation of the pinholes is dominated by the SMC flow behavior. Thus, flow behavior analysis is very important to understand and improve the pinhole defect.

PINHOLES AND SHEAR RATE

Figure 5 shows the shear rates under the various molding conditions. The shear rate reaches its maximum as mold closing, then decreases quickly as the mold approaches full closure. Shapes of shear rate curves vary with the conditions. In order to analyze the shear rate behavior during molding at a distance of r = 155 mm from the mold center where the most of pinholes were observed, the shear rate when SMC reaches the pressure transducer at 155 mm from the mold center and the maximum shear rate were defined as γ_{Ini} and γ_{Max} respectively (Figure 6). Table 4 shows measured shear rates and the corresponding pinhole count.

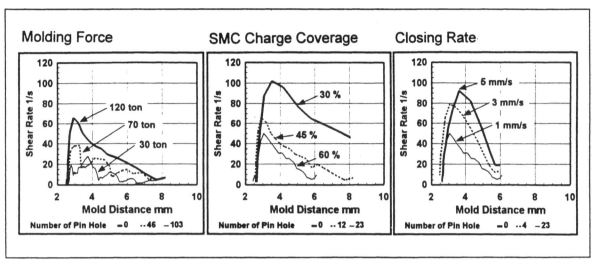

Figure 5. Shear Rate Curves Under Various Molding Conditions

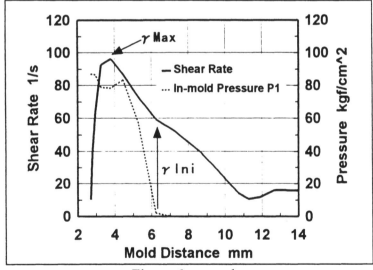

Figure 6. γ_{Ini} and γ_{Max}

Table 4. Shear Rate and Pinholes

Molding Force, ton	γ_{Ini}	γ_{Max}	Pinholes
120	31	59	0
70	11	43	46
30	11	25	103
SMC Charge Coverage, %			
30	73	92	0
60	7	46	23
Closing Rate, mm/s			
5	23	83	0
3	13	74	4
1	7	46	23

The shear rates increases as the molding force increases, SMC charge coverage decreases, and the closing rate increases. The number of pinholes decreases as the shear rate increases.

Figure 7 shows the result from the regression analysis for the number of the pinholes with the log transformed γ_{Ini} and γ_{Max}. The number of pinholes decreases as either the γ_{Ini} or γ_{Max} increase.

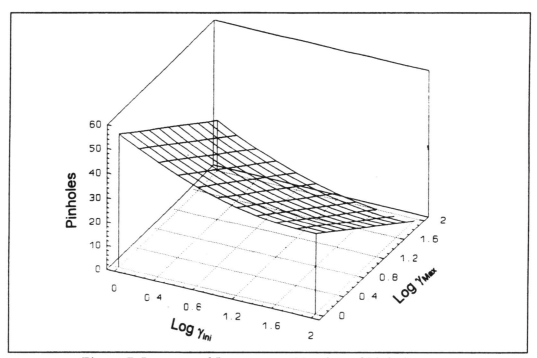

Figure 7. Log γ_{Ini} and Log γ_{Max} vs. Number of Pinholes

In order to observe the SMC flow behavior with the different shear rates, a bicolored SMC charge (upper layer charge: white, lower layer charge: pink) was molded to visualize the flow. Table 5 shows the molding conditions, γ_{Ini}, γ_{Max}, and the number of pinholes.

Table 5. Molding Conditions and The Number of Pinholes

	High Shear Rate	Low Shear Rate
Upper Mold Temperature, °C	140	140
Molding Force, ton	120	30
Closing Rate, mm/s	5	1
SMC Charge Coverage, %	45	45
γ_{Ini}, 1/s	31	11
γ_{Max}, 1/s	59	25
Number of Pinholes	0	103

Figure 8 shows the molded products.

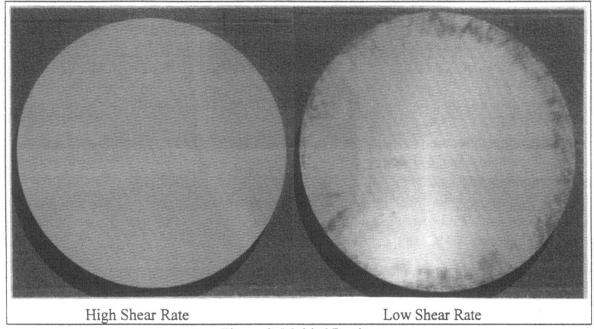

High Shear Rate   Low Shear Rate

Figure 8. Molded Product

The number of pinholes depends on the shear rates. Pinholes were not formed under the high shear rate. The upper and the lower layer of SMC charge were observed only on the surface and on the back side of the molded product respectively. On the other hand, many pinholes formed under the low shear rate. Also the lower layer of charged SMC was observed on the edge of the molded surface. It appears that the shear rate means SMC packing rate. SMC is deformed with both elongational deformation caused by the mold closing and melting flow caused by the heat from the mold surfaces during the SMC packing. Under the high shear rate condition, the SMC packing rate is high, i. e. SMC deforms elongationally with low deformation resistance (Figure 9. Flow Model a).

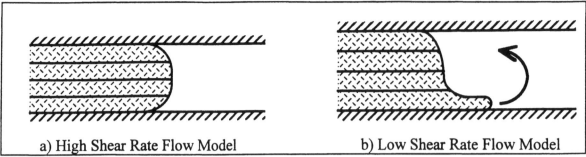

| a) High Shear Rate Flow Model | b) Low Shear Rate Flow Model |

Figure 9. SMC Flow Model

Under the low shear rate condition, the SMC packing rate is low, i. e. SMC deforms elongationally with high deformation resistance and the melting flow dominates the packing (Figure 9, Flow Model b). Since the lower layer of charged SMC was observed on the molded surface, the flow of the SMC at low shear rate is assumed to behave as follows: When the SMC is charged on the lower mold, the lower layer of the SMC is melted by the heat. Then the melted SMC rolls upward and traps the air as the flow front moves outward.

The SMC flow behavior can be explained according to the shear rate.

SMC COMPOSITION AND PINHOLES

Figure 10 shows the shear rate curves of SMC sample C, E, and G molded under the molding conditions #6 and #7.

Many pinholes formed under the low shear rate, whereas few pinholes formed under the high shear rate. However, the number of pinholes can not be explained only by the shear rate for different SMC compositions.

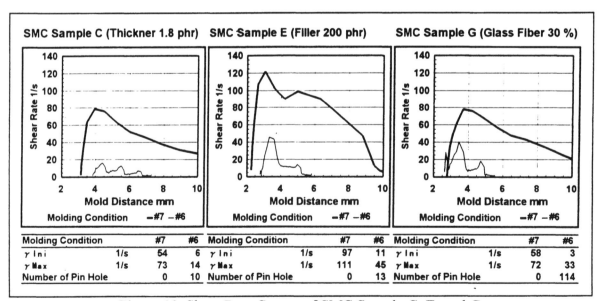

Figure 10. Shear Rate Curves of SMC Sample C, E, and G

Generally, the relationship of the shear rate and shear stress can be represented by a power equation (3). The apparent viscosity is represented by Equation (4). Substituting Equation (3) in Equation (4) gives Equation (5).

$$\sigma = a \times \gamma^n \tag{3}$$

$$\eta = \sigma / \gamma \tag{4}$$

$$\eta = (a \times \gamma^n) / \gamma = a \times \gamma^{n-1} \tag{5}$$

The equation (5) is log transformed as follows:

$$\log \eta = (n - 1) \times \log \gamma + \log a \tag{6}$$

where,
a: constant
σ: shear stress
n: structural viscosity index
γ: shear rate
η: apparent viscosity

There is a linear relationship with slope (n-1) between the log transformed shear rate and apparent viscosity (Equation 6). The structural viscosity index, n, is an indicator of the rheological characteristics. When the structural viscosity index = 1, the apparent viscosity is constant and the fluid is called a Newtonian fluid. When the structural viscosity index $\neq$ 1, the apparent viscosity depends on the shear rate and the fluid is called a non-Newtonian fluid.

The log transformed shear rate γ_{Ini} and apparent viscosity η under the molding conditions #1 to #7 were plotted to obtain the structural viscosity index. Figure 11 shows the shear rate γ_{Ini} versus apparent viscosity η and the number of pinholes at the three levels of thickener content.

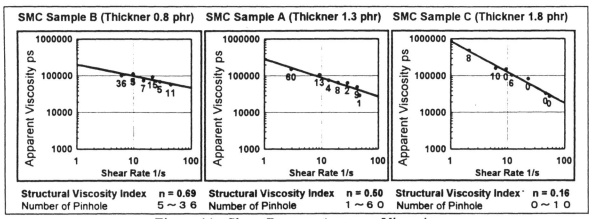

Figure 11. Shear Rate vs. Apparent Viscosity

The structural viscosity index decreases with the increase in thickener content. The number of pinholes decreases significantly with the increase in shear rate. Structural viscosity indexes and shear rate ranges without pinholes are shown in Table 6.

Table 6. Viscosity Index and Shear Rate Range

SMC Sample	Thickener			Filler			Glass fiber		
	0.8 phr	1.3 phr	1.8 phr	100 phr	150 phr	200 phr	20%	25%	30%
	B	A	C	D	A	E	F	A	G
Structural Viscosity Index, n	0.69	0.50	0.16	0.33	0.50	0.02	0.53	0.50	0.36
Shear Rate Range without Pinholes	none	none	> 20	none	none	> 30	none	none	> 50

The structural viscosity index of sample C, SMC with the highest thickener content, is 0.16. Pinholes did not form when the shear rate was more than 20 1/s. This SMC performs over a wide range of shear rates without pinholes, i.e. it can be molded under a wide range of conditions (#1, 2, 3, and 7) without pinholes. The viscosity index of sample E, the SMC with the highest filler content, is 0.02. Pinholes did not form when the shear rate was more than 30 1/s. Glass fiber content did not affect the number of pinholes.

The effects of SMC composition on the pinholes are as follows:

1) Thickener and filler content affect the number of pinholes significantly.
2) Structural viscosity index decreases as the thickener and filler content increase. This means that the degree of non-Newtonian characteristics of the SMC becomes high, and the apparent viscosity during the molding decreases with the increase in shear rate significantly.
3) SMC with low structural viscosity index can be molded under a wide range of molding conditions without pinholes.

CONCLUSIONS

- The shear rate curves are dominated by the molding conditions. SMC in-mold flow behavior can be predicted from the shear rate curves.
- The number of pinholes can be reduced by the increasing the shear rate.
- SMC flow is dominated by the elongational deformation under high shear rate, whereas the flow is dominated by the SMC melting flow under low shear rate.
- In-mold SMC flow is non-Newtonian. The structural viscosity index, dependence of the apparent viscosity on the shear rate, varies with SMC composition.
- Pinholes did not form in SMC which has a low structural viscosity index, i. e. the SMC can be molded under the wide range of molding conditions without pinholes.

REFERENCES

1. Mihata, I., et al., 1989. "SMC/BMC Flow and Cure Behavior in Compression Molding," 44th Annual SPI Conference, Session 1-C
2. Menges, G., et al., 1981. "Flow Testing of Sheet Molding Compound and Its Practical Application," 36th Annual SPI Conference, Session 23-C
3. Burns, R., et al., 1977. "The Rheology of Glass Fiber Reinforced Polyester Molding Compounds," 32nd Annual SPI Conference, Session 7-C
4. Silva-nieto, R. J., et al., 1980. "Rheological Characterization of Unsaturated Polyester Resin Sheet Molding Compound," 35th Annual SPI Conference, Session 2-E

5. Lee, L. J., et al., 1981. "The Rheology and Mold Flow of Polyester Sheet Molding Compound," SPE 39th ANTEC, p.334

6. Kau, H. T., et al., 1986. "Experimental and Analytical Procedures for Flow-dynamics of Sheet Molding Compound (SMC) in Compression Molding," SPE 44th ANTEC, p.1345

7. Marker, L., et al., 1977. "Rheology and Molding Characteristics of Glass Fiber Reinforced Sheet Molding Compounds," 32nd Annual SPI Conference, Session 16-E

8. Schmeizer, E., et al., 1985. "SMC-processing up the Learning Curve," 40th Annual SPI Conference, Session 16-B

9. Costigan, P. J., et al., 1985. "The Rheology of SMC during Compression Molding, and Resultant Material Properties," 40th Annual SPI Conference, Session 16-E

BIOGRAPHIES

Kenichi Hamada

Mr. Hamada received his M.S. degree in applied chemistry from Waseda University. He has been with Dainippon Ink and Chemicals, Inc. since 1996, and is now a research engineer at Kansai Research and Development Center in Sakai, Japan.

Tetsuya Harada

Mr. Harada received his B.S. degree in applied chemistry from Doshisya University. He has been with Dainippon Ink and Chemicals, Inc. since 1987, and is now a research engineer at Kansai Research and Development Center in Sakai, Japan.

Takashi Tomiyama

Mr. Tomiyama received his M.S. degree in polymer chemistry from Kyoto University. He has been with Dainippon Ink and Chemicals, Inc. since 1985, and is now a research engineer at Kansai Research and Development Center in Sakai, Japan.

Hirokazu Yamada

Mr. Yamada received his M.S. degree in organic chemistry from Osaka University. He is a group manager for the press molding group at Dainippon Ink and Chemicals, Inc. at Kansai Research and Development Center in Sakai, Japan.

Iosipescu Shear Testing of Sheet Molding Composites (SMC)

E. M. ODOM, T. A. KVANVIG, T. P. VANHYFTE AND T. H. GRENTZER

ABSTRACT

The focus of this study is on the measurement of in-plane shear properties of Sheet Molding Composite (SMC) materials using the Iosipescu shear test as described in the V-Notched Beam Method ASTM D 5379-93. To test the SMC materials of this study, a new Iosipescu shear fixture was designed and fabricated. The advantages of this fixture are the ability to test specimens without tab procedures in order to insure gauge section failures. These advantages translate into less time required for sample preparation, increased precision during testing, and, therefore, higher productivity. Testing of interlaminar shear properties using the ASTM D 2344-84 short beam shear test are included for comparative purposes. The materials of interest in this study were two FRPs, one with a polyester matrix and the other with a vinyl ester matrix. The in-plane shear strength of some of these FRP formulations is remarkable. The values ranged from 16 to 22 ksi and are higher than a unidirectional AS4/3501-6 carbon fiber/ epoxy composite. Some of the values reported in this study approach those of 6061-T6 aluminum. Additionally, the effect of reinforcement orientation and the effect of hollow glass spheres on shear properties were investigated. The test results suggest that the Iosipescu shear test provides highly repeatable values for in-plane shear strength and modulus and will be of great value to design and material engineers in a variety of industries.

INTRODUCTION

There is increasing use of Sheet Molding Composite (SMC) Materials in automotive applications. As the applications move from the aesthetic to the structural, the need arises for engineering properties such as tensile, compressive and shear moduli and strengths. While mechanical test methods for tension and compression are very well developed, test methods for measuring shear properties are not. The main reason for this lack of development is that most composite materials, of which SMCs are a subset, are used in sheet form, while the most advantageous form for shear property measurement would be thin walled tubes, which could be subjected to torsional loading. To overcome this problem of material form, a variety of in-plane shear tests applicable to planar specimens have been suggested, e.g., the two and

E. M. Odom, University of Idaho.
T. A. Kvanvig, T. P. VanHyfte and T. H. Grentzer, Ashland Chemical Company.

three rail shear test, the [±45] tensile, the 10° off-axis tensile and the Iosipescu shear test. This paper examines the applicability of the Iosipescu shear test, as well as its superiority to the short beam shear test.

IOSIPESCU SHEAR TEST

Recently, the Automotive Composites Consortium (ACC) adopted the Iosipescu in-plane shear test as outlined in "Test Method for Shear Properties of Composite Materials by the V-Notched Beam Method" (ASTM D 5379-93) as part of their testing protocol. A sketch of the loading fixture and the specimen shear and moment diagram is shown in Figure 1.

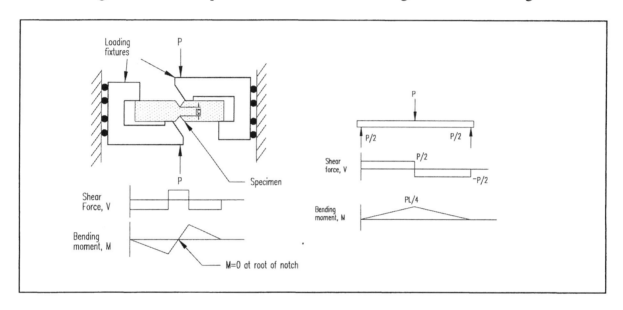

Figure 1   Sketch of Iosipescu (left) and SBS Shear (right) Test Specimens Along with Their Shear and Moment Diagrams.

This test method was originally proposed for metals in 1967 and modified for use on composite materials in the early 1980s (Adams and Walrath 1982). During the past 15 years of continuous use and development, this test method has been controversial. The controversy involves a variability of reported accuracy and precision for measured modulus values. It was found that the primary cause of this variability is due to specimen twisting during testing. Recently, a new fixture was designed that alleviates specimen twisting (Conant and Odom 1993). Another problem associated with this test method is unacceptable crushing failures, which most often occur at the inner load point of the fixture rather than shear failures in the gauge section. The ASTM standard recommends specimen tabbing to prevent these failures. However, specimen tabs are time consuming to fabricate, bond to specimens and have been shown to be only partially effective in preventing load point failures. To alleviate this problem, a new fixture design was developed and fabricated to test the SMC materials in this study. This new fixture is shown in Figure 2. As can be seen, this fixture has inner loading points that can rotate and therefore distribute the load over a larger area, which reduces the compressive contact stresses in the specimen.

SHORT BEAM SHEAR TEST

The interlaminar shear strength is measured using the short beam shear test as described in ASTM D 2344-84. A sketch of the specimen and shear and moment diagrams are included in Figure 1. This test is most often used for quality assurance due to the advantages of a small easy to prepare specimen geometry and its sensitivity to changes in matrix properties and matrix-fiber adhesion. The Standard does not recommend the use of the results of this test method for design purposes, since the stress state at the mid-span of the specimen is a combination of shear, compressive and flexural stresses. However, the Standard does indicate that data can be used for quality control and specification design.

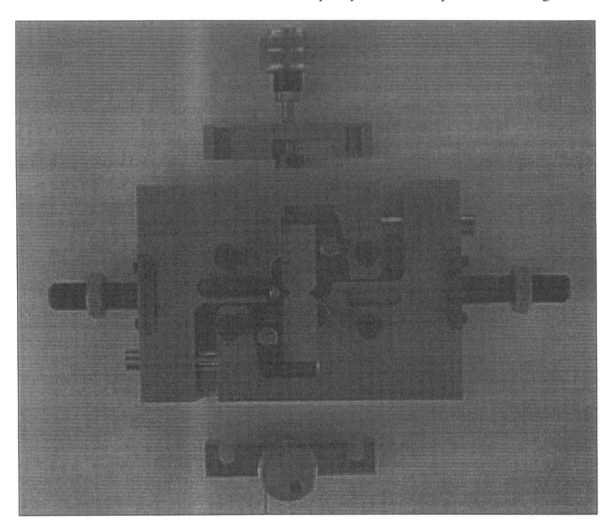

Figure 2: Design of Iosipescu Shear Test Fixture Designed for Testing SMC Materials. (Note semicircular inner loading points).

TEST PROCEDURE

The Iosipescu shear test specimens were fabricated using a procedure developed at the University of Idaho and tested per the procedures outlined in the ACC's *Test Procedures*. After the specimens were fabricated, Micro Measurement N2A-00-C032A-500 strain gauges were mounted on the specimens. These gauges have a grid that spans the entire width of the gauge section and therefore sense the average shear strain in the gauge section. Five specimens were tested for each material variation and orientation. After each test was concluded, the specimen was inspected using the guidelines in the ACC's *Test Procedures* to determine the mode of failure and the shear stress-shear strain data were reduced to obtained the shear strength and shear moduli.

For completeness, the specific gravity of each specimen was determined using Archimede's method of weighing the specimen in air, W_{air}, and water, W_{H2O}, and then calculating the density using

$$\rho_c = \frac{W_{air}}{W_{air} - W_{H_2O}} \rho_{H_2O}$$

where ρ_c is the density of the composite specimen and ρ_{H2O} is the density of water.

The short beam shear tests were conducted in accordance with ASTM D 2344. The samples were tested on Instron 4202 and 4204 model mechanical testers. The cross head speed was set at 0.05 inches/minute and the span to depth ratio was 5 to 1.

MATERIAL DESCRIPTION

The first SMC material set investigated the influence of resin type and level on the measured mechanical properties. The material components of each of the different materials are given in Table I. The PET- based unsaturated polyester base resin was combined with additional styrene monomer, acrylic-based resin, and inhibitors to complete the formulation. The system was catalyzed with t-butyl perbenzoate. Zinc stearate was added as an internal mold release agent. The A-paste was blended with a Cowles type mixer to 85° F. B-side paste was blended in (to 90° F) and the paste was processed across the SMC machine. Approximately 15% continuous fiberglass roving and 45% two-inch chopped fiberglass roving were used during processing. The sheet was run on a 36-inch SMC machine. The doctor blades were set to give different amounts of resin concentrations during processing. Sample sets B1-B2 and C-E differ in the ratio of resin to fiberglass reinforcement due to doctor blade settings. Molding was done on a 12" x 12" Class-A tool heated to 300 °F (top 305 °F and bottom 295 °F). The press was set for 75 tons (1000 psi), at a closure speed of 5-9 seconds.

The second SMC material tested used chopped fiberglass roving in a matrix of unsaturated vinyl ester base resin that was catalyzed with t-butyl perbenzoate and t-butyl peroxy-2-ethylhexanoate. Magnesium oxide was used as the thickening agent and

TABLE I
COMPONENTS OF 1ST SMC SERIES MATERIALS

Major Components of Each Material Designation	Material Designation			
	B1	B2	C	E
PET- based unsaturated polyester resin (phr)	90	90	90	90
Acrylic-based resin (phr)	10	10	0	0
saturated polyester resin (phr)	0	0	10	10
glass fiber (%)	61	62	65	60

calcium carbonate was used as the filler. Sample A has 150 parts of filler; whereas, Sample B has 50 parts of filler and 15 parts of hollow glass bubbles. Molding was done on a 12" x 12" Class-A tool heated to 300°F (top 305 °F and bottom 295 °F). The press was set for 75 tons (1000 psi), at a closure speed of 5-9 seconds. To explore the repeatability of the material processing, material shear properties, and the Iosipescu shear test method, four thicknesses of material were used.

TEST RESULTS

Tables II indicates the mean specific gravity, Iosipescu in-plane shear strength and shear modulus (Iosipescu), and short beam shear interlaminar shear strength of the four chopped-continuous-chopped glass fiber/unsaturated polyester matrix FRPs tested. The numbers in parentheses are the standard deviation of the measurements. The data indicate very low variability in the shear moduli, strength and specific gravity. In fact, the coefficient of variation (standard deviation/mean) expressed as a percentage for all the specimens tested is 1.3% for the specific gravity, 6.5% for the Iosipescu shear modulus, 7.7% for the Iosipescu shear strength, and 4.4% for the Short Beam Shear strength. The Iosipescu shear modulus value is even more remarkable when using a rule of mixtures: It can be shown that a 1.3% variation in specific gravity corresponds to about 2.5% change in fiber volume, which can be shown to effect the shear modulus by about 3.5%. This means that the coefficient of variation in Iosipescu shear modulus attributable to the test method is 3%, based on specimens tested from the same molded plaque, and to material variation 3.5%. Both of these values are excellent. Furthermore, it should be noted that this is the same magnitude of variability observed when using this test method and fixture on aerospace FRPs.

The variability of both the Iosipescu shear strength and short beam shear strength is small. Unfortunately, the contributions of variation of specific gravity on the measured shear properties cannot be quantified due to the present impreciseness of failure theories.

The in-plane shear strength of the FRPs reported in Table II is remarkable. For example, the in-plane shear strength measured using the Iosipescu shear test for a 6061-T6 aluminum was 28 ksi, and the shear strength for a unidirectional AS4/3501-6 carbon fiber/ epoxy composite was 14.0 ksi. Values for the SMC materials of this study range from 18-22 ksi. Another important observation from the data in Table II is that the Iosipescu shear modulus and strength observed for each material tested in both the 0° and 90° specimen orientations are equivalent. These observations are in accordance with basic assumptions made in the study of the mechanics of deformable bodies.

TABLE II EXPERIMENTAL OBSERVATIONS FOR
POLYESTER SMC MATERIAL

Material and Test Orientation		Specific Gravity	Iosipescu Shear		SBS Strength ksi
			Modulus Msi	Strength ksi	
5891-156-B1	0°	1.76(.01)	0.67(.08)	18.6(2.8)	5.6(.24)
	90°	1.84(.03)	0.75(.06)	20.6(2.0)	3.1(.53)*
5891-156-B2	0°	1.71(.01)	0.69(.03)	19.6(0.4)	3.9(.23)
	90°	1.70(.04)	0.67(.04)	19.8(2.2)	2.7(.37)*
5891-156-C	0°	1.86(.01)	0.81(.03)	22.7(1.6)	5.5(.17)
	90°	1.80(.03)	0.82(.11)	22.2(3.1)	3.5(.36)*
5891-156-E	0°	1.82(.01)	0.74(.02)	21.5(1.4)	5.6(.25)
	90°	1.77(.03)	0.70(.05)	20.2(1.4)	3.1(.20)*

xx(y) indicates the mean and standard deviation of the five specimens tested.
* specimens did not fail in shear.

TABLE III EXPERIMENTAL OBSERVATIONS FOR
VINYL ESTER SMC MATERIAL

Material Thickness and Matrix Formulation		Specific Gravity	Iosipescu Shear		SBS Strength ksi
			Modulus Msi	Strength ksi	
.100 mil	A	1.91(.01)	0.67(.03)	16.6(1.4)	3.7(.43)*
.125 mil	A	1.88(.08)	0.70(.02)	18.3(0.8)	3.7(.27)*
	B	1.55(.01)	0.70(.07)	16.6(1.5)	3.3(.16)*
.187 mil	A	1.92(.01)	0.71(.04)	18.3(0.8)	3.5(.31)*
	B	1.58(.02)	0.66(.04)	17.8(0.6)	3.0(.16)*
.250 mil	A	1.92(.02)	0.67(.03)	19.4(1.2)	3.5(.22)*

xx(y) indicates the mean and standard deviation of the five specimens tested.
* specimens did not fail in shear.

Table III indicates the mean specific gravity, Iosipescu in-plane shear strength and shear modulus (Iosipescu), and short beam shear interlaminar shear strength of the vinyl ester based SMC material. While the addition of hollow glass beads to the matrix significantly changes the specific gravity of the material, it does not appreciably change the shear modulus or shear strength. Additionally, the observed values in Table III for specific gravity and shear modulus indicate that the processing is highly repeatable over a range of material thicknesses.

Finally, the SBS test does not obtain true shear failures for all of the materials tested. When this occurs the standard indicates the values should be reported and noted as non-shear failures. This is not surprising since the matrix and fiber matrix interface of the material under test are subjected to the shear stresses induced by the short beam shear test SBS while the Iosipescu shear test measures in-plane shear properties, which are fiber dominated. The data between the two test methods cannot be correlated.

CONCLUSIONS

A new Iosipescu shear test fixture was designed and fabricated to test SMC materials. The goal of the fixture was to be able to test specimens without tabs and still have gauge section failures rather than loading point induced failures. The fixture design met this goal. The Iosipescu shear test and the SBS test were then used to measure the shear properties of two classes of SMC materials. The average coefficients of variation of the shear strength and moduli measured were about 5%. The average Iosipescu in-plane shear strengths for the two SMC materials tested ranged from 18 to 22 ksi. Material orientation in the test fixture as well as the effect of hollow glass spheres on shear properties were measured, and the results suggest that these conditions do not affect the shear strength and moduli.

REFERENCES

D. E. Walrath and D. F. Adams, "The Iosipescu Shear Test as Applied to Composite Materials," *Experimental Mechanics,* Vol. 23, No. 1:105-110.

N. R. Conant and E. M. Odom, "An Improved Iosipescu Shear Test Fixture," *Journal of Composites Technology & Research,* Vol. 17, No. 1:50-55.

BIOGRAPHIES

Edwin Odom is an Associate Professor of Mechanical Engineering at the University of Idaho. He received his B.S. in 1974, M.S. in 1982, and PhD degrees in 1991 from the University of Wyoming. Prior to the University of Idaho, he worked for General Electric, the University of Wyoming and the U.S. Army Material Test Directorate.

Tom Kvanvig received his M.E.M.E. degree from the University of Idaho in 1995 and served an internship with the Analytical Services and Technology Section of the Ashland Chemical Company, a Division of Ashland Inc.

Terry VanHyfte is Development Manager with the Composite Polymers Division of Ashland Chemical Company, a Division of Ashland Inc. His current responsibilities include development of existing products to fit customer applications. He received his B.S. degree from Illinois State University in 1984. Prior to joining Ashland Chemical Company, he worked for McDonnell Douglas, Rockwell International, and Owens Corning.

Thomas H. Grentzer is a Senior Staff Scientist with Ashland Chemical Company in the Analytical Services and Technology Section. His current responsibilities involve the rheological, dynamic mechanical, and thermal analysis characterization of polymers and composite materials in order to support research and development efforts. He also is the Group Coordinator for the Materials Characterization Group, which includes mechanical testing, thermal analysis, rheology, and gel permeation chromatography. Tom received his M.S. degree from the University of Akron in 1983 and his B.S. degree from Case Western Reserve University in 1979.

Shrinkage Control of Low Profile Unsaturated Resins Cured at Low Temperature

WEN LI AND L. JAMES LEE

ABSTRACT

With the development of new manufacturing processes, such as low pressure compression molding of sheet molding compound (SMC) and resin transfer molding (RTM) and SCRIMP processes, low shrinkage molding compounds which can be processed at low temperatures have attracted considerable interest from the composite industry. In this paper, an integrated kinetics-rheology-morphology-dilatometry study on a unsaturated polyester resin (UPE) mixed with different low profile additives (LPA) was carried out to investigate the shrinkage control mechanism of LPA under low temperature cure. The reaction rate was determined by a differential scanning calorimeter (d.s.c.), while a scanning electron microscope (SEM) and a Rheometrics dynamic analyzer (RDA) were employed to follow the morphological and rheological changes respectively. The volume change of the resin mixture during the curing process was measured by a dilatometer. It was found that the shrinkage behavior of the resin mixture strongly depends on the competition of the shrinkage induced by the resin polymerization and the expansion induced by microvoid formation. The results also showed LPAs with higher molecular weight works better at low temperatures, and low LPA content seems to work very well under low temperature cure. Several series of molding experiments were conducted to verify the dilatometry results.

INTRODUCTION

Low profile additives (LPAs) are thermoplastic materials that are generally served as non-reactive additives in unsaturated polyester (UPE) and vinyl ester resins. They are initially soluble or form a stable dispersion in the styrene and resin mixture before cure, but become incompatible with the cured resin during the curing process. Common LPAs include poly (vinyl acetate), poly (methyl methacrylate), thermoplastic polyurethanes and polyesters. LPA has been found highly effective in eliminating the polymerization shrinkage of UPE resins in high temperature molding processes such as compression molding of sheet molding compounds (SMC) and injection molding of bulk molding compound (BMC).

The effect of LPA type, molecular weight, and concentration on resin shrinkage, surface quality and dimensional control of molded polymer composites have been studied by many researchers (Pattison et al., 1974; Atkins, 1978, 1993; Melby and Castro, 1989). The effect of cure conditions on LPA behavior, including temperature, pressure, and thermal history, have also been investigated (Ruffier, 1993; Kinkelaar et al., 1992, 1994). Most

Wen Li and L. James Lee, Department of Chemical Engineering, The Ohio State University, Columbus, OH 43210.

studies of LPA mechanism focused on the curing at high temperatures, since LPAs found most of their applications in high temperature and high pressure processes. Although the detailed LPA mechanism is still a subject of controversy, it is now generally agreed that the most important features for LPA to work in high temperature processes are thermal expansion, phase separation and inversion between LPA and cured UPE resin, and microvoid formation along the interface or inside the LPA phase (Bartkus and Kroekel, 1970; Atkins, 1978, 1993; Bucknall et al., 1985; Suspene et al., 1991, Hsu et al., 1991). During molding, the compound is first heated to the mold temperature. The resin and LPA would thermally expand in this stage. The free-radical copolymerization of UPE molecules and styrene monomers is initiated by the decomposition of initiators. The increase in molecular weight and the change in polarity of the reacting UPE resins cause the once homogeneous system to become locally heterogeneous. The reacting UPE tends to phase out and a second phase is formed. The thermal expansion of unreactive LPA partially compensates the polymerization shrinkage. As the curing process goes on, a phase inversion would occur where the increasing amount of reacted UPE becomes the continuous phase and the LPA is forced to a discrete phase (in some cases, a co-continuous phase structure may form). The LPA rich phase contains mostly LPA and styrene with some unreacted UPE resin. As the polymerized material continues to shrink, the stress between the two phases is induced. Microvoids may form at the interface or throughout the LPA rich phase. The stress is relieved and polymerization shrinkage compensated. More microvoids may form during demolding to compensate the thermal shrinkage.

Recently, because of the growing interest of new manufacturing processes such as low pressure/low temperature SMC molding, resin transfer molding (RTM) and vacuum infusion resin transfer molding like Seemann composites resin infusion molding process (SCRIMP), low shrinkage molding compounds with the ability to be processed at low temperature and low pressure have attracted considerable interest from the composite industry. Therefore, further understanding of the low profile mechanism at low temperature and low pressure cure is important.

Since the resin thermal history in the high temperature processes such as SMC compression molding and in the low temperature processes such as SCRIMP is totally different, the behavior of LPA may vary from process to process. In SMC processing, the cure cycle includes heating the compound from room temperature to the mold temperature, followed by a strong reaction exotherm, then the cooling in the mold and during demolding. To achieve the maximum shrinkage control in SMC processing, large thermal expansion of LPA during heating and reaction and microvoids formation during cooling are essential. Therefore, a desirable LPA for high temperature applications should have a large thermal expansion coefficient and a low glass transition temperature. In contrast, SCRIMP is conducted at room temperature and there is little temperature change during curing. Obviously, thermal expansion of LPA can no longer be counted for in low temperature cure.

The objective of this study is to determine LPA performance and to provide a better understanding of low profile mechanism under low temperature cure.

EXPERIMENTAL

MATERIALS

The unsaturated polyester (UPE) resin used in this study is Q6585 from Ashland Chemical, which is a 1:1 mixture of maleic anhydride and propylene glycol with an average 10.13 vinylene groups per molecule and an average molecular weight of 1580 g mole^{-1}, containing 35% by weight styrene. The low profile additives used are three polyvinyl acetate based thermoplastic (named LPA-A, LPA-B, and LPA-C) with different molecular weight (MW=190,000, 160,000, and 90,000 g mol^{-1}, repectively) from Union Carbide.

All the samples being tested were formulated to provide a styrene (St) double bond to UPE double bond ratio of 2.0. 1.5 percent methyl ethyl ketone peroxide (MEKP) with 0.5 percent cobalt octoate were used as the initiator. The compositions and cure conditions are listed in Table I, and Fig. 1 shows the ternary phase diagrams of UPE/LPA/St, which were obtained by observing the cloudy point when various amounts of styrene were added to the mixture at a constant temperature (35°C).

INSTRUMENTATION AND PROCEDURES

A differential scanning calorimeter (d.s.c., DSC2910, TA Instruments) was used to measure reaction kinetics. The sample was sealed in a volatile aluminum sample pan which may stand 2 atm internal pressure. Isothermal runs were followed by scanning the cured sample from room temperature to 250°C to determine the residual heat with a heating rate of 10°C. The scanning run was then repeated to obtain the baseline.

For the rheological measurements, a Rheometrics Dynamic Analyzer-700 was employed in the oscillation mode to test the viscosity change during reaction. The frequency used was 1 radian/sec, and the strain was set at 10%.

Sample volume change during reaction was measured by a dilatometer developed earlier in our laboratory (Kinkelaar and Lee, 1992) at 100 psi. The sample was sealed in a PE pouch, then degassed under vacuum. A small hole was poked at the edge of the pouch, and air bubbles inside the pouch, formed under vacuum, were squeezed out. The pouch was heat sealed again, and placed inside the sample chamber of the dilatometer. The sample chamber contained both the sample pouch and the encapsulating fluid (Dow Corning 550 fluid). First the cure heating cycle was performed on the fresh sample, and then the same heating cycle was repeated on the cured sample to determine the thermal response of both the cured sample and the dilatometer. The percentage volume change due to the polymerization shrinkage was determined by subtracting the second curve from the cure curve, counting for the sample volume and also the dilatometer geometry. More details regarding the dilatometer design and operation can be found elsewhere (Kinkelaar and Lee, 1992).

The sample cured in the dilatometer, without etching by any solvent, was gold-coated for morphological measurements. The scanning electron microscopy (SEM) used was Hitachi S-510 with 25 kV power.

Samples cured in the dilatometer were also subjected to the BET internal surface area measurement (Kinkelaar and Lee, 1992). Here, a Micromeritics 2100E Accusor system was used. Krypton was used as the absorbate, and the sample was outgassed around 35°C for at least 72 hours before the surface area measurement was performed.

RESULTS AND DISCUSSION

KINETICS AND RHEOLOGY

Although the addition of LPA may significantly influence the volume change and microstructure of the UPE resin, its effect on kinetic and rheological changes appeared to be minor. As shown in Fig. 2a, the molecular weight of LPA seemed to have no influence on the conversion profile. An increase of LPA concentration slightly decreased the final conversion, as shown in Fig. 2b. Figure 2c shows the temperature effect on reaction. Final conversion increased as the temperature was increased as expected.

The effect of LPA molecular weight on gel time is little as shown in Fig. 3a. The increase of LPA concentration slightly delayed the gelation, as shown in Fig. 3b. In Fig. 3c, it is found that the increase of temperature shortened the gel time as expected.

DILATOMETRY EXPERIMENTS

EFFECT OF LPA MOLECULAR WEIGHT

Figure 4 is the plot comparing the percent volume change versus time for the UPE resin with 3.5% different LPAs cured at 35°C. For all the cases, the samples showed an initial expansion when the heating cycle started, followed by a sharp polymerization shrinkage. The shrinkage curves in these two regions showed only a slight difference, since their reaction kinetics were similar, as shown in Fig. 2a.

The sample with 3.5% highest molecular weight LPA (i.e. LPA-A) showed an expansion at 314 minutes, which continued until the end of the cure. The sample with 3.5% LPA-B behaved the similar way, except that the shrinkage control performance was slightly less than that of 3.5% LPA-A. No second expansion was observed for the sample with 3.5% LPA-C and this sample had the highest shrinkage. It is clear that the higher molecular weight LPA provided the better shrinkage control. The sample appearance was also quite different for the three samples. The two cured samples with the second expansion were stark white (opaque) in appearance and the one without second expansion was translucent. A good low profile performance corresponds to the stark white appearance.

To further investigate the relationship between shrinkage control and sample opacity, the dilatometry of sample with 3.5% LPA-A was repeated. Here, the metal lower part of the dilatometer was replaced by a transparent one made by Plexiglas. The opacity change of the sample during cure was observed and videotaped. Three stages of opacity change were found during reaction: transparent in the beginning, turning translucent during the shrinkage period, and turning opaque when the second expansion occurred. In the first stage, as shown in Fig. 5a, the transparent sample suggests that the UPE resin was compatible with LPA and styrene. Figures 5b and 5c correspond to the second stage. In this stage, the copolymerization of UPE molecule and styrene monomer was initiated. The increase in molecular weight and the change in polarity of the reacting UPE caused the separation of the resin mixture into two phases, a UPE-rich phase and a LPA-rich phase. The sample became more and more translucent. We, however, were not able to detect the phase inversion point. In the beginning of stage 3, which was around 314 minutes, the sample turned opaque. As recorded by the videotape, a white spot first emerged at 314 minutes, and in less than one minute, it covered the whole sample (Figs. 5d-f). The volume jump measured by dilatometer matched very well with such morphological change. The opacity increased as the expansion continued.

The SEM microphotographs in Fig. 6 show the final morphology of the three samples. For all three samples, a coexistence of two distinct regions: a flake-like region and a particulate region was found. According to the literature, the flake-like structure corresponds to a high concentrated UPE region. The particulate region, on the other hand, is a co-continuous structure formed by UPE and LPA. The SEM micrograph of the sample with 3.5% LPA-B, as shown in Fig. 6b, is similar to that of LPA-A (Fig. 6a), except that the relative area of the particulate region is smaller in the latter case. For the sample with LPA-C, Fig. 6c shows the particulate phase formed a dispersed region, while the UPE network formed the continuous matrix. This is probably why LPA-C has a relatively poor shrinkage control.

As mentioned earlier, microvoid formation plays an important role in LPA shrinkage control. Opacity measurement and SEM morphological measurement can not provide quantitative information about the microvoid formed inside the sample. A direct way is to use BET surface area measurement technique.

BET technique is most commonly used to measure the internal surface area of catalytic particles. It requests the pores inside the sample to be continuous, so that the adsorbate gas may penetrate into the pores. The results of the BET studies are summarized in Table II. The shrinkage results, sample appearance and the internal surface area data are well correlated. That is the sample with good shrinkage control has a high surface area measured by BET, and is always opaque. This observation confirms that the opacity increase during sample expansion is the result of microvoid formation.

EFFECT OF LPA CONCENTRATION

To investigate the effect of LPA concentration, dilatometry runs of samples with 3.5, 6, and 10% LPA-A were carried out. Figure 7 shows the shrinkage behavior of various resin mixtures cured at 35°C. The volume change followed nearly the same path for all three samples, except that expansion of the sample with 3.5% LPA-A started at 314 minutes and that of the sample with 6% LPA-A started at 460 minutes. No expansion of the sample with 10% LPA-A was observed in the experiment. Cured samples with 3.5% and 6% LPA-A were opaque, while that with 10% LPA-A was mostly translucent. After stored at room temperature overnight, the same sample, however, turned completely opaque. This phenomenon implies that the sample with 10% LPA would also expand except the time would be far longer than the experiment time.

The effects of varying LPA concentration on the resin morphology are shown in Fig. 8. At the 3.5% level, as shown in Fig. 8a at a low magnification (x100), the structure was distinctly inhomogeneous. Flake-like areas, consisting mainly cured UPE resin, alternated with particulate-like co-continuous structures formed by UPE and LPA. The particles inside the particulate region were up to 7-8 μm and loosely packed. At 6% LPA, the structure was still quite inhomogeneous. The size of the flake-like region, however, was much smaller, in comparison with that in Fig. 8a, and the two regions formed another co-continuous structure. The particle size of the particulate phase decreased to 2-3μm. Further increasing the LPA concentration to 10%, only particulate structure was found. The particles were coagulated and tightly packed.

The BET results shown in Table II correlate well with the dilatometry results (i.e. Samples 1, 4 and 5).

EFFECT OF CURE TEMPERATURE

The effect of cure temperature on shrinkage control was conducted for the resin with 3.5% LPA-A. As shown in Fig. 9, the time that the second expansion occurred was 10, 35, 314 minutes for 80, 55, 35°C respectively. Obviously, the second expansion period shifted strongly with temperature. The final shrinkage results did not vary too much: 4.6% for 35°C cure, 5.5% for 55°C cure, and 4.7% for 80°C cure. These results agreed well with the BET data listed in Table II (i.e. Samples 1, 6 and 7).

Comparing Fig. 9 with Fig. 2c, it is noted that for all three cases, the resin conversion at which the second expansion occurred is around 55-60%. Checking the other two cases, 3.5% LPA-B and 6% LPA-A cured at 35°C, the conversion at which the second expansion occurred is also within this range. This implies that a certain resin reaction must be reached before the microvoid formation could start.

Increase of cure temperature caused a marked change in sample morphology. Unlike the two-phase structure at 35°C cure (Fig. 10a), a homogeneous particulate structure was resulted for the sample cured at 55°C (Fig. 10b). Further increase the cure temperature to 80°C caused a great change in appearance (Fig. 10c). The particles were highly packed and deflated.

EFFECT OF GLASS FIBER PRESENCE

Glass fiber is one of the principle ingredients of composites materials. It is used to achieve the necessary dimensional control and mechanical properties, reduce coefficient of thermal expansion, and increase heat distortion temperature of molded parts. To study the effect of glass fiber presence on LPA behavior, dilatometry of two samples, one with 20% fiber by weight and the other 30% fiber by weight, was compared with that of neat resin mixture. The fiber used was a carbon fiber mat (plain weave). All other aspects of the formulation and sample preparation were unchanged as Sample 3, and the same cure conditions were followed.

Figure 11 shows the volume change versus time based on the neat resin mixture. The general trend is the same. The second expansion was still observed, except it occurred earlier with the presence of fiber mats. During polymerization, fibers resist cure shrinkage, thereby setting up stresses at the interface of the fiber and the polymer matrix. Such stresses may have caused an early formation of microvoids. As observed in the experiment using the transparent mold, the microvoids first grew along fibers, and then migrated into the resin phase as cure progressed. Changing of fiber content does not seem to have strong influence on volume change, since the two curves almost overlapped with each other as shown Fig. 11.

MOLDING EXPERIMENTS

Two series of room temperautre molding experiments were conducted using the same formulations listed in Table I. The first series followed the ASTM standard for linear shrinkage measurement (ASTM D2566-86). The mold dimension was 10 inches long with 0.5 inch internal radius. The resin was cured under room temperature and the change of specimen length was measured by a caliper gauge. In the second series of molding tests, a vacuum infusion technique similar to the SCRIMP process was used. Before molding, four layers of QM6408 dry glass fiber mats were laid upon a steel plate which has a chrome coated surface finish. The fiber stack was covered by a vacuum bag, and the outer edges of the bag were sealed. A resin inlet and a vacuum outlet on each end of the mold were also formed. During molding, vacuum was applied through the outlet, which forced the bag to press tightly against the fiber stack. The liquid resin was introduced to the inlet via a supply line. After mold filling and curing, the molded part was removed by peeling away the vacuum bag.

The surface quality of the molded samples was measured by a profilometer, Federal's Surfanalyzer 4000. It provides the profile of surface, roughness, and waviness. Most commonly, a roughness average (Ra) is used as a quantitative standard to compare the surface quality. It is the arithmetic average height of roughness irregularities measured from the mean line within the sample length. The results are listed in Table III.

CONCLUSION

The LPA performance was investigated under low temperature cure in the view of kinetics, rheology, morphology and volume change. In the study of volume change, it was found that a second expansion occurred for the samples with good shrinkage control. The results indicated that microvoid formation at a later stage of curing is a critical factor for shrinkage control. The mechanism of microvoid formation under isothermal conditions (i.e. no thermal expansion and shrinkage effects) is still not known and needs to be further studied. The shrinakge result, sample opacity and the internal surface area were well correlated. Two series of molding experiments were also performed to compare the compound linear shrinkage and the composite surface quality with the dilatometry results. The correlation is quite good. In general, the results showed that high molecular weight LPAs works better than the low molecular weight ones, and relatively low LPA concentration seems to be able to provide good shrinkage control. More experiments, however, need to be carried out to further verify this observation.

ACKNOWLEDGMENT

The authors would like to thank Union Carbide Corporation for financial support and LPA donation, and Ashland Chemical Company for resin donation.

REFERENCES

Atkins, K. E. 1978. "Polymer Blends" (Eds. D.R. Paul and S. Newman), chapter 23, Academic Press, NY.

Atkins, K. E. 1993."Low Profile Additives: Shrinkage Control Mechanism and Applications" in "Sheet Molding Compound Materials: Scienec and Technology" (Ed. H. Kia), Hanser Publisher.

Bartkus, E. J. and Kroekel, C. H. 1970. *J. Appl. Polym. Sci.* 15:113.

Bucknall, C. B. Davies, P. and Partridge, I. K. 1985. *Polymer*, 26:109.

Hsu, C. P., Kinkelaar, M., Hu, P. and Lee, L. J. 1991. *J. Polym. Eng. Sci.*, 31(20):1450.

Kinkelaar, M., Wang B. and Lee, L. J. 1994. *Polymer*, 35(14):3011.

Kinkelaar, M. and Lee, L. J. 1992. *J. Polym. Eng. Sci.*, 45:37.

Melby, E. G. and Castro J. M. 1989. "Comprehensive Polymer Science", vol. 7, Pergamon Press, Oxford.

Pattison, V. W., Hindersinn, R. R. and Schwartz, W. T. 1974. *J. Appl. Polym. Sci.*, 18:1027.

Ruffier, M., Merle, G. and Pascault, J. P. 1993. *Polymer Engineering and Science*, 33(8):466.

Suspene, L. Fourquier, D. and Yang, Y. S. 1991. Polymer, 32(9):1595.

BIOGRAPHIES

Wen Li

Wen Li is a Ph. D. candidate in the Department of Chemical Engineering at The Ohio State University. She received her B.S. degree in 1991 and M.S. degree in 1994 from Zhejiang University.

L. James Lee

L. James Lee is a professor in the Department of Chemical Engineering at The Ohio State University. He received his B.S. degree from National Taiwan University in 1972, and his Ph.D. degree from University of Minnesota in 1979. Before joining The Ohio State University, he was a research scientist at GenCorp Research Division from 1979 to 1982. His research interests include composite manufacturing, polymer processing, polymerization engineering, and polymer rheology.

Table I. Formulations used in this study (based on weight)

Materials/ Conditions	1 3.5% LPA-A	2 3.5% LPA-B	3 3.5% LPA-C	4 6% LPA-A	5 10% LPA-A	6 3.5% LPA-A	7 3.5% LPA-A
Q6585	63.63	63.64	63.64	61.98	59.34	63.63	63.63
LPA-A	9.46	-	-	16.22	27.03	9.46	9.46
LPA-B	-	8.75	-	-	-	-	-
LPA-C	-	-	8.75	-	-	-	-
MEKP	1.5	1.5	1.5	1.5	1.5	1.5	1.5
cobalt octoate	0.5	0.5	0.5	0.5	0.5	0.5	0.5
benzoquinone	0.03	0.03	0.03	0.03	0.03	0.03	0.03
temperature($^{\circ}$C)	35	35	35	35	35	55	80
pressure(psi)	100	100	100	100	100	100	100

Table II Summary of BET surface measurements

Sample	1	2	3	4	5	6	7
Surface area (m^2/g)	0.433	0.226	0.047	0.415	0.359	0.319	0.419
Sample appearance	opaque	opaque	translucent	opaque	opaque	opaque	opaque

Table III. Surface roughness and linear shrinkage measurements

	No LPA	3.5% LPA-C	3.5% LPA-A	6% LPA-A	10% LPA-A
Ra (mm)	1.9	0.65	0.375	0.425	0.6
$\Delta L/L$(%)	2.395	1.826	~0	—	1.00

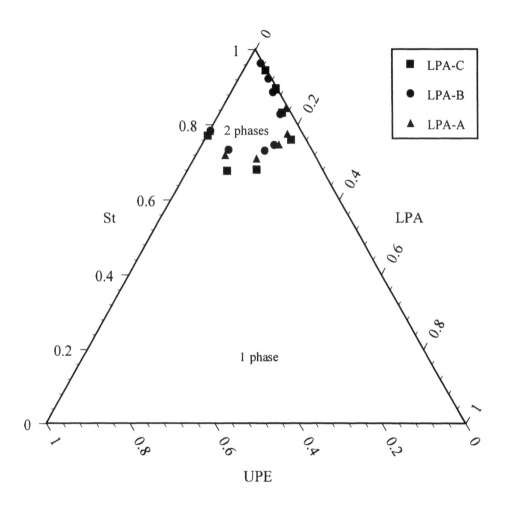

Figure 1. Phase diagram of styrene-UPE-
LPA ternary systems.

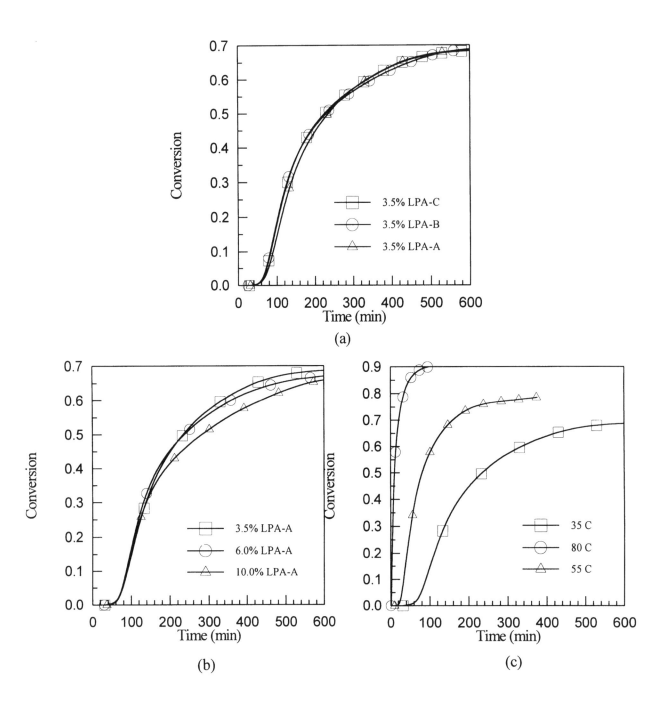

Figure 2. Conversion vs. time data obtained from d.s.c. measurement: (a) samples with different LPAs cured at 35°C, (b) samples with various LPA concentrations cured at 35°C, (c) temperature effect on the sample with 3.5% LPA-A.

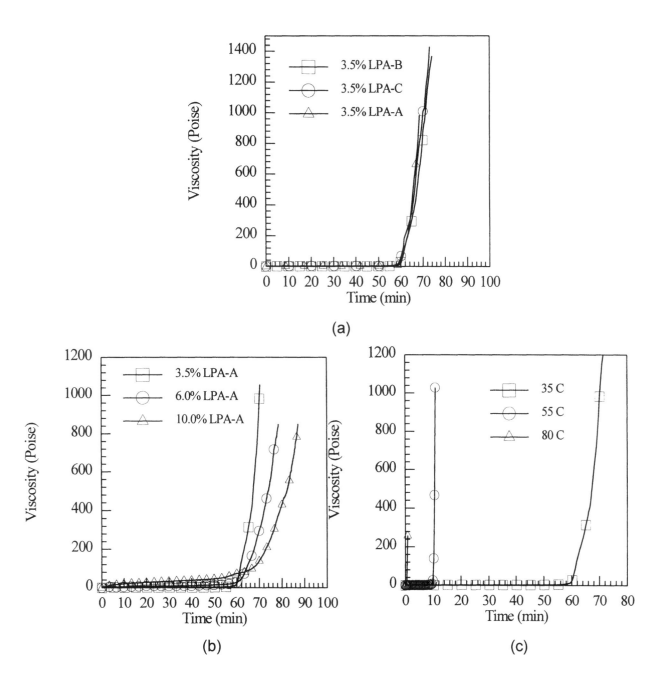

Figure 3. Viscosity vs. time data obtained from RDA measurement: (a) samples with different LPAs cured at 35°C, (b) samples with various LPA concentrations cured at 35°C, (c) temperature effect on the sample with 3.5% LPA-A.

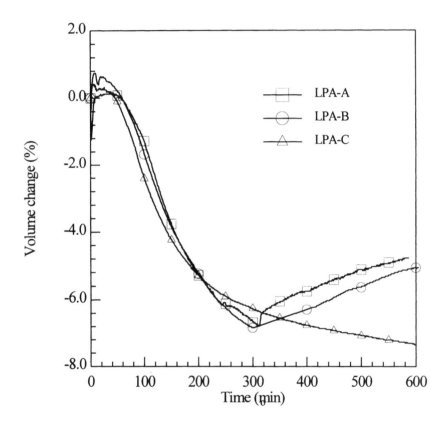

Figure 4. Comparison of volume change versus time of
samples with 3.5% different LPAs cured at 35°C

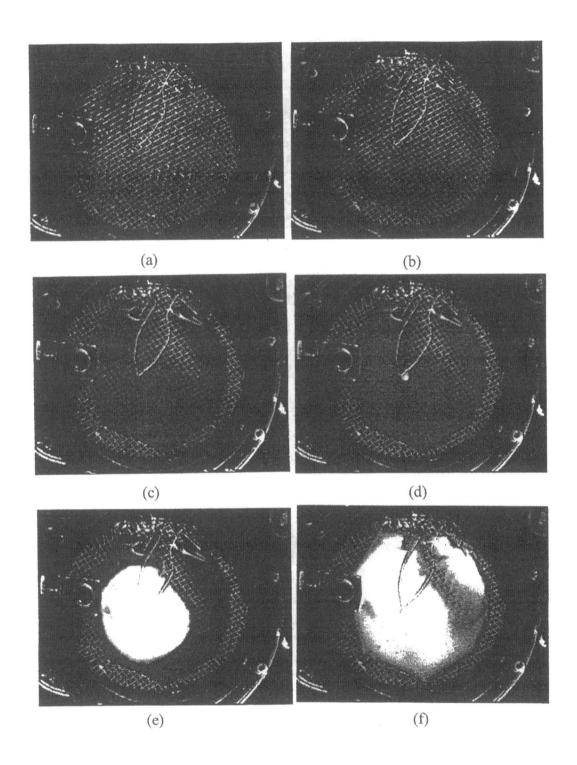

Figure 5. Change in opacity during cure of 3.5% LPA-A sample at 35°C: (a) 0 min, (b) 51.67 min, (c) 272.1 min, (d) 314 min, (e) 314.15 min, (f) 314.33 min.

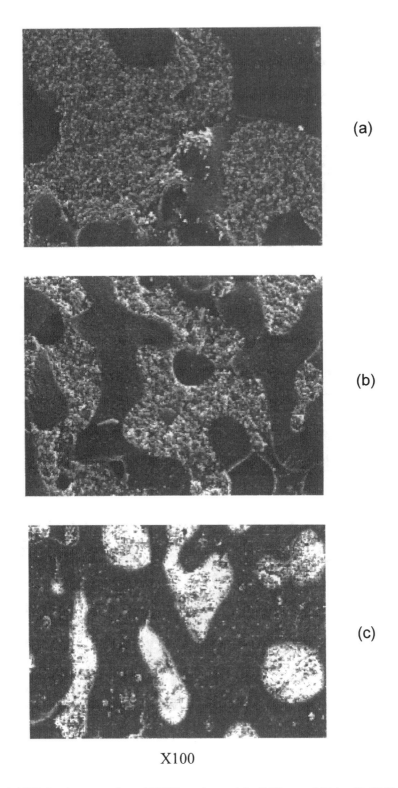

X100

Figure 6 SEM micrographs of UPE resins with different LPAs (3.5%) cured at 35ºC: (a) LPA-A, (b) LPA-B, (c) LPA-C

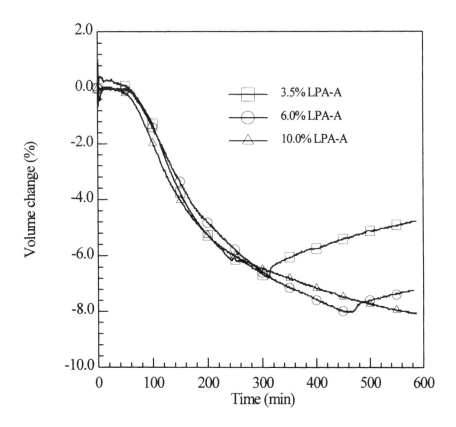

Figure 7. Comparison of volume change versus time of
samples with various LPA-A concentrations cured at 35°C.

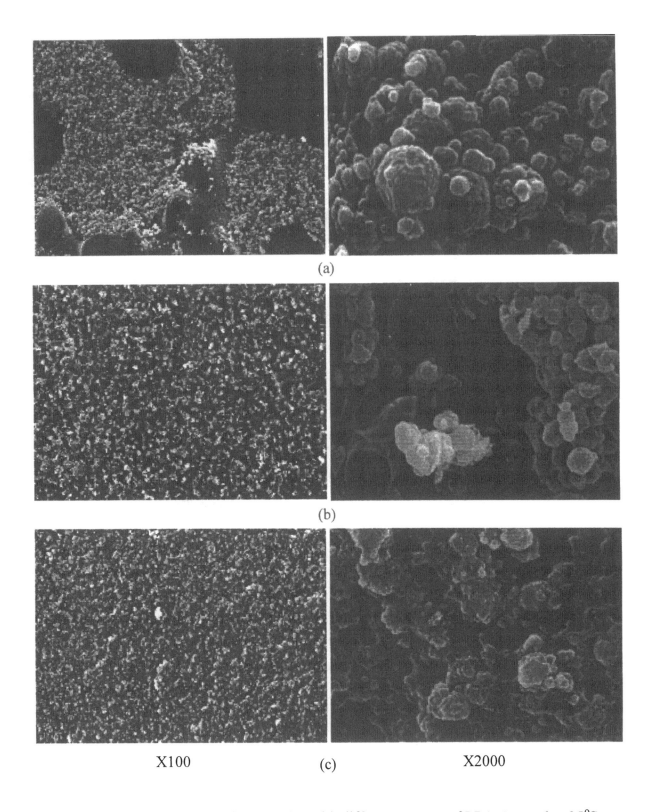

X100   (c)   X2000

Figure 8 SEM micrographs of UPE resins with different content of LPA-A cured at 35°C:
(a) 3.5%, (b) 6.0%, (c) 10.0%.

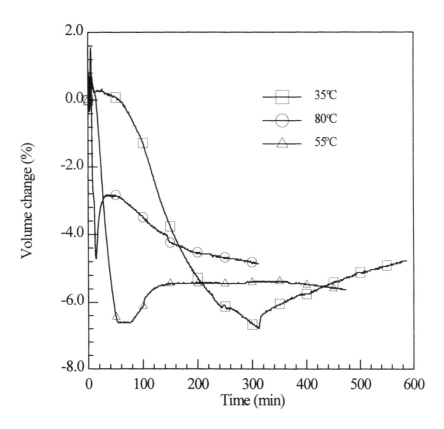

Figure 9. Effect of cure temperature on shrinkage behavior of

the sample with 3.5% LPA-A.

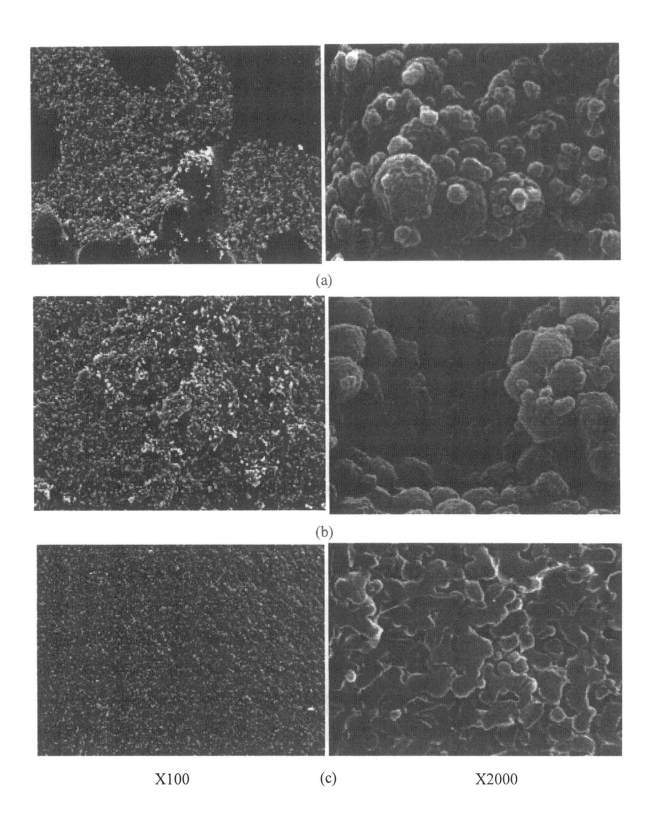

X100   (c)   X2000

Figure 10 SEM micrographs of UPE resins with 3.5% LPA-A cured at different temperatures:
(a) 35°C, (b) 55°C, (c) 80°C

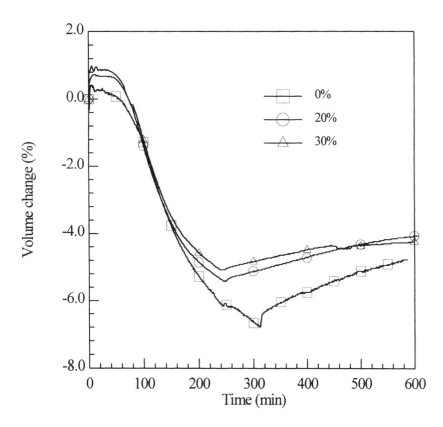

Figure 11. Effect of the presence of glass fiber on shrinkage

behavior of the sample with 3.5% LPA-A.

An Investigation of Post-Fill Curing Behavior of Compression Molded SMC Parts

H. U. AKAY, O. SELCUK, O. GURDOGAN, M. REVELLINO AND L. SAGGESE

ABSTRACT

Curing analysis of a compression molded SMC part, performed using Technalysis' compression molding simulation program PASSAGE®/Compression, is presented in this paper. A truck panel is modeled with two different SMC formulations. The curing behavior of the two formulations are compared.

INTRODUCTION

Fiber orientation and curing characteristics of compression molded SMC parts are of interest for determination of warpage and structural integrity of such parts. Studies pertaining to mold filling and fiber orientation analysis of compression molded parts using a computer program developed by Technalysis were presented elsewhere, e.g., (Technalysis, 1996; Reifchneider and Akay, 1994; Reifschneider, et al., 1995).

In this paper, we summarize our recent studies conducted for curing simulation of SMC parts. Since the thermoset SMC undergoes chemical reactions when subjected to heat during compression molding processes, a complete curing analysis of such materials requires the solution of flow and heat transfer equations coupled with the curing kinetics of the material during the molding process. The material gains rigidity as the chemical reaction takes place with temperature. Hence, the time needed for the molded material to reach the rigid state is of importance in terms of deciding when to remove the part from the mold. In the present work, coupled fluid flow and heat transfer equations are solved during the filling stage, while coupled heat transfer and curing kinetics equations are solved until the material reaches its gel point following the filling stage.

GOVERNING EQUATIONS

Simulation of compression molded parts can be performed in two stages which can be described as filling and post-filling stages. For thermoset materials such as SMC, compression flow equation is solved coupled with the effects of heat transfer and curing kinetics. For most processes, however, the filling stage is much faster then the post-filling stage hence the curing kinetics may be neglected during filling. The effects of energy and curing kinetic equations are dominant during the post-filling stage. Flow and energy

---

H.U. Akay, O. Selcuk and O. Gurdogan, *Technalysis Inc., Indianapolis, IN - USA*
M. Revellino and L. Saggese, *Iveco, SpA, Torino, ITALY*

equations are solved during the filling stage can be found elsewhere, e.g., (Barone and Caulk, 1986; Tucker, 1987; Reifschneider and Akay, 1994).

POST-FILLING STAGE

Heat transfer and curing kinetics effects are coupled during the post-filling stage as follows:

$$\rho C_p \frac{\partial T}{\partial t} - \frac{\partial}{\partial z}\left(k\frac{\partial T}{\partial z}\right) = \rho Q_{tot} \frac{\partial c}{\partial t}$$

(1)

where $c = Q/Q_{tot}$ is the degree of cure $(0 \le c \le 1)$, Q is the amount of heat released until time t, and Q_{tot} is the total heat evolved during curing of the thermoset. Through the thickness variations of the degree of cure is calculated by using same number of layers of finite elements as in the energy equation.

For the SMC materials investigated here, the kinetic model in the form

$$\frac{dc}{dt} = d_1 c^m (1-c)^n$$

(2)

was used, which is a subset of a kinetic model proposed by Kamal and Ryan, 1987. In the above equation, c is the degree of cure, $d_1 = ae^{-b/RT}$, R is the universal gas constant, T is the absolute temperature, and m and n are constant. The constant coefficients a, b, m and n as well as Q_{tot} are obtained to fit Eq. (2) from a series of DSC measurements performed by DSM Research in Netherlands (Riedel and Kroezen, 1996).

The finite element method is used for the discretization of the above equations. The resulting equations are solved with a volume-of-fill method for filling and backward-Euler time-integration scheme for energy and curing kinetic equations. Automatic time-stepping is used for time-dependent equations. The final temperature distribution at the end of filling is used as an initial condition for the post-filling computations.

NUMERICAL RESULTS AND DISCUSSION

This investigation was conducted jointly by Iveco and Technalysis for modeling of Iveco's truck panels by using Technalysis' compression molding program. Two different SMC formulations labeled, SMC 933 and SMC 950 containing 33% and 27% fibers, 20% and 15% unsaturated polyester resin, 8.8% and 10.7% thermoplastic resin contents, respectively, were considered for comparisons. The measured properties of these SMC's are listed in Table 1.

Figure 1 shows the initial charge locations and number of SMC layers used, each having a thickness of 2.8 mm. Since the part is symmetric, only the half of it is modeled. Shown in Figure 2 is the finite element mesh of the part which consists of 2,180 nodal points and 4,192 triangular elements. The location of nodes 77, 762, 1811 are marked on the part which will be used to monitor the variation of temperature and curing. Indicated on the same figure are three regions, R1, R2 and R3, corresponding to 4.5 mm-, 2.0 mm- and 3.0 mm- thick regions of the part. For a press speed of 8.0 mm/s the calculated temperature distribution at the end of filling is shown in Figure 3 for SMC 950. The mold and initital charge temperatures were set at 430 K and 300 K, respectively. For the curing simulations, the temperature distribution obtained at the end of filling was used as the initial temperature in each case.

TABLE 1. MEASURED CURING PROPERTIES OF TWO SMC's.

Property	SMC 933	SMC 950
Q_{tot} (J/kg)	91,200	100,000
R ($J/K \cdot mol$)	8.31441	8.31441
a ($1/s$)	3.325×10^{14}	1.368×10^{24}
b (kJ/mol)	121	193.0
m	0.991	1.978
n	4.94	4.14

Figures 4 and 5 represent the temperature and degree of cure histories at different layers through the thickness of selected nodes, respectively. For this analysis, 5 layers of finite elements were employed to model the temperature and degree-of-cure calculations through the half thickness of the part. Layer 1 corresponds to mold wall and layer 5 corresponds to the center line of the part.

As it may be observed in Figure 4, the peak temperatures of SMC 950 are higher than SMC 933. For example, at node 77, the peak temperatures are 525 K and 460 K for SMC 950 and SMC 933, respectively. This indicates that SMC 950 has a stronger exothermic reaction. Also, the curing histories shown in Figure 5 support this observation. For both materials at the peak region, the temperature of the layers which are closer to the mold wall remain cooler. Thus, the degree of cure for these layers are lower than the center of the part thickness.

Finally, Figure 6 shows bulk degree of cure magnitudes of 0.9 and 0.7 at $t = 85$ seconds after filling for SMC 950 and SMC 933, respectively. This figure supports the decision made for selection of SMC 950 due to its superior curing characteristics for the production of this front panel. The methodology presented here can be used for determination of curing characteristics and the optimum time for ejection of the part.

REFERENCES

Barone, M.R. and D.A. Caulk, 1986, "A Model for the Flow of a Chopped Fiber reinforced Polymer Compound in Compression Molding," *J. Applied Mechanics*, 53: 361-371.

Kamal, M.R. and M.E. Ryan, 1987, "Thermoset Injection Molding," *Injection and Compression Molding Fundamentals*, Edited by A.I. Isayev, Marcel Dekker, New York, NY, 329-376.

Reifschneider, L.G. and H.U. Akay, 1994, "Applications of a Fiber Orientation Prediction Algorithm for Compression Molded Parts with Multiple Charges," *Polymer Composites*, 15 (4): 261-269.

Reifschneider, L.G., H.U. Akay, G. Molina and A. Ajmar, 1995, "An Experimental Verification of Fiber Orientation Predictions for Compression Molded SMC Parts," *Proceedings of ANTEC '95*, Boston, MA, 1905-1909.

Riedel, R. and T. Kroezen, 1996, "Trials Determination Barone-Caulk Model Experiments and Kamal Kinetic Experiments," *DSM Research Report Submitted to Technalysis*, Geleen, The Netherlands.

Technalysis Inc., 1996, "PASSAGE®Compression - A Finite Element Program for Compression Molding with Fiber orientation and Post-Fill Curing - Version 3," *A User's Manual, Technalysis, Inc.*, Indianapolis, IN.

Tucker, C.L., 1987, "Compression Molding of Polymers and Composites," *Injection and Compression Molding Fundamentals*, Edited by A.I. Isayev, Marcel Dekker, New York, NY, 481-565.

BIOGRAPHIES

Hasan U. Akay

Dr. Akay is the Vice President and Technical Director of Technalysis. He is also Professor of Mechanical Engineering at Purdue School of Engineering and Technology, IUPUI, Indianapolis, IN. He received his Ph.D. from the University of Texas at Austin in 1974. His research interests include the applications of the finite element method in fluid and solid mechanics areas.

Oguzhan Gurdogan

Dr. Gurdogan is the Director of Engineering Software Development Group at Technalysis. He received his Ph.D. from Ohio State University in 1984. He joined Technalysis in 1986 where he directs the software development activities of a number of software packages for fluid flow and heat transfer problems, including compression molding and metal casting applications.

Ozan Selcuk

Ozan Selcuk is a member of Engineering and Software Development Group at Technalysis. He received his Master's degree in Mechanical Engineering from Purdue University at IUPUI in 1994. He has been with Technalysis since 1994 where he is involved with development and applications of a number of software packages, including compression molding and metal casting applications.

Mario Revellino

Dr. Revellino is the Manager of the IVECO Plastics Department. He received his Master's degree in Chemistry from Torino University. He has been with IVECO since 1983 where he is involved in testing and new technology development.

Luigi Saggese

Luigi Saggese is a member of the IVECO Plastics Department. He received his Master's degree in Chemical Engineering from Napoli University. He joined IVECO in 1987 where he is involved with structural and flow analysis of compression molded SMC parts.

Figure 1. Initial charge locations and number of layers (each layer 2.8 mm thick).

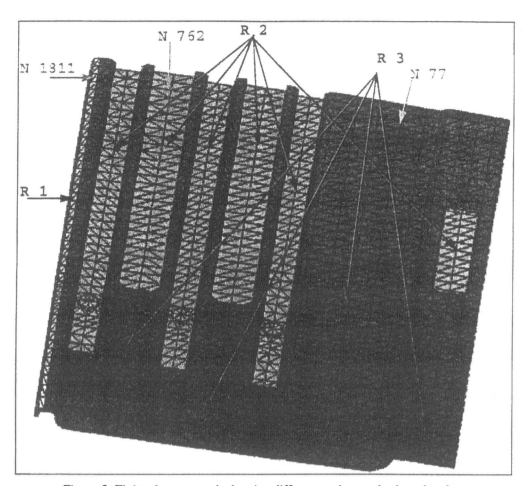

Figure 2. Finite element mesh showing different regions and selected nodes.

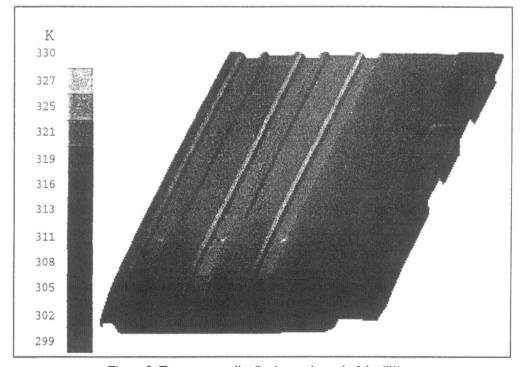

Figure 3. Temperature distribution at the end of the filling.

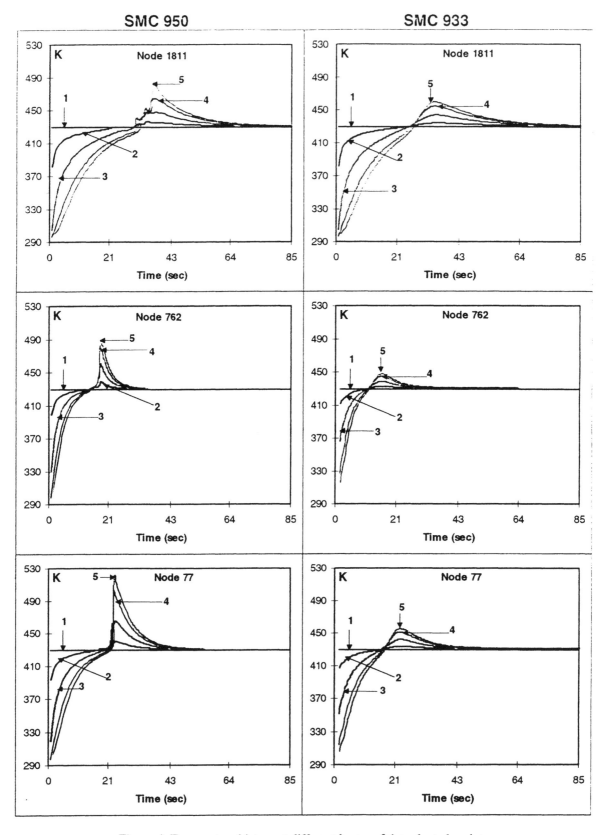

Figure 4. Temperature history at different layers of the selected nodes.

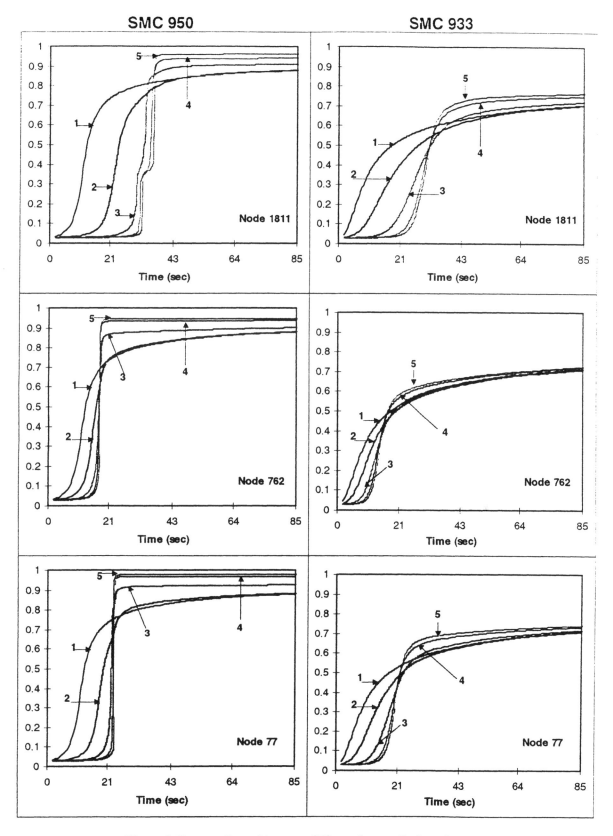

Figure 5. Degree of cure history at different layers of selected nodes.

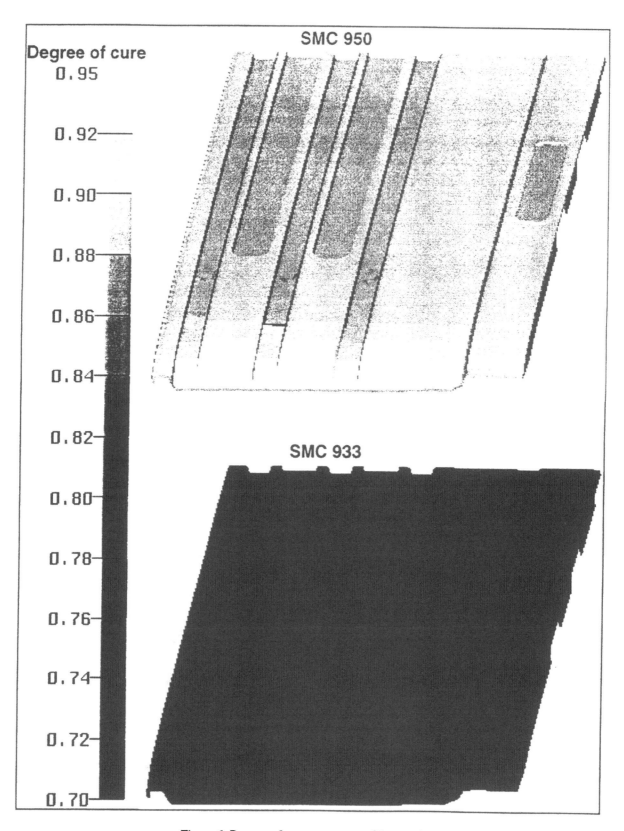

Figure 6. Degree of cure contours at 85 seconds.

SESSION 12

Resins, Coatings, Fibers and Foams—Partners in Composites Formulation

Viscosity Increase of High Molecular Weight Unsaturated Polyester Resin through Use of Filler and/or Thickener

EIICHIRO TAKIYAMA, YOSHITAKA HATANO AND FUMIO MATSUI

I. summary

A new method has been developed by which number average molecular weight (Mn) of unsaturated polyester resin (UP) can be increased to a level (5,000-15,000) far above the current level (1,500-2,500).

The end groups of the obtained high molecular weight unsaturated polyester resin (HUP) are mostly hydroxyl groups. By substituting these hydroxyl groups with carboxyl groups in whole or in part through reaction with an acid anhydride, HUP was obtained.

Influence of the types of end groups on HUP viscosity has been studied by investigating viscosity changes of HUP mixed with a filler, a thickener, and a filler and a thickener by changing the types of end groups. The thickeners used were metal oxide and hydroxide, the fillers, silica powder, alumina trihydrate and calcium carbonate.

The viscosity of all of the mixtures with fillers increased with an increase in the number of end groups substituted with carboxyl groups, while the tensile strength of cured resin exhibited a tendency to decrease.

The viscosity of a mixture with a thickener was affected by the HUP molecular weight, the kind and ratio of end groups, and the kind of thickener. The influence of water content on the viscosity was also substantial.

II. Introduction

The molecular weight (Mn) of UP ranges roughly between 1,500 and 2,500, which it can be said is in the range of an oligomer.

In general, UP contains an equal ratio of carboxyl groups and hydroxyl groups as end groups, which resulted from the synthesis of such UP carried out through an esterification reaction.

When UP is used in preparation of SMC (sheet molding compound) or BMC (bulk molding compound) the end groups of such UP are left as they are and no substitution to change those end groups to carboxyl groups or to hydroxyl groups is usually done.

Eiichiro Takiyama, Yoshitaka Hatano, Fumio Matsui

SHOWA HIGHPOLYMER Co., Ltd., Kanda Chuo Bidg. Kanda Nishiki-Cho 3-Chome, Chiyoda-ku, Tokyo, Japan

It is presumed that such polymer came to possess a large number of end groups affected by a large number of molecules. If the end groups are substituted in whole with the same groups, such substitution will significantly affect the resin properties.

A new technology has been developed by SHOWA HIGHPOLYMER Co.,Ltd. to synthesize a high molecular weight unsaturated polyester (HUP) with Mn being as high as 5,000-15,000. The technology utilize a deglycol reaction and increases the molecular weight, while conversely decreasing the number of end groups.

The end groups of HUP obtained through the deglycol reaction are mostly hydroxyl groups. Such hydroxyl end groups can be substituted with carboxyl groups in a reaction with an acid anhydride, and furthermore, the number of end groups can be freely controlled as needed.

The influence of the number and the kind of end groups, the ratio of end groups (carboxyl and hydroxyl) and the use of filler, thickener, and filler/thickener on the resin viscosity has been investigated. As a result, useful information has been developed for molding of HUP.

III. Raw Materials

The following commercially available materials were used as raw materials without any treatment.

Neopenthyl glycol, Isophthalic acid, Fumaric acid

Styrene, hydroquinone, tetraisopropyl titanate

Magnesia, Magnesium hydroxide, Silica powder, Alumina trihydrate, calcium carbonate

IV. Sample Preparation and Measuring Procedure

IV-1. Synthesis of HUP

1,091g of neopenthyl glycol, 829g of isophthalic acid, and 2.5g of tetraisopropyl titanate were charged into a 3,000cc separable flask equipped with an agitator, a partial condenser, a gas induction tube and a thermometer, followed by an esterification reaction under nitrogen gas flow at 190-195°C until the reactant acid value reached 27, then 580g of fumaric acid was added at 150°C to the flask, and the esterification reaction was continued at 195-200°C until the reactant acid value reached 31. Then a deglycol reaction was carried out under 5 torr until the molecular weight of HUP (Mn) reached 6,100. At this point a part of the reactant was taken out as sample. The reaction was continued until the molecular weight (Mn) of the reactant (HUP) reached 12,000 at which point the acid value was zero.

Changes of the molecular weight (Mn,Mw) of HUP during the deglycol reaction at given hours were as given in Fig.2.

IV-2. Molecular Weight Measurement

Equipment: Shodex GPC(Showa Denko K.K.)

Eluent: C_4H_8O(THF:tetrahydrofuran)

Sample column: KF-805Lx2

Column temperature: 40°C

Flow rate: 1.0ml/min.
Detector: Shodex RI-71
Control sample: Polystyrene(Showa Denko Shodex Standard)
Calibration curve: See Fig.1

IV-3. Viscosity Measurement
3-1. Sample with Filler
A commercially available helical stand type viscometer for high viscosity measurement (Toki Sangyo K.K.) was used.

A sample was first charged into a glass bottle with a screw stopper and the bottle was kept in a thermostat at 40°C for the number of days as needed for the tests and viscosity measurements were made at 40°C.
3-2. Sample with Thickener
The above-mentioned viscometer with a different rotor was used for measurement of the viscosity of the samples with fillers.

The highest viscosity measurable was one million poise and the measurement was made at 40°C. The samples were kept in the thermostat in the manner described earlier.
3-3. Samples with Thickener and Filler
Viscosity measurement of the samples with a thickener a filler was made in the same manner as described above.

IV-4. Control of Water Content
The water content of HUP containing a 50wt.% styrene solution was 300ppm, which was reduced to 50ppm with the help of a molecular sieve equal to 20wt.% of the HUP. Samples containing 0.2wt.% of water were prepared by adding an amount of equivalent to 0.2% of the HUP.

Commercially available thickeners were used without any treatment.

Calcium carbonate containing 0.08wt% of water was dried at 120°C for 12 hours to 0.01wt.%.

Water content was measured by the Karl Fischer method.

IV-5. Treatment of End Group
The following is an example of carboxylation of all of the hydroxyl end groups of HUP (Mn:12,000) with trimellitic anhydride.

500g of the HUP was charged into a 1,000cc flask equipped with an agitator, a reflux condenser, a thermometer and a dripping funnel. The HUP was first melted at 200°C, then 16g of trimellitic anhydride was added to the melted HUP, and the reactant was agitated for 30 minutes.

The molecular weight (Mn) distribution of the HUP at the end of the esterification, before and after treatment, is given in Fig.4. Even after all of the hydroxyl groups of the HUP had been substituted with carboxyl groups, no change was observed in the molecular weight (Mn) distribution of the HUP. As a result, it was concluded that the substitution reaction had been completed because no trace of trimellitic acid was detected on the low molecular weight side. After the reaction temperature was decreased to 150°C, the amount of styrene necessary to prepare a 50wt.% resin solution was added and investigation was made as described in V.

V. Investigation

V-1. Sample

The following samples were prepared and evaluated.

V-1-1. Sample with filler

1-1-1. HUP

Two types of HUP synthesized as described in IV-4 were used, as follows:

  (i)Mn= 6,100   Water content=0wt.%

  (ii)Mn=12,000   Water content=0wt.%

1-1-2. End Group Type

The end groups were treated trimellitic anhydride (TA) and succinic anhydride (SA) as shown below:

  (i) Hydroxyl 100 mol%   Carboxyl   0 mol%

  (ii) Ditto   50 wt.%   Ditto   50wt.% treated w/TA

  (iii) Ditto   50 wt.%   Ditto   50wt.% treated w/SA

  (iv) Ditto   0 wt.%   Ditto   100wt.% treated w/TA

  (v) Ditto   0 wt.%   Ditto   100wt.% treated w/SA

1-1-3. Filler

The following chemicals were used as filler.

  Silica powder, Alumina trihydrate, Calcium carbonate

In total 30 types of samples were prepared as indicated in Table 1 and Table 2.

V-1-2. Sample with Thickener

1-2-1. HUP

HUP samples were synthesized based on the procedure described in IV-1 with molecular weight of 5,000 and 10,000. The water content was controlled at 0.0 and 0.2wt.% as described in IV-4. They were as follows:

  (i)Mn= 5,000   Water content=0.0wt.%

  (ii)Mn= 5,000   Water content=0.2wt.%

  (iii) Mn=10,000   Water content=0.0wt.%

  (iv) Mn=10,000   Water content=0.2wt.%

1-2-2. End Group Type

The same as the fine described in V-1-1-2.

1-2-3. Thickener

The following chemicals representing 1wt.% of HUP were used as thickener.

  Magnesia, Magnesium hydroxide, Zinc oxide, Calcium hydroxide

In total 80 types of samples were prepared.

V-1-3. Sample with Thickener and Filler

1-3-1. HUP

The same types of HUP described in V-1-2-1 were used with their Mn being 5,000 and water content,0.0 and 0.2wt.%.

1-3-2. End Group Type

All of the end groups of the above-mentioned types of HUP were substituted with carboxyl groups of trimellitic anhydride or succinic anhydride, and those not substituted remained to be of hydroxyl groups.

1-3-3. Thickener and Filler

A combination of magnesia and calcium carbonate widely used in preparation of SMC's and BMC's was selected and six samples were prepared in all.

The samples prepared for the investigation came to a grand total of 126.

V-2. Effect of Filler on Viscosity

The samples were kept for 28 days in a thermostat and their viscosity changes were measured. The results were given in Table 1 and 2. For further understanding, the changes are represented graphically in Fig.5-10. In these and ensuing figures the term TA means the samples were treated with trimellitic anhydride, and the term SA, with succinic anhydride.

V-2-1. Effect of Silica Powder

220 parts of silica powder were used per 100 parts of HUP.

(i) HUP Mn=6,100 (Fig.5)

A sample of HUP whose end groups were all carboxylated with trimellitic anhydride registered the highest viscosity increase in this group.

(ii) HUP Mn=12,000 (Fig.6)

A sample of HUP treated as in (i) above registered the highest viscosity increase in this group.

V-2-2. Effect of Alumina Trihydrate

250 parts of alumina trihydrate were used per 100 parts of HUP.

(i) HUP Mn=6,100 (Fig.7)

A sample of HUP whose end groups were all carboxylated with trimellitic anhydride and a sample HUP whose end groups were all carboxylated with succinic anhydride showed large increases. The viscosity increase of the sample treated with succinic anhydride was especially high. On the contrary, untreated samples whose end groups were 50% carboxylated with trimellitic anhydride registered viscosity decrease.

(ii) HUP Mn=12,000 (Fig.8)

A sample of HUP whose end groups were all carboxylated with succinic anhydride registered a great increase in viscosity, while other samples registered only slight changes in viscosity.

V-2-3. Effect of Calcium Carbonate

250 parts of Calcium carbonate were used per 100 parts of HUP.

(i) HUP Mn=6,100 (Fig.9)

A sample of HUP whose end groups were all carboxylated with trimellitic anhydride registered the highest viscosity increase , followed by a sample treated with trimellitic anhydride at the rate of 50%.

In general, treated samples registered great and steeper increases whthin a short period of time but subsequent changes were not so steep and rapid.

(i) HUP Mn=12,000 (Fig.10)

Unlike the case in (i) above, only samples treated with succinic anhydride showed great viscosity changes, however, samples treated with trimellitic anhydride did not show such changes.

V-3. Effect of Filler on Mechanical Properties

Viscosity changes were greatly affected by the molecular weight, the type of end groups and the kind of filler. Relationships between viscosity changes and mechanical properties are a very interesting subject to study.

V-3-1. Flexural Strength

Samples (Mn=6,100) treated with calcium carbonate registered greater increases in viscosity within a short period of time than untreated ones. Measurements taken at 40°C during 28 days were as given in Table 3. The Table also shows the flexural strength and flexural modulus of the samples on the 28 day.

A sample whose end groups were all carboxylated with trimellitic anhydride registered >6kg/mm^2, however, this is thought to have resulted from residual foams caused by the rapid viscosity increase.

The flexural strength improved with an increase of carboxyl end groups.

V-3-2. Tensile Strength, Elongation and Tensile Modulus

The effects of the kind of filler, the type of treatment and HUP molecular weight on mechanical properties were as indicated in Fig.11-16.

(i) Tensile Strength (Fig.11 and 12)

A tendency for tensile strength to decrease with an increase in carboxyl groups was observed. The effect of the molecular weight and the types of end groups on the tensile strength of the samples with silica powder, alumina trihydrate and calcium carbonate was as indicated in Fig.17, 18 and 19.

(ii) Elongation (Fig.13 and 14)

No clear tendency in elongation was observed, presumably because of great variations in the effect of fillers.

(iii) Tensile Modulus (Fig.15 and 16)

No noticeable variations in tensile modulus between samples were observed.

V-4. Effect of Thickener on Viscosity

Samples were prepared as described in V-1-2 and kept at 40°C for 20 days. One part of a filler was used per 100 parts of HUP. Viscosity changes during the period were measured and the results were as given in Fig.20-28.

V-4-1. Magnesium Oxide

(i) HUP Mn=5,000 (Fig.20)

A sample of HUP with its end groups all carboxylated with trimellitic anhydride and whose water content was 0.2wt.% and an untreated sample with water content of 0.2 wt.% showed some viscosity increase.

(ii) HUP Mn=10,000 (Fig.21)

only an untreated sample of HUP with water content of 0.2wt.% showed an increase in viscosity, however, it was a small increase about a quarter of those of (i) above. The type of polyester that registered an increase in viscosity both in (i) and (ii) above had end groups that were all hydroxyl groups, however, the degree of the increase was small.

V-4-2. Magnesium Hydroxide

(i) HUP Mn=5,000 (Fig.22)

Only a sample of HUP with 50% of its end groups carboxylated with succinic anhydride and whose water content was 0.2wt.% showed a considerably viscosity increase.

(ii) HUP Mn=10,000 (Fig.23)

As in (i) above, only a sample of HUP with 50% of its end groups carboxylated with succinic anhydride and whose water content was 0.2wt.% registered a remarkably high viscosity increase in spite of the decrease in the number of end groups.

When magnesium hydroxide is used as thickener, carboxylation of 50wt.% is believed to be effective in increasing viscosity.

V-4-3. Zinc Oxide

(i) HUP Mn=5,000 (Fig.24)

A sample of HUP with its end groups all carboxylated with trimellitic anhydride and whose water content was 0.0wt.% and a sample with 50% of its end groups carboxylated with succinic anhydride and whose water content was 0.2 wt.% registered slight increase in viscosity.

(ii) HUP Mn=10,000 (Fig.25)

Samples of HUP with all of their end groups carboxylated with succinic anhydride and whose water content was 0 and 0.2wt.% , and a sample with 50% of its end groups carboxylated with succinic anhydride and whose water content was 0.2 wt.% registered a small increase in viscosity.

V-4-4. Calcium Hydroxide

(i) HUP Mn=5,000 (Fig.26)

Unlike the earlier cases, an unsaturated sample with water content of 0.0wt.% registered a remarkable viscosity increase.

(ii) HuP Mn=10,000 (Fig.27)

An unsaturated sample containing 0.0wt.% of water showed a very slight increase in viscosity.

There is a significant difference between (i) and (ii) above.

Among the samples with thickeners, the following three samples were picked as they registered remarkable changes (increases) in viscosity as indicated in Fig.28.

(i) HUP

    Mn   10,000
    End groups   50% carboxylated with succinic anhydride
    Water content   0.2wt.%
    Thickener   Mg(OH)$_2$

(ii) HUP

    Mn   5,000
    End groups   50% carboxylated with succinic anhydride
    Water content   0.2wt.%
    Thickener   Mg(OH)$_2$

(iii) HUP

    Mn   5,000
    End groups   untreated
    Water content   0wt.%
    Thickener   Ca(OH)$_2$

V-5. Effect of Thickener and Filler on Viscosity

Samples were prepared as described in V-1-3. Viscosity changes of the samples with magnesium oxide and calcium carbonate were investigated. These systems are most commonly used in preparation of SMC's and BMC's from conventional UP. The Viscosity changes at 40°C during 20 days were measured and the results were as given in Fig.29.

All of the samples except an unsaturated one containing no water showed noticeable viscosity increases. Especially those treated with trimellitic or succinic anhydrides registered remarkable increases in viscosity when water content was 0wt.%.

When water content was 0wt.%, the effect of treatment with trimellitic anhydride on viscosity increase was greater than that of succinic anhydride treatment.

It is worthy of note that even an untreated sample (all hydroxyl end groups) containing 0.2wt.% water exhibited an increase in viscosity. It is in general safe to say that water content greatly influences viscosity increase.

VI. Conclusion

(1) Water content

Regardless of the kind of thickener, the viscosity of HUP is greatly affected by water content, suggesting the necessity for a highly precise control of water content.

(2) Thickener

The following three resin systems registered remarkable increase in viscosity during 20 days.(Fig.28)

   (i)   HUP (Mn,5,000; end group, 50% carboxyl groups of succinic anhydride; and water content,0.2wt.%)

      Thickener: Magnesium hydroxide

      Viscosity: 146 poise (after 20 days at 40°C)

  (ii) HUP (Mn,10,000; end group, 50% carboxyl groups of succinic anhydride; and water content,0.2wt.%)

      Thickener: Magnesium hydroxide

      Viscosity: 992 poise (after 20 days at 40°C)

 (iii) HUP (Mn,5,000; end group, 100% hydroxyl groups of; and water content, 0.0wt.%)

      Thickener: Magnesium hydroxide

      Viscosity: 456 poise (after 20 days at 40°C)

(3) The effect of combined use of a thickener and a filler was investigated with magnesium oxide used as a thickener and calcium carbonate, as filler.

Samples of HUP (Mn=5,000) treated with trimellitic anhydride registered viscosity increases higher than one million poise regardless of water content. The effect of water content on viscosity was outstanding when HUP was treated with succinic anhydride. A sample with water content of 0.2wt.% registered a viscosity increase beyond one million poise, however, a sample containing no water remained at 15,000 poisse.

Although HUP has a much smaller number of end groups than does conventional UP, it is highly possible to establish an effective system to increase viscosity for HUP through optimum selection of end group treatment, molecular weight and precise control of water content. There is a similar system commonly used for the purpose of increasing conventional UP's viscosity.

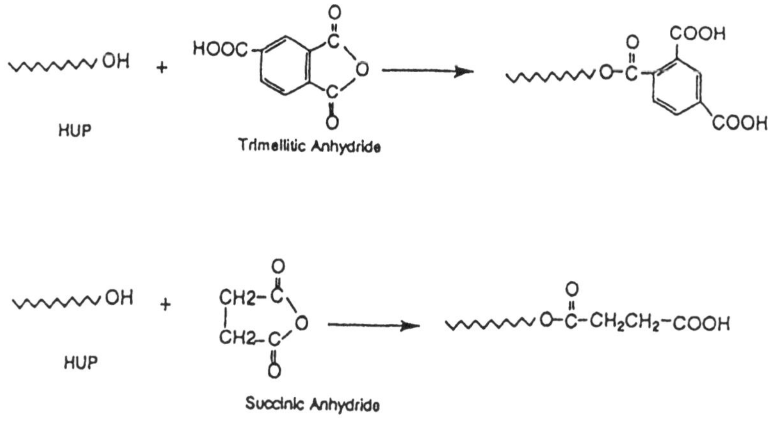

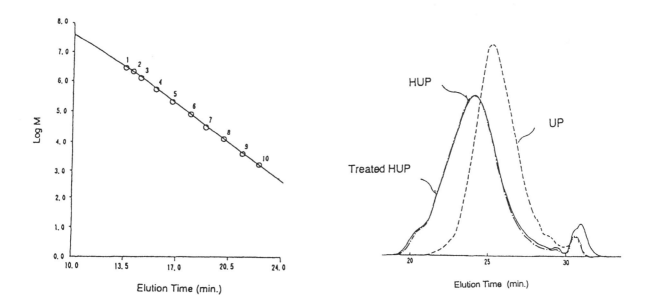

Fig.1 Substitution of End Group

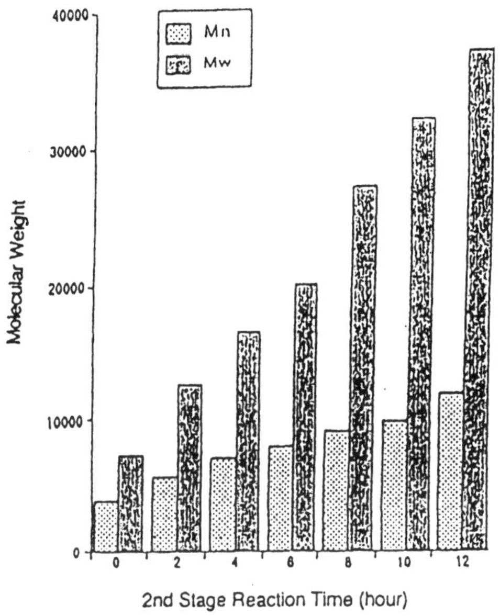

HUP

Trimellitic Anhydride

HUP

Succinic Anhydride

Fig.2 HUP Molecular Weight Changes

Molecular Weight

2nd Stage Reaction Time (hour)

Mn
Mw

Fig.3 GPC Calibration Curve

Log M

Elution Time (min.)

Fig.4 Molecular Weight Distribution Changes

HUP

UP

Treated HUP

Elution Time (min.)

Tab.1 Viscosity Changes (1)
Mn=6100 (poise at 40°C)

Substitution of End Group	Ratio (%)	Filler	Parts	Days					
				1	3	7	14	21	28
Untreated	0	Silica Powdered	220	460	560	690	720	800	850
		Alumina Trihydrate	250	540	500	380	340	300	210
		CaCO₃	250	58	58	51	45	51	51
Trimellitic Anhydride (TA)	50	Silica Powdered	220	900	1060	1260	1200	1570	1570
		Alumina Trihydrate	250	750	500	620	340	290	270
		CaCO₃	250	420	400	460	590	690	820
	100	Silica Powdered	220	1260	1500	3090	3870	4510	4420
		Alumina Trihydrate	250	2270	3090	5220	6130	10690	23460
		CaCO₃	250	3140	5440	8560	9730	10720	12260
Succinic Anhydride (SA)	50	Silica Powdered	220	830	980	1120	1180	1200	1280
		Alumina Trihydrate	250	1040	1070	1150	1020	1020	1170
		CaCO₃	250	160	160	160	140	160	160
	100	Silica Powdered	220	1070	1150	1250	1300	1360	1540
		Alumina Trihydrate	250	1820	2820	9040	15600	28640	58720
		CaCO₃	250	300	270	240	210	210	190

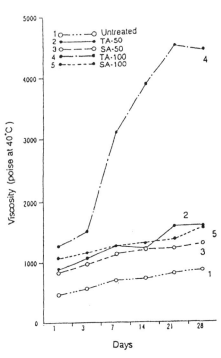

Fig.5 Viscosity Changes
Silica Powdered, Mn=6100

Tab.2 Viscosity Changes (2)
Mn=12000 (poise at 40°C)

Substitution of End Group	Ratio (%)	Filler	Parts	Days					
				1	3	7	14	21	28
Untreated	0	Silica Powdered	220	1460	1650	1810	2270	2640	2830
		Alumina Trihydrate	250	1830	1410	1420	1600	1810	1950
		CaCO₃	250	880	910	980	1040	1100	1220
Trimellitic Anhydride (TA)	50	Silica Powdered	220	2110	2350	2660	3260	3570	3570
		Alumina Trihydrate	250	3760	3360	2780	3090	2780	2980
		CaCO₃	250	850	740	740	900	1010	1220
	100	Silica Powdered	220	2160	2530	3200	3680	4690	5090
		Alumina Trihydrate	250	4720	4860	5520	6210	6580	9100
		CaCO₃	250	1060	980	850	770	800	800
Succinic Anhydride (SA)	50	Silica Powdered	220	2420	2500	2820	3260	4060	4130
		Alumina Trihydrate	250	3500	3220	3630	4450	4850	6270
		CaCO₃	250	1980	2000	2020	2160	2320	2510
	100	Silica Powdered	220	2320	2610	2750	3140	3790	4640
		Alumina Trihydrate	250	4690	6860	7380	12060	17860	33040
		CaCO₃	250	2940	2800	2800	2880	2960	3300

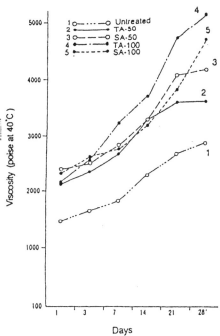

Fig.6 Viscosity Changes
Silica Powdered, Mn=12000

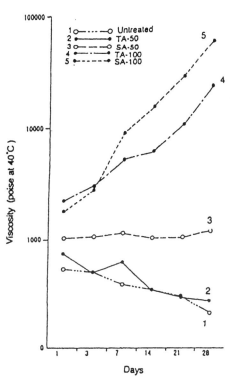

Fig.7 Viscosity Changes
Alumina Trihydrate, Mn=6100

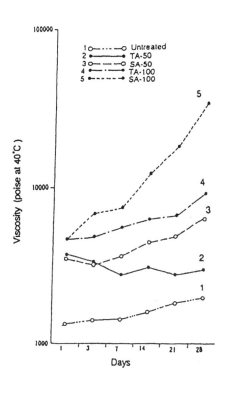

Fig.8 Viscosity Changes
Alumina Trihydrate, Mn=12000

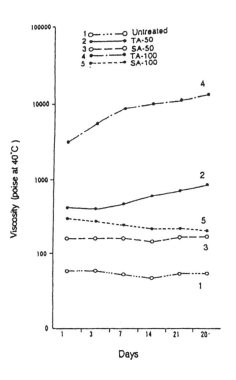

Fig.9 Viscosity Changes
CaCO₃, Mn=6100

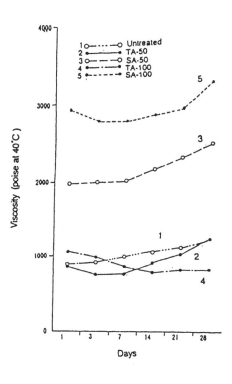

Fig.10 Viscosity Changes
CaCO₃, Mn=12000

Tab.3 Effect of CaCO$_3$ on Viscosity and Property
Mn=6100 (poise at 40°C)

Viscosity (poise at 40°C)

Substitution of End Group (%)	Days					
	1	3	7	14	21	28
Untreated	58	58	51	45	51	51
TA-50	420	400	460	590	690	820
SA-50	160	160	160	140	160	160
TA-100	3140	5440	8560	9730	10720	12260
SA-100	300	270	240	210	210	190

Strength After 28Days

Substitution of End Group (%)	Flexural Strength (kg/mm2)	Flexural Modulus (kg/mm2)
Untreated	7.6	1370
TA-50	9.0	1310
SA-50	>6*	1400
TA-100	7.8	1370
SA-100	8.6	1330

*Void Included

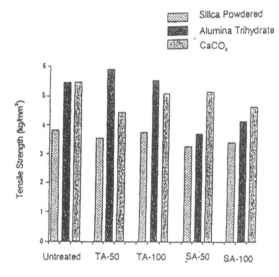

Fig.11 Tensile Strength
Mn=6100

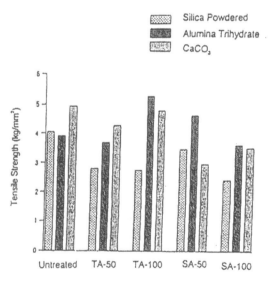

Fig.12 Tensile Strength
Mn=12000

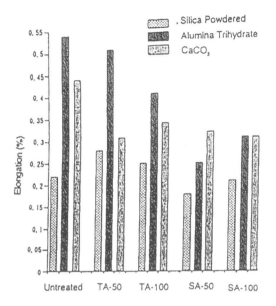

Fig.13 Elongation
Mn=6100

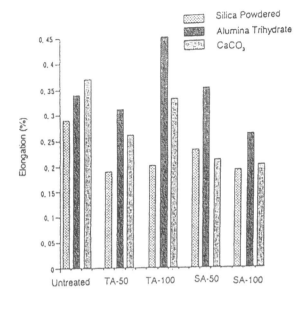

Fig.14 Elongation
Mn=12000

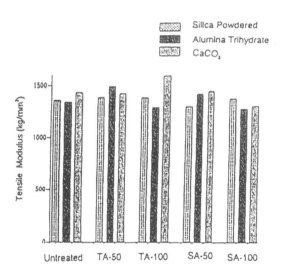

Fig.15 Tensile Modulus
Mn=6100

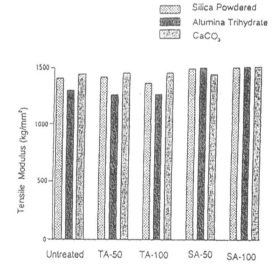

Fig.16 Tensile Modulus
Mn=12000

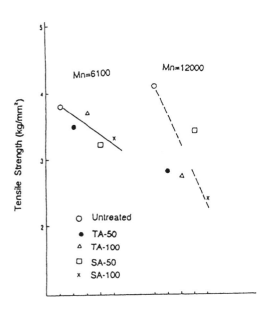

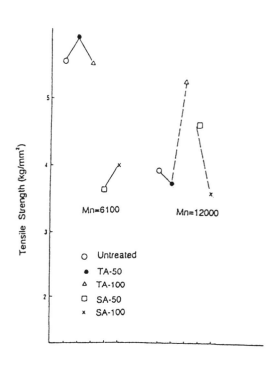

Fig.17 Effect of End Group on Tensile Strength
Silica Powdered

Fig.18 Effect of End Group on Tensile Strength
Alumina Trihydrate

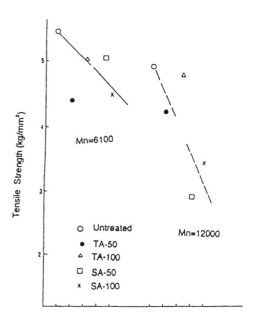

Fig.19 Effect of End Group on Tensile Strength
CaCO$_3$

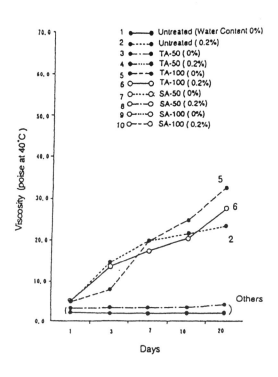

Fig.20 Viscosity Changes
MgO, Mn=5000

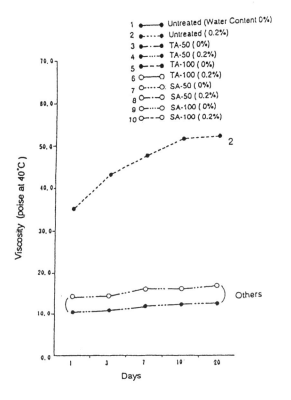

Fig.21 Viscosity Changes
MgO, Mn=10000

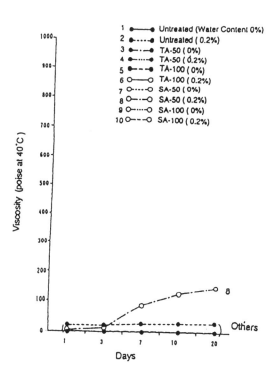

Fig.22 Viscosity Changes
Mg(OH)$_2$, Mn=5000

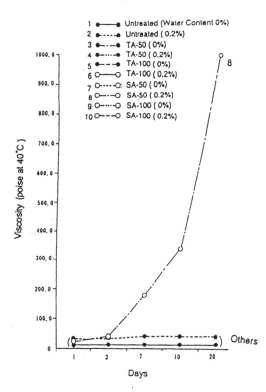

Fig.23 Viscosity Changes
Mg(OH)$_2$, Mn=10000

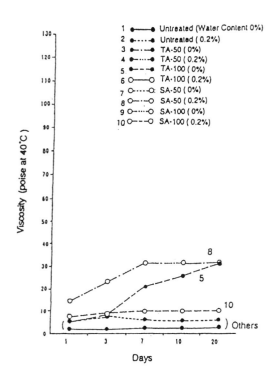

Fig.24 Viscosity Changes
ZnO, Mn=5000

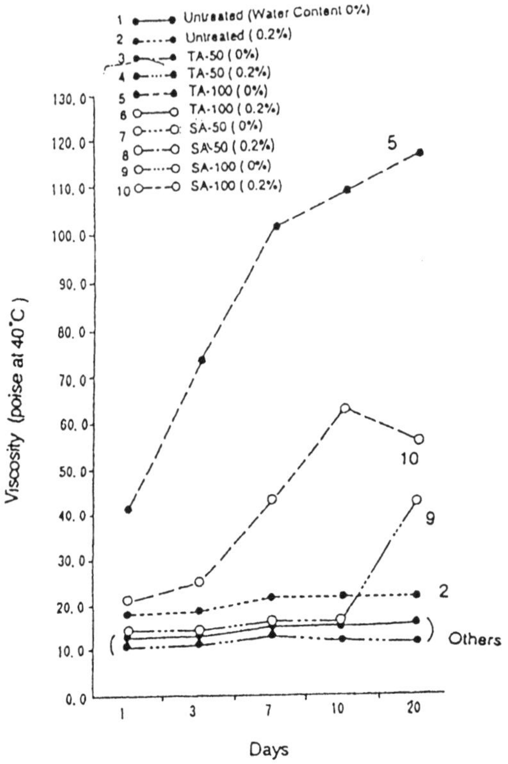

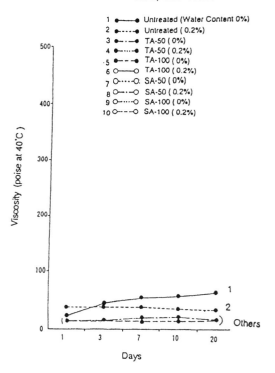

Fig.25 Viscosity Changes
ZnO, Mn=10000

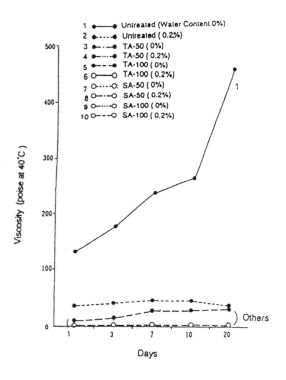

Fig.26 Viscosity Changes
Ca(OH)$_2$, Mn=5000

Fig.27 Viscosity Changes
Ca(OH)$_2$, Mn=10000

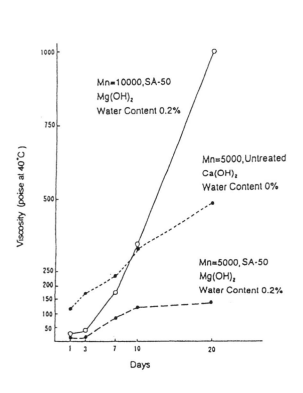

Fig.28 Viscosity Changes

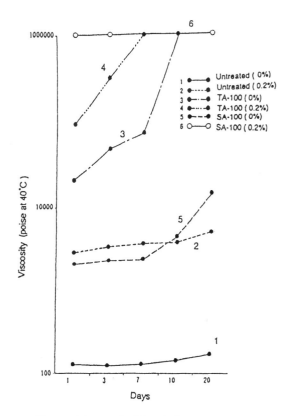

Fig.29 Effect of CaCO3/MgO on Viscosity
Mn=5000

High-Performance Synthetic Foams

HARRY S. KATZ AND RADHA AGARWAL

ABSTRACT

Syntactic foams are low density composites with relatively high compressive strength. Commercially available products usually have densities ranging from 25 to 35 lb/cu.ft., 2000 psi compressive strength and compressive modulus from 100 to 150 ksi. This paper describes an investigation that has produced room-temperature-curable high performance syntactic foams with the unique combination of 20 pounds per cubic foot density, compressive strengths at the 2000 psi level and over 170 ksi compressive modulus. These syntactic foams were prepared by compounding hollow microsphere fillers, short fiber reinforcements and coupling agents in a high-strength epoxy matrix. Packing concepts were used to optimize the packing of various size spheres. The addition of short fiber reinforcements was investigated in order to increase the compressive strength and modulus of the foam. Various coupling agents were studied to obtain improvements in composite properties and reduced water absorption, and to reduce the viscosity of the uncured mixture. Mixing and fabrication methods were evaluated. These syntactic foams were tested for density, compressive modulus and strength, and water absorption.

OBJECTIVE

The objective of this work was to develop a syntactic foam that will have a density of 20 lbs./cu. ft. in combination with a compressive strength of 2000 psi., compressive modulus of 170 ksi and very low water absorption when immersed in water at 1000 psi. Another objective was to obtain these properties with a room-temperature-curable system.

Harry S. Katz and Radha Agarwal, Utility Development Corp., 112 Naylon Ave., Livingston, N.J. 07039

BACKGROUND

Foam materials have many marine, military, commercial and industrial applications. Foam cores are used in sandwich structures and can provide advantages over honeycomb structures. Syntactic foams, which consist of hollow spheres in a matrix, have been specified for many critical applications because of their good combination of high compressive strength at relatively low densities. Commercially available products usually have densities ranging from 25 to 35 lb/cuft, 2000 psi compressive strength and compressive modulus from 100 to 150 ksi. High performance, low density syntactic foams are candidates for many future aerospace and marine applications.

EXPERIMENTAL

Our main approach toward achieving the objectives of this work was the study of packing concepts, The reason is that effective packing of different size spheres can provide low interstitial void space. Thus less of the resin, which has a much higher density than the hollow spheres, will be needed to fill the voids and a minimum density composite can be obtained.

Fig. 1. shows how the interstitial void volume of spheres can be reduced by packing an appropriate size of small diameter spheres into larger diameter spheres. Uniform diameter spheres, of any size, will pack with a void volume of about 38.5% of the total volume of the container. The combination of small spheres, in a volume fraction of 0.85 large to 0.15 small reduces the void content to about 28% A ratio of 0.72 large to 0.28 small can result in a void content of only 15%. Thus, very little resin matrix will be required to fill the interstitial voids and a small excess of the resin will provide a flowable or easily processed uncured mix that can be molded to produce the final product.

Fig. 2. is a good visual demonstration that a proper ratio of the diameter of spheres and fibers will result in interstitial packing and a minimum void content. Note that when the proper ratio of sphere to fiber diameter (0.5) is used, the void volume is only 35% . In contrast, when a poor ratio (4) is used, there is a high void volume of 50% that would result in a high resin demand. Also note in Figure 2 that both containers have the same volume of solids, but that proper packing will produce a high performance composite whereas poor packing will result in poor performance for most composite systems. [1] There are many different types of hollow spherical fillers [2] and short fibers that are good candidates for evaluation in this application.

MATERIALS

Among the many hollow spherical fillers that were evaluated during this program were glass products from PQ Corporation, 3M, and HGS Corporation. Phenolic spheres from Union Carbide and polyvinylidenechloride (Saran) spheres from Nobel Industries and Pierce and Stevens were also included in our trial formulations. Typical final candidates were Q2116 from PQ Corporation, 3M Co.'s S15 hollow glass microspheres, HGS-60 from HGS Corporation and Expancel 091 DE80 from Nobel Industries.

The factors involved in our choice of the epoxy matrix candidates were room-temperature-curable two-component systems with high compressive strength and capable of safe handling without a major concern of dermatitis or other potential problems. Formulation ingredients were bisphenol A and bisphenol F type epoxy resins, blends of these resins, reactive diluents, and different curatives. These were procured from a number of companies, including Shell Chemical, Ciba-Geigy , Air Products and Dow Chemical Company.

A number of silane coupling agents, which were added to the resin as an integral blend, and titanates were evaluated. Among these were A1100 and A187 from Union Carbide; Z6040 and Z6032 from Dow Corning and a number of titanates from Kenrich Corporation.

Short fibers evaluated included carbon-graphite fibers from a number of manufacturers, polyester fibers, Spectra (polyethylene fibers) from Allied Signal and Kevlar from Dupont.

CONCEPT ILLUSTRATION

Figure 3 illustrates the concept of packing small spheres between larger spheres to obtain minimum density and also interstitial fibers to increase the compressive strength and modulus.

MIXING

The final physical properties of a syntactic foam part, molded from a specific formulation, can show a wide difference when mixed or fabricated by different methods. Our initial trials were conducted by hand mixing and simply hand pressing the trowelable mixture into high durometer silicone rubber molds. Later mixes were made on a Ross Mixer with double planetary blades. Typical batches ranged in size from one quart to 2 gallons in volume. Aluminum cylindrical molds were used to fabricate the test samples, and diameters ranged up to 6-inches. In our effort to improve the uniformity of each mix, we came to the conclusion that the most effective method was the use of mix mulling equipment, as manufactured by Simpson Technologies and

Lancaster Corporation. This type of equipment was our preferred choice for quick and thorough mixing of our syntactic foam batches. Figure 4. is a photo of this type of equipment. The unit we used was capable of mixing about a 2-gallon volume of foam, and much larger models are available.

FABRICATION METHODS

Figure 5. shows a fabrication set-up that was used to fabricate many of our syntactic foam samples. In this fabrication method, vacuum and pressure were applied simultaneously and the pressure was varied to determine an optimum level.

RESULTS AND DISCUSSION

Compressive strength and modulus was tested in accordance with ASTM D695. Some of the hollow spherical fillers and their effect on density and compressive strength are shown in Figure 6. 3M Company's S15 and PQ Corporation's 2116 hollow microspheres had a good combination of compressive strength and density. Expancel 091 DE 80 from Nobel Industries was found to be the lowest density hollow microspheres, but the compressive strength of these hollow spheres was very poor.. Therefore, our approach was to use it in a small weight ratio of the total fillers as an interstitial filler between larger spheres.

The selection of short fibers was based on strength and modulus. Glass fibers would have been a major candidate, if density was not an important consideration. A number of grades of silane treated E-glass and S-glass fibers were studied. Lower density fibers were evaluated, including graphite fibers, polyester fibers, Spectra (polyethylene fibers) and Kevlar. The length and diameter of these fibers were crucial choices in order for them to act as reinforcements, without disturbing efficient packing of the hollow spheres.

The interface bonding of spheres or fibers with the resin matrix is a very important factor that must be optimized in order to obtain a high performance syntactic foam for critical applications. The use of coupling agents or special surface treatments for fillers and reinforcements have been found to be very effective to improve the interface bonding and the performance and durability of composite materials. A number of coupling agents at various concentration levels were evaluated. Figure 7 shows the effect of coupling agent on the compressive strength. These coupling agents were used at 2% by weight in the formulation mix. All the three coupling agents (A1100, A187 and Z6040) in this evaluation showed a significant improvement in compressive strength over the same system without the integral blend coupling agent.

Hydrostatic tests were run by use of a hydraulic cylinder, which was filled with water. Syntactic foam samples were immersed in the water. 1000 psi pressure was applied for about 100 cycles and each cycle was about 15 min. Various foams gave a great variation in water absorption. It is apparent that some of the hollow spheres may fracture and become filled with water. The better test values that we obtained ranged between 0.5% to 2.0% water absorption.

CONCLUSIONS

This work proved that the concept of hollow sphere/fiber packing in an epoxy matrix provides a high performance syntactic foam that met the objectives of this program, and results in a product that will be suitable for many different applications.

ACKNOWLEDGEMENT

The authors thank the U. S. Department of the Navy for the funding to carry out this work.

REFERENCES

1. Milewski, J.V. 1987, Ch. 3., Packing Concepts in the Use of Fillers and Reinforcement Combinations in HANDBOOK OF REINFORCEMENTS FOR PLASTICS, Milewski, J.V. and Katz, H.S., Editors, Van Nostrand Reinhold, New York.

2. Ruhno, R.A, and Sands, B.W.,1987, Ch. 22 Hollow Spheres in HANDBOOK OF FILLERS FOR PLASICS, Katz, H.S. and Milewski, J.V., Editors, Van Nostrand Reinhold. New York.

BIOGRAPHIES

Harry, S. Katz is the president and founder of Utility Development Corporation, a firm that has been involved for over 25 years in consulting for industrial firms and government agencies in the fields of plastics and composite materials. He is a chemist with graduate education in polymer chemistry. He has been involved with many technical societies, including membership in SPI, ACS, SPE and SAMPE.

Radha Agarwal received a Ph.D. in polymer chemistry from the India Institute of Technology, New Delhi, India in 1985. She has been working with Utility Development Corporation, as Research Director, for nine years. She has been involved in projects dealing with polymer coatings, adhesives, fillers and composites.

1.0 LARGE	0.85 LARGE	0.72 LARGE
0.0 SMALL	0.15 SMALL	0.28 SMALL
Void Content	Void Content	Void Content
37.5 %	28 %	15%

FIGURE 1. Packing of Large and Small Spheres

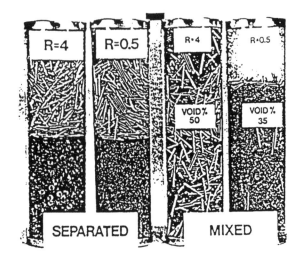

FIGURE 2. Packing of Spheres and Fibers

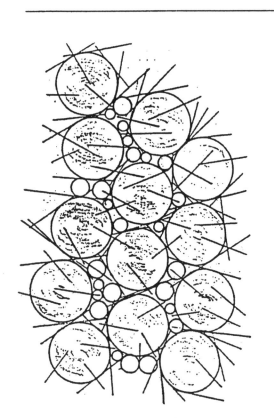

FIGURE 3. Syntactic Foam Concept Sphere and Fiber Packing

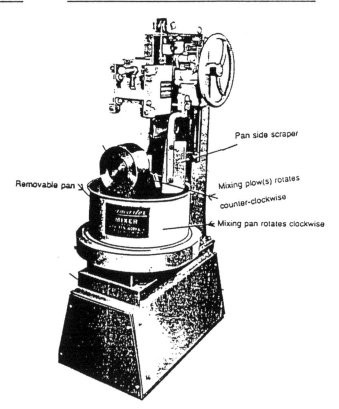

Pan side scraper

Mixing plow(s) rotates counter-clockwise

Removable pan

Mixing pan rotates clockwise

FIGURE 4. Mix-Muller Equipment

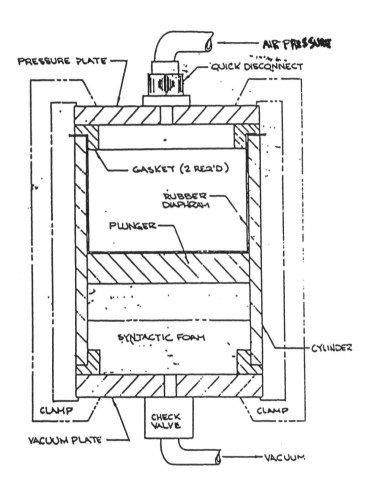

FIGURE 5. Fabrication Set-Up for Syntactic Foam Test Cylinders

Compressive Strength vs Density for Various Hollow Microspheres

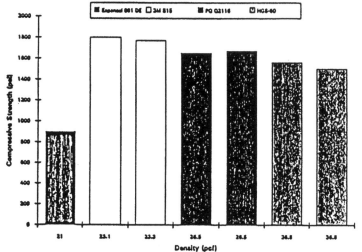

FIGURE 6. Compressive Strength vs.
Hollow Microspheres

Compressive Strength vs Density with Various Coupling Agents

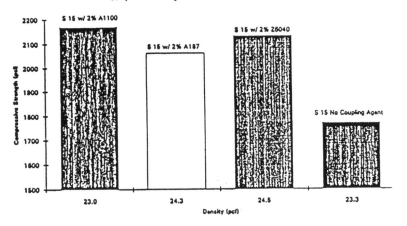

FIGURE 7. Effect of Various Coupling
Agents on Compressive Strength

Influence of Surface Treatment of Filler on Hydrothermal Aging of Particle Filled Polymer Composite

TOHRU MORII, NOBUO IKUTA AND HIROYUKI HAMADA

ABSTRACT

This study dealt with the influence of silane treatment of filler on hydrothermal aging of particle filled polymer composite (artificial marble). Glass and aluminum hydroxide with and without silane treatment were prepared, and various kinds of particle filled composites were molded by using these fillers. The specimens were immersed in hot water, and weight change and change of color were evaluated. Moreover the degradation mechanism was discussed from the microscopic observation. The silane treatment affected the degradation of the interface due to the long-term immersion, and it affected little the degradation due to the short-term immersion. The silane treatment restrained the formation of void and debonding, and the degradation of the interface occurs little.

INTRODUCTION

In recent years, particle filled polymer composite called as "artificial marble" has been used as a main component of bath tub and a top board of kitchen counter. Artificial marble has excellent water resistance, transparency and gloss. Artificial marble is usually made of glass or aluminum hydroxide as filler and cross-linked polyester or vinylester resin as matrix. In general, 100-200phr particle is filled in the matrix, and as a result, there exist many interfaces between filler and matrix. It is well-known that degradation of not only matrix resin but also filler/matrix interface affects the reduction of various properties if it is used in hydrothermal environment such as bath tub. Some research have been done for the weight change of the particle filled composites in water environment (Janas and McCullough, 1987), however, the degradation of the interface between filler and matrix has never been understood well.

The authors have studied the degradation of fiber reinforced plastics (FRP) in water environment (Morii et al., 1993a, 1993b, 1994a, 1994b). Weight change measurement, bending test and microscopic observation have been conducted, and it has been clarified that the degradation of the fiber/matrix interface significantly affects the long-term properties of FRP. In this paper our evaluation method was applied to the evaluation of degradation of the various kinds of artificial marble in hydrothermal environment. Glass and aluminum hydroxide with and without silane treatment were used as filler particle, and the influence of silane treatment on the degradation was discussed from the experimental results of weight change measurement, color measurement and microscopic observation.

Tohru Morii and Nobuo Ikuta, Shonan Institute of Technology, Tsujido-Nishikaigan, Fujisawa, Kanagawa 251, Japan
Hiroyuki Hamada, Kyoto Institute of Technology, Matsugasaki, Sakyo-ku, Kyoto 606, Japan

EXPERIMENTAL PROCEDURE

Materials used were two kinds of matrix resin and four kinds of filler. Matrices were terephthalic cross-linked polyester and vinylester resins and fillers were glass and aluminum hydroxide particles. Fillers with and without silane treatment were prepared for both particles. The mean particle diameters were 50μm in glass particle and 25μm in aluminum hydroxide particle. Using these materials six kinds of specimens were molded by casting. Compositions of the specimens were summarized in Table I. Casted compound was cured at 65°C for 30min and was post-cured at 95°C for 60min in the temperature controlled oven.

In order to discuss the degradation in hydrothermal environment, all the specimens were immersed in distilled water at 95°C. Before immersion all the specimens were dried to equilibrium in the vacuum oven in order to parch the absorbed moisture.

Weight change measurement was conducted by weighing the specimen with 50mm square and 7.5mm thickness before and after immersion. Before immersion, the weight of dry specimen (W_o) was measured, and after that the specimen was immersed. The immersion times were 3, 10, 30, 100, 300, 1000 and 3000hours. After immersion for the fixed period, the weight of wet specimen (W_w) was measured, and the specimen was re-dried in vacuum oven. After equilibrium in weight, the weight of re-dried specimen (W_d) was measured. From these weighing, the net weight gain (M_g) and the weight loss (M_l) were evaluated by Equation 1.

$$M_g = \frac{W_w - W_d}{W_o}, \qquad M_l = \frac{W_o - W_d}{W_o} \qquad (1)$$

Here, M_g means the weight gain due to water absorption and M_l means the weight loss due to dissolution from the specimen.

Change of the tone of color was evaluated by measuring L, a, and b values of the weight change measurement specimen. From these values E value was calculated by

$$E = \sqrt{L^2 + a^2 + b^2} \qquad (2)$$

From the obtained E values ΔE value was defined as the difference of E values between the virgin and immersed specimens.

RESULTS AND DISCUSSION

WEIGHT CHANGE

Figure 1 shows the changes of the net weight gain, M_g, with the square root of immersion time. M_g increased linearly against the square root of immersion time to 30hours independent of the specimen. M_g's of the specimens with the treated filler gradually approached constant value, and those showed a Fickian diffusion behavior. On the other

TABLE I   DETAILS OF USED MATERIALS

Specimen ID	Matrix	Filler	Silane treatment
UP/AT	Cross-linked polyester	Aluminum hydroxide	Yes
UP/AN	Cross-linked polyester	Aluminum hydroxide	No
UP/GT	Cross-linked polyester	Glass	Yes
UP/GN	Cross-linked polyester	Glass	No
VE/AT	Vinylester	Aluminum hydroxide	Yes
VE/AN	Vinylester	Aluminum hydroxide	No

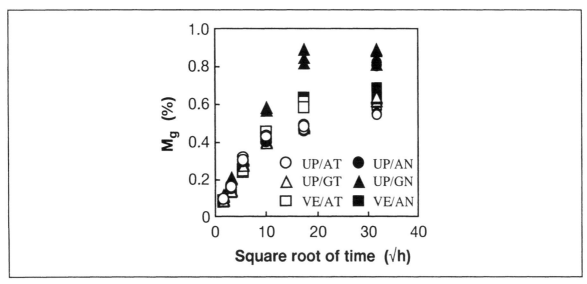

Figure 1   Changes of the net weight gain (M_g) with the immersion time.

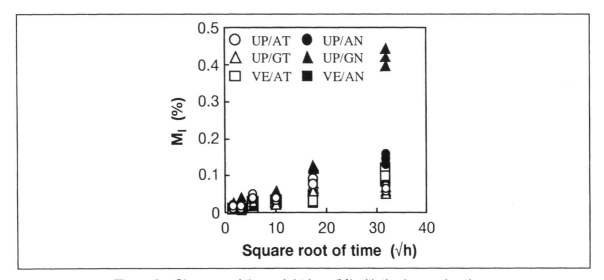

Figure 2   Changes of the weight loss (M_l) with the immersion time.

hand, M_g's of the specimens with the non-treated filler showed continuous increases to 1000hours.   Therefore the influence of the silane treatment of filler appears after 30hours. The specimens with the treated filler showed almost the same change in M_g independent of the kind of filler.   This suggests that the weight gain is mainly caused by the water absorption of matrix resin and that the filler/matrix interface affects little the weight gain behavior in the specimens with the treated filler.   In the specimens with the non-treated filler the changes of M_g after 30hours were different due to the filler , however, those were independent of the matrix resin.   In UP/AN and VE/AN the change of M_g showed almost same tendency with that of UP/AT and VE/AT to 300hours.   At 1000hours M_g's of UP/AN and VE/AN increased again, and therefore it is considered that the degradation of the interface in occurred after 300hours in these specimens.   On the other hand, M_g of UP/GN showed significant increase after 30hours.   Therefore it is considered that the degradation of the filler/matrix interface easily occurs in the glass filled specimen.   Figure 2 shows the changes of the weight loss, M_l, with the square root of immersion time. M_l's were almost equal to zero to 100hours independent of the filler.   In the specimens with treated filler M_l's indicated gradual

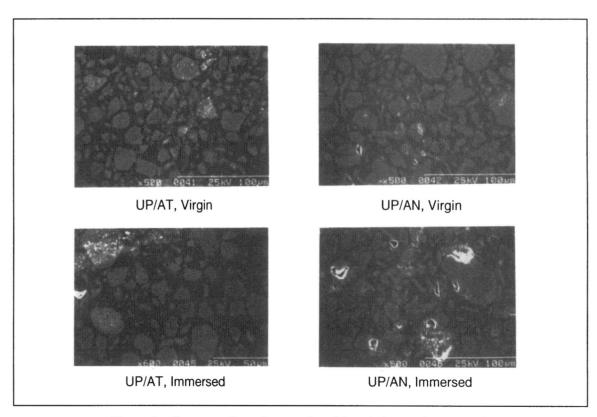

UP/AT, Virgin

UP/AN, Virgin

UP/AT, Immersed

UP/AN, Immersed

Figure 3   Cross-section micrographs of the virgin specimens and
the specimens immersed for 1000hours.

increase after 100hours, and they were same tendency independent of filler.   However, the changes of M_l in the specimens with the non-treated filler were different due to the kind of filler. M_l's of the specimens with the non-treated filler were quite larger than those of the specimens with the treated filler.   In particular, M_l of UP/GN increased remarkably after 100hours, and it reached about 7 times of M_l value of UP/GT at 1000hours.   From these results the time when M_l began to increase corresponded to the time when the difference of M_g appeared between the treated and non-treated specimens.   Therefore the increase of M_l is closely related to the increase of M_g in the longer immersion time.   This fact suggests that the increase of M_l induces the remarkable increase of M_g.

Figure 3 is the cross-section micrographs of the virgin specimens and the specimens immersed for 1000hours for UP/AT and UP/AN.   In the virgin UP/AT any defect could not be observed on the cross-section, and this indicated effective bonding between filler and matrix, and such state on the cross-section never changed even after immersion in water.   In the virgin UP/AN, however, some voids and debondings could be observed at the filler condensed region on the cross-section.   This suggests the poor bonding between filler and matrix.   In the immersed UP/AN the larger voids and debondings could be observed.   It is considered that such larger voids were produced by the enlargement of the small voids due to water immersion.   From this result it is clear that the increase of M_l is caused by the enlargement of the voids and that the increase of M_g is caused by the water penetration into such voids.   In the treated specimen the enlargement of the voids never occurs, and therefore the remarkable increase of M_l never occurs.   From these observations it is concluded that the silane treatment of filler affects the formation and enlargement of voids.

CHANGE OF COLOR

Figure 4 shows the changes of ΔE due to the water immersion.   In the specimens with

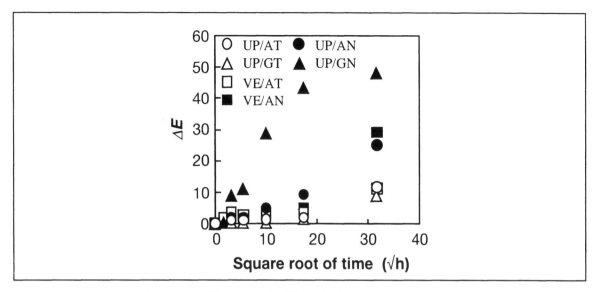

Figure 4   Changes of ΔE due to hydrothermal aging.

the treated filler the changes of ΔE showed the same tendency, and they increased after 300hours.   On the other hand, ΔE increased remarkably after 100hours in UP/AN and VE/AN and after 3hours in UP/GN.   The changes of ΔE were quite similar to the changes of M_l.   It suggests that the main cause of the change of color is the increase of the voids and debondings.   The color of the specimens with the treated filler turned to yellow due to water immersion.   Therefore it is considered that the changes of color in these specimens are caused by the color change of the matrix resin due to water absorption.   On the contrary, the color of the specimens with the non-treated filler turned to white and the transparency decreased drastically.   In these specimens the voids and the debondings increased due to water immersion.   Such defects inside the material refracted the light, and this phenomenon led to the significant whitening of the specimen.   Such whitening resulted in the significant increase of ΔE.   Consequently it is derived that the change of color is a index of the increase of the defects such as void and debonding due to water immersion.

DEGRADATION MECHANISM

From the experimental results it is obvious that the degradation process of the specimen is different due to the silane treatment of filler.   Figure 5 represents the degradation mechanism of the specimens with the treated filler and with the non-treated filler.   In the first process only the matrix absorbs the water, and the silane treatment never affects this process.   However, the following degradation process is different due to the silane treatment.   In the specimen with the treated filler gradual degradation of the interface occurs at longer immersion time (second process), and microscopic debonding (which cannot be observed by microscope) occurs between filler and matrix.   Such debonding induces a little increase of M_l and its following M_g.   According to these process the color of the specimen turns to yellow. In the specimen with the non-treated filler, the voids enlarges more and more due to water immersion and debonding is caused between filler and matrix in the second process.   The increase of void and debonding induces the remarkable increase of M_l.   In this process the water penetrates into the void and debonding, and as a result, M_g increases remarkably. Such degradation of the interface whitens the color of the specimen.   Therefore the silane treatment influences only the degradation of the long term immersion.

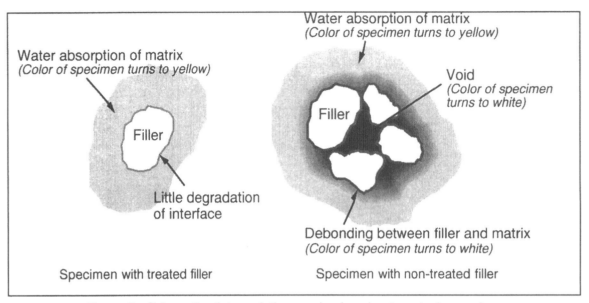

Figure 5   Schematic of degradation mechanism due to water immersion.

CONCLUSION

This study dealt with the influence of silane treatment on the hydrothermal aging of artificial marble.   The silane treatment restrained the formation and enlargement of the voids and the debondings due to the long-term immersion.   The specimens with the treated filler showed the degradation of only matrix resin, while the specimens with the non-treated filler showed the degradation of filler/matrix interface in addition to the matrix resin.   These differences in the degradation mechanism affected the weight change behavior and the change of color.

REFERENCES

Janas, V. F. and R. L. McCullough. 1987. "Moisture Absorption in Unfilled and Glass-filled, Cross-linked Polyester," *Composites Science and Technology*, 29:293-315.

Morii, T., T. Tanimoto, H. Hamada and Z. Maekawa. 1993a. "Degradation Behaviour of Glass Fibre Mat Reinforced Polypropylene Immersed in Hot Water," *Polymers and Polymer Composites*, 1:175-182.

Morii, T., H. Hamada, Z. Maekawa, T. Tanimoto, T. Hirano and K. Kiyosumi. 1993b. "Weight Changes of a Randomly Oriented GRP Panel in Hot Water," *Composites Science and Technology*, 49:209-216.

Morii, T., H. Hamada, Z. Maekawa, T. Tanimoto, T. Hirano, K. Kiyosumi and T. Tsujii. 1994a. "Weight Changes of the Fibre/Matrix Interface in GRP Panels Immersed in Hot Water," *Composites Science and Technology*, 50:373-380.

Morii, T., H. Hamada, Z. Maekawa, T. Tanimoto, T. Hirano and K. Kiyosumi. 1994b. "Fracture and Acoustic Emission Characteristics of Glass Fiber Reinforced Plastics Panel in Hot Water," *Polymer Composites*, 15:206-216.

The Role of Constituent Materials in the Development of Syntactic Foam

THOMAS J. MURRAY AND NOEL J. TESSIER

ABSTRACT

Syntactic foams, the unique blend of ordered reinforced air and matrix material, have found many uses in both subsea and aerospace industries. The material was originally developed in the late 1950's for use as a buoyancy material for drilling platforms but has current applications which include; void filler for both the 688 and SEAWOLF Class submarines, man-rated stabilization for deepwater research and recovery craft, insulation for subsea pipelines, plug assist tooling for thermoforming, core materials for damage tolerant and fire resistant structures, and large subsea mooring buoys with expected 30 year life cycles. In each of these applications the material development and design phase optimized the combination of hollow glass or reinforced epoxy spheres, thermoset or thermoplastic matrix material, and composite skin to meet specific product and processing requirements. In this paper, an overview of the development of some of these unique applications will be presented, with specific insight as to the role that the constituent materials play in the final part.

INTRODUCTION

Syntactic foams constitute a unique family of composite material systems that have been developed for niche market applications. Syntactic foams are defined as materials that contain hollow spherical particles. This encapsulation of air into a material by controlling the density and strength of the hollow spheres results in a material system that can be designed for each application. They are typically more expensive on a per pound basis than other "blown" foams or lightweight materials and therefore are used in performance driven applications where the added material cost is justifiable.

Recent advances in the technology of syntactic foam design and processing has brought them to the threshold of broader use for their unique properties. This paper follows the natural development process for a new syntactic foam. It begins with a discussion of the constituent materials, manufacturing processes and performance of the syntactic foams in the traditional market of subsea buoyancy. Then, from this knowledge

Thomas J. Murray and Noel J. Tessier, Emerson and Cuming Composite Materials, Inc., 59 Walpole Street, Canton, MA 02021 base, it follows the development of three new products with different requirements in; 1) void filling, 2) insulation, and 3) low coefficient of thermal expansion applications.

CONSTITUENT MATERIALS

Syntactic foams are defined as materials that contain hollow spherical particles. The subsea buoyancy syntactic foams contain a resin matrix, hollow glass microspheres, hollow composite shell macrospheres and protective skin. Materials selection criteria, properties and typical applications follow.

RESIN MATRIX

The important selection criteria for a resin matrix is its specific strength and stiffness, the ability to bond to the untreated hollow spheres, the rheology of the matrix during processing, the ability to control the cure, hydrolytic stability, and cost. Table I displays the typical matrices used for subsea applications. Important properties are, density, compressive strength and modulus and processing characteristics. Epoxies dominate in this market because of their high strength, bonding characteristics, hydrolytic stability, low shrinkage, low viscosity, controllable exotherm and cure, and costs.

HOLLOW SPHERES

The selection criteria for hollow microspheres (10 to 140 microns diameter) are specific strength and modulus, hydrolytic stability, processing characteristics and cost. Table II presents their important properties of density, strength and size. Hollow Glass microspheres dominate in the subsea buoyancy market because of their high compressive strength at low densities, hydrolytic stability when encapsulated in the matrix, processing characteristics and cost.

The selection criteria for hollow macrospheres (1.5 mm to 100 m diameter) is similar to that of hollow microspheres with the addition of size selection. Table III presents the important properties of macrosphere density and strength. Composite shell macrospheres dominate the market because of their diameters, moderate compressive strength at low densities, hydrolytic stability when encapsulated in the matrix, processing characteristics and cost.

SYNTACTIC FOAM PROPERTIES

The combination of different resin matrices and hollow spheres of different diameters, densities and strengths leads to changes in performance and processing characteristics that allow the materials engineer much latitude for any application.

Table IV presents the properties of hollow microsphere syntactic foams with changes in hollow microsphere type with the same resin matrix and changes in resin matrix with the same hollow microshpere. From Tables IV and II, it can be seen for RG resin that a doubling of microsphere strength (SCOTHCLITE™ K1 Glass Bubbles to K2 Bubbles) results in a syntactic foam strength increase of 29%. From Tables IV and I, it also shows that a 17% increase in resin strength (RG to MR) results in a 113% increase in syntactic foam strength.

Table V presents the properties of hollow macrosphere and microsphere syntactic foams. From Table V it can be seen that macrosphere strength dominates the mechanical properties. Here RG K1 to MR K1 strength increase of 113% translates to only a 10% increase in strength with EPG16 macropsheres.

PROTECTIVE SKIN

The selection criteria for protective skin materials are mechanical properties, abuse resistance, hydrolytic stability, density, load transitioning and cost. Table VI presents the important mechanical properties of two different skins. Fiberglass composite skins dominate the subsea buoyancy market because of their mechanical properties at low density, toughness, hydrolytic stability, load transitioning and costs.

PROCESSING CHARACTERISTICS FOR BUOYANCY APPLICATIONS:

The processing of syntactic foams for buoyancy is characterized by the following:

1. Ability to maintain high loadings of hollow spheres with consistent processing and superior performance with a variety of final part geometries.
2. Controlled rheological properties in mixing through mold filling in quantities up to 100 cubic feet in volume.
3. Elimination of air voids that lead to inconsistent processing and performance.
4. Ability to protect the syntactic foam from abuse at the lowest density and creation of an interface that prevents syntactic foam to skin delamination.
5. Ability to achieve consistent cure with low to no residual stresses due to shrinkage or thermal and degree of cure gradients.

NEW PRODUCT DEVELOPMENTS

VOID FILLER FOR SUBMARINE CONTROL SURFACES

This application required a new set of materials and process conditions over traditional subsea buoyancy products. The most important change was the ability to process the material into a prefabricated, large steel structure. This filling process needed to be adaptable to parts transported to the filling site or filled in the field. In addition, the performance requirements on the materials were 25PCF density, high compressive strength and modulus, excellent fatigue life over 20 years of expected life and low moisture absorption.

The knowledge base of traditional buoyancy materials identified areas that needed development. These developments were high strength macrosphere at low density 16 PCF, ambient temperature cure resin system of low shrinkage and low heat of reaction and a filling process with the ability to be performed at any location.

The syntactic foam developed is called VF-25S. It has been used in filling complicated steel cavities of up to 120 Cubic feet. It has a density of 25.5 PCF on average, with a compressive strength of 2600 PSI, a compressive modules of 124,000 PSI and a bulk modules of 199,000 PSI. Water absorption is very low with 1000 cycle to 1000 PSI and 30 day soaks at 1,000 PSI showing less than 1% weight gain.

LOW CTE MATERIAL FOR MRI SUPERCONDUCTING MAGNETS

This application was totally removed from the subsea market and required much different material performance. The most important requirements were the ability to mold and cure in-situ on a prefabricated stainless steel part, the ability to be easily machined, the ability to stay bonded to stainless steel at 4°K and to be processed and sold at low cost.

Syntactic foam was determined to be a suitable material able to meet all the requirements. Developments were needed for thermal shock resistance, lowest CTE possible, controlled cure, processing in-situ and higher mechanical properties than traditional foams.

The syntactic foam developed stayed bonded to stainless steel at 4°K, had good thermal shock resistance, was capable of in-site processing and maintained consistent high mechanical properties.

INSULATION FOR FLEXIBLE PIPE

The subsea production of oil at water depths greater than 1000 feet has created a problem due to the loss of oil temperature over the increased flowline lengths. As the oil cools from wellhead to production facility, paraffins persipitate out of solution in the lines and cause reduced production rates or stoppage. One way to combat this problem is to insulate the flexible flowlines to reduce the heat loss. Unfortunately, conventional insulation systems comprised of blown or crosslinked foams could not survive the deep-sea environment. Working in conjunction with one of the industries flexible pipe manufacturers, (Wellstream Corp.), a program was established to develop a syntactic based insulation which was capable of survival at depths of 3000 feet without loss of thermal resistance, had low thermal conductivity, could be applied using available cost effective pipe wrap technology, and could withstand the harsh acidic outgassing of flexible pipe interiors.

For this application, a continuous form of the material was desirable. This led to the development of an extrusion process capable of efficiently mixing hollow glass microspheres into a polypropylene based thermoplastic material under low shear conditions. A twin screw extrusion process using a proprietary screw design was developed. A small, (less than 50 micron average diameter), microsphere was selected for the application. Development of multistage addition process for the balloon was crucial to the program. Further, tight controls of the addition process for both the resin and microspheres was necessary in order to minimize head pressure fluctuations to produce a finished product directly from the twin screw. Also, in order to get sufficient bonding of the glass spheres to the polypropylene, a coupling agent was added to the spheres along with a grafting agent to the resin. The final product of the development program was a continuously extruded profile with properties given in Table VI.

Table I
Typical Resin Properties

Type	Density (PCF)	Compressive Strength (Psi)	Compressive Modulus (Psi)	Application	Characteristic
RG	67.4	8,750	290,000	Macrosphere Foam ≤ 4000 Feet	Low Viscosity Elevated Cure Temp
FL	69.9	7,630	264,000	Macrosphere Foam ≤ 3000 Feet	Room Temp Cure Moderate Pot Life
MR	69.9	10,300	360,000	Macrosphere Foam ≤ 9000 Feet	Low Viscosity Elevated Cure Temp
ET	68.6	5,700	230,000	Macrosphere Foam for Military Void Filler	Room Temp Cure Long Pot Life
EL	77.3	22,600	537,000	High Performance Block to 20,000 Feet	Low Viscosity High Modulus
DS	70.5	20,700	536,000	High Performance Block to 20,000 Feet	High Modulus Low Density
EI	55.5	600	17,000	Flexible Pipe Insulation	Extrudable Thermoplastic

Table II
Typical Microballoon® and Glass Bubble® Properties

Designation	Nominal Density (PCF)	Minimum 80% Survival Strength (Psi)	Median Diameter (Microns)	Application
K1	7.5	250	69	Macrosphere Foam ≤ 3000 Feet
K15	9.4	300	67	Macrosphere Foam ≤ 4000 Feet
K20	12.5	500	61	Macrosphere Foam ≤ 7000 Feet
FTD-200	12.5	600	60	High Performance Block
K25	15.6	750	52	Macrosphere Foam ≤ 9000 Feet
FTD-202	15.6	700	55	High Performance Block
SDT-28	17.5	1300	16	High Performance Block
K37	23.1	2000	50	Flexible Pipe Insulation
FTD-235	23.1	2200	40	High Performance Block
SDT-40	25.0	>4000	12	Flexible Pipe Insulation

Table III
Typical Macrosphere® Properties

Designation	Nominal Density (PCF)	Average 50% Survival Strength (Psi)	Typical Application Foam Depth (Feet)
HG7	7.3	400	1000
EPC11	11.3	850	2000
EPG14	14.3	1200	2500
EPG16	16.3	1775	3000
EPB16	16.3	2500	NA
EPG20	20.3	2460	4500
EPG27	27.3	3500	6000
EPG30	30.3	4500	7000

Table IV
Typical Microballoon® Foam Properties

Resin Type	Micro Balloon Type	Loading Level (Vol %)	Foam Density (PCF)	Compressive Strength (Psi)	Compressive Modulus (Psi)
RG	K1	53	35.6	3,350	185,000
RG	K15	53	36.6	3,400	187,000
RG	K2	53	38.0	4,330	224,000
ET	K1	50	38.0	4,275	230,000
MR	K1	53	36.8	7,160	366,000
DS	Blend	>70	30.5	11,800	518,000

Table V
Typical Macrosphere® Foam Properties

Resin Type	Micro Balloon Type	Macro Sphere Type)	Foam Density (PCF)	Compressive Strength (Psi)	Compressive Modulus (Psi)
RG	K1	EPC11	21.0	1,625	66,000
RG	K1	EPG16	24.0	1,775	75,500
RG	K2	EPG20	27.4	2,260	84,500
RG	K2	EPG27	31.6	2,440	98,600
MR	K1	EPG16	24.5	1,950	87,000
MR	K2	EPG27	32.2	3,300	116,000

Table VI
Typical Properties of Eccohide® Skin with Two Fabrics

Property	Type I Fabric	Type II Fabric
0° Tensile Strength (psi)	36,400	30,500
90° Tensile Strength (psi	18,000	7,000
0° Compressive Strength (psi)	29,500	29,000
90° Compressive Strength (psi)	21,000	17,000

Table VII
Typical Properties of ECCOTHERM® Insulation

Thermal Conductivity	0.081 BTU/hr-ft-°F
Operating Temperature	155°F
Density	43.7 PCF
Service Depth	3300 Feet
Moisture Absorption (Weight gain 24 hrs at Depth)	<.5%
Compressive Strength	775 Psi
Compressive Modulus	25,000 Psi
Tensile Strength	600 Psi
Tensile Modulus	13,000 Psi

Newly Developed Flexible Vinyl Ester Resin with Air-Drying

TOMOAKI AOKI, KAZUYUKI TANAKA, KOUJI ARAKAWA
AND KANEMASA NOMAGUCHI

ABSTRACT

This paper covers a newly developed vinylester resin used mainly for concrete overlay. The surfaces of concrete structures such as building roof, floor and drainage in food processing plants and chemical plants are damaged by water, boiling water, corrosive detergents, disinfectants and other chemicals. Therefore, the surfaces must be protected by resin overlay.

Traditionally, the matrix resins used for this application have been unsaturated polyester resin and vinylester resin with air drying additives which are mainly paraffinwax . However although the paraffinwax improves the surface curing property of these resins, it often involves a delamination between intermediate layer and top-coat in the overlay.

The authors solved the problem of this delamination by developing a flexible vinylester resin with air drying property (developed FVE). Also this newly developed resin is optimum in its performance balance; chemical resistance, mechanical and curing property compared with a conventional unsaturated polyester and vinylester resins.

INTRODUCTION

In Japan, the application of concrete overlay to waterproof and prevent degradation of concrete structures is increasing year by year. An unsaturated polyester resin is used for waterproofing and a vinylester resin is used for preventing degradation of the surface.

The general method of overlay onto the concrete substrate is as follows; At first, a primer commonly a urethane resin is applied onto the substrates, and then an intermediate layer which is fiber reinforced plastics(FRP) or paste(resin with filler) is overlapped. When the surface of intermediate layer dries, usually it takes about a few hours, top-coat is covered onto it. We call a formulation of these three layers "overlay".

Tomoaki Aoki, Kazuyuki Tanaka, Kouji Arakawa and Kanemasa Nomaguchi, Hitachi Chemical Co. Ltd., Yamazaki Works, 4-13-1, Higashi-cho, Hitachi-shi, Ibaraki, JAPAN, 317

A shorter laydown time has been desired in this application. To make laydown time short, the surface curing time of each layer has to be short.Moreover, the overlay has to follow when cracks occur in the concrete substrate. For this reason, the matrix resins used for the overlay are flexible. The surface curing property is dependent on its bulk curing property. Reactivity of radical curable flexible resin is usually lower than that of other rigid resin. The reason why the flexible resin has less reactivity is the less quantity of double bond in its polymer chain to make it flexible. As oxygen easily inhibits its polymerization, the stickiness remains on the surface. This remnant of stickiness makes the laydown time longer. Therefore, the air drying additive such as paraffinwax is added to improve this problem. But this improvement involves a delamination between intermediate layer and top-coat in the overlay. To prevent the delamination, the authors investigated on synthesizing a fraction of air drying into a polymer chain and modification to flexible. Consequently, the flexible vinylester resin with air drying property has been newly developed.

Presented in this paper is a general description to evaluate this resin's performance compared with other radical curable resin generally used in concrete overlay.

EXPERIMENTAL

Materials (Resin and Filler)

The developed FVE was evaluated comparing with three other resins which were commonly used as matrix resin for concrete overlay. The resins compared were a conventional flexible vinylester resin(conventional FVE), a general flexible unsaturated polyester resin(general FUP) and a general vinylester resin(general VE). These resins are summarized in TABLE 1. The filler used 250 PHR in the intermediate layer was the aggregate of commercial grade round grained silica sand. A formulation of filler is shown in TABLE 2.

TESTING PROCEDURES AND METHODS

Liquid Resin Properties

Liquid resin properties including viscosity and pot life were measured according to JIS(Japanese Industrial Standard) K 6901. Surface curing time was determined at 25°C according to JIS K 5400, and was defined as tack-free. In this determination, the film specimens were coated 200 μ m in thickness on a glass plate. Liquid resins were cured by means of 55% methylethylketone peroxide (1.0PHR) and 6% cobalt octoate (0.5PHR).

Cured Resin Properties

Appling to the evaluation of cured resin properties, these resins were cured at ambient temperature by means of above mentioned polymerization techniques in a metal mold, and then post cured at 60°C for three hours.

Mechanical Property

Tensile properties were determined according to JIS K 6301. The shape of specimens were type 1 dumbbell and test speed was 10 mm/min. The destructive energy was calculated an area surrounded by tensile stress and elongation line recorded in the tensile test chart.

Abrasion Resistance

The film specimens were 1mm in thickness and determined their abraded quantities by means of Taber's abrasion tester (Toyo Seiki Co.) according to JIS K 6902. The load was 530 g and the examination turns were 250.

Chemical Resistance

The type 1 dumbbell shape specimens specified by JIS K 6301 were immersed in various solutions for thirty days at 40°C. Subsequently, the retention of tensile strength were measured the same to above mentioned in tensile property.

Overlay Properties

The specimen was prepared from a concrete slab (300mm×300mm×60mm), on which a urethane resin was primed at first and the paste consisting of a formulation of resin and filler was trowelled 3mm in thickness(intermediate layer). After surface cured(tack-free), each specimen was covered in 0.3mm thickness with top-coat which was the same as the resin to the intermediate layer. To set them the same surface curing(almost four hours constant), the paraffinwax was added to conventional FVE, general FUP and general VE. The applied condition of each intermediate layer is shown in TABLE 3. All the specimens were cured at ambient temperature for three days.

Adhesion Property

Peeling test [1] was applied to evaluate the adhesion property between intermediate layer and top-coat as a function of overlapped time from 2 to 24 hours after troweling intermediate layer. It was observed whether delaminated or not between these layers.

Thermal Shock Resistance

To evaluate the toughness of the overlay, the thermal shock test was made by air chamber at a temperature of 80 or 90°C and at various lower temperature from 20°C to -30°C, and maintained an hour at each temperature. Every condition cycled five times. The specimens were prepared by troweling the above mentioned compound onto a concrete slab, on which a surface of polyethylene terephthalate film(100mm×100mm) had already been placed so that the central part of the trowelled layer was free from concrete slab surface. As thermal stress could concentrate onto this part of the overlay, cracks would appear on its surface if it had less toughness. After each condition test, the delamination which occured between the overlay and the concrete slab, and cracks were observed.

RESULTS AND DISCUSSION

Liquid Resin Properties

As shown in TABLE 4, the developed FVE's surface curing time was shortest, whereas its pot life was almost the same as other resins. From this result, it was recognized that the developed FVE had air drying property. On the other hand, as the other flexible resins did not contain any air drying additives, their surfaces did not cure over twelve hours.

Cured Resin Properties

As shown in TABLE 5, the destructive energy was 750 MJ/m^3 and shore D hardness was 47 respectively for the developed FVE. As compared with other flexible resin, both properties of developed FVE were higher. It could be said that developed FVE is tough. Moreover, abraded quantity of the developed FVE was low. From these result, this developed FVE seemed to be suitable for more loaded overlay application.

The chemical resistance of developed FVE is shown in TABLE 6. It seemed that its chemical resistance was better than that of conventional FVE and general FUP.

Overlay Properties

Adhesion Property

All resin's surface curing times were adjusted in three or four hours to set workability condition constant. This condition is shown in TABLE 3. Peeling test was run to investigate the adhesion strength between intermediate layer and top-coat. The test result is shown in TABLE 7. However the delamination occurred within eight hours with conventional FVE, general FUP and general VE, developed FVE was not observed over 24 hours. This result means the paraffinwax covered their surfaces completely within eight hours. So, in case of covering top-coat as times goes by after troweling intermediate layer, the adhesion strength goes down and the possibility of delamination occurring increases. Traditionally, to prevent this delamination, intermediate layer had been covered with top-coat before the surface had been fully cured, or the surface had been sandblasted in case of as times goes by after troweling intermediate layer. From this result, this developed FVE seems to solve the delamination in various overlay applications. It could be said that sandblastless laydown will be possible.

Thermal Shock Resistance

As developed FVE's modulus was relatively higher than that of the other flexible resin, decreasing the adhesion strength was concerned between the overlay and the concrete substrate in hot and cold cyclic atmosphere. TABLE 8 shows the result of the thermal shock test. No delamination of the developed FVE was observed and no cracks occurred on the top-coat under the temperature deviation condition from 90℃ to -30℃. Concerning general VE, it was rigid, the delamination was observed the temperature under deviation condition from 80℃ to 0℃. Therefore, this developed FVE is considered to be suitable for concrete overlay matrix resin where hot water and some other chemical substances are continuously poured with heat cycle condition.

CONCLUSIONS

The advantages of this newly developed flexible vinylester resin with air drying are as follows,
1)   The surface curing property is excellent without any air drying additives. It could be possible to solve the problem of delamination in the overlay.
2)   The cured resin seems to be more tough and harder than that of a flexible unsaturated polyester resin and a conventional flexible vinylester resin.
3)   Chemical resistance seems to be superior to that of a flexible unsaturated polyester resin.

Because of these advantages, this resin is very suitable for concrete structures such as roof, floor, drainpipe and drainage in food processing plants and chemical plants damaged by water, boiling water, corrosive detergents, disinfectants and other chemicals for waterproofing and preventing degradation.

REFERENCES

1)Uragami.,et al. 1992. "Adhesive Strength of FRP Lining on Concrete", J. Soc. Mat. Sci., Japan, 41 (467) : 1305 - 1310.

BIOGRAPHIES

Tomoaki Aoki
Tomoaki Aoki is a researcher with Hitachi Chemical at Yamazaki Works. He received his M.S. degree in mineral resources and material engineering from Waseda University in 1991.

Kazuyuki Tanaka
Kazuyuki Tanaka is a section manager of composite production development with Hitachi Chemical at Yamazaki Works. He received his B.A. degree in applied chemistry from Kyoto University in 1977.

Kouji Arakawa
Kouji Arakawa is a senior researcher with Hitachi Chemical at Yamazaki Works. He received his B.A. degree in polymer chemistry from Osaka Prefecture University in 1969.

Kanemasa Nomaguchi

Kanemasa Nomaguchi is an assistant to works manager with Hitachi Chemical at Yamazaki Works. He received his B.A. degree in organic chemistry from The Gakushuin University in 1959.

TABLE 1　Materials (Resin)

Resin	Type
developed FVE	Bisphenol A based
conventional FVE	Bisphenol A based
general FUP	Isophthalic acid based
general VE	Bisphenol A based

TABLE 2　Materials　(Filler formulation)

Silica Sand No.5*	30　wt%
Silica Sand No.7*	30　wt%
under 400mesh Silica Powder	40　wt%

*Sand No. is according to JIS G 5901

TABLE 3　Applied　Condition　of　Intermediate　Layer

Item	developed FVE	conventional FVE	general FUP	general VE
Surface Curing Time　(hour)	4	4	4	3
Addition of wax	no	yes	yes	yes

TABLE 4　Liquid　Resin　Properties

Properties	Unit	developed FVE	conventional FVE	general FUP	general VE
Viscosity	mPa·s	280	250	350	300
Pot Life	min	26	28	26	29
Surface Curing Time	hour	4	>12	>12	6

TABLE 5   Cured Resin Properties

Item	Unit	developed FVE	conventional FVE	general FUP	general VE
Tensile Strength	MPa	14	8	3	60
Tensile Modulus	MPa	220	130	88	2800
Elongation at Break	%	140	110	104	2.7
Destructive Energy	MJ/m³	750	255	130	-
Abraded Quantity	mg/100turns	85	120	75	150
Hardness(Shore D)	-	47	30	27	84

TABLE 6   Chemical Resistance   (% Retention of Tensile Strength)

Solutions	developed FVE	conventional FVE	general FUP	general VE
30% Sulfuric Acid	◎	◎	△	◎
30% Sodium Hydroxide	○	○	×	◎
5% Ethanol	◎	○	△	◎
Saturated Calcium Hydroxide	◎	◎	○	◎
5% Sodium Hypochlorite	◎	◎	◎	◎
Tap-Water	◎	◎	◎	◎

◎ : >90%   ○ : 90-70% △ : 70-50%   × : less than 50%

TABLE 7   Adhesion Property   (Peeling Test)

Time to Overlap (hour)	developed FVE	conventional FVE	general FUP	general VE
2	○	○	○	○
4	○	○	○	×
8	○	×	×	×
24	○	×	×	×

○ : Not Delaminated   × : Delaminated

TABLE 8   Thermal Shock Resistance

Condition (cycle number)	developed FVE	conventional FVE	general FUP	general VE
80℃/20℃ (5)	○	○	○	○
80℃/ 0℃ (5)	○	○	○	×
80℃/-20℃ (5)	○	○	○	× (Cracks)
90℃/-30℃ (5)	○	×	○	× (Cracks)

○ : Not Delaminated   × : Delaminated

Composites—Leading the Offensive in Corrosion-Resistant Duty for Large Industrial Vessels

NOTE: No formal papers were submitted for this session.

Large Diameter FRP Applications in FGD Systems

JOHN C. MCKENNA AND **A. HERBERT**

Presentation will consist of a general discussion of large diameter chimney liners, slurry tanks and scrubber systems. Detailed historical information from recent stack liner survey will be demonstrated showing positive results in the use of composites for FGD applications. The field winding process and other considerations will be discussed including the examination of the material handling concerns for large structures.

John C. McKenna and A. Herbert, Ershigs, Inc., P.O. Box 1707, Bellingham, WA 98227

The Suitability of Isopolyester FRP for Water Treatment and Sewer Applications

H. R. DAN EDWARDS AND BEN BOGNER

For more than three decades, unsaturated polyesters based on isophthalic acid have been the workhorse matrix resin for fiber reinforced polymer (FRP) pipe, tanks and structures used in the transmission and treatment of water. Applications extend far beyond sewer pipe, the best known and largest use, to include reservoir pipe, power plant outfalls, municipal sewer pipe relining, and tanks for the storage of water treatment chemicals to name a few. In this paper many of these applications will be profiled and the technical reasons justifying the choice of isopolyesters will be reviewed.

While good field experience with isophthalate polyesters is most compelling, laboratory testing is convenient for isolating the resistance to specific chemicals. Total immersion of relatively thin laminates and elevated temperature are effective means to accelerate permeation as long as the test temperature is below the glass transition temperature of cured matrix resin. Increasing chemical concentrations beyond that seen in actual service can produce misleading results. New techniques to estimate actual service life expectancies from laboratory test results will be discussed.

H.R. Dan Edwards and Ben Bogner, Amoco Chemical Company, 150 West Warrenville Road, Naperville, IL 60563

Case Histories of ASME RTP-1 Stamped Vessels Designed by Subpart A and Subpart B

ALFRED L. NEWBERRY

In some cases RTP-1 stamped FRP vessels may be designed with the rules of formulas found in Subpart A. The stress analysis method found in Subpart B may be used for any vessel but must be used for vessels not covered by Subpart A rules.

The first case history is a vessel which was designed by Subpart A rules. The second case history is a vessel which could not be designed by Subpart A rules and had to be designed by Subpart B stress analysis. In this case, as in most cases, this analysis was performed using Finite Element Analysis. Both vessels were fabricated and stamped by Tankinetics Inc., Harrison, AR.

The differences in the design methods will be discussed. Special attention will be given to the methods and details involved in performing a Subpart 3B design.

Alfred L. Newberry, FEMech Consulting, 110 Rose Street, Harrison, AR 72601

The 0° and 90° Laminate—An Advanced Fabrication Method for FRP Corrosion Resistant Vessels and Equipment

JOE LASSANDRO

The intent of the paper is to present a method of laminating Fiberglass Reinforced Plastic cylinders for corrosive service which offers high strength in both the hoop and axial directions and excellent corrosion resistance. The intent is to provide an understanding of the technique, its capabilities and its advantages over present methods.

Joe Lassandro, Viatec, Inc., Road #3, Box 714, Plum Creek Road, Schuylkill Haven, PA 17972

SESSION 13-D / 1

Dual Laminate Piping

GARY ALAN GLEIN

Many alternatives are available for piping in chemical or high purity applications. Dual Laminate is constructed with an inner thermoplastic liner which is bonded to fiber reinforced plastic for mechanical strength.

This presentation will overview the design, construction, and use of dual laminate piping

Topics covered will include the description of products available, typical chemical and high purity applications, methods of design and construction and a cost/benefits analysis compared to fiberglass and metallic alternatives.

Information will be developed from reviews of past literature as well as interviews of users, fabricators and material suppliers.

Gary Alan Glein, Norcore Plastics, 1144 Thorne Road, Tacoma, WA 98421

SESSION 14

Design and Testing of Pultruded Composite Structures

ASCE/PIC Standard for Design of Pultruded Composite Structures

RICHARD E. CHAMBERS, MAX L. PORTER AND ASHVIN A. SHAH

ABSTRACT

This paper summarizes Phase 1 of an American Society of Civil Engineers' project to develop a national consensus standard for structural design, fabrication and erection of the pultruded fiber reinforced plastic products of structural shapes, plates and rods. The paper describes the literature survey conducted to help develop an outline of a draft standard or a prestandard. The paper presents a recommended outline of the proposed prestandard including scope and organization of individual chapters. Rationale for the design philosophy proposed for the prestandard are presented, including design criteria needed by structural engineers to design structural members of pultruded FRP products. Standards that are needed from the industry are also proposed.

INTRODUCTION

The American Society of Civil Engineers (ASCE) has been involved in development of engineering of composites in civil-engineering type applications since the 1960's; it produced the ASCE Structural Plastics Design Manual in 1984 (ASCE, 1984). The more recent national consensus standards development program for composites in construction began in the early 1990's. The Pultrusion Industry Council (PIC) of the Society of the Plastics Industry (SPI) of the Composites Institute (CI), has the long-range mission to develop accepted standards for structural design, fabrication and erection of fiber reinforced plastic (FRP) pultruded structural products, specifically, rod, plates, and shapes for use as mainstream components of construction. In 1995, the PIC sponsored at the ASCE, the development of a design draft standard or a *prestandard* with a view to process the prestandard upon completion as an ASCE national consensus standard in accordance with the rules of the American National Standards Institute.

ASCE convened a steering committee comprising pultrusion industry representatives and prominent structural and materials experts in the fields of structures and composites to

Richard E. Chambers, P.E., Chambers Engineering P.C., 2356 Washington St., Canton, MA 02021

Max L. Porter, Ph.D., P.E., Iowa State University, Ames, IA 50011

Ashvin A. Shah, P.E., Old Army Road, Scarsdale, NY 10583

oversee the development of the ASCE/PIC prestandard. In late 1995, ASCE retained Chambers Engineering P.C. to undertake Phase I, a one-year startup and planning phase of the program. Phase 1 included the following scope:

- Literature database survey of relevant documents, as background to the prestandard program, and as a basis for developing an outline of the prestandard and identifying gaps in knowledge that might impede promulgation of the standard.

- Prestandard outline -- the first step in defining approach to the development of the standard including the recommended design philosophy and relationship of the ASCE design standard with other material or industry standards.

LITERATURE DATABASE

The database of literature developed on this project contains some 350 documents published since circa 1980 that were screened for their relevance from a much larger initial selection. Dr. A. S. Mosallam, California State University at Fullerton, was retained by Chambers to assist in developing the database. The database is annotated with a capsule summary that serves as a brief-form abstract of the content of each document to aid development of the prestandard. The information is sorted into the following seven categories, that represent key broad issues in structural design of engineered composites in construction.

1) Materials -- properties and behavior of the composite.

2) Environmental Durability -- effects of exposure to thermal and chemical loads.

3) Time Effects -- creep deformation and static and cyclic fatigue.

4) Section & Member Performance -- analysis, design and tests of performance of members and structure.

5) Buckling -- local and general instability and interactions thereof.

6) Connections -- adhesive, mechanical fasteners alone or in combination.

7) General -- texts, manuals, and handbooks, standards and specifications.

The literature search revealed a newly published document that closely parallels the mission of the ASCE/PIC prestandard. The EUROCOMP Design Code & Handbook (Spon, 1996) is of interest since it *"represents the views of a wide body of designers, academic and manufacturing organizations as to what is considered to be current good practice."* It has neither legal standing nor the consensus of a standards-writing body, however.

The European code references the ASCE Structural Plastics Design Manual (SPDM) (ASCE, 1984), and some of its approaches are consistent with the SPDM. For example, it adopts the limit states approach as the structural design protocol as recommended in the SPDM. As another example, local and lateral torsional buckling of open thin-walled beam sections is treated the same as in the SPDM.

PRESTANDARD OUTLINE

A comprehensive outline of the prestandard has been developed that is annotated to provide a glimpse of the detail of the provisions. The broad outline follows the format found in other contemporary structural design standards, whereas the provisions are tailored to the characteristics of pultruded composites.

Basis for Outline

The review and outline development was guided by the following criteria: accommodate the contemporary load and resistance factor design (LRFD) protocol used for civil engineering structures (described in the next section); include commercial pultruded rod, sheet and structural shapes; reflect properties, behavior and flexibility of configuration and microstructure architecture furnished by FRP; consider methods of approach to dealing with problems that are common to both traditional materials and composites; and seek formats that are user-friendly to encourage acceptance by design professionals and code authorities.

The combination of two contemporary structural LRFD standards, one for engineered wood construction and the other for steel construction, proved to offer a framework for the overall format of this prestandard for pultruded structural shapes. That is, the characteristics of both the *materials* and the *configuration* of pultruded FRP *members* were reflected by combining the ASCE-LRFD specifications for wood in developing the framework for the *materials* characteristics covered in Chapters 1 and 2 (ASCE, 1995a). The concepts provided in AISC-LRFD specifications (AISC, 1994) for steel reflect specific structural characteristics of pultruded FRP members dealing with issues that are related to the *member* and the *configuration* of the product.

Organization & Scope of Chapters

Following are the 8 Chapters that comprise the standard, together with a capsule description of the scope of each chapter. Detailed provisions will be tailored to performance of pultruded FRP structures.

Chapter 1 -- General Provisions: Presents objectives of the standard and general provisions that apply to the standard as a whole. Goal is to provide design criteria for engineering grades of FRP pultruded rod, plates, and shapes used as structural members in construction. The scope applies to the structural design of buildings, and other structures and fabricated components where public safety is a consideration and rational design criteria are required.

Chapter 2 -- Design Requirements: presents design provisions that are common to the standard as a whole, such as reference documents, requirements such as stability and lateral support, and materials and product oriented provisions such as *reference* conditions for establishing *reference resistance. The chapter includes* general serviceability criteria; refer to future applications-specific design appendices and codes for serviceability criteria.

Chapter 3 -- Tension Members: Applies to members subjected to concentric axial tension and portions of members subjected to significant local tension arising from connection details. However, this chapter refers to Chapter 5 for members loaded in combined bending and axial tension, and refers to Chapter 7 for tension in connector regions.

Chapter 4 -- Compression Members and Bearing: Covers members subjected to concentric axial compression and to localized compression in locations of bearing. This chapter refers to Chapter 5 for members in combined bending and axial compression, including members with eccentric axial loads and to the Appendix for special cases of single-angle columns.

Chapter 5 -- Members In Bending and Shear: Covers both compact and non-compact prismatic members subjected to flexure, shear and torsion. Provisions provide for both flexural bending and flexural shear. This chapter refers to Chapter 6 for members loaded in biaxial bending and/or combined bending and axial tension or compression, refers to Chapter 2 for serviceability criteria, and refers to Chapter 8 for single angles.

Chapter 6 -- Members Under Combined Forces and Torsion: The provisions of this chapter apply to prismatic members subjected to axial force and bending about one or both principal axes, with or without torsion, and/or torsion only.

Chapter 7 -- Connections, Joints and Fasteners: applies to the design of mechanical and bonded joints, or combinations thereof, for connecting FRP pultruded members, including beam-to-column, beam-to-beam, and column-to-base connections. Mechanical fasteners include threaded rods, bolts, washers and nuts made from either FRP composites or metal. Bonded connections include both wet lay-up laminated and adhesive-bonded single and double lapped, and shop and field fabricated butt and strap joints.

Chapter 8 -- Special Considerations: This chapter covers special considerations in the design of pultruded FRP members, including such issues as strength design of flanges and webs subjected to *single* or *double* concentrated forces, design of single-angle members.

LIMIT STATES DESIGN PROTOCOL -- LRFD

Design Philosophy

The ASCE/PIC design standard that emerges from this program should be flexible and applicable to a broad variety of pultruded structural members of various shapes and laminate constructions on a routine basis. This generic approach does not exclude the potential for creation of innovative structural solutions that will be required in order to penetrate the mainstream civil engineering construction market in a significant way. The design philosophy embraced in the prestandard, (termed LRFD for Load and Resistance Factor Design) is the contemporary protocol used in the structural design of civil-engineering-type structures. It allows fine-tuning of the structure to carry applied loads efficiently and safely, based on the structural performance of the structural member. The ASCE Structural Plastics Design Manual (ASCE, 1984) summarizes the contemporary philosophy of structural design of civil engineering structures:

> *The purpose of design is the achievement of acceptable probabilities that the structure being designed will not become unfit for the use for which it is required, i.e., that it will not reach a limit state.*

This philosophy has to be the guide in the framing of this prestandard in order to offer an economical structure while also furnishing reasonable assurance that reliability and safety obtained are consistent with expectations for any mainstream civil-engineering structure. This demands a *technically sound* approach based on research, existing performance data, experience, and sound engineering rationale. Information on materials and section properties must permit calculation of stress, strain, deformations and strength under any loads or load combinations applied to the member.

Features of LRFD vs. ASD

The traditional Allowable Stress Design (ASD) approach utilizes a single safety factor (**SF**) to account for the variability of both load effects and material or member strength, as follows:

Calculated Stress ≤ Allowable Stress

Sum of Loads ≤ Resistance (strength) + Factor of Safety

$$\Sigma\, Q_i \leq R + SF$$

The result is that the strength of the material is reduced to an *allowable stress* to compensate for the potential for *both* an increase in load above the design value and a decrease in member strength below the design value. Thus, the ASD approach lumps together the effects of load and member characteristics and does not distinguish between the two effects for their contribution to reliability and structural performance.

Load and Resistance Factor Design -- LRFD

LRFD protocol was developed as a refined design procedure that would better estimate the effects of loads (stresses) applied to a structure and the strength (resistance or capacity) offered by the structure. The general approach is as follows:

Factored Load Effects ≤ Factored Resistance

$$\Sigma\, \gamma i \times Q_I \leq \phi \times R^A$$

where: $R' = R*C1*C2*...Cn$

Notation:

γ_I = *Load factor* to account for deviations of the load-related uncertainties and variability, specified in ASCE 7 (ASCE, 1995b) or in Codes having project jurisdiction.

Q_i = *Nominal load* is load specified in ASCE 7 (ASCE, 1995b) or in Codes having project jurisdiction.

ϕ = *Resistance factor* reflects variabilities in materials properties, and mode and consequences of failure, and uncertainties introduced by analyses used in determining structural effects.

R^A = *Adjusted resistance* reflects effects outside of reference conditions, e.g. load duration end-use temperature and moisture environments.

R = *Reference resistance* is material strength or section capacity (or modulus or stiffness) under standard reference conditions, including specific values or ranges for load duration, moisture environment and temperature.

$C_1, C_2...C_n$ = *Adjustment Factors* account for a conditions outside of standard reference conditions, e.g., temperature and moisture environment, and load duration effects on apparent modulus and strength.

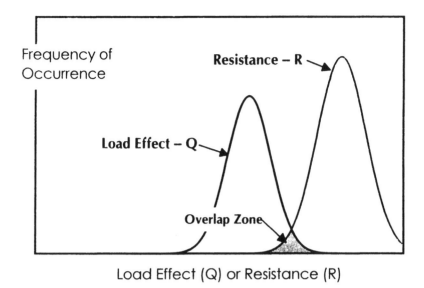

Load Effect (Q) or Resistance (R)

Figure 1 -- Frequency distribution of load effect Q and resistance R.

The essence of the LRFD approach is illustrated in Fig. 1, taken from the AISC Manual (AISC, 1994), which shows generalized frequency distributions resulting from variabilities in resistance and in load effects. The LRFD approach utilizes two factors -- the load and resistance factors -- to assure that the failure zone, as defined by the overlap in curves in Fig. 1, is acceptably small.

CONSIDERATIONS FOR MATERIALS AND PRODUCT STANDARDS

The standards for materials characteristics covered in Chapters 1 and 2 of the outline and for the product member properties needed for design in Chapters 3 to 8 are not in the scope of this current effort to develop the ASCE/PIC-LRFD prestandard on design. The material and product standards are, however, key components of the structural design system. Thus, observations and needs that were developed from the present study are discussed below.

Although there are several ASTM standards that deal with some aspects of pultruded structural shapes, there are no standards that characterize either their structural performance or the materials from which they are manufactured. The properties of materials and of the pultruded structural products are required in a detail sufficient to provide acceptable probabilities that the structure will perform as intended. Properties and factors called for in the LRFD equation need to be determined by a standardized approach. Thus, test methods for determining the material properties and statistical procedures for analyzing test results must be either drawn from existing standards or specially developed in order to assure consistency and reproducibility of the product performance. This performance would include: (a) test methods for elastic properties of modulus and Poisson's ratio, and reference resistance (strength) properties, all obtained for various stress modes usually in the three orthogonal axes of these orthotropic materials, and (b) analytical procedures to compute statistically valid reference resistance values and working values for elastic constants from test results obtained on coupons or members to achieve an acceptable overlap of the R-curve, Fig. 1.

(See ASTM reference (ASTM, 1993)) resistance computations for wood as an example of protocol for reducing test data to properties needed for LRFD.)

A classification system that defines, and is limited to *structural grades* of materials and laminate constructions may prove to be an effective means for dealing with a variety of laminate constructions that the industry may wish to furnish. The beginnings of such a system have been proposed by Evans (SPI, 1992) and the SPI Composites Institute (Evans, 1986).

Standardized procedures are required for determining section properties for use in strength and stiffness determinations under axial, shear, and bending forces. Consider as examples: shear lag, local buckling and compact sections, slenderness ratio limits, lower-bound strength limits of flange-web junction in wide flange sections.

Finally, adjustment factors are needed to modify reference values to reflect anticipated departures from reference conditions in the specific design project. Adjustment factors (C_n) would be determined to account for such end-use conditions as temperatures, moisture, load duration and other environmental conditions outside of the reference values established in the ASCE/PIC-LRFD standard. The number of adjustment factors required is significant.

CONCLUSIONS AND RECOMMENDATIONS

In order to promote widespread application of pultruded fiber reinforced polymer products in the civil engineering construction market, a national consensus design standard is required. Its objective is to reasonable assurance that reliability and safety are consistent with expectations for mainstream civil-engineering type structures. Accordingly, this will encourage acceptance of the FRP products by design professionals and code officials. ASCE recommends that the design philosophy of the proposed design standard be the Load and Resistance Factor Design, which is the contemporary protocol used in the national consensus standards for structural design of wood and steel products. ASCE has proposed an outline of a draft standard or a prestandard based on a literature survey of publications dealing with research on the design criteria for FRP products.

The next steps in the program require review of results of research, reduction of relevant approaches to the format required for standards provisions, development of actual provisions, and development of a commentary that gives background and basis for the provisions for reference by the user. This program must provide input to the ancillary standards for the product and for the properties and factors needed to characterize the products that the industry intends to furnish. These standards, vital components of the structural design system, are on the critical path for timely implementation of the ASCE/PIC-LRFD standard.

Detailed development of the prestandard is warranted, and should commence -- at an accelerated pace with expanded support and budget. At some point, full implementation of the structural design system will also require development for standards for fabrication and erection as envisioned in the long-range mission for the ASCE/PIC program.

REFERENCES

ASCE Structural Plastics Design Manual. 1984. American Society of Civil Engineers, NY, NY.

Structural Design of Polymer Composites - EUROCOMP Design Code and Handbook. 1996. E & FN Spon, Chapman and Hall, London.

LRFD Specification for Engineered Wood Construction - Draft Markup. 1995. American Society of Civil Engineers, New York, NY.

Load & Resistance Factor Design - Vol. 1. 1994. Structural Members, Specifications, & Codes, Second Edition, American Institute of Steel Construction.

Minimum Design Loads for Buildings and Other Structures. 1995. ASCE 7, American Society of Civil Engineers, New York, NY.

ASTM D5457-93. 1993. Standard Specification for Computing Reference Resistance of Wood-Based Materials and Structural Connections for Load and Resistance Factor Design Components, and Connections for Load and Resistance Factor Design, Annual Book of ASTM Standards, Vol. 04.09, American Society of Testing & Materials, West Conshohocken, PA.

Evans, D. J. 1986. Classifying Pultruded Products by Glass Loading Proc. of 41st Annual Conference, Session 06-E, SPI.

Society of Plastics Industry, Composites Institute. 1992. Recommended Specification for Materials Used in Pultruded Structural Shapes Pultrusion Industry Council of the Composites Institute, Society of the Plastics Industry, New York, NY.

BIOGRAPHIES

Richard E. Chambers is the Principal Investigator for the ASCE/PIC project. He is a Registered Professional Engineer in Massachusetts, and the Principal and President of Chambers Engineering P.C. He holds an S.B. in Building Engineering & Construction, and S.M. in Engineering (Materials & Structures) from Massachusetts Institute of Technology. His career covers structural design of buildings, research and consulting in plastics, composites and other construction materials. He is an author of the ASCE Structural Plastics Design Manual.

Max L. Porter, P.E., Ph.D., Parl., is a Professor of Civil Engineering at Iowa State University where he has taught over 30 different courses, conducted over 70 research projects principally on various kinds of composite structures, reinforced concrete, and masonry, and published over 70 papers and 200 reports. He is a member of the Board of Governors of the new Structural Engineering Institute (of ASCE), Chair of the ASCE Codes and Standards Activities, past Chair of the Executive Committee of the Building Standards Division, Chair of the Structural Composites and Plastics Standards Committee, Chair of the Steering Committee for the ASCE/PIC-LRFD Standards, Chair of the Masonry Standards Joint Committee, and past Chair of the Composite Deck Standards Committee, as examples of over 50 such leadership or officer positions in several professional organizations.

Ashvin A. Shah is a practicing engineer specializing in structural engineering, natural hazards, water resources, energy resources, and consensus standards development. Since July 1, 1996, Mr. Shah has provided technical support to the staff and committees of the American Society of Civil Engineers (ASCE) on several of ASCE's grants and contracts work. Currently Mr. Shah provides technical support to ASCE staff and the ASCE Project Steering Committees for various ASCE projects funded by industry or government agency including the Pultrusion Industry Council project on the development of a Prestandard for the Design of Pultruded Composites Structures. Prior to July 1, 1996, Mr. Shah was employed by ASCE as Director of Engineering. In this capacity, Mr. Shah supervised ASCE's Codes and Standards program, and programs in structural engineering, materials engineering, engineering mechanics, energy, and aerospace engineering, lifeline earthquake engineering, forensic engineering, and cold regions engineering. He assisted over 200 technical committees involves in developing technical publications, publishing technical journals, and organizing specialty conferences. Mr. Shah served as the secretary to the ASCE 7 standards committee on Minimum Design Loads for Buildings and Other Structures during its 1995 revision cycle. Prior to joining ASCE in 1990, Mr. Shah gained over twenty-five years experience in project management, design and construction of power plant facilities, industrial facilities, aerospace facilities, and commercial buildings. Education: BE, MSCE; Professional Engineer License in New York.

Design Considerations for Pultruded Composite Beam-to-Column Connections Subjected to Cyclic and Sustained Loading Conditions

AYMAN S. MOSALLAM

ABSTRACT:

Pultruded composite frame connections have a different behavior as compared to other conventional metallic connections. Two major issues need to be addressed when designing such structures, namely; the cyclic and creep behavior. This paper presents results of a study on investigating the behavior of beam-to-column exterior composite connections subjected to both cyclic and sustained loading conditions. This study aimed at understanding the behavior of pultruded fiber reinforced polymer (PFRP) frame structures under repeated and long-term loads. The general purpose of the equi-amplitude cyclic tests was to quantify the cyclic moment/rotation (M/Θ) characteristics of such structures and associated low-cycle fatigue behavior of RTM molded composite Universal Connector (UC). Results from full-scale connection specimens tested under cyclic loads are reported in this paper. A common mode of failure was observed for all specimens. This was a combination of bolt thread shaving and flexural fatigue-type failure of pultruded threaded rods. Other local failures to the pultruded thin-walled beam sections were observed near the ultimate moment. Minor cracks at the corners of the universal connectors were also observed. For the creep test, an identical exterior composite beam-to-column connection was fabricated with the same dimensions, details, and loading configuration to both quasi-static and cyclic specimens. Experimental creep data for the first 1,400 hours is presented. Relative rotation, and compression strain creep curves are reported. Test results indicated that during the first 420 hours, an increase of 46% in the relative rotation was recorded. This indicated the importance of considering the connection creep in the design of any PFRP frame structure. A simplified design expression to include the effect of creep and its effect on the composite frame connection moment capacity is proposed.

Civil and Mechanical Engineering Departments,
California State University, Fullerton
Fullerton, California 92834

INTRODUCTION:

Understanding the behavior of frame connections for pultruded fiber reinforced plastic (PFRP) structures is an essential key to satisfy both the safety and efficiency requirements of such structures. Unlike steel structures, limited number of full-scale experimental data is available on PFRP framing connections. This represents one of the major obstacles limiting the efficient and wide spread use of the material. There are several reasons behind the limited information on PFRP connection. First, lack of awareness of the manufacturers and fabricators of the importance and the seriousness of this issue. The experimental results have demonstrated the danger of adopting the recommended framing connections details provided in the pultruders "design" manuals [1], [2], and [3]. Second, the limited available appropriate connecting elements (e.g. Universal Connectors) for frame connections. Third, the inherent structural deficiencies of the majority of commercially produced open-web pultruded profiles (e.g., H, I, C, L). Fourth, due the absence of authoritative unified design code, structural designers of most of the PFRP structures that were built in the last decade, or so, have utilized and continue to utilize these inadequate and, in most cases, unsafe steel-like connection details. For these reasons, work in this area will require involvement not only in the connection characterization, but also in developing both appropriate frame connectors and providing stiffening details of existing PFRP open-web structural sections. This concept was discussed in detail by Mosallam [1].

MOTIVATION AND OBJECTIVES:

Previous full-scale studies on the performance of FRP portal frames, indicated the major impact of connection behavior on the overall performance of these structures including buckling and post-buckling capacity, the premature localized failure of open-web members, the ultimate strength, as well as the overall creep behavior of the thin-walled FRP structures. In order for pultruded composites to be considered as structural materials for civil engineering applications, these materials must prove their structural reliability, and higher efficiency under different loading conditions (e.g., dead, live, wind, earthquake, etc.) during the expected useful life of the structure. For composite connection, connection efficiency can be expressed as:

$$\eta = \frac{P_j}{P_m} \qquad \text{...............................Eq. (1)}$$

Where: η = Connection Efficiency (%),

P_j = Load producing failure to connection (lb.), and

P_m = Load producing failure to member (lb.)

For this reason, reliable information on the behavior of pultruded composites under these loading regimes must be available to the structural engineers. The majority of the previous composite connections studies have focused mainly on the static behavior of these FRP connections with little work on the dynamic, and seismic behavior of these joints [3]. Furthermore, due to the fact that the PFRP are viscoelastics, structural engineers must include the long-term (creep) effects under ambient and other varying environmental conditions.

PFRP CONNECTIONS RELATED WORK:

In the past few years, several research studies have been conducted in the area of characterizing the structural behavior of pultruded composite connections [5], [6], [1], [7], [8], [9], and others).

In 1984, Morsi et al. conducted an experimental study on a PFRP column- to-base connection using both composite and metal connecting elements [10]. Between 1984 and 1990, no work has been reported on the FRP frame connections. The first comprehensive full-scale study was conducted by Mosallam [5]. In this study, detailed theoretical and experimental investigations on the short and long-term behavior of PFRP frame structures were conducted. The effect of connections' details made of pultruded FRP composites, and their semi-rigid behavior have been studied. Based on this work, Bank, Mosallam, and Gonsior conducted theoretical and experimental investigations on the performance of PFRP connections [11]. Bank and Mosallam presented results of an experimental study for three different PFRP connection details [12]. In 1992, Bank, Mosallam, and McCoy extended this work by performing a study on different types of PFRP connections using existing structural profiles as connecting elements [13]. In 1994, Bruneau and Walker conducted an experimental study on the cyclic behavior of PFRP interior beam-to-column "rigid" connection [3]. The connection detail in this study was identical to those for steel connection with no consideration to the anisotropic nature of the materials and the fiber orientation of the connecting element (in their case the "T" stub-connectors). In their details, an expected premature failure occurred at an early stage of the experiment. This was because the fibers of the T-stubs were running in the wrong direction. This was described in details in several papers (e.g. [5], [2], [1]). This was reflected in the non-uniform hysteresis of the load/deflection curves produced in this study. Similar studies with steel-like connection details for both interior and exterior PFRP beam-to-column connections were reported by Bass and Mottram [14]. In their study, and in an effort to increase the flexural strength of these connections, double unidirectional cleat angles were used. Again, the fibers in both angles were running in the wrong direction, and the only addition was the increase of the matrix cross-section by doubling the thickness of the cleat angles (in their case isophthalic polyester). For the same reason, the same expected delamination failure of the top cleat occurred. This common type of failure for steel-like connections was described in details by Mosallam [5]. The same stiffening approach was adopted by Sanders et. al. [4] in their experimental study on the behavior of adhesively bonded "steel-like" pultruded connections. In 1996, Smith et. al. [16] presented a study on the behavior of exterior beam-to-column connections using both pultruded rectangular and H-profiles. Again, same premature failure was observed.

THE UNIVERSAL CONNECTOR:

In the present study, another approach was adopted by using appropriate composite connecting elements. In the development of these connectors, mixture of the past experience, available research and design data, and knowledge of the anisotropic behavior of the composite materials to develop an optimum design for FRP connecting element(s) were used to come up with an optimize geometry and fiber architecture of the connecting elements. The design criteria of the composite connector included: 1) proper fiber orientation, 2) ease of erection and duplication, 3) geometrical flexibility and suitability to be used in connecting a large variety of commercially available pultruded shapes and, 4) maximizing both the overall connection stiffness and ultimate capacity. The FRP connector (universal connector or UC) presented in this study was designed and fabricated from E-glass/vinylester composition. The UC element (Figure 1) can be used for the majority of PFRP connections for different shapes, e.g., exterior and interior beam-to-column connection, column-base connections, continuous beam connections, beam-to-girder connections, and others.

Figure (1) The Universal Connector

DESCRIPTION OF THE EXPERIMENTAL PROGRAM:

General

As it was mentioned earlier, the experimental program was composed of two major parts. First, experimental program on cyclic behavior of PFRP connections. Second, the performance of PFRP beam-to-column connections under sustained loads (ambient room conditions). For both testing programs, identical exterior connections detail was employed. The specimen geometry represents a portion of a typical moment-resisting frame structure (see Figure (2)).

Commercially produced PFRP H-profiles (4" X 4" X 1/4" E-glass/polyester) were used for both beam and column sections. The connecting element used in these connections were namely a combination of the following: high-strength epoxy adhesive (828 Epoxy with Versamid 140/V-40 Curing Agent), 1/2" pultruded threaded rods, molded square composite nuts, UC # 4 (#4 is a designation which was proposed by Mosallam for Universal Connector of 4" width [1]). The details of connection specimens for both the cyclic and the creep tests are shown in Figure (3).

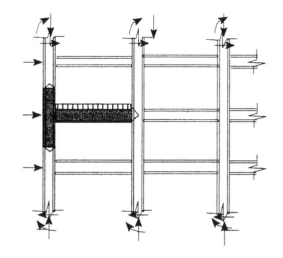

Figure (2): Typical Moment-resisting Frame Structure

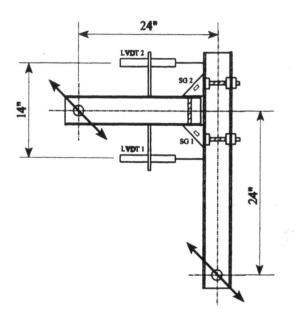

Figure (3): Connection Specimens Dimensions and Instrumentation

CYCLIC LOAD PROGRAM:

The objective of these tests were to: i) identify both the rotational stiffness and strength of each connection under cyclic loading, ii) identify the connection failure modes under cyclic loading and compare these modes to those observed from the quasi-static tests performed by the principal author in previous studies [1], [7], iii) evaluate the performance of the modified molded connectors under low cycles fatigue conditions, iv) and finally, to investigate the impact of using high-strength epoxy adhesives (in combination with pultruded threaded rods and molded nuts) on strength, stiffness, and ductility of these connections as compared to bolted-only connection details.

Test Description:

In performing experimental work on the seismic performance of frame connections, progressively increasing cyclically reversing loads are applied. In earthquake situation, the rate at which such loadings occur on individual members is relatively slow, and for this reason, it is possible to perform experiments by subjecting the connection specimen to a slowly cyclically reversing forces or displacements.

All connections were tested in a Closed-loop Electro-hydraulic MTS Testing System. Figure (4) shows the test setup for the cyclic tests. All data were collected using a computerized data acquisition system.

Figure 4: Cyclic Test Setup

In order to evaluate the flexural/rotation (M/Θ) characteristics for each specimen, two measurements must be obtained from each test. The first quantity is the applied moment at the connection, and the second quantity is the relative rotation between the beam and the column. Due to the fact that the tested connection specimens are statically determined, the first quantity (M) can be calculated at any load level (P) by knowing the moment arm. The loads were obtained directly from the control panel of the MTS machine which was connected directly to the data acquisition system and the moments were calculated during the analysis. The relative rotation was measured by two deflection readings at a fixed distance (14") using a rotational rigid metal bar fixture as shown in Figure (3). The two deflection readings were obtained using two ± 2" stroke linear variable differential transducers (LVDT) which were mounted at the ends of the rotational bar fixture. The LVDT's cores were bonded to the two column fixed point. At the beginning of each test, the LVDT's cores were positioned to approximately the middle of their stroke. The LVDT's signals were stored in a computer through a data acquisition system. In addition, two strain gages were mounted diagonally at the exterior diagonal faces of both the top and the bottom Universal Connectors at a distance of 0.5" from the edge of the UC to monitor the behavior and stress history of the connecting elements. These data were collected using a data acquisition system via a Vishay conditioning box. The reading frequency for all data was 12 Hz. At the start of each test, initial data were recorded, and a loading protocol was followed. Figure (5) shows the loading history which was used for these connections. This loading history was selected to simulate the behavior of these connections under cyclic loading due to an earthquake and associated after shocks. However, the results of this study maybe different under different cyclic loading history.

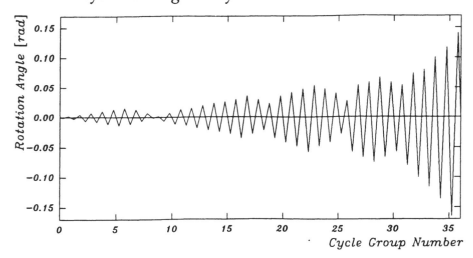

Figure 5: Loading History

The loading history consists of 36 cycle groups. Each cycle group, represented in Figure (5) as one compressed cycle, represents five cycles with the same amplitude. The relative rotation was gradually increasing for the first six cycle groups (total of 30 cycles). The load was decreased for the next three cycle groups (7,8& 9). Following that, the load was gradually increased again for eight cycle groups (total of 40 cycles) up to cycle group # 17 with an amplitude of (0.04 rad. The maximum amplitude was achieved at cycle group # 36 (total number of cycles=180) with maximum rotation value of about 0.16 rad. For all specimens, the test was conducted under displacement control, i.e., over the course of the 5 cycles, the rotation angle (Θ) was held at a certain value. Reversal cyclic displacements were applied to each connection detail. The shape of the loading signal was selected to be a negative sine wave (i.e., compression first, then tension). The loading frequency for all tests was 0.1 Hz. Load, rotation angle (Θ), and strain were measured continuously throughout the experiment. For each test, a total of 36 cycle groups, each 5 cycles, have been investigated (see Figure (5)). Cycle groups 23, 27, 31, and 35 were selected for more detailed evaluation as it will be discussed later.

CYCLIC BEHAVIOR OF SPECIMEN(BO):

For the bolted-only specimen (BO), the first "crackling" sound was noticed at the fifth group cycle (after 25 cycles of loading) at the 0.012 rad. amplitude. This crackling sound continued and became louder, especially at cycle group # 6, with a maximum amplitude of 0.0145 rad. After one hundred (100) cycles (beginning of group cycle # 21), a creaking sound was heard. This was the result of the beginning of shaving of the threads which allowed some sliding between the UC's bases and the flanges. After about 135 cycles (cycle group # 27), hair cracks were visible at the corners of the UC's and loud noise was heard. This continued until the major local failure of the threaded rods and nuts connecting the bottom UC in the side of the beam occurred. The main cause of the sudden failure of the threaded rods was the excessive decrease in the cross section (about 30 %) of the rod due to the failure of the outermost glass fibers because of the cyclic bending load with superimposed tension. Figure (6) shows a close-up photograph to one of the failed rods. From this figure, one can notice the fiber degradation of the threaded rod, and the reduction of the effective cross section of the composite bolt. The figure shows also the shaved thread mode of failure as described earlier.

CYCLIC BEHAVIOR OF SPECIMEN (BA):

For the combined bolted/adhesive specimen, BA, the first crackling sound was heard as early as cycle group 3 and continued until cycle group # 5 (= ±0.008 to ±0.012 rad.). The first crack in the adhesive material was clearly visible at cycle group # 5 (after 25 cycles). The crack line started from the edge of the H-beam top flange. As the load increases, hair cracks were observed at the corner of the UC, as well as at the bolts. For the bolted-only

the edge of the H-beam top flange. As the load increases, hair cracks were observed at the corner of the UC, as well as at the bolts. For the bolted-only connection specimen, BO, similar behavior was observed. However, higher relative moments were measured compared to the specimen BO, especially at the early stage of the test. A major failure of the adhesives occurred at around cycle group # 14 (± 0.022 rad. amplitude). At this stage the UC was practically not bonded by the adhesive to both the column and beam flanges anymore. However, the stiffness was still relatively higher than that for specimen BO. As the experiment progresses, and after the failure of the adhesive of specimen BA, the behavior of both specimen became more and more similar to each other and both specimens started to exhibit similar, relatively, lower rotational stiffness. Starting with cycle group # 31, the connection behaved merely as a bolted-only with an average rotational stiffness of 250 kip-in/rad for both specimens (see Table 1). The major local failure of the threaded rod/nut system occurred at cycle group # 35 (same as for specimen BO). At this stage, both bolts connecting the lower UC to the beam section failed in a flexural fatigue failure mode. As a result, the rotational stiffness was decreased, especially in compression and hence the connection moment capacity dropped dramatically. Figure (7) shows the moment/rotation hysteresis for the three cycle groups, 23, 27, and 31 for both specimen. From this graph, one can see that these three cycle groups had the same amplitude of ±0.07 rad.

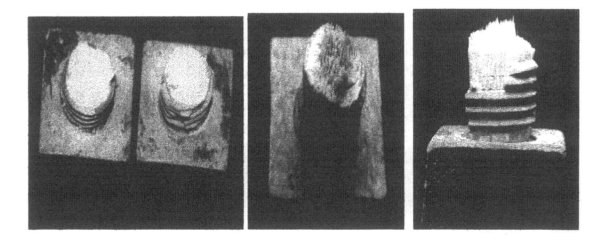

Figure 6: Close-up Photographs of the Typical Failure
Modes of Composite Threaded Rods

The response of the two specimens on cycle groups 23 and 27 was not very different, and in the case of specimen BO it was even identical. This indicates that the two specimens have not been significantly damaged in between. Cycle group 31, however, shows an obvious loss in stiffness coupled with a relative strength degradation. The apparent damages under the repeated load increase gradually. This happened in the form of hair cracks at the UC and progressive failure of the adhesive, progressive shaving of the rods, and progress of the cracks in the rods. The maximum moment was achieved at cycle group 23 for specimen BA and at cycle group 35 for BO (see Table 1). For the BA specimen, and following the major local failure of the threaded rods, the maximum moment resisted in compression was less than 50%, and the moment in tension was about 10% compared to the moment before the local failure occurred. This can be seen in Figure (8) where a large drop in the rotational stiffness occurred.

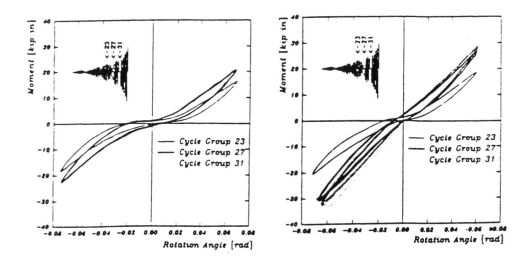

**Figure 7: Hysteresis-Curves for Specimens BO & BA
for Cycle Groups # 23, 27, and 31**

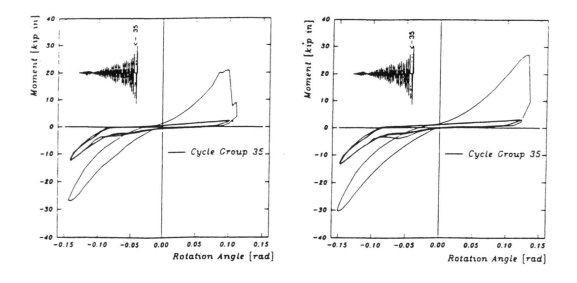

Figure 8: Hysteresis-Curve for BO and BA at Cycle Group 35

Cycle Group	θ [rad]	Bolted only (BO) max.Moment [kip-in] M⁺	M⁻	k [kip-in / rad]	max. Strain [%] tens.	comp	Bolted with Adhesive (BA) Max Moment [kip-in] tens.	comp	k [kip in / rad]	Max. Strain [%] tens.	comp
2	±0.004	2.61	-2.2	430	0.08	-0.16	6.30	-6.4	2200	0.26	-0.26
23	±0.07	22.08	-23.41	320	0.49	-0.61	29.38	-33.48	580	0.85	-1.13
27	±0.07	20.88	-22.60	260	0.52	-0.70	26.8	-30.90	470	0.51	-1.45
31	±0.07	16.36	-18.40	220	0.86	-1.01	18.77	-20.59	260	0.14	-1.65
35 before failure	±0.14	23.93	-27.64	180	1.03	-1.22	27.34	-30.41	190	0.40	-1.86
35 after fail.	±014	2.33	-12.54	150	---	-1.10	3.10	-13.21	160	---	-1.56

Table (1) Summary of Experimental Cyclic Results

The rotational stiffness dropped from 185 kip-inch/rad. to approximately 155 kip- inch/rad. This allowed the beam to rotate, relative to the column, by a large angle under lower moment. As the thread shaving progresses, the nuts moved about 1/8". This slip was blocked because the shaved material filled the undamaged threads. This made it harder for the composite nut to move further. At this stage, the threaded rods were under reversal bending and the outer fibers of these rods started to fail progressively. This was continued until the remaining cross section of the threaded rod was insufficient to carry the applied cyclic load. Another interesting mode of local failure was also observed, that is the appearance of delamination hair cracks in the composite nuts. The behavior of the threaded rods and nut system requires more in-depth evaluation. For this reason, more experimental studies are required to investigate other possible modes of failures by changing the geometry, locations of the threaded rods, as well as loading history. The commercially available threaded rods have major design problems. The first type of these rods are made of pultruded smooth rods and the threads are cut similar to steel threaded bolts. However, in this case the longitudinal fibers are damaged with the depth of the thread and become discontinuous because of this process. The other commercially produced composites threaded rods are composed of smooth pultruded rods, and the threads are made from another material which is molded on the outer skin of the smooth pultruded composite rod. The later type was used in this experiment. For this reason, it was expected to see the thread-shaving mode of failure due to the lack of bond and homogeneity of the rod core and the outer threaded shell. This is also the main reason of the limited torque which can be applied to the bolted joints. The "maximum safe" torque recommended by the manufacturer for this size was applied to all bolts (T_{max}= 15 lb-ft). According to the manufacture, the 1/2" diameter bolts, used in this program, have an ultimate thread shear strength is 4,000 (lb/inch of thread engagement). The maximum design tensile load is 2,000 lb, and the compression, and flexural strength values are 55,000 psi and 60,000 psi, respectively.

PFRP CONNECTION RESPONSE UNDER SUSTAINED LOADS (CREEP TESTS):

Due to the viscoelastic nature of pultruded polymer composites. It is essential to evaluate the effect of long-term loading of these connections. Several studies were conducted on the creep behavior of PFRP structures. Results of comprehensive experimental and theoretical study on creep behavior of PFRP portal frames was presented by Mosallam [1]. A detailed review on this subject can be found in Mosallam, and Chambers [7]. Connection creep is a function of time and temperature, connection configuration, stress level, orientation of the composite adherend, and types of adhesive. For adhesively bonded joints, the adhesive's rate of creep follows exponential laws similar to expressions used in stress-rupture analysis. In general, the adhesive creep rate (C_r) is given by:

$$C_r = A e^{\frac{-Q}{RT}} \quad \text{...Eq. (2)}$$

where A and Q are constants based on both the stress level and material properties, R is the gas constant, and T is the connection temperature.

EXPERIMENTAL CREEP PROGRAM:

As it was mentioned earlier, no experimental connections creep data for pultruded frame structures are available. In order to model and include the rotational creep effects, full-scale creep tests are required to precisely predict the actual behavior of each class of composite connections. For this reason, a pilot creep experimental program was initiated which included testing several adhesively bonded and bolted coupon specimens. In addition, a bolted-only full scale connection detail was subjected to a sustained load. The specimen was identical to those used in the cyclic loads testing program. The dimensions and test setup is described in Figure (3). The connection was tested under ambient environmental conditions. The average temperature was 75°F, and an average relative humidity of 73%. The specimen was instrumented with six 120 ohms electrical strain gages. All strain gages were coated by M-coating for environmental protection, and strain gages were connected to a strain indicator and switch box system. As for the static and the cyclic test specimens, the relative rotation angle (Θ) was measured using two deflection creep readings. All readings were taken manually. Both rotational and strain readings were taken every hour, for the first few hours, and twice a day thereafter. Both temperature and humidity were recorded at each reading. Creep data up to 1,400 hours are reported. The load was applied at the two ends using a special 20,000 lbs capacity lever arm creep Tester (see Figure 9).

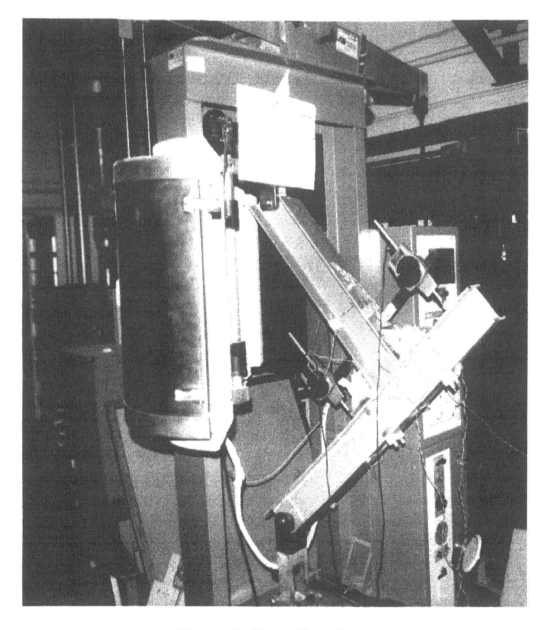

Figure 9: Creep Test Setup

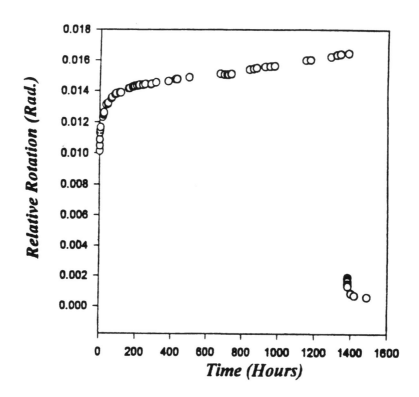

Figure 10: Rotation Angle (Θ) vs. Time
(Creep & Recovery)

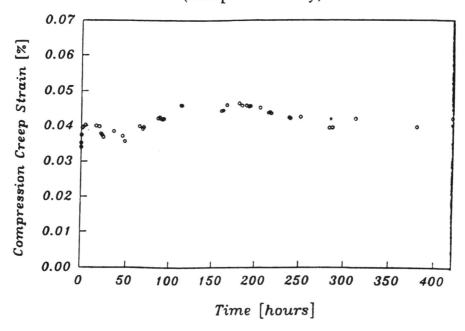

Figure 11: Compression Strain Creep Curve

$$K_v = \frac{K_o K_t}{K_t + K_o (t / t_o)^{n_k}} \quad \ldots\ldots\ldots\ldots\ldots\ldots\ldots \text{Eq.(3)}$$

Where: $K_o = M/\Theta_o$, $K_t = M/m_k$, $t_o =$ unit time (hours), $M =$ applied moment $n_k =$ material-dependent rotational creep parameter, and $m_k =$ stress-dependent creep parameter. The two creep coefficients can be determined experimentally for different stress levels by plotting the creep curves on a log-log scale (see Figure 12).

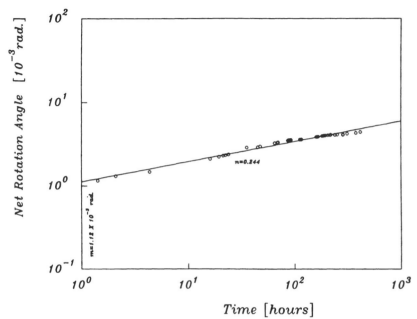

Figure (12): Determination of Creep Parameters m_k & n_k

This topic was not covered in this paper. However, it is highly recommended to initiate more in-depth studies in this area in order to evaluate the creep behavior of composite connections using short-term creep tests.

CONCLUSIONS AND RECOMMENDATIONS:

This study presents experimental results on cyclic and long-term behavior of exterior PFRP beam-to-column connections. Cyclic test results showed that the PFRP connections exhibits ductile failure and enough warnings are given before the ultimate failure of connection. In all cases the major local failure occurred at the lower UC. Hence, since the upper UC was still able to carry load, local failure occurred, and the structure was still capable of carrying the loads. The test results also indicated that the combined (bolted with adhesive) connection have a relatively higher initial stiffness. This is valid until the epoxy adhesive has failed. A common mode of failure was observed for both bolted and bolted/adhesive connections. This was in the form of threaded rods bending fatigue failure. The inefficient design of commercially produced threaded rods was identified through the "shaving" failure mode of the threaded shell. The design of a composite element or structure must not be a copy of a steel structure. The properties are too different and in order to be able to make more efficient designs, the anisotropic material properties must be considered. One has to "think composites" to come up with a proper connection design. It should be noted that these results were obtained from only four specimens with a specific details and loading history. A larger number of specimens are required to consolidate these results.

The creep test results indicated also that moment connection rotational creep cannot be ignored. In a period of 1,400 hours, the rotational stiffness dropped over 70% from the initial stiffness. This will affect the moment capacity, end restraint modeling, and stress distribution of any frame structure built from this pultruded composite material. Subsequently, this can affect the stability, long-term deflection, and, ultimately, the failure mode of the frame structure, especially when higher moments are applied. More work is required to understand the long-term behavior of such connections when subjected to higher stress level where non-linear creep behavior is expected.

ACKNOWLEDGMENT:

This study was supported by the National Science Foundation (NSF) under grant CMS-9415053. The support and encouragement of Dr. Ken P. Chong, program Director is highly appreciated. Thanks goes to Mr. H. Schmitz for his assistance in the laboratory effort. Pultruded composites were donated by Bedford Reinforced Plastics, Inc., Bedford, PA.

REFERENCES:

1. Mosallam A.S., (1994) "*Connections for Pultruded Composites: A Review and Evaluation*", Proceedings, the Third Materials Engineering Conference, ASCE, San Diego, CA, November 13-16, 1001-1017.
2. Mosallam, A.S., and Bank, L. C., (1992) "*Short-Term Behavior of Pultruded Fiber- Reinforced Plastic Frame,*" Journal of Structural Engineering, The American Society of Civil Engineers, Vol. 118, No. 7, July, pp. 1973-1954.
3. Bruneau, M., and Walker, D. (1994) "*Cyclic Testing of Pultruded Fiber-Reinforced Plastic Beam-to-Column Rigid Connection*" Journal of Structural Engineering, The American Society of Civil Engineers, Vol. 120, No. 9, September, pp. 2637- 2652.

4. Smith, S.J., Parsons, I.D., and Hjelmstad, K.D. (1996) "*A Study of the Behavior of Joints in GFRP Pultruded Rectangular and I-Beams*," Proceedings, The First International Conference on Composites in Infrastructure (ICCI'96), Tucson, AZ, January 15-17, pp. 583-595.

5. Mosallam, A.S., (1995) "*Connection and Reinforcement Design Details for Pultruded Fiber Reinforced Plastic(PFRP) Composite Structure*" Journal of Reinforced Plastics and Composites. Vol. 14, No. 7, July, pp. 752-784.

6. Mosallam, A.S., (1993), "*Stiffness and Strength Characteristics of PFRP UC/Beam-   to-Column   Connections*." Composite Material Technology, PD-   Vol. 53, pp. 275-   283.

7. Mosallam, A.S. and Abdelhamid, M.K., (1993), "*Dynamic Behavior of Pultruded PFRP Structural Sections*" Composite Material Technology, PD- Vol. 53, pp 37-44.

8. Zahr, S., Hill, S., Morgan, H., (1993), "*Semi-Rigid Behavior of PFRP/UC Beam-To-Column Connections*," Proceedings, The ANTEC Conference, Society of Plastic Engineers, New Orleans, Louisiana, May 13, 1496-1502.

9. Prabhakaran R., Razzaq, Z., and Devara, S. (1995). "*Block Shear Failure Modes in Pultruded Composites Plate Connections*." Proceedings, The Composites Institute 50th Annual Conference & Expo, The Society of Plastics Industry, Inc., Cincinnati, Ohio, Jan. 30- Feb. 1, paper # 10-F.

10. Morsi, A., Wissman, D., and Cook, J. (1984). "*Column-base Connections of Fiber Reinforced Plastics*," Proceedings, ASCE Structures Congress, Atlanta, Georgia, May.

11. Bank, L.C, Mosallam, A.S., and Gonsior, H. (1990). "*Beam-to-Column Connections for Pultruded FRP Structures*", Proceedings, First ASCE Materials Engineering Congress, Denver, CO. , August, 13-15, 804-813.

12. Bank, L.C., Mosallam, A.S. (1991) "*Performance of Pultruded FRP Beam-to-Column Connections*," Proceedings, 9th ASCE Structures Congress, ASCE, NY, NY, 389-392.

13. Bank, L.C., Mosallam, A.S., and McCoy, G.T. (1992). "*Make Connections   Part of Pultruded Frame Design*," Modern Plastics Magazine, Ed. P. A. Toensmeier, McGraw-Hill, August, pp 65-67.

14. Bass, A.J., and Mottram, J.T. (1994) "*Behavior of Connections in Reinforced Polymer Composite Section*," The Struct, Engrg., 72 (Autumn).

15. Structural Plastics Design Manual, (1984), Report # 63, Chapter 4, ASCE, NY, NY.

16. Sanders, D.H., Gordaninejad, F., and Murdi, S. (1996) "*FRP Beam-to-Column Connections Using Adhesives*," Proceedings, The First International Conference on Composites in Infrastructure (ICCI'96), Tucson, AZ, January 15-17, pp. 583-595.

Professor Ayman Mosallam, Ph.D., P.E.

Dr. Mosallam is a Research Professor in the Civil, and Mechanical Engineering Departments at California State University, Fullerton, Fullerton, California. He received his Ph.D. Degree in the area of structural Engineering from the Catholic University of America and his undergraduate degree from Cairo University in Egypt. He has previously been affiliated with the George Washington University (GW), The Engineering Research Institute (ERI) of the University of District of Colombia (UDC), and the Catholic University of America, Washington, D.C. He has over eighteen years of experience in the area of structural engineering. Professor Mosallam is a registered Professional Structural Engineer at the District of Colombia, Washington, D.C. For the past eight years, he has been heavily involved in the area of full-scale testing of PFRP structural systems. He is a member of the ASCE/SCAP Committee, and serves as the Secretary of the SCAP Subcommittee on the ASCE Structural Plastics Design Manual revision. He is also a member of the ASCE Standards for Structural Composites and Plastics, and serves as the Chairman of the Standards subcommittee on Composites connections. He has played a major role in developing the data base for pultruded composites and the Pre-Standard project of the SPI/PIC/ASCE project in 1996. He is the Chairman of the International Composites Engineering Infrastructure Committee. Dr. Mosallam is a Editorial Board Member of Composites Journal (Part B: Engineering) of Elsevier, and is the Guest Editor of Infrastructure Issues of the Journal. He has published over than fifty technical papers on characterization, performance and design of composite structures. Professor Mosallam is the recipient of a number of national and international awards, including the Best Design Paper Awards from the Composites Institute in 1995 and 1993, the Best Overall Paper Award from Modern Plastics Magazine, and the Best Commercial Innovation Paper Award from Plastics World Magazine in 1993. He is listed in Who's Who in Technology (7th edition).

Simulation of Progressive Failure of Pultruded Composite Beams in Three Point Bending Using LS-DYNA3D

DAVID W. PALMER, LAWRENCE C. BANK AND T. RUSSELL GENTRY

ABSTRACT

An analytical procedure to simulate the progressive failure of pultruded composite beams loaded in flexure is described in this paper. In the testing portion of the work, pultruded composite beams were loaded in three point bending and subjected to a midspan deflection large enough to cause failure. From this test, the force vs. displacement behavior of the beam is measured. The development of a methodology to numerically simulate the three point bending test has been initiated. While several failure mechanisms are observed during testing, including matrix cracking, fiber breakage, crushing and delamination, this study focused on the progressive tearing failure unique to thin wall pultruded beams of rectangular cross section. In this context, tearing refers to the separation of the vertical and horizontal members at the corners of the cross section. Tearing typically begins directly under the loading head and progresses toward the ends of the beam as vertical deflection increases. Using the LS-DYNA3D explicit finite element software package, a modeling technique capable of capturing the aforementioned tearing action was developed. Comparison of numerical force vs. displacement results and experimental data showed that at the current level of development, the finite element model captures the overall behavior of the beam with sufficient accuracy to warrant its application to the design process.

INTRODUCTION

As part of an effort to study the potential application of pultruded composite materials to highway guardrail systems, a program of testing has been undertaken to study the large deflection behavior of rectangular cross section pultruded composite beams loaded in flexure [1]. In the flexure tests, the beams were loaded in quasi-static displacement controlled three point bending. The results of the tests took the form of a load force vs. beam displacement curve, providing a measure of both the load carrying capacity of the beam and its energy absorbing characteristics. As outlined in [2], the quasi-static test is a very useful tool in designing guardrails because it provides an order of magnitude estimate of the true impact performance of the rail. With the intent of expediting the design process, the present study will attempt to determine whether an explicit finite element code, specifically LS-DYNA3D [3], can be used to numerically simulate the quasi-static three point bend test.

David W. Palmer, Lawrence C. Bank and T. Russell Gentry, Department of Civil Engineering, The Catholic University of America, Washington, D.C. 20064

Figure 1 Pultruded Composite Beam Mounted in Quasi-Static Test Bed

QUASI-STATIC TESTING OF PULTRUDED COMPOSITE RECTANGULAR CROSS SECTION BEAMS

The quasi-static test configuration in use at Catholic University is shown in Figure 1. As shown in the figure, the beam to be tested is mounted at both ends on 381 mm sections of W6x8.5 steel I beam spaced 1905 mm apart. The test proceeds by lowering the load head of the universal test machine at a specified rate, typically in the range of 5 to 10 millimeters per minute, and recording the force reported by the load cell. As reported in [1], several cross sectional configurations have been considered to date. This study however, will focus on a particular configuration, namely a 76.2 mm square cross section pultruded composite beam with 6.25 mm wall thickness.

When considering all test results collected to date, it was noted that the force vs. deflection curve from each beam followed a stiffening-softening-stiffening trend. The first phase of the beam response was elastic, producing a positive, stable relationship between the load head force and the beam deflection. The test specimens then typically showed a sudden, drastic loss of stiffness at the end of the elastic phase, resulting from the rupture of the pultruded material at the corners of the beam cross section directly under the loading head. After this initial stiffness loss, plastic deformation of the I-beam blockouts and progressive failure of the composite material were observed simultaneously in the second stiffening phase. In all cases, tests continued until the composite specimen experienced complete failure, breaking into two pieces, or until the displacement limit of the test machine was reached.

In the course of testing pultruded beams with single or multi-cell closed rectangular cross sections, a very particular type of damage unique to composite materials has been observed. Specifically, at the corners of rectangular sections, out-of-plane shearing failures occurred in which the vertical and horizontal members of the beam were torn apart from one another. The tearing is referred to as out-of-plane because the shearing action occurs in a plane perpendicular to the horizontal plane. The horizontal members of the cross section were typically seen to separate from the vertical members in a manner that preserved the

Figure 2 Progressive Tearing Failure of Square Cross Section Pultruded Beam

inner width of the horizontal members. That is, the tearing occurred in the plane of the inner surface of the vertical members of the cross section so that the dimensions of the vertical members of the beam were not reduced in any way. This type of failure was consistently observed in single and multi-cell rectangular cross section beams, and was the predominant mode of observable damage in most cases. Tearing progressed along the length of the beam, beginning at the load head and moving toward the ends, as the displacement of the load head increased. An illustrative example of this tearing behavior is shown in Figure 2, which displays two sequential photos taken during the quasi-static test of a 76.2 mm square cross section beam with 6.25 mm thick walls.

Failure of this type is particularly useful in guardrail applications, as the remaining intact portions of the beam are still capable of carrying load and producing the tension field effect necessary to stop the colliding vehicle. Cases where very little tearing was observed did occur. However, these observations were limited to those cases where relatively thick wall beams were being tested. Based on test observations, the existence and extent of tearing damage appears to be a function of the wall thickness, with less tearing being observed as wall thickness increased. This suggests that beyond a certain ratio of cross sectional dimension to wall thickness the tearing behavior would be completely lost, resulting in a single catastrophic failure event rather than progressive failure of the beam.

NUMERICAL MODELING OF QUASI-STATIC TESTING

An explicit finite element code such as LS-DYNA3D is ideal for modeling the quasi-static three point bending test, as the tests involve very large deflections and progressive material failure with discontinuously nonlinear response. The numerical portion of this study focused on a particular beam configuration, namely a 76.2 mm square cross section with 6.35 mm wall thickness and a length of 1905 mm. These dimensions correspond to those of the beam pictured in Figures 1 and 2. The mesh used in simulating the quasi-static test is displayed in Figures 3a and 3b, from isometric and side views. The beam was fixed to the

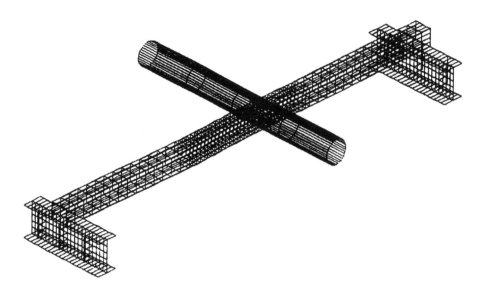

Figure 3a Isometric View of Quasi-Static Test Finite Element Mesh

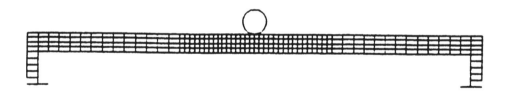

Figure 3b Side View of Quasi-Static Test Finite Element Mesh

blockouts at both ends, which had the material and geometric properties of the blockouts used in the test configuration of Figure 1. The elastic constants of the composite material were supplied by the manufacturer of the beam, and are listed in Table 1 as well as reference [4]. The load head is represented as a rigid tube 76.2mm in diameter. The simulation run proceeded by specifying a displacement rate for the load head and allowing it to travel 360 mm. Various values of head speed were used.

The challenge in accurately modeling the large deflection three point bend testing of pultruded square cross section beams lies in the capture of the out of plane tearing phenomena. To model this behavior in LS-DYNA3D, the square cross section beam was modeled as a collection of four single layer orthotropic elastic flat strips joined at the edges. The LS-DYNA3D code allows for the simulation of spotwelds, which were used to connect the strips in forming the beam. The spot weld is realized in the model as a short rigid beam connecting two nodes. The beam is rigid in the sense that nodal rotations and displacements are rigidly coupled, however, brittle failure of the beam can be specified. It is the brittle failure option that will allow the model to capture the sudden rupture of the material under the load head as well as the progressive tearing along the length of the beam. In the current model, the length of the spotweld is set at 1mm. The failure of the spotweld is governed by the following relationship:

$$\left(\frac{|f_n|}{S_n}\right)^N + \left(\frac{|f_s|}{S_s}\right)^M \geq 1$$

where f_n and f_s are the current normal and shear forces in the weld, and S_n and S_s are the normal and shear forces at spotweld failure. The above expression will yield an elliptical failure envelope that may be adjusted with the manipulation of the exponents N and M. Both of these parameters are set to 1 in this study.

The estimates of parameters S_n and S_s were based on the material strength values provided by the manufacturer of the beam and were calculated as

$$S_n = \frac{Y l t}{q} \qquad\qquad S_s = \frac{T l t}{q}$$

where Y is the strength of the material transverse to the fiber direction, T is the shear strength of the material, l is the length of the beam over which the calculation is made, t is the wall thickness and q is the number of nodes under consideration. As can be seen in Figure 3b, the density of the mesh varies in the axial direction, thus yielding different values of S_n and S_s for the center and end regions. The values of Y and T used in this study are given in Table 1.

E_{xx}	20.69 GPa
$E_{yy}=E_{zz}$	6.89 GPa
G_{xy}	2.89 GPa
G_{xz}	0.1 GPa
G_{yz}	0.031 GPa
$\nu_{yx} = \nu_{zx}$	0.12
ν_{zy}	0.4
Y	0.055 GPa
T	0.041 GPa

Table 1 Elastic Constants of Composite Material used in Simulations

LS-DYNA3D RESULTS AND DISCUSSION

Four values of load head displacement rate were considered in the simulation, namely 5, 50, 100 and 500 cm/s. The load head force vs. load head displacement plots output from LS-DYNA3D for the above displacement rates are shown in Figures 5 through 8 respectively. The load head force is inferred by monitoring the vertical reaction forces under the I-beam blockouts. The load head displacement is calculated by multiplying the displacement rate by the time. Additionally shown in Figures 5 through 8 are the experimental force and displacement data taken from the quasi-static test of the 76.2 mm square cross section pultruded beam with 6.35 mm wall thickness and length of 1905 mm. The load head speed in the experimental test was 5 mm per minute.

The run times required to complete the simulations are plotted in Figure 4, where they are seen to rapidly fall off as the displacement rate increased. All simulations were completed on a Gateway 2000 G6, with 64 Megabytes of RAM and a single Pentium Pro processor running at a clock speed of 200 MHz. The slowest run, which corresponded to a load head displacement rate of 5 cm/s, took 49 hours and 57 minutes, while the fastest run, which corresponded to a load head displacement rate of 500 cm/s, took 30 minutes. In each run, the displacement rate of the load head is specified by entering the desired displacement of the head and the amount of time in which the head must traverse the given distance. The load head traveled 360 mm in each run, but was brought down in less time with each successive run, so that the displacement rates corresponded to those listed above. The decreasing run times are a result of the fact that the integration time step used in the program remained constant as the load head displacement rate increased. As the head travel times decreased, fewer cycles of the program were required, resulting in shorter run times.

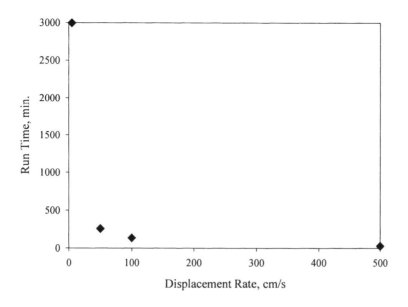

Figure 4 LS-DYNA3D Run Time vs. Load Head Displacement Rate

As can be seen in Figures 6 and 7, load head speeds of 50 and 100 cm/s produced results which most closely resembled the test data, with each force vs. displacement curve showing a large drop in stiffness at approximately 0.06 m of load head displacement. These sudden drops in stiffness were a result of several spotwelds breaking along the uppermost corners of the beam directly under the load head. The extent of splitting observed in the post processed animation, approximately 3 inches to either side of the load head, was consistent with that observed in the experiment. In both the 50 and 100 cm/s cases, a less severe stiffness loss was observed prior the large stiffness drop, at approximately 0.04 m of load head displacement, as shown in Figures 6 and 7. In both cases, these reductions in stiffness were caused by buckling of the vertical walls of the beam directly under the load head. As is clear from Figures 6 and 7, this initial stiffness reduction was not observed in the experiment. Spotwelds were observed to break progressively as the runs continued past the large stiffness drop, and in both the 50 and 100 cm/s cases tearing was limited to approximately 6 inches to either side of the load head, and occurred only along the upper spotwelded joints. In this respect, the 50 and 100 cm/s cases were inconsistent with the laboratory observations, in that the laboratory tested beam exhibited tearing along both the top and bottom corners, with most damage occurring on the underside of the beam. Figure 2 displays the tearing damage observed during the experiment.

The slowest head speed, 5 cm/s, resulted in a much more gradual loss of stiffness, with each successive reduction corresponding to small groups of spotwelds breaking. The 5 cm/s run and the 50 and 100 cm/s runs are similar in the sense that the spotwelds broke in groups. That is, stiffness losses occurred as sudden drops rather than steady declines. The same runs are clearly different when considering the number and severity of stiffness drops, although the rates of stiffness gain later in the runs are quite similar. A second likeness among the slowest three speeds was that almost all of the tearing behavior was observed along the upper spotwelds, which is inconsistent with the experimental observations. A slight amount of tearing was observed along the lower corners in the three slower runs, but only near the end of the simulation, at deflections greater than 300 cm. Only the fastest run, at 500 cm/s, produced a significant amount of tearing along the bottom corners. Although the load head force vs. load head displacement plot from the fastest run is quite unsteady, the tearing behavior observed in this case was closest to that observed in the experiment. The final mesh configuration for the 500 cm/s run is shown in Figures 9a and 9b, in isometric and side views.

It is clear in viewing Figures 5-8 that the response of the beam is highly dependent on the load head displacement rate. It is also apparent in viewing Figures 5-8 that as load head speed increased, dynamic effects became more pronounced. This is most clearly observed in Figure 8, which shows wide fluctuations in force as load head displacement increased. Dynamic effects are additionally evident in Figures 5, 6 and 7, which show that as load head speed increased, additional time was required for the force fluctuations resulting from welds breaking to die out. This is particularly visible when comparing the results in Figures 6 and 7 in the displacement range of 0.06 to 0.10 m. All results in Figures 5-8 have been filtered at 500 Hz in the LS-DYNA3D post processor. Future study will consider the effects of the impulsive loads applied to the beam by the brittle fracture of the spotwelds. The effects of variation in spotweld length and failure envelope parameters N and M will be considered in future work as well.

At the time of this writing, the dependence of the beam response on the load head displacement rate is not certain. A fundamental discrepancy exists in that as the conditions in the model approach those in the experiment, specifically regarding the speed of the load head, the resulting force-displacement behavior does not show a tendency to converge to the experimental data. The variety of results presented in Figures 5-8 strongly suggest that an optimum combination of mesh density, time integration step size and load head speed exists. Future efforts will investigate the effects of changes in these variables on the convergence rate and simulation run times.

Generally speaking, the model captured the tearing behavior observed in the laboratory, although the pattern of weld breakage and the associated levels of axial and shear force at the time of breakage were not consistent among the four simulations. It must be noted that only one experimental run was completed, and the elastic constants used in the simulation are those provided by the manufacturer. Additionally, data was collected in 5 second intervals during the experimental investigation, which precluded the recording of small fluctuations in load head force. Included among the recommendations for future study is the completion of additional experimental investigations, with data taken at smaller time intervals, as well as an independent assessment of the material properties.

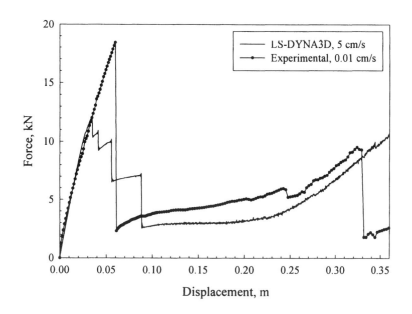

Figure 5 Load Head Force vs. Load Head Displacement, 5 cm/s

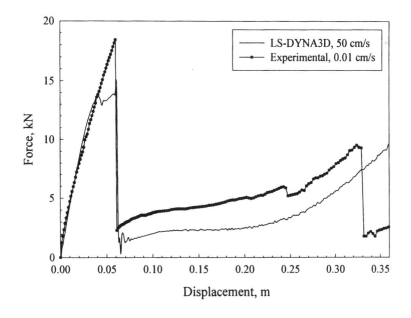

Figure 6 Load Head Force vs. Load Head Displacement, 50 cm/s

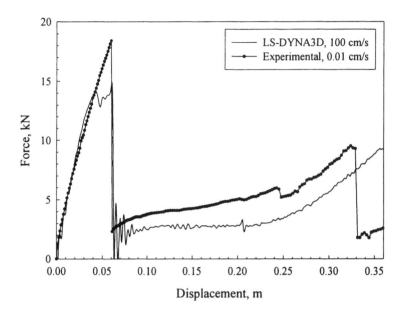

Figure 7 Load Head Force vs. Load Head Displacement, 100 cm/s

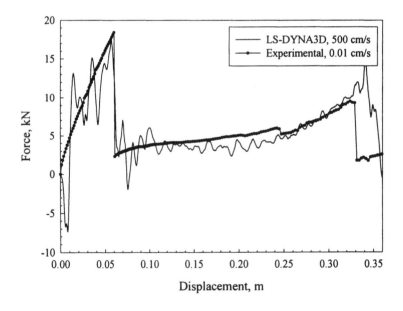

Figure 8 Load Head Force vs. Load Head Displacement, 500 cm/s

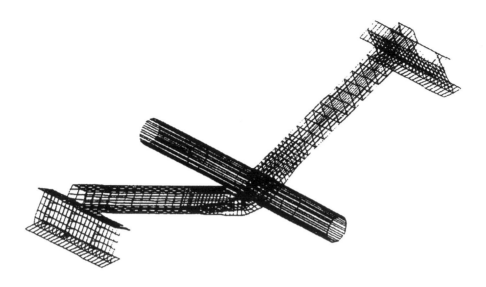

Figure 9a Isometric View of Quasi-Static Test Finite Element Mesh
Final Configuration at Load Head Speed of 500 cm/s

Figure 9b Side View of Quasi-Static Test Finite Element Mesh
Final Configuration at Load Head Speed of 500 cm/s

CONCLUSION

It must be emphasized that the results presented in this article are quite preliminary in nature and will require much additional study before the output generated by LS-DYNA3D is fully understood. However, the authors feel that the progress to date is quite promising, and they are confident that the modeling technique presented herein can be applied in preliminary design situations.

ACKNOWLEDGEMENTS

This material is based on work supported by the Federal Highway Administration under Cooperative Agreements DTFH61-92-X-00012 and DTFH61-95-X-00024. Any opinions, findings and conclusions or recommendations expressed in this publication are those of the Authors and do not necessarily reflect the view of the Federal Highway Administration.

REFERENCES

1. Gentry, T.R., Bank, L.C., Yin, J. and Lamtenzan, J.D., 1996. "Damage Evolution and Progressive Failure in Composite Material Highway Guardrails," to appear in the Proceedings of the 5th Symposium on Occupant Protection and Crashworthiness, International Mechanical Engineering Congress and Exposition.

2. Bank, L.C., Gentry, T.R., Yin, J. and Lamtenzan, J.D., 1996. "DYNA3D Simulation of Pendulum Impact Tests on Steel Guardrails," to appear in the Proceedings of the 5th Symposium on Occupant Protection and Crashworthiness, International Mechanical Engineering Congress and Exposition.

3. LS-DYNA3D, Livermore Software Technology Corporation, 2876 Waverley Way, Livermore, CA 94550, USA.

4. Creative Pultrusions Inc., 1996. *Creative Pultrusions Design Guide.*

An Investigation of the Influence of Tightening Torque and Seawater on Bolted Pultruded Composite Joints

R. PRABHAKARAN, S. DEVARA AND Z. RAZZAQ

ABSTRACT

Pultruded FRP composites are important structural materials that are finding increasing use in civil engineering applications. Mechanical joints, in particular bolted joints, are the preferred method of joining structural elements. Extensive work needs to be done, both in the area of testing and in the area of developing standards, in order to make bolted joints in pultruded composite structures efficient and safe. This paper is intended to further this effort by providing information regarding test results and the application of Load and Resistance Factor Design (LRFD) procedures.

In this investigation, two failure modes (block shear and net tension) were studied for five bolt-hole geometries; the tests consisted of four phases, namely dry finger-tightening, dry torqued, saltwater soaked finger-tightening and saltwater soaked torqued conditions. The test results were processed by LRFD-type equations which are used in the case of bolted steel connections. The emphasis here is on the development of design procedures, rather than on actual test results.

INTRODUCTION

Bolted joints in steel structures are designed according to LRFD procedures of the American Institute of Steel Construction (AISC, 1994). For designing bolted joints in composite structures, similar design procedures need to be developed. There have been many investigations of bolted joints in FRP composites. The stress distributions and failure modes in aerospace grade composites have been studied using finite element analysis (Hart-Smith, 1976; Crews, 1981). The stress distributions around loaded holes and the contact angle between the bolt and the composite specimen have been studied (Prabhakaran and Naik, 1986a, 1986b, 1987a, 1987b). Work on bolted joints in pultruded composites has been limited (Rosner and Rizkalla, 1992; Mosallam et al., 1994). The structural efficiency of bolted connections in pultruded glass FRP composites has been reported (Sotiropolous et al., 1994). A preliminary attempt to extend the application of LRFD procedures to pultruded composite plate connections has been reported (Prabhakaran et al., 1995). In the present investigation, previous work is extended to explore the effects of tightening torque and saltwater exposure on the strength of bolted joints of several bolt-hole geometries; these effects are interpreted using LRFD type equations.

R. Prabhakaran, S. Devara and Z. Razzaq, Old Dominion University, Norfolk, VA 23508.

OBJECTIVES

This paper presents the results from a series of tests on bolted joints in pultruded composites. The block shear and net tension failure modes are investigated for five bolt-hole configurations. Four test conditions are investigated: (i) dry, finger-tightening, (ii) dry, torqued, (iii) saltwater soaked, finger-tightening, and (iv) saltwater soaked, torqued. In each case, the failure loads are processed using LRFD type equations.

EXPERIMENTAL PROGRAM

The test specimens were fabricated from 1/2 in. thick E-glass fiber reinforced isophthalic resin matrix pultruded composite sheet (Series 1500 manufactured by Creative Pultrusions, Inc.). The composite has several layers of glass roving and continuous roving mat. The specimen configurations and dimensions are shown in Figure 1. These dimensions were chosen to highlight the block shear and net tension modes of failure. The specimens had one, two or four holes and were labeled type A, B, C, D, and E, as shown in the figure. Three specimens of each type were tested in a double-lap configuration, as shown in Figure 2. High strength steel bolts, 5/8 in. diameter, and washers were used to connect the composite specimens with the steel fixture-plates. The holes in the pultruded composite specimens were drilled carefully, to produce a very small clearance (of the order of 0.005 in.). The specimens were tested in tension on a 400,000 pounds capacity Tinius-Olsen testing machine.

As mentioned earlier, four sets of specimens were tested. The first set was simply finger-tightened. The second set was torqued to 196 ft. lbs, which corresponds to 80 percent of the recommended maximum torque for the particular steel bolts that were used. In a multiple bolted joint, the same torque level was used for all the bolts. The third set of specimens were soaked in seawater in a tank. The water was heated to 52°C for four and a half months, to accelerate the effect of salt water on the composite. The specimens were then taken out, dried and the bolts were finger-tightened before testing. The fourth set of specimens were soaked in seawater, as described earlier, and then the bolts were torqued to 196 ft. lbs. before testing.

RESULTS

The results for the dry, finger-tight case were reported earlier (Prabhakaran et al., 1995). For the sake of completeness, the failure loads and failure modes of the different types of specimens are given in Table I. The table shows that there is some scatter in the failure loads and the failure mode in a given type of specimens is the same, except in two cases (types B and E).

The results for the dry, torqued case are shown in Table II. While there is scatter in the test results, it is observed that specimens of each type exhibit the same failure mode. This Table also shows the average failure loads for each type of specimen, for the dry finger-tight case as well as the dry torqued case; the percentage increase due to the applied torque is shown in the last column. It is seen that the largest increase in strength is in the type A and type C specimens; for all the other types of specimens, where the failure mode is net tension, the increases are small.

The results for the seawater soaked and finger-tight cases are shown in Table III. Again, in spite of the scatter in the failure loads of the same type of specimens, the failure modes are the same for each specimen type. The last column shows the percentage reduction in the failure load due to the seawater soaking, as compared to the dry case.

Only specimens with four holes (type D and type E) were tested under both seawater soaked and torqued conditions. These results are presented in Table IV. For both types of specimens, the overall trends are the same: seawater immersion, even with the application of a torque, results in a loss of strength compared to the dry cases; the tightening torque partially offsets the detrimental effect of salt water exposure.

An observation of the failed specimens showed that for the dry torqued, seawater soaked finger-tight and seawater soaked torqued cases, specimens within each type (type A, type B, etc.) failed in the same mode. However, it should be noted that for the block-shear or the net tension failures of the staggered four-hole specimens, there are a few possibilities within each mode. The different possibilities for the net tension failure of seawater soaked E-type specimens are shown in Figure 3: the seawater soaked finger-tight specimens failed as shown in Figure 3(a) and the seawater soaked torqued specimens failed as shown in Figure 3(b). The dry finger-tight and dry torqued E-type specimens, shown in Fig. 4, failed in the same type of net tension failure mode. Thus, the failure of the seawater soaked torqued specimens in net tension involving only one hole can be attributed to the combined effect of salt water and the torque on the pultruded composite. The unnotched tensile strength of the composite showed a decrease of 6 percent due to saltwater exposure and this may have a bearing on the observed failure modes.

ANALYSIS OF RESULTS

As mentioned earlier, the specimen dimensions and the bolt hole configurations were selected to promote block shear and net tension failure modes.

The block shear test results for the seawater soaked finger-tight specimens A-1, A-2, A-3, C-1, C-2 and C-3 are summarized in Table V. The experimental ultimate loads for these specimens are given in the second column as R_{uE}. The R_{uE} values and the dimensions of the specimens are used to establish a block shear resistance factor ϕ_{BS} by the following expression:

$$R_{uE} = \phi_{BS}R_n = \phi_{BS}(\tau_u A_{ns} + \sigma_u A_{nt}) \tag{1}$$

where the material ultimate shear and normal stresses, τ_u and σ_u, are taken as 5,500 psi and 33,000 psi, respectively, and R_n is the maximum possible block shear resistance given by the expression in parentheses. The net shear area, A_{ns}, the net tension area, A_{nt}, and the R_n values for the specimens are tabulated in the third through the fifth columns of Table V. The ϕ_{BS} values calculated using Equation (1) are presented in the sixth column. The mean value of the six different ϕ_{BS} values is found to be 0.44. As seen from the seventh column of the table, if this mean ϕ_{BS} value is utilized, then $\phi_{BS}R_n/R_{uE}$ values for the six specimens range from 0.93 to 1.06. Adopting a reduced ϕ_{BS} value of 0.41, however, leads to $\phi_{BS}R_n/R_{uE}$ values ranging from 0.86 to 0.98, as seen from the last column of the table.

The net tension test results for specimens B-1, B-2, B-3, D-1, D-2, D-3, E-1, E-2 and E-3 are summarized in Table VI. The experimental ultimate loads are given in the second column as T_{uE}. The T_{uE} values and the dimensions of the specimens are used to establish a net tension resistance factor, ϕ_{NT}, by the following expression:

$$T_{uE} = \phi_{NT}T_n = \phi_{NT}(\sigma_u A_{nt}) \tag{2}$$

where the material ultimate normal stress σ_u is 33,000 psi and T_n is the maximum possible net tensile resistance given by the expression in parentheses. The net tension area, A_{nt}, and the T_n values for the specimens are tabulated in the third and fourth columns of Table VI. The ϕ_{NT} values calculated using Equation (2) are presented in the fifth column. The mean value of ϕ_{NT} is 0.43 but it leads to nonconservative values of $\phi_{NT}T_n/T_{uE}$. Therefore, a smaller value, 0.37, is used for ϕ_{NT}.

It should be noted that the ϕ_{BS} and ϕ_{NT} values proposed here are based on limited experimental results and therefore are tentative. Results for other testing conditions were analyzed in a similar manner and the resistance factors for various cases are summarized in Table VII.

CONCLUSIONS

Based on the experimental results and the LRFD type analysis of these results, the following principal conclusions are drawn:

- Application of a bolt-tightening torque improves the strength of dry bolted joints — more when the failure mode is cleavage or block shear and less when the mode of failure is net tension.

- Application of a bolt-tightening torque improves the strength of seawater soaked bolted joints — significantly when the failure mode is net tension.

- When the bolts are not tightened, seawater soaking diminishes the joint strength appreciably for all failure modes.

- Application of a bolt-tightening torque does not compensate fully for the damage done by seawater soaking.

- LRFD type equations can be used for block shear and net tension failure load predictions.

- Resistance factors for block shear and net tension failure modes are proposed on a tentative basis.

REFERENCES

American Institute of Steel Construction, Load and Resistance Factor Design Specification. 1994. Volume 1.

Crews, J. H., Jr. 1981. "Bolt-Bearing Fatigue of a Graphite/Epoxy Laminate." Joining of Composite Materials. ASTM STP 749:131–144.

Hart-Smith, L. J. 1976. Bolted Joints in Graphite-Epoxy Composites. NASA CR-144899.

Mosallam, A. S., Abdelhamid, M. K., and Conway, J. H. 1994. "Performance of Pultruded FRP Connections Under Static and Dynamic Loads." Journal of Reinforced Plastics and Composites, 13:386–407.

Naik, R. A. and Prabhakaran, R. 1987b. "Failure Modes for Polycarbonate Under Clearance-Fit Pin-Loading." Polymer Engineering and Science, 27:1681–1687.

Prabhakaran, R. and Naik, R. A. 1986a. "Measurement of Contact Angle in a Clearance-Fit, Pin-Loaded Hole." Experimental Techniques, 10:25–27.

Prabhakaran, R. and Naik, R. A. 1986b. "Investigation of Non-Linear Contact for a Clearance-Fit Bolt in a Graphite/Epoxy Laminate." International Journal of Composite Structures, 6:77–85.

Prabhakaran, R. and Naik, R. A. 1987a. "A Fiber-Optic Technique for the Measurement of Contact Angle in a Clearance-Fit Pin-Loaded Hole." Experimental Techniques, 11:20–22.

Prabhakaran, R., Razzaq, Z, and Devara, S. 1995. "Block Shear Failure Modes in Pultruded Composite Plate Connections." Proceedings of 50th Annual Conference, Composites Institute, SPI:10F1–10F7.

Rosner, C. N. and Rizkalla, S. H. 1992. Proceedings of Advanced Composite Materials in Bridges and Structures, Canadian Society for Civil Engineering.

Sotiropoulos, S. N., Ganga Rao, H. V. S., and Allison, R. W. 1994. "Structural Efficiency of Pultruded FRP Bolted and Adhesive Connections." Proceedings of 49th Annual Conference, Composites Institute, SPI:8A1–846.

BIOGRAPHIES

Dr. R. Prabhakaran

Dr. R. Prabhakaran is an Eminent Professor in Mechanical Engineering at the Old Dominion University in Norfolk, Virginia. He has published extensively in the areas of experimental mechanics and composite materials. He is the Technical Editor of the international journal, Experimental Mechanics, and is a Fellow of the Society for Experimental Mechanics.

Satish Devara

Satish Devara has completed his Master of Science degree in Mechanical Engineering at the Old Dominion University. He has worked and published in the area of pultruded composites.

Dr. Zia Razzaq

Dr. Zia Razzaq is a Professor in Civil and Environmental Engineering at the Old Dominion University. He has published extensively on various topics in structures. He is a Fellow of the American Society for Civil Engineering.

TABLE I–FAILURE LOADS AND MODES FOR DRY FINGER TIGHT SPECIMENS

Specimen	Failure Mode	Failure Load, lbs
A-1	cleavage	12,770
A-2	cleavage	12,820
A-3	cleavage	12,050
B-1	net tension	21,530
B-2	net tension	23,280
B-3	block shear	23,470
C-1	block shear	27,000
C-2	block shear	24,260
C-3	block shear	23,510
D-1	net tension	45,600
D-2	net tension	46,100
D-3	net tension	43,600
E-1	net tension	43,200
E-2	net tension	45,500
E-3	block shear	43,600

TABLE II–FAILURE LOADS AND MODES FOR DRY TORQUED (196 FT.LBS) SPECIMENS

Dry Torqued				Dry Finger Tight: Average Load, lbs	Percent Increase
Specimen	Failure Mode	Failure Load, lbs	Average Load, lbs		
A-1	cleavage	19,170			
A-2	cleavage	16,770	18,277	12,547	46
A-3	cleavage	18,890			
B-1	net tension	25,120			
B-2	net tension	24,920	24,770	22,760	9
B-3	net tension	24,270			
C-1	block shear	32,170			
C-2	block shear	32,330	33,100	24,923	33
C-3	block shear	34,800			
D-1	net tension	50,200			
D-2	net tension	49,200	49,833	45,100	11
D-3	net tension	50,100			
E-1	net tension	47,000			
E-2	net tension	45,400	46,800	44,100	6
E-3	net tension	48,000			

TABLE III–FAILURE LOADS AND MODES FOR SEA WATER SOAKED FINGER-TIGHT SPECIMENS

Specimen	Failure Mode	Failure Load, lbs	Average Load, lbs	*Percentage Decrease
A-1	cleavage	10,460		
A-2	cleavage	10,150	10,487	16
A-3	cleavage	10,850		
B-1	net tension	19,320		
B-2	net tension	19,480	18,220	20
B-3	net tension	15,860		
C-1	block shear	19,780		
C-2	block shear	20,220	19,693	21
C-3	block shear	19,080		
D-1	net tension	31,600		
D-2	net tension	29,390	32,073	29
D-3	net tension	35,230		
E-1	net tension	34,470		
E-2	net tension	31,380	34,650	21
E-3	net tension	38,100		

*Compared to dry finger-tight case.

TABLE IV–FAILURE LOADS AND MODES FOR SEAWATER SOAKED AND TORQUED SPECIMENS

Seawater Soaked & Torqued				Percent Change, Compared to		
Specimen	Failure Mode	Failure Load, lbs	Average Load, lbs	Finger Tight Dry	Torque Applied Dry	Sea Water Soaked Finger Tight
D-1	net tension	39,350				
D-2	net tension	38,300	38,583	-14%	-23%	+20%
D-3	net tension	38,100				
E-1	net tension	41,300				
E-2	net tension	39,900	40,567	-8%	-13%	+17%
E-3	net tension	40,500				

TABLE V–BLOCK SHEAR ANALYSIS RESULTS FOR SEA WATER SOAKED FINGER TIGHT CASE

Test Specimen	Exp. Ultimate Load, R_{uE} (lbs)	A_{ns}(in^2)	A_{nt}(in^2)	R_n (lbs)	Φ_{BS}	$\Phi_{BS}R_n/R_{uE}$ with	
						$\Phi_{BS}=0.44$	$\Phi_{BS}=0.41$
A-1	10,460	0.594	0.594	22,869	0.457	0.96	0.90
A-2	10,150	0.594	0.594	22,869	0.444	0.99	0.92
A-3	10,850	0.594	0.594	22,869	0.474	0.93	0.86
C-1	19,780	1.188	1.188	45,738	0.432	1.02	0.95
C-2	20,220	1.188	1.188	45,738	0.442	1.00	0.93
C-3	19,080	1.188	1.188	45,738	0.417	1.06	0.98

TABLE VI—NET TENSION ANALYSIS RESULTS FOR SEA WATER SOAKED FINGER TIGHT CASE

Test Specimen	Exp. Ultimate Load, T_{uE} (lbs)	A_{nt} (in^2)	T_n (lbs)	Φ_{NT}	$\Phi_{NT}T_n/T_{uE}$ with	
					$\Phi_{NT}=0.43$	$\Phi_{NT}=0.37$
B-1	19,320	1.188	39,204	0.493	0.87	0.75
B-2	19,480	1.188	39,204	0.497	0.86	0.75
B-3	15,860	1.188	39,204	0.405	1.06	0.92
D-1	31,600	2.376	78,408	0.403	1.07	0.92
D-2	29,390	2.376	78,408	0.375	1.15	0.99
D-3	35,230	2.376	78,408	0.449	0.96	0.82
E-1	34,470	2.469	81,477	0.423	1.02	0.88
E-2	31,380	2.469	81,47	0.385	1.12	0.96
E-3	38,100	2.469	81,477	0.468	0.92	0.79

TABLE VII–LRFD RESISTANCE FACTORS FOR PULTRUDED FRP BOLTED JOINTS SUBJECTED TO VARIOUS CONDITIONS

Bolted Joint Type	Resistance Factors for	
	Block Shear	Net Tension
Finger tight and dry	0.50	0.50
Torque applied and dry	0.70	0.55
Sea water soaked finger tight	0.41	0.37
Sea water soaked torque applied	—	0.44

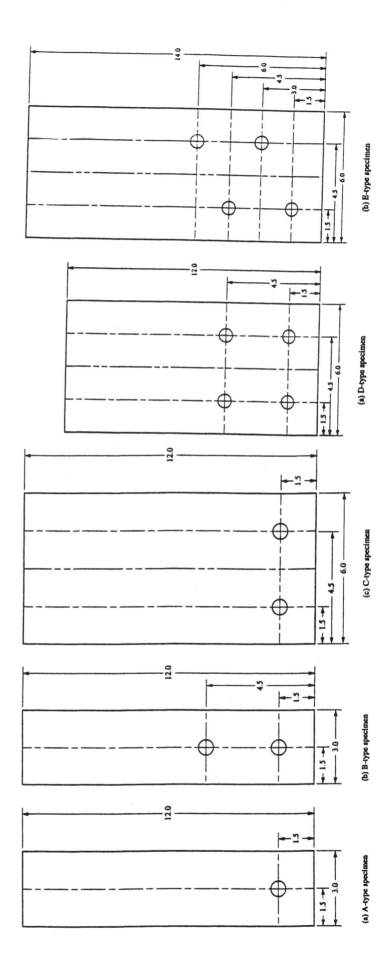

Figure 1. Dimensions of the tested bolted joint specimens

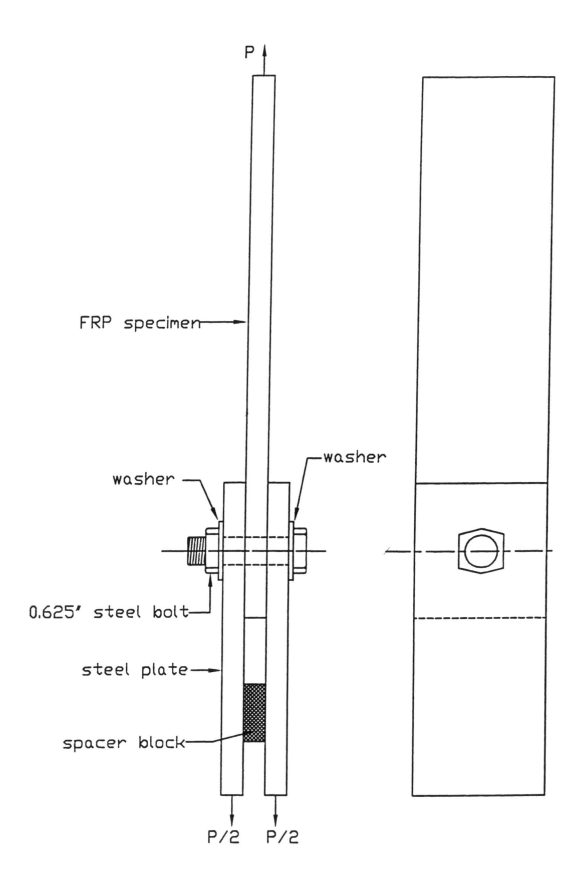

Figure 2. Bolted joint test setup

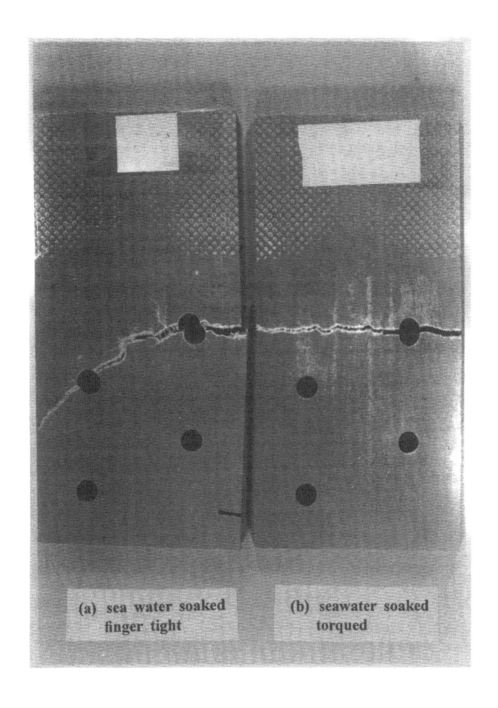

(a) sea water soaked
finger tight

(b) seawater soaked
torqued

Figure 3. Failed E-type specimens

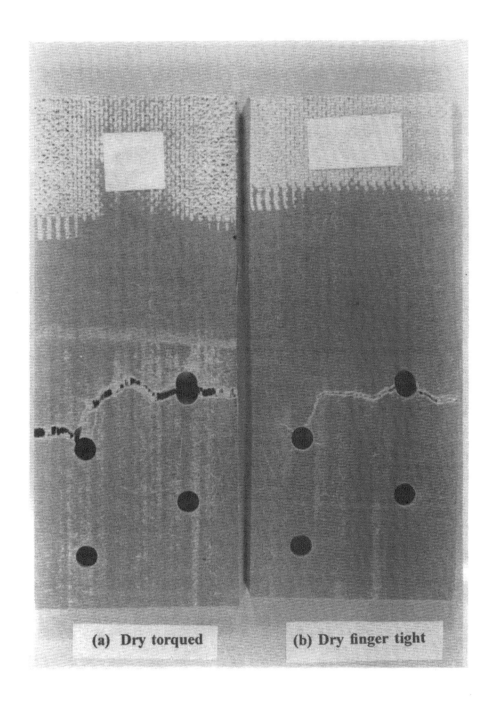

Figure 4. Failed E-type specimens

An Experimental Investigation of Bolt-Preload Relaxation in Pultruded Composite Joints

R. PRABHAKARAN, R. SRINIVAS AND Z. RAZZAQ

ABSTRACT

Pultruded composites are important infrastructure materials. They utilize polymers as matrices and are therefore susceptible to time-dependent effects. This paper is aimed at supporting the development of design guidelines and sound engineering practice related to bolted joints in pultruded composites.

Thick pultruded composites are susceptible to entrapment of air and the resulting voids contribute to the time-dependent behavior. In this paper, specimens from 1/2 in. thick pultruded sheets were first tested under compression in the thickness direction. Under different initial loads, the relaxation in the load and the changes in the specimen strain were monitored and were used to develop empirical relations.

Next, steel bolts were instrumented with strain gages and calibrated in tension. These bolts were used (singly) to apply preloads to composite specimens. Tests were conducted at three torque levels: 75, 150, and 215 ft. lbs. At each torque level, the relaxation in the preload was monitored and used to develop empirical relations.

In the third phase, the relationship between the tightening torque and the bolt preload was investigated at the three torque levels. The torque coefficient (K) was statistically determined for 10 repetitions of tightening at the highest and lowest torque levels, for unlubricated and lubricated conditions. The 'K' factor was obtained for the initial as well as the stabilized preloads.

In the fourth phase of the investigation, the effect of repeated bolt-tightening on the open-hole tensile strength and the pin-loaded tensile strength was investigated for one particular single-hole specimen geometry. The information presented in this paper is aimed at providing a better understanding of pultruded composite bolted joints.

INTRODUCTION

Pultruded composites are becoming important civil engineering structural materials because of their light weight, ease of manufacture, continuous production and other beneficial characteristics. Mechanical joints, such as bolted joints, represent an important method of joining pultruded composites. At present, the design of bolted joints in pultruded composites appears to be based on an extension of the procedures developed for steel structures, complemented by results from specific tests conducted on bolted pultruded composite joints.

R. Prabhakaran, R. Srinivas and Z. Razzaq, Old Dominion University, Norfolk, VA 23508.

The importance of bolt-preload on the performance of bolted joints is well known; because of this importance, the preload is calculated and the joint is usually torqued to produce the predetermined preload. But the preload does not remain at the initial value. Even in steel joints (steel plates clamped by steel bolts) at room temperature, the bolt tension has been observed to drop due to the elastic recovery that takes place when the wrench is removed; creep and yielding in the bolt due to the high stress level at the root of the threads also result in a preload relaxation (Kulak, Fisher and Struik, 1987). At high temperatures, the bolt preload relaxes due to viscoelastic effects, necessitating retightening at specific intervals.

In composites, the beneficial effects of bolt clamp-up have been recognized (Shivakumar and Crews, 1983). In the case of pultruded composites, the bolted joint strength has been found to increase significantly with the bolt-tightening torque (Sotiropoulos, Ganga Rao and Allison, 1994). Bolt tightening torque has been found to increase the dry as well as the seawater soaked joint strength (Prabhakaran et al., 1997). However, the strength improvement decreases over the long term due to relaxation. Polymeric matrix-based composites are viscoelastic even at room temperature and the time-dependent effects are more pronounced at high temperature and moisture levels. Most of the FRP materials, unless they incorporate reinforcements in the thickness direction, are more vulnerable to viscoelastic effects along the thickness, which is the direction in which the bolt preloads act. It is therefore necessary to investigate the mechanical behavior of composites in the thickness direction and also to study the time-dependent behavior of bolt preloads so that design guidelines can be developed for bolted joints. The present investigation is a step in this direction.

OBJECTIVES

This paper presents the results from tests conducted on pultruded composite specimens subjected to compressive loads in the thickness direction as well as specimens subjected to bolt-preloads. Load relaxations and strain variations under constant deformation conditions are reported, along with empirical relations. The influence of dry and lubricated bolt conditions on the torque coefficients, as well as the statistical nature of the important variables are reported. Finally, the effect of bolt clamp-up forces on the strength of the composite material is shown.

EXPERIMENTAL PROGRAM

In this investigation, the test specimens were fabricated from 1/2 in. thick E-glass fiber reinforced isophthalic resin matrix pultruded composite sheet (Series 1500 manufactured by Creative Pultrusions, Inc.). The composite has several layers of glass rovings and continuous roving mat. The 1/2 in. thick sheets have been observed to have a significant number of voids. The experimental program consisted of four phases. In the first phase 1 in. square specimens were tested under compression in the thickness direction. Under constant deformation conditions, the relaxation in the load and the specimen strain (in the direction of the applied load) was monitored. In the second phase, a 5/8 in. diameter hole (oversize by 0.005 in.) was machined in 4 in. by 1.75 in. composite specimens and a 5/8 in. diameter steel bolt with washers was tightened with torque values of 75, 150 and 215 ft. lbs. The bolt was instrumented with an electrical resistance strain gage, as shown in Figure 1, and was calibrated. At each torque level, the relaxations in the preload were monitored.

In the third phase 4, in. by 1.75 in. specimens were subjected to bolt preloads at the highest and lowest torque levels of the second phase. The statistical nature of the relationship between the torque and the preload was investigated for dry and lubricated bolt conditions. The differences between the initial and the stabilized conditions were also investigated. In the fourth phase of the investigation, the effect of repeated bolt-tightening on the open-hole tensile strength and the pin-loaded tensile strength was investigated.

RESULTS

The monotonic stress-strain behavior in the thickness direction was monitored for 1 in. by 1 in. composite specimens in different load ranges — 950 lbs, 4750 lbs, 9500 lbs and 19,000 lbs. The behavior is nonlinear, even in the smallest load range. Typical results for the 19,000 lb. load range are shown in Figure 2. The material, containing voids, appears to become stiffer due to the collapse of the voids and the stress-strain curve tends to stabilize after approximately 5 cycles. Load and strain relaxations in the thickness direction, under various initial loads, were monitored under constant deformation conditions. The load relaxation with time is shown in Figure 3; at all the three initial load levels, the magnitude of the compressive load decreased with increasing time. The variation of the compressive strain in the thickness direction with time is shown in Figure 4 for the same three initial load levels. From this figure, it is seen that at the 950 lb. initial load, the magnitude of the compressive strain decreased (corresponding to the decrease in the compressive load); at the higher load levels, the magnitude of the compressive strain increased (in spite of the decrease in the compressive load), possibly due to the gradual collapse of voids in the material.

Four steel bolts were instrumented with electrical resistance strain gages, as shown in Figure 1, and were calibrated in tension. The calibration constants ranged from 8.1 lb/$\mu\epsilon$ to 10.7 lb/$\mu\epsilon$. Composite specimens with holes were subjected to bolt tightening torques of 75 ft-lbs., 150 ft-lbs., and 215 ft.lbs. and the resulting preloads were monitored as a function of time. The bolt preload relaxation is shown in Figure 5 for all the three torque levels. The preload is seen to decrease significantly, especially at the higher torque levels, and then tends to stabilize after several hours.

Composite specimens with holes were subjected to bolt tightening torques of 75 ft-lbs and 215 ft-lbs. Each specimen was subjected to the same values of the torque ten times; the initial preload immediately after torque application and the stabilized preload after a suitable time were recorded. After each preload stabilization, the bolt was removed and the specimen was allowed to recover sufficiently, before retorquing. The mean and standard deviations for the stabilized preloads are shown in Table I. These values are shown for dry and lubricated bolt-thread conditions. It is seen that the standard deviations are appreciable for both dry and lubricated bolt conditions. Also, the mean preload under the lubricated condition is larger than the corresponding preload under dry conditions.

In the fourth phase of this investigation, 8 in. long, 2 in. wide specimens with a central 5/8 in. diameter hole were pulled to failure under different conditions. A set of three specimens were tested in tension. Another set of three specimens were subjected to a bolt-tightening torque of 215 ft. lbs ten times, with suitable intervals. This was done to simulate service conditions where the same structural member is used and reused again. Then, with the bolt removed, the specimens were tested in tension. The mean failure loads are shown in Table II. Similar tests were performed under pin-loaded conditions. Finally, three specimens, which were subjected to ten cycles of preloading and unloading, were tested in tension, with the bolt in place under a torque of 215 ft. lbs. The results show that prior bolt tightening leads to a small increase in the failure load, possibly due to the collapse of voids (which may otherwise act as damage nuclei).

ANALYSIS OF RESULTS

The load relaxation results from the first two phases were analyzed using the empirical relation:

$$P = P_o t^{-B} \tag{1}$$

where P_o is the initial load, P is the load at time t and B is a constant. The relaxation exponent, B, is shown as a function of the initial load in Figure 6. It is seen that the values of B for the two phases show similar trends and very small differences — as is to be expected.

The tightening torque and the resulting bolt preload are related by the equation:

$$T = KFd \tag{2}$$

where T is the torque, F is the bolt preload, d is the nominal bolt diameter and K is the torque coefficient. The torque coefficient depends on the friction in the bolt assembly and other parameters (Shigley and Mischke, 1989). From the above equation, it is clear that the smaller the value of K, the greater the preload for a given tightening torque. The influence of thread lubrication on K and the statistical variation of K have been investigated in the case of steel joints (Blake an Kurtz, 1965).

In the present investigation of pultruded composite joints, the variation of K with the tightening torque, T, for lubricated and unlubricated conditions, is shown in Figure 7. The figure shows the torque coefficient based on the initial preload as well as the stabilized preload. The same results are tabulated in Table III. In the light of Equation (2), it is seen that for a given tightening torque, the peload is greater under lubricated conditions; also, the preload decreases with increasing time, due to viscoelastic effects.

CONCLUSIONS

Based on the experimental results presented here, the following principal conclusions can be drawn:

- Pultruded composites, with polymeric matrices, exhibit time-dependent behavior; the voids in thick materials contribute an additional influence on the mechanical behavior.

- The behavior of the pultruded composite under a bolt preload can be correlated with the behavior under a direct compressive load in the thickness direction.

- The concept of a torque coefficient (K) can be extended to pultruded composite joints. However, additional work needs to be done to establish numerical values under different conditions.

- Repeated bolt tightening and dismantling (corresponding to reuse of composite structural elements) leads a collapse of the internal voids in the material, which in turn tends to improve the strength; however, actual loads on composite joints may induce various kinds of damage which need to be considered.

REFERENCES

Blake, J. C. and H. J. Kurtz. 1965. "The Uncertainties of Measuring Fastener Preload." Machine Design, 37:128–131.

Kulak, G. L., J. W. Fisher, and J. H. A. Struik. 1987. Guide to Design Criteria for Bolted and Riveted Joints. John Wiley and Sons.

Prabhakaran, R., S. Devara, and Z. Razzaq. 1997. "An Investigation of the Influence of Tightening Torque and Sea Water and Bolted Pultruded Composite Joints," Proceedings of the SPI Composites Institute Conference.

Shigley, J. E. and C. R. Mischke. 1989. Mechanical Engineering Design. McGraw-Hill Book Company.

Shivakumar, K. N. and J. H. Crews, Jr. 1983. "Bolt Clamp-up Relaxation in a Graphite/Epoxy Laminate," ASTM STP 813:5–22.

Sotiropoulos, S. N., H. V. Ganga Rao and R. W. Allison. 1994. "Structural Efficiency of Pultruded FRP Bolted and Adhesive Connections," Proceedings of the SPI Composites Institute Conference:8A1–8A6.

BIOGRAPHIES

Dr. R. Prabhakaran
Dr. R. Prabhakaran is an Eminent Professor in Mechanical Engineering at the Old Dominion University in Norfolk, Virginia. He has published extensively in the areas of experimental mechanics and composite materials. He is the Technical Editor of the international journal, Experimental Mechanics, and is a Fellow of the Society for Experimental Mechanics.

Ravi Srinivas
Ravi Srinivas is a graduate student in the Department of Mechanical Engineering at the Old Dominion University and is working in the area of pultruded composites for his Master's thesis.

Dr. Zia Razzaq
Dr. Zia Razzaq is a Professor in Civil and Environmental Engineering at the Old Dominion University. He has published extensively on various topics in structures. He is a Fellow of the American Society for Civil Engineering.

TABLE I–STATISTICAL VARIATIONS OF STABILIZED BOLT PRELOADS

CONDITION	TORQUE (ft-lb)	PRELOAD (lbs)	
		MEAN	STANDARD DEVIATION
DRY	75	4143	704
	215	15740	3201
LUBRICATED	75	5506	997
	215 (fresh)	16530	2808

TABLE II–FAILURE LOADS OF OPEN HOLE AND PIN LOADED
SPECIMENS, WITH AND WITHOUT REPEATED PRIOR TORQUING

Condition	Mean failure load, lbs
Open Hole (No Torque)[a]	13,800
Open Hole (prior torque)[b]	14,600
Pin Loaded (No Torque)[a]	13,300
Pin Loaded (prior torque)[b]	13,900
Under Torque	15,500

a. No previous torque applied on the specimen.
b. A 215 ft-lb torque applied 10 times prior to testing.

TABLE III–INITIAL AND STEADY TORQUE COEFFICIENT
UNDER DRY AND LUBRICATED CONDITIONS

CONDITION	TORQUE (ft-lb)	MEAN TORQUE COEFFICIENT	
		INITIAL	STEADY
DRY	75	0.35	0.46
	215	0.27	0.34
LUBRICATED	75	0.27	0.32
	215	0.26	0.29

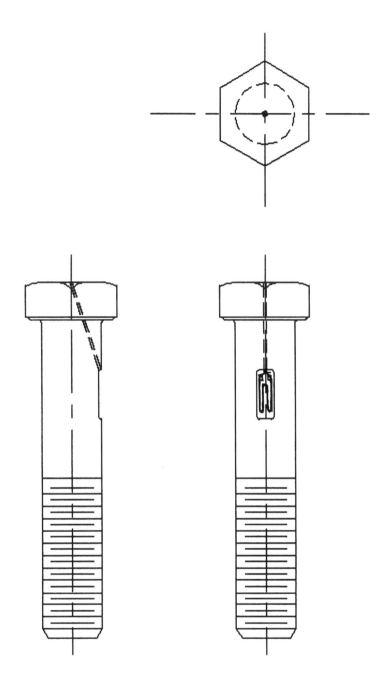

FIGURE 1. Instrumented bolt

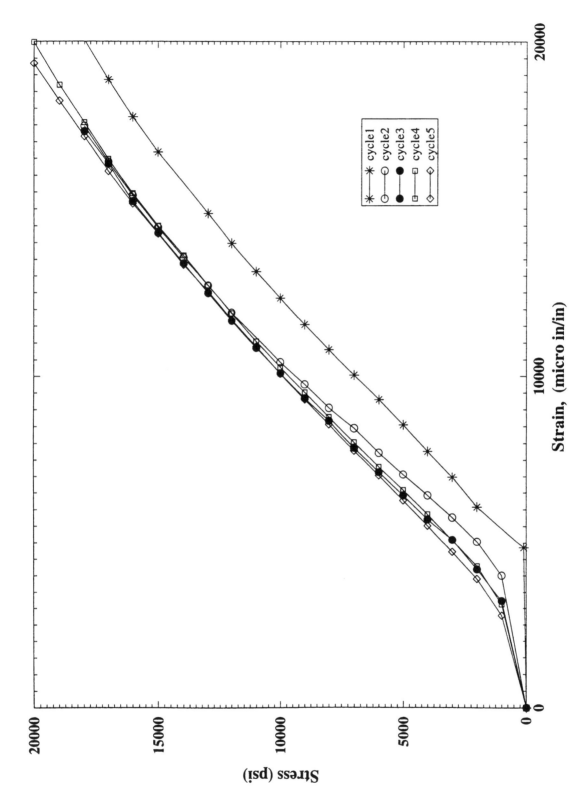

FIGURE 2. Stress-strain behavior in the thickness direction for the 19,000 lbs. load range

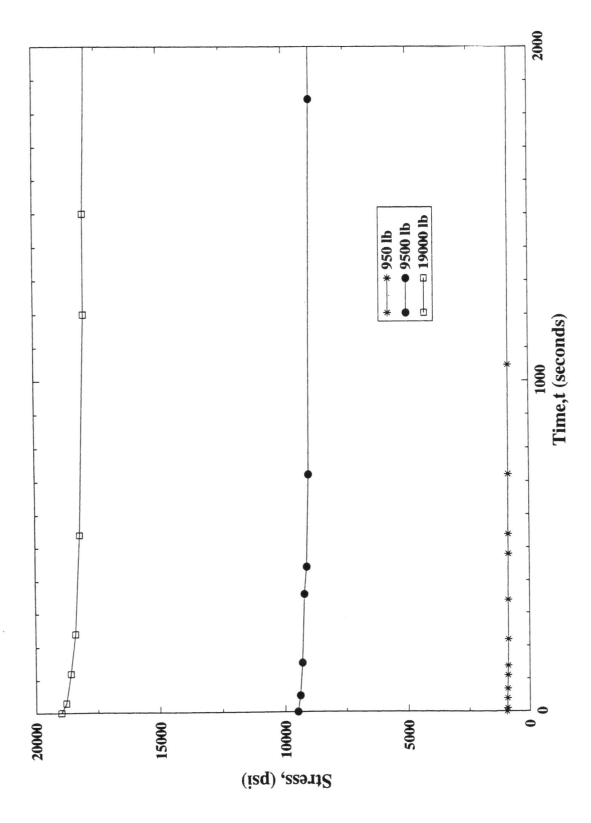

FIGURE 3. Relaxation of compressive load under constant deformation

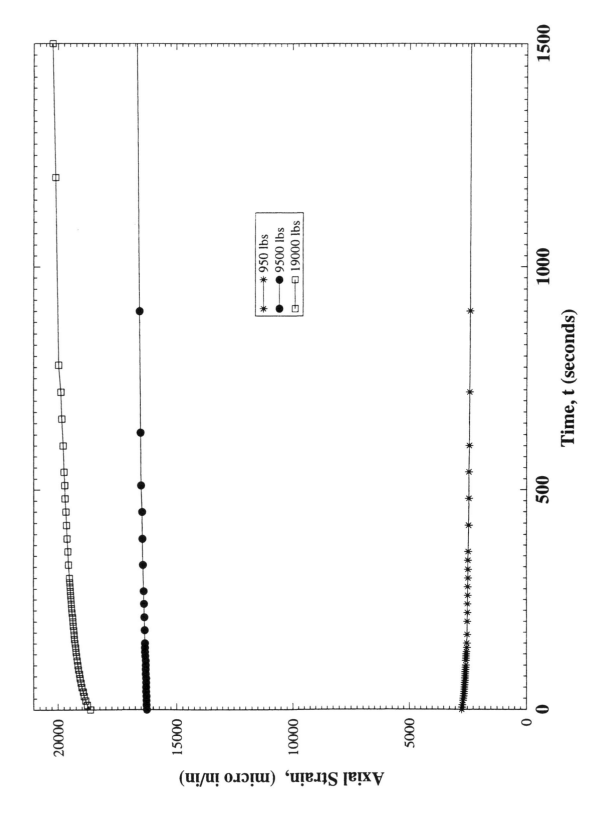

FIGURE 4. Variation of compressive strain under constant deformation

Legend:
* 950 lbs
● 9500 lbs
□ 19000 lbs

Axis labels:
- Time, t (seconds): 0, 500, 1000, 1500
- Axial Strain, (micro in/in): 0, 5000, 10000, 15000, 20000

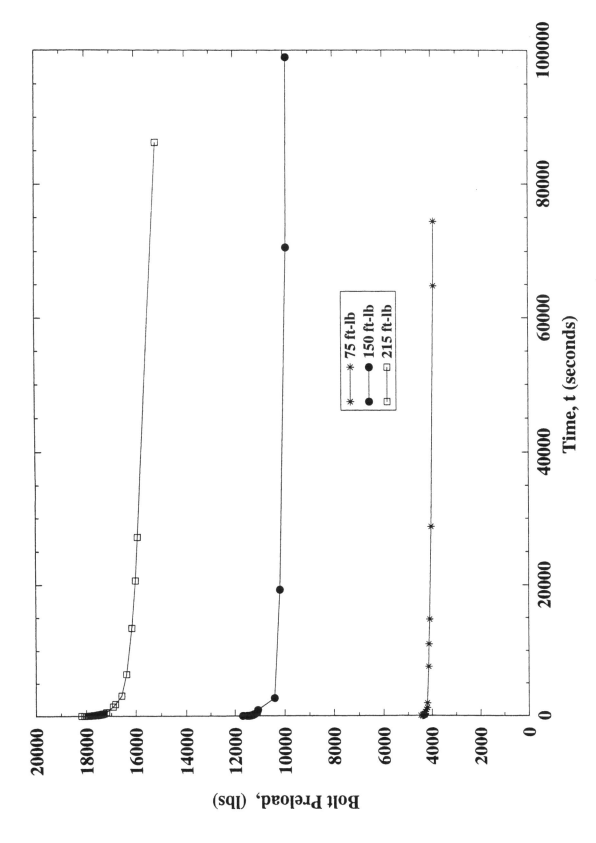

FIGURE 5. Bolt preload relaxation at three torque levels

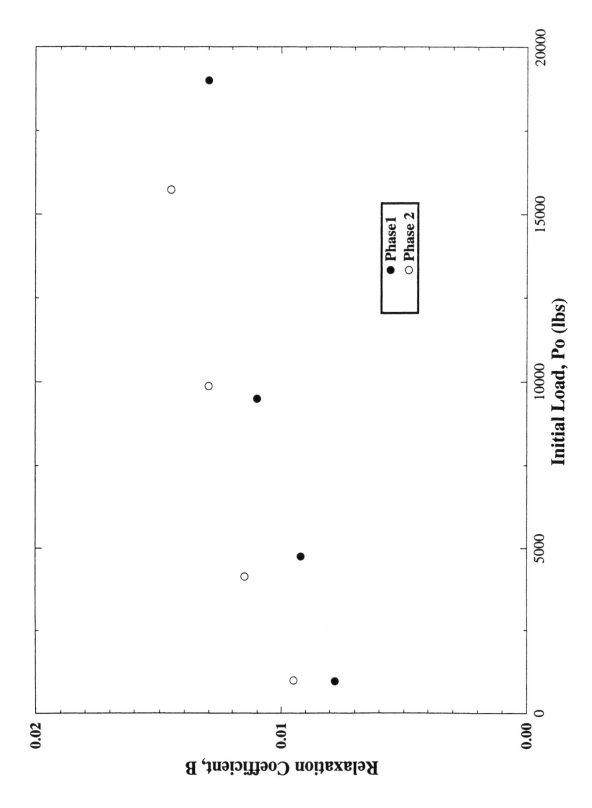

FIGURE 6. Variation of relaxation exponent, B, with initial load, P_o

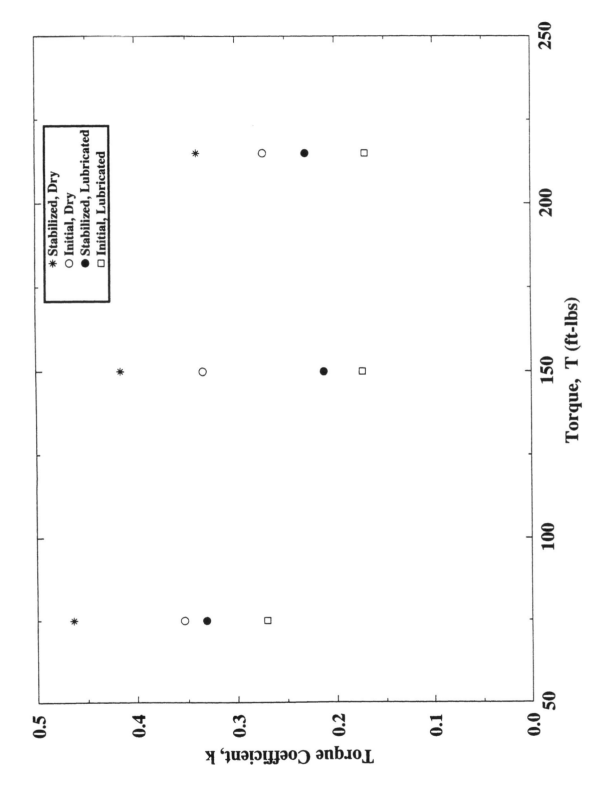

FIGURE 7. Variation of torque coefficient with applied torque

The Effect of Environmental Exposure on the Behavior of Pultruded Mechanical Connections

ROBERT L. YUAN AND SHANE E. WEYANT

ABSTRACT

This paper discusses an experimental investigation to examine the effects of moisture and temperature on the behavior of FRP composite materials used for civil engineering structures. The study included moisture absorption evaluation and the bearing capacity of mechanical connections made from pultruded FRP flat sheet. Various temperatures and wet environmental conditions were investigated. During the experimental testing, the maximum moisture content and the bearing load-displacement relations were recorded. Contingent upon the experimental results, a reduction of the ultimate bearing capacity FRPs is determined, with the intent that mechanical connections might be designed for exposure to elevated temperature and moisture.

INTRODUCTION

In recent years, the fiber-reinforced plastics (FRP) composites have emerged as an effective material for use in civil engineering structures because of their strength, light weight, and high resistance to corrosive environments (ASCE Report, 1984). The corrosion resistance, not found to this extent in any other structure use material, makes FRP especially suitable for structures such as cooling tower systems, chemical storage tanks, and transportation structures. These types of structures typically are exposed to high temperatures, excessive humidity, and wet environments (Morris, 1984). There is at present a scarcity of data on the effects of these conditions on the strength properties of FRP composites in general, and in particular, on the effects of the combination of elevated temperatures and moisture on the bearing capacity of FRP composites. Bearing capacity is an important design parameter, as the mechanical connections are designed to preclude a catastrophic shear or splitting tension failure by providing adequate edge and end distances (Rosner and Rizkalla, 1995). Accordingly, they are designed so that the bearing strength of FRP controls the mechanical connection strength.

Robert L. Yuan, Department of Civil Engineering, University of Texas at Arlington, Box 19308, Arlington, TX 76019-0308

Shane E. Weyant, Creative Pultrusions, Inc., Pleasantville Industrial Park, Box 6, Alum Bank, PA 15521

The objectives of this research program are:

(1)　　To study the moisture absorption behavior and the maximum moisture content and to compare the results with the maximum moisture absorption currently recommended in the design guide; and

(2)　　to investigate the bearing strength of FRP mechanical connections with the combined effects of temperature and moisture absorption.

EXPERIMENTAL PROGRAM

A testing program was devised to determine the effects of moisture and temperature by first subjecting test specimens to a simulated extreme exposure environment, and then testing each specimen for bearing capacity. Specimens were immersed in different temperature water baths for varied lengths of time and were tested for bearing capacity at each immersion time and temperature combination.

The specimen size was selected in order to preclude a shear or tear-out failure. The thickness of the coupon specimens used in this investigation was 3/8 in. (9.5 mm); the specimens had a width of 4 in. (101.6 mm) and a length of 13 5/16 in. (338 mm) and the bolt diameter was 9/16 in. (14.3 mm). A width-to-hole diameter ratio (w/d) of seven (7) and an edge-distance-to-hole diameter ratio (e/d) of five (5) were maintained to ensure bearing failure for the test programs. In each test, a 9/16 in. (14.3 mm) high-strength stainless steel bolt was used. The bolt length was selected to ensure that the composite plate in each connection intercepted the shank and not the threads of the bolt.

The bearing strength tests were conducted according to ASTM D953, Procedure A, Standard Test Method for Bearing Strength of Plastics (ASTM, 1995). A photograph of the test setup is shown in Figure 1.

Figure 1 - Mechanical Connection Test Setup

The connection coupon specimen consisted of a single composite plate, which was loaded by a single bolt in double shear at the lower end of the plate. Two dial gages, reading in 0.0001 in. (0.00254 mm) were used for measuring the net displacement of the bolt with the relation to the free end of the specimen in a tension loading condition. The loading rate for this test maintained at a rate of 0.0003 in./sec. (0.08 ± 0.03 mm/sec).

The temperature parameter in the program includes 73°F (23°C), 100°F (38°C), 125°F (52°C), and 150°F (66°C). The maximum moisture content varies from 0.5% to 1.7% by weight depending upon immersion temperature in the study. The immersion times at each temperature vary from one day to sixty days. The specimens were immersed in different temperature water baths. The absorption parameters were determined first by immersing a "witness coupon" in each temperature bath and monitoring the moisture absorption over time.

EXPERIMENTAL RESULTS AND COMMENTS

Moisture Absorption

The moisture content of a specimen was measured by weighing a sample. By monitoring the weight gain, the percentage moisture content by weight of the material was determined. Figure 2 illustrates the moisture content by weight percentage gain as a function of time at various temperatures. The results indicate that the time to the maximum moisture content as well as the maximum moisture content itself varies with temperature. The maximum moisture contents ranged from 0.5% at room temperature to 1.7% at 150°F (66°C). It is also apparent that the maximum moisture content increases most dramatically above 100°F (38° C).

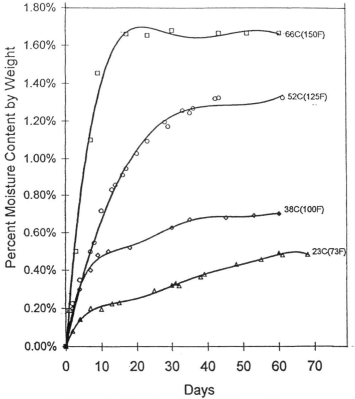

Figure 2 - Moisture Content Versus Time

Analysis of the absorption behavior is required to predict the moisture content at a particular immersion time. The absorption behavior of FRP composites was analyzed in this study based on the Springer Methods of computing the Fickian Absorption parameters (Springer, 1988).

The theoretical moisture absorption curve and the observed moisture contents were plotted as a function of the square root of time for temperature T = 150°F (66°C) in Figure 3. By comparing the theoretical curve with the test data curve, it can be concluded that the absorption behavior of the FRP used in this study can be approximated using Fick's law, especially at high temperatures.

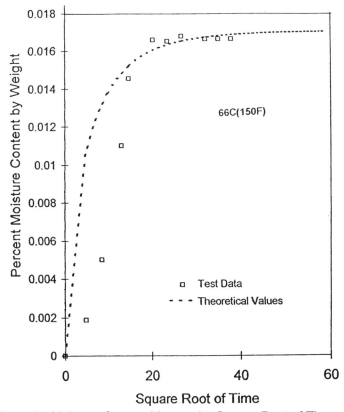

Figure 3 - Moisture Content Versus the Square Root of Time

Bearing Strength

The bearing capacity has been defined by ASTM D953 to be the load at which there is a corresponding four percent (4%) deflection of the original bolt hole diameter. The bolt hole diameter of the specimens used in this study was kept constant at 9/16 in. (14.3 mm); the corresponding failure deflection, accordingly, is 0.022 in. (0.57 mm).

From observation of the load test and analysis of the load versus deflection curves, it is demonstrated that the four percent (4%) deflection and corresponding load usually coincide with the point on the load versus deflection curve where the relationship deviates from linear behavior. This point is characterized by a decrease in the load during testing with a corresponding increase in deflection which causes the relationship curve to drop. In addition, a significant, audible "crack" is also heard at this point. Because of these observations, it is conjectured that this point represents the incipient internal "failure", where the parallel fibers buckle and the surrounding resin crushes. This point may be defined as the bearing capacity of the connection because of the beginning failure of the internal structure. After reaching this load, the material is subjected to a progressive failure as the load continues and the failure pattern continues.

Because this point represents thc incipient failure load that has been reached in each test during the investigation and this point is usually at or below the load at the four percent (4%) bolt hole deflection, it will be considered the bearing capacity for the purposes of this study.

The bearing strength appears to decrease due to exposure to combined moisture and temperature. For comparison of bearing load values, the bearing load found from tests at 0.00% moisture content was approximately 6400 lb. This value is very close to the calculated bearing load of 6750 lb. Calculated from the given bearing capacity value provided in the manufacturer's design manual of 32,000 psi.

The average percent bearing strength retention versus moisture content at different temperatures is graphed in Figure 4. The results indicate the decrease in bearing capacity as the moisture content increases.

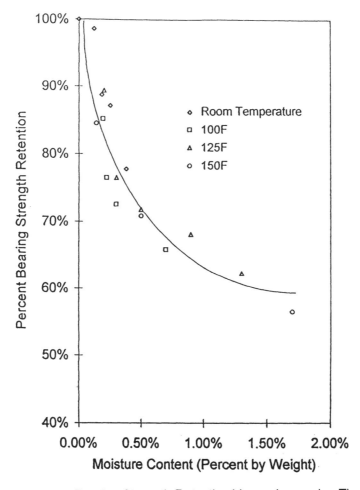

Figure 4 - Bearing Strength Retention Versus Immersion Time

CONCLUSIONS

Based on the measured absorption and bearing behavior of the specimens with various moisture and temperature conditions, the following conclusions can be made.

1. Moisture absorption of the FRP specimens was generally consistent with Fickian diffusion behavior, especially at high temperatures.

2. The maximum moisture content of a specimen increased with an increase in temperature. The maximum moisture absorption given by the FRP design manual of 0.6% by weight appears to be close to the value found at room temperature. At higher temperatures, the maximum moisture content increases.

3. There appears to be a greater reduction in FRP bearing capacity due to exposure to combined temperature and moisture than to exposure to moisture alone.

4. The design of mechanical connections in a FRP structure should take into account the effect of combined moisture and temperature on the bearing capacity. Exposure to moisture alone at room temperature could reduce the bearing capacity by nearly thirty percent (30%).

REFERENCES

American Society of Civil Engineers, 1984, "Structural Plastics Design Manual," *Reports on Engineering No. 63.*

American Society of Testing and Materials, 1995, "Standard Test Method for Bearing Strength of Plastics," ASTM D 953-93, *Annual Book of ASTM Standards Part,* Philadelphia, PA; 208-212.

Morii et al., 1994, "Hydrothermal Effects on Mechanically Fastened Glass/Polypropylene Composite Joints," *Polymer Composites,* Vol. 15, No. 6' 408-417.

Rosner and Rizkalla, "Bolted Connections for Fiber-Reinforced Composite Structural Members," *ASCE Journal of Materials*; 223-230.

Springer, G.S., "Environmental Effects," *Environmental Effects on Composite Materials*, Vol. 3; 18-28.

BIOGRAPHIES

Robert L. Yuan is a Professor at The University of Texas at Arlington in the Department of Civil and Environmental Engineering.

Shane Weyant is an Assistant Sales and Marketing Director for Creative Pultrusions, Inc., in Alum Bank, PA.

SESSION 15

Material Substitution with Thermoplastic Composites

NOTE: No formal papers were submitted for this session.

Replacing Metal with Engineering Thermoplastics

DAN VOTURE

The presentation discussed some of the key design considerations to be made when replacing metal parts with engineering thermoplastics. In a metal-to-plastic conversion, engineers need to evaluate whether or not a thermoplastic part can be designed to meet the performance requirements of the existing metal part. Although the material strength and stiffness of reinforced thermoplastics are lower than most metals, it is often feasible to design plastic parts to replace metal design and achieve the necessary structural requirements. Depending on the safety factors employed in the metal design, typical metal-to-plastic conversions will involve design modification. Engineers also need to consider whether or not a thermoplastic part can be molded to meet the specified part tolerances. Knowledge of the variables affecting molding tolerances and limitations associated with the injection molding process is key to understanding attainable tolerances in thermoplastic parts. A successful conversion should provide either a cost reduction, a weight reduction, or enhanced product performance. Parts and illustrations were used to highlight design considerations as well as some of the advantages engineering thermoplastics exhibit over metals.

Dan Voture, LNP Engineering Plastics, 475 Creamery Way, Exton, PA 19341

High Performance Nylon Composites in Automotive

CLINT CHRISTIAN

Over the past 30 years or so, the Automotive industry - driven by CAFE standards and other regulatory measures - has adopted the use of thermoplastics extensively in the automobile. Every system of the automobile has been affected, including Powertrain. In Powertrain, nylon has been particularly evident in its growth in helping the OEM's to solve problems as they move to smaller and lighter vehicles and to vehicles with enhanced quality and lower cost.

The presentation on "High Performance Nylon Composites in Automotive" dealt with the nylon success mentioned above - specifically in powertrain, and specifically where glass and mineral reinforcement, and toughening technologies have married with nylon to create a very user-friendly solution for engineering and business professionals throughout the Automotive Value Chain. Focus will be on application successes that have occurred, that occur as we speak, and a look to the future.

Clint Christian, DuPont Automotive, 950 Stephenson Highway, Troy, MI 48007-7013

The Northstar Air Intake Manifold—Conversion from Metal to Plastics

KEN BARAW

When it was first introduced, the air intake manifold for the General Motors Northstar engine was an assembly of parts made from aluminum, magnesium, plastic, steel, brass, and silicon rubber. Utilizing techniques recently developed, it became possible to "re-make" this assembly into a one piece plastic part with an intricate interior design. This process, known as lost-or lost-core molding involves the casting of alloy cores. These cores are assembled and placed into an injection molding tool and overmolded. The alloy is removed using a melt-out bath. This complete redesign resulted in a part that was lighter, less expensive, quieter, and enabled the engine to produce more power. Due to its integral design, the total number of parts required to produce a finished assembly was drastically reduced. Other functions were also incorporated into the integral manifold unit that were previously not possible. This 1995 SPE award-winning project opened the door for the conversion of metal mainfolds to plastic in the US automotive industry.

Ken Baraw, BASF Corporation, 1609 Biddle Avenue, Wyandotte, MI 48192

Polyphenylene Sulfide Composites: Applications in Automotive Fuel Systems

JONAS ANGUS

Traditionally, metal is the material of choice for most automotive fuel system components. Over the years, in the automotive industry, the trend for lighter components, cost savings and the mandate by both the California and Federal governments for oxygenated fuels is generating a lot of interest for high performance thermoplastic materials. Oxygenated or alternate fuel helps to meet the challenge for cleaner burning engines and the reduction of pollution. Fortron POLYPHENYLENE SULPHIDE (PPS), a high performance linear thermoplastic polymer exhibits excellent oxygenated fuel resistance in terms of dimensional stability, negligible weight and volume change at both 60°C and 120°C. Test results of the compatibility of Fuel C, auto-oxidized fuel, aggressive fuel and fuel methanol/aggressive fuel will be discussed in detail and how they relate to components such as fuel rails, fuel sensors, throttle bodies, bobbins, fuel pumps and turbines. The benefits of thermoplastic processing, as it relates to injection molding in a single operation (eliminating machining,) welding, straightening, corrosion protection processes and all the hand assembly operations (as in the case of metal fabricated fuel components for elevated temperature applications) will be discussed.

Jonas Angus, P. Eng. Program Executive, Hoechst Technical Polymers, 1195 Centre Road, Auburn Hills, MI 48326

Advanced Thermoplastic Composites for Office Seating

JAMES HURLEY

Due to high demands in terms of strength, flexibility, aesthetic appeal and low manufactured cost, office seating has usually been the domain of traditional materials such as metal and wood. With this unique dynamic Purpose Chair, Kimball International, in collaboration with BASF, has made new advances in seating comfort, design flexibility, part consolidation and affordability. The main structural component of the Purpose Chair is a molded plastic seat/back shell. In order to minimize weight and part thickness, the shell was designed with a ribbed channel, which increases the load bearing capabilities while maintaining the flexibility needed to accommodate user stretching, turning and reclining. The shell is produced using Ultramid B3ZG6 Q660, a glass fiber reinforced and impact-modified nylon 6 from BASF Plastic Materials. In addition to it's excellent surface finish and easy processing, this material combines high flexural strength with the superior elongation and impact properties necessary to meet the demands of the ANSI/BIFMA standards.

James Hurley, BASF Corporation, 1609 Biddle Avenue, Wyandotte, MI 48192

Advances in *in-situ* Fiber Placement of Thermoplastic Matrix Composites for Commercial Applications

MICHAEL J. PASANEN

ABSTRACT   Eight years of developing fiber placement technologies at Automated Dynamics Corporation has resulted in a robust manufacturing process for both commercial and aerospace applications. *In-situ* consolidation technologies for thermoplastic matrix composites have been adapted to the commercial market to include recreational products, piping and industrial rolls and shafts. High performance thermoplastics such as PEEK, nylon and PPS have been the candidate materials for these applications. Increased performance of these materials over thermosets in the areas of damage tolerance, fatigue resistance, and chemical resistance has been well documented. The limitation of thermoplastic matrix composite products from broad scale commercialization to date has been cost. This paper describes new approaches which have been investigated to increase throughput of *in-situ* thermoplastic fiber placement to reduce manufacturing costs for thermoplastic composite components.

INTRODUCTION

The cost of composite structures needs to come down in order for wide scale use in the commercial arena. Three areas can be targeted: reduction in the composite material costs, reduction in the cost of expendable materials and reduction in direct labor costs.

Of these costs, the greatest single contributor to overall cost is that of labor. The effort at Automated Dynamics has been to reduce labor costs through the use of *in-situ* bonded fiber placement of thermoplastic matrix materials. This method of on-line consolidation of the composite material eliminates the requirement of vacuum bagging and autoclaving of the component.

It is well known that the consolidation of melt processable thermoplastics is governed by the relationship between temperature, pressure and time, and although there can be trade offs among these three parameters to achieve the same results, (i.e. increasing the process temperature and pressure allows a reduction in the time at elevated conditions which is required to achieve complete consolidation) the exact relationship between these variables is not fully understood. This is particularly true when considering the very short times under elevated conditions required of *in-situ* thermoplastic fiber placement.

Michael J. Pasanen, Automated Dynamics Corporation, 407 Front Street, Schenectady, New York 12305

In general, *in-situ* consolidation requires that as much energy as possible be transferred to the composite very rapidly. The elevated temperature of the composite will in turn reduce the viscosity of the matrix allowing the matrix molecules to flow and migrate across boundaries. When the energy level entering the material becomes too great, and degradation of the material begins, the processing rate can be increased to bring the energy per unit volume of material back to a level which does not induce degradation.

This paper describes current fiber placement technology at ADC and efforts which are under way to increase the throughput of the fiber placement equipment to better meet the needs of the commercial market.

CURRENT FIBER PLACEMENT TECHNOLOGY

FIBER PLACEMENT HEAD

The primary function of the fiber placement head is to guide one or more in coming tapes / tows from a material creel to the point of application on the substrate as well as provide heat and pressure to bond the new material to the substrate. Secondary functions of the head include providing cut, queue and restart functions for the in coming materials.

A schematic illustration of ADC's fiber placement head is shown below.

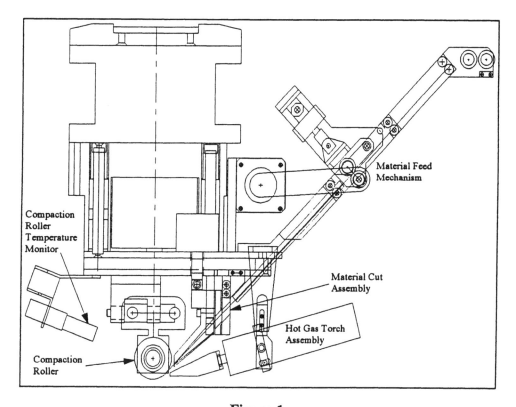

Figure 1

The method which ADC uses to supply energy to the in-situ fiber placement process is a combination of forced hot gas convection at and around the nip region and conduction through a heated pressure roller. This method of energy input has been selected due to the relatively high heat density which can be achieved with hot gas, the ability to easily direct the gas stream to a desired target, compact size, low weight and low cost. The compaction roller is integrally heated in order to prevent the relatively large mass of the roller from becoming a heat sink, and pull a large portion of the energy out of the composite material prematurely.

A slide assembly is pneumatically actuated to provide a compaction force to the bond zone at the roller nip point. This also allows vertical compliance to maintain a constant compaction load while allowing variations in the surface height.

PROCESSING CONDITIONS

Processing conditions have been established for several material systems. The conditions were determined through manufacture of a series of NOL style rings, each fabricated at different manufacturing conditions. The conditions which produced the best results for interlaminar shear strength, based on ASTM D2344, were selected as the baseline conditions to be used for in-situ fiber placement for the given material. The table below shows process conditions and interlaminar shear strength for AS4 / PEEK, AS4 / PPS and AS4 / nylon 6.

Table I

Material	Process Rate (in/sec)	Gas Stream Temperature (°C)	Gas Stream Flow Rate (SLPM)	Compaction Roller Temp. (°C)	Applied Compaction Load (pounds/in.)	Apparent Interlaminar Shear Strength (psi)
AS4/PEEK	3.0	950	200	450	300	11,300
AS4/PPS	3.0	950	110	435	200	9,500
AS4/nylon 6	3.0	925	50	*	50	7,500

INCREASING THROUGHPUT

A process rate of 3.0 inches per second yields a throughput rate of between $\frac{1}{2}$ and 2 pounds of completed part per hour. The exact throughput rate will depend on the width and thickness of the specific tape or tow being used. In order to increase the throughput of the fiber placement equipment to a level which will make thermoplastic composites more economical in commercial markets, different forms of the fiber placement head are being investigated. There are two major components to the head which will effect the rate at which the fiber placement equipment may operate: the compaction device and the energy input device.

---

* Compaction temperature roller not actively controlled or monitored with this version of fiber placement head.

COMPACTION DEVICES

Several configurations for a compaction device have been investigated. These include a pressure shoe, pressure roller, track device and tension. Because many of the parts which are fabricated with fiber placement are complex in shape, the track device and tension systems are not practical. They cannot match complex contours and concave regions and maintain intimate contact with the substrate material. It has been experimentally determined that a pressure shoe can cause excess damage to the fibers of both the incoming material and the substrate material. A pressure roller combines both the compact size of a pressure shoe with the rolling contact of a track device. For these reasons, a compaction roller has been the method of choice.

HEAT / ENERGY SOURCES FOR IN-SITU PROCESS

The hot gas method of heating employed at ADC limits the amount of heat which can be transferred to the material, given the maximum power output of a hot gas torch and the processing rate at which it is desired to operate. A secondary energy source, used in addition to the hot gas is one avenue being investigated to increase the heat-up rate of the materials during *in-situ* fiber placement.

Heating the materials at the nip point can be achieved by many different methods. Ultrasonic, focused IR, laser, microwave, resistance, induction, open flame, conduction and hot gas have all been investigated by various sources for thermoplastic *in-situ* fiber placement. Each of these methods has advantages and disadvantages. A brief review of each follows along with an assessment of its applicability as a complimentary energy source for fiber placement.

Hot Gas Heating

This has been the primary heat source for ADC's thermoplastic fiber placement technology. It was selected due to its relatively high energy density, fast response time, ability to easily direct the energy stream, light weight and low cost. The use of an inert gas as the carrier for the heat energy adds the additional benefit of acting as a shielding gas against oxidation of the polymer at high temperature.

A disadvantage is that only a small amount of the heat energy is absorbed by the composite material at the nip region.

Ultrasonic Heating

Ultrasonic heating for welding of plastic materials is well established. This is primarily for the welding together of fixed shapes through seams or spot welds. Some work in the area of ultrasonic bonding of thermoplastic composites has also been performed. In-situ fiber placement requires a continuous process. Some work has been performed in this area, however (5). The major difference between this heating method and others is that in addition to heat, shear energy is introduced to the bond interface through mechanical vibration. It is envisioned that this mechanical agitation aids in the formation of the interfacial bond.

Focused IR

Work using focused infrared energy as a primary energy source for *in-situ* thermoplastic fiber placement been performed by Oak Ridge National Lab. (9) and is on going at NASA Langley. A fixed focal point requires that the point of application is precisely fixed with respect to the IR source. This can be difficult to achieve in the dynamic environment of fiber placement, particularly if a complex shaped part is being fabricated.

Laser

Lasers provide the highest available energy density of the methods outlined. The efficiency is high in the welding zone due to the high emissivity of graphite and the beam can be shaped to the desired profile and directed to the bonding zone with wave guides.

One of the advantages of using laser energy is its highly localized nature. This can be used to only partially melt the incoming tape and the substrate. This will virtually eliminate fiber movement in the consolidated structure and result in high quality laminates, if the initial prepreg is of high quality and contains no voids. If, on the other hand, the prepreg contains a significant amount of voids, the power can be increased to fully melt the incoming tow / tape and attempt to drive out the voids.

Two types of lasers have been used for in-situ thermoplastic fiber placement:

1. Nd:YAG This operates in the infrared wavelength range of around 1500 nm. Fiber optics with glass lenses and wave guides can be used to shape and direct the beam.

2. CO_2 This laser operates in the wavelength range of 10 μm. At this wavelength normal optical glass is opaque, so special lenses need to be used to shape the beam. Also, fiber optics are not usable at this wavelength, so positioning the beam in a moving robotic system is extremely difficult.

The major disadvantages of lasers is the cost and safety issues.

Microwave

A microwave applicator has been developed at McDonnell Douglas which applies energy to a composite laminate through the thickness of the material (15,16). In the case of carbon fiber thermoplastic composites the 2.4 Ghz frequency of the microwaves penetrates to a depth of 0.0065 cm, which is very close to one ply thickness. Heating of the material to processing temperatures has been found to be rapid (approximately 0.7 seconds for graphite / PEEK) using a power density of 50 W/cm^2.

The microwave applicator which was developed covers an area of approximately 100 cm^2 and is circular. This allows a fairly high throughput rate, however, because of the wide area covered it limits the application to only flat or very large radius of curvature parts. Also, one of the assumptions in the system is that there is good thermal contact between the plies being bonded. It is rarely the case for two composite prepreg surfaces to be in such good thermal contact before they are melted. It seems that this heating approach would work well as a through the thickness energy source used as a secondary consolidation step after intimate contact is achieved by a primary heat / pressure system.

Resistance

Some attempts at using resistance heating without a separate resistance element in an in-situ process have met with some success. This method uses two electrodes: one on the incoming tape and one on the part. Two methods have been used for the part electrode. The first used a metallic mandrel as the electrode, the second used a compliant "top of component electrode". Investigation has shown that the contact resistance and variability between the electrodes and the graphite fibers is greatly reduced when contact is made in the melt region. Reference 13 details this information.

Disadvantages include constant dragging of the electrodes over the material surfaces is bound to cause some fiber damage and material build up in the electrodes if they are located in the melt region of the matrix. Also, this method will work with only conductive fibers such as graphite. The top of component electrode method requires a ceramic or other non-conductive mandrel to prevent the electrical path from grounding through the part.

Experiments have shown some positive results making rings with less than 1.5 % void content for graphite / PEEK.

Inductive

Inductive heating is caused by alternating magnetic fields creating eddy currents in conductive media. The eddy currents heat the material by either hysteresis losses (in magnetic materials) and/or resistive losses. This method is commonly used to melt or heat treat metals.

Advantages of induction heating include a high response rate to control of the heat source and a localized area of heating, with a properly designed coil.

Disadvantages include: For composites, the fibers are limited to conductive fibers only, no heating would be evident in glass fiber reinforced materials without additional susceptors. Also, the coil shape, geometry of the materials being heated and the position of the coil relative to the materials being heated cause significant changes in the resultant heating patterns. This makes the inductive heating technology not well suited to a continuous process such as fiber placement, where the geometry of the target materials are constantly changing.

Open Flame

A hydrogen / oxygen flame has been used as a heat source to filament wind T700 / nylon 6 at rates up to 5.25 inches per second (7,8). Although some positive results were obtained, the process was described as "challenging". This, added to the inherent danger in an open flame, makes this method a low priority for a secondary heat source.

PROCESS CONSIDERATIONS

It is generally recognized that thermoplastic matrix composites are consolidated in two steps. The first is intimate contact and wetting between the two surfaces being bonded. The second is autohesion or healing, in which molecular diffusion across the bond line occurs. The specific process parameters required to achieve these conditions need to be determined for each material.

For thermoplastic processing, general results have been obtained as illustrated in figure 2, below. The different color areas represent interlaminar shear strength as measured by ASTM D2344. Process parameter values are compared to apparent interlaminar shear strength to determine the best conditions to use for processing a particular material.

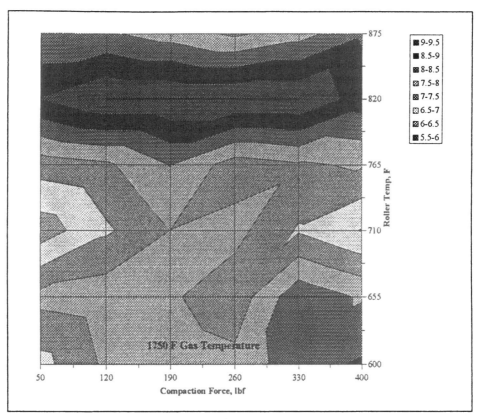

Figure 2

As can be seen from this plot, and in general it holds for all the thermoplastic composite materials which have been investigated, the application pressure in the region of reasonably good consolidation is not critical to the process. At both very low and very high pressures, the interlaminar shear strength of the specimens starts to drop off, but there is a wide window where the pressure does not have a large effect on the part consolidation.

One of the biggest processing problems has been lofting of the material behind the compaction roller. This phenomenon is caused by the lifting force associated with the material tending to partially stick to the roller. At a point part way around the roller, the tension in the tow / tape coming from the material that is stuck to the substrate overcomes the adhesive force of the material to the roller, and the material comes free of the roller. The problem is that the lofted material is not in contact with the substrate when it releases from the roller, and intimate contact between plies of material is not achieved. Lofting is in figure 3.

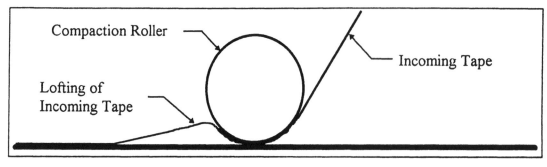

Figure 3

The amount of lofting which occurs in process can be controlled by altering the process control parameters. Different materials have behaved differently in this regard. In most cases, altering the process conditions by a slight amount from those determined to give best properties can greatly change the degree to which lofting occurs.

Another factor which has an effect on the degree to which lofting occurs is the smoothness of the underlying ply of material. A smoother substrate ply will yield better intimate contact between the in coming material and the substrate, which in turn reduces the amount of lofting which occurs.

MATERIAL CONSIDERATIONS

Investigations have indicated that the exact nature of the prepreg material has an impact on its processability during *in-situ* fiber placement. Although the research is not at this time conclusive, further investigations into the resin and morphology are on going.

CONCLUSION

The energy sources described above were subjectively compared for their suitability for use with ADC's existing fiber placement equipment. The options were assessed on the maturity of the technology, availability, safety, system cost, and the type of energy which is being added to the system. For example, it is the opinion of the author that the hot gas and heated roller system which is currently in use at ADC is sufficient to bring the materials to melt temperature in the times required of fiber placement.

One approach which is currently being investigated adds ultrasonic energy to the materials down stream of the compaction roller. In this way, heat energy is being added to the system at a point which will effectively increase the time which the materials are under elevated temperature conditions. In addition, shear energy will be added to the materials which will aid in the intermingling of the molecules at the bond interface.

Although thorough testing of this approach has not been concluded, preliminary results show that there is some promise of increases in material through put while maintaining equivalent resultant laminate properties using this approach. This, combined with investigations into other aspects of fiber placement such as raw material characteristics and robot motion algorithms, should yield results which enable throughput rate to be increased by several factors and bring the cost of thermoplastic composite structures down accordingly.

REFERENCES

(1) Lin, Miller, Buneman, "Predictive Capabilities of an Induction Heating Model for Complex-Shape Graphite Fiber / Polymer matrix Composites", 24th International SAMPE Technical Conference, October 1992.

(2) Border, Salas. "Induction Heated Joining of Thermoplastic Composites Without Metal Susceptors", 34th International SAMPE Symposium, May 1989.

(3) Howes, Loos, Hinkley. "The effect of Processing on Autohesive Strength Development in Thermoplastic Resins and Composites", *Advances in Thermoplastic Matrix Composite Materials, ASTM STP 1044*, G. M. Newaz, Ed., American Society for Testing and Materials, Philadelphia, 1989, pp. 33 - 49.

(4) Dara, Loos, "Thermoplastic Matrix Composite Processing Model", NASA-CR-176639, 1986.

(5) Bullock, Boyce, "On-line Consolidation of Thermoplastic Matrix Composite Tape Using Ultrasonic Heating", 39th International SAMPE Symposium, April 1994.

(6) Strong, D. Johnson, B. Johnson, "Variables Interactions in Ultrasonic Welding of Thermoplastic Composites", SAMPE Quarterly, January 1990, pp. 36 - 41.

(7) Funck, Neitzel, "Challenge of Thermoplastic Composites for Bicycle Frames", Proceedings of the Forth Japan International SAMPE Symposium, 1995, pp. 1309 - 1314.

(8) Funck, Ostgathe, Breuer, Neitzel, "Mass Applications for Composites: The Demand for New Processing Techniques", JEC Composites Manufacturing, April 1996.

(9) Norris, Campbell, Tobin, "In-situ Thermoplastic Consolidation System", Oak Ridge National Laboratory Technical Report, Contract DE-AC05-84OR21400, March 1993.

(10) Güçeri, "On-Line (In-Situ) Consolidation of Thermoplastic Composites - A Review", Proceedings of 8th Thermoplastic Matrix Composites Review, January 1991.

(11) Towell, Johnston, Cox, "Thermoplastic Fiber Placement Machine for Materials and Processing Evaluations",

(12) Holmes, Don, Gillespie, "Integrated Process Model for Resistance Welding of Thermoplastic Composites", 39th International SAMPE Symposium, April 1994 pp. 1855 - 1867.

(13) Miller, Van den Nieuwenhuizen, "Full In-Situ Consolidation of Graphite Fiber Thermoplastic Matrix Composites by Direct Electric Heating", Proceedings of the Tenth Thermoplastic Matrix and Low Cost Composites Review, February 1993.

(14) Hinkley, Working, Marchello, "Graphite / Thermoplastic Consolidation Kinetics", Proceedings of 39th International SAMPE Symposium, April 1994 pp. 2604 - 2611.

(15) Lind, Wear, Kurz, "Microwave Heating for Fiber Placement Manufacturing of Carbon-Fiber Composites", Proceedings of the Symposium on Microwaves: Theory and Application in Materials Processing held during the 93rd Annual Meeting of the American Ceramic Society, April 1991.

(16) Lind, Wear, "Design Considerations for Manufacturing Carbon-fiber Thermoplastic Composites Using Microwave Heating", Polymer Materials Science Engineering, 1995.

Optimum Beam Design for Automotive Pedal Application

CHUL S. LEE AND BRIAN IRWIN

Plastics materials are replacing metals in many of the automotive pedal applications. In order to achieve structural stiffness and strength of metallic pedals various beam sections, such as I beam, Channel beams or box shaped beams, are being utilized in the design of the plastics pedals. Various rib patterns are also introduced to add stiffness both in bending and torsional loadings.

Unfortunately, there is no guidance on which beam sections or which rib patterns to use for a given loading condition. The presentation evaluated efficiencies of various beam sections under bending load and torsional loadings. The study also included optimum rib patterns for each of the two loadings conditions. It also included effect of rib thickness on the performance per weight of the pedal design.

The presentation contained a chart showing relative performance of various beam sections and also rib patterns for both bending load condition and also for the torsional loading condition.

Finally examples of rib pattern generation using a topology optimization program are shown, using accelerator pedal example.

Chul S. Lee, Allied Signal, Inc., 20500 Civic Center Drive, Suite 4000, Southfield, MI 48085 and Brian Irwin, University of Michigan, 711 Arch Street, Ann Arbor, MI 48104

SESSION 16

Ways to Accelerate Composites' Acceptance in the Civil Infrastructure Market: How Your Company Can Benefit from Current Projects

Composite Repair/Upgrade of Concrete Civil Engineering Structures

ORANGE S. MARSHALL, JR., PAMALEE A. BRADY AND JOHN P. BUSEL

ABSTRACT

The U.S. Army Construction Engineering Research Laboratories, in partnership with the Composites Institute, has initiated a cooperative research project to develop, test, demonstrate, and commercialize fiber reinforced composite materials systems for in-place strengthening, repair or upgrade of existing reinforced concrete civil engineering structures. The goal of this work is the integration and promotion of fiber reinforced polymer composite technologies into the construction industry for the repair of concrete columns, beams and decking. The major thrusts include identification of mechanical and physical requirements and cost targets for civil engineering applications, development of composite materials systems for concrete structures, definition of design guidelines for repair and upgrade of reinforced concrete with advanced composite material systems, laboratory experimentation and testing to determine short term and long term performance, and field demonstrations of advanced composites for repair and upgrade of concrete structures. The approach taken in this work has been the development of two working groups focused on 1) composite plate and wrap technologies, and 2) composite post-tensioning technologies for repair of reinforced concrete structures. This paper presents an overview of the programs of each of these working groups. Laboratory testing includes long term thermal cycling effects of composites bonded to concrete, shear strengthening of concrete beams, and bend angle effects on FRP post-tensioning tendons. The demonstrations include post-tensioning of concrete beams, repair of concrete beams to improve their flexural capacity, strengthening of under-reinforced concrete walls, and repair of concrete columns with corroded rebar. The development of design guidelines utilizing the results of this research and others, will also be discussed.

1. Orange S. Marshall, Jr., Principal Investigator, U.S. Army Construction Engineering Research Laboratories, PO Box 9005, Champaign, IL 618216-9005.
2. Pamalee A. Brady, Principal Investigator, U.S. Army Construction Engineering Research Laboratories, PO Box 9005, Champaign, IL 61826-9005.
3. John P. Busel, Manager, Market Development, The Composites Institute of The Society of the Plastics Industry, Inc., 355 Lexington Avenue, New York, NY 10017.

INTRODUCTION

Much of the nation's infrastructure is deteriorating due to problems associated with reinforced concrete. Factors contributing to the infrastructure deterioration include the effects of the environment (harsh climate, de-icing salts, seismic activity), the increase in both quantity and weight of loads on structures from truck traffic, changes in use of structures, and the under design of older structures. In many cases, funds for replacement are not available. Repair or upgrade using traditional materials and methods may be expensive, difficult to install, or both. Fiber reinforced polymer (FRP) composites are being explored worldwide as a promising solution to these problems. The use of externally applied composites to rehabilitate, increase the strength, and provide seismic resistance of beams, columns, and decks have many benefits. These include high strength-to-weight ratio, resistance to corrosion, relative ease of application, minimum disruption of traffic, durability, and minimal maintenance requirements compared to traditional materials.

The goals of this project are to identify candidate materials systems, establish material performance specifications to include the use of glass, aramid, and/or carbon fiber reinforced polymers, and develop design protocols and guidelines to be used in specifications and standards for each type of strengthening system. Demonstration projects will field test composites materials systems on typical civil engineering structures. Performance of these systems will be monitored. This paper gives a general overview of current activities to investigate and field demonstrate composites products.

PROJECT PARTICIPANTS

The U.S. Army Construction Engineering Research Laboratories (USACERL) and the Composites Institute (CI) of the Society of Plastics Industries entered into a cooperative research project to develop, test, and demonstrate composites materials systems for in-place strengthening, repair or upgrade of concrete civil engineering structures. Organizations, participating in this project working with the CI include the American Concrete Institute, the American Society of Civil Engineers, the Civil Engineering Research Foundation, the Ohio Department of Transportation and the California Department of Transportation. USACERL is working with the U.S. Army Cold Regions Research and Engineering Laboratory, the U.S. Army Waterways Experiment Station, the Naval Facilities Engineering Service Center, and the U.S. Air Force Wright Laboratories.

CURRENT DEMONSTRATION ACTIVITIES

Demonstration projects utilizing existing material systems are being conducted as a part of the project. The purpose of these demonstrations are to document the materials applications and to acquire data needed in order to develop industry consensus guidelines/standards for the use of composite materials for the repair or upgrade of concrete structures. This paper describes the status of these ongoing projects.

POST-TENSIONED FRP TENDONS TO STRENGTHEN DOUBLE T-BEAMS

In a condominium in Lantana, Florida, over 200 reinforced double T-beams located on the first floor were deteriorated due to a damp marine environment. This created extensive concrete cracking, spalling, and exposure of the rebars on the stem of the beams. The building owners and engineers decided to restore the structural integrity of the building by strengthening the beams using steel post tensioning tendons. Structural Preservation Systems (SPS), Inc. working with Chalaire and Associates, Inc., the project engineers, provided the opportunity to evaluate carbon tendons on this project. In April 1996, five of the damaged beams were strengthened with ten carbon Leadline® tendons, supplied by Mitsubishi Chemical, and installed by Structural Preservation Systems, Inc. One of the beams was instrumented to monitor long term behavior. Instrumentation is monitoring creep of the tendon and slip in the anchorage. Strain gages are mounted on the tendons to measure the forces, tendon slip, and strains, the end anchorages are instrumented with transverse and longitudinal strain gages to monitor the pressure exerted by the grout material and the deflection of the beam is measured with a deflectometer.

The effects of bend angles of carbon tendons is a critical issue for this type of application, as is the performance of the anchorage. In a parallel effort, USACERL will perform laboratory testing of the tendons used in this demonstration to determine their strength characteristics at different bend angles and bend radii simulating installation conditions.

EXTERNALLY BONDED FRP SYSTEM FOR SHEAR STRENGTHENING OF BEAMS

Prestressed concrete joists known to be deficient in shear reinforcement were experimentally tested at USACERL to determine the improved performance possible using an FRP repair. Some highway bridges constructed in accordance with American Concrete Institute (ACI) codes of the 1950's and 1960's have been found to have less shear capacity than flexural capacity; FRP composites may provide a cost effective solution to this problem for state DOT's. Beams were a hybrid design consisting of a prestressed tapered reinforced concrete web with openings and a site cast top flange, Fig. 1. Two beams were tested each under four point bending; two symmetrical point loads with a 9 ft - 8 in. span length and a 11 ft - 2 in. shear span. One beam was damaged to a predetermined level prior to repair (Fig. 2) and the other was repaired without damage. A glass/aramid/epoxy composite system called Tyfo-S Fiberwrap™ System, supplied by Hexcel-Fyfe Co was used as an external wrap. Commercially available two component epoxy glue was used to bond the fiberglass to the concrete. Experimental data on the strength, stiffness, steel strain, FRP strain, deflection and mode of failure of the repaired beams were obtained and comparisons were made with unrepaired control beams. Preliminary force vs. displacement results show the increased strength and displacement capacity of the repaired beams over the control beam, Fig. 3. This plot also highlights the importance of properly applying the FRP to address the performance problem. A premature horizontal shear failure initiated at the web/slab interface of one beam end of the repaired joist due to a lack of FRP at this location. Overall, the FRP showed the potential to significantly increase capacity and avert a brittle shear failure.

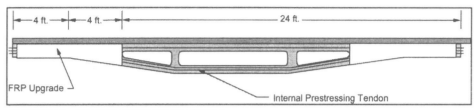

FIGURE 1: BEAM CONFIGURATION

FIGURE 2: DAMAGED BEAM

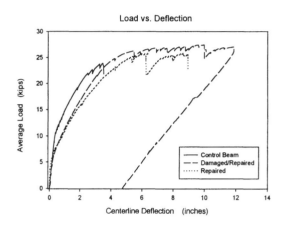

FIGURE 3: FORCE VS. DISPLACEMENT COMPARISONS

EXTERNALLY BONDED FRP SYSTEM STRENGTHENING OF STORAGE TANK WALLS

In 1994, Structural Preservations Systems, Inc. (SPS) and Gannett Fleming applied a carbon/epoxy sheet material to the inside wall of a concrete sewage treatment tank in Hollidaysburg, PA, to evaluate composites as a repair material on an experimental basis. The treatment tank was damaged shortly after construction when a nearby river overflowed its banks and floated the tank off of its foundation. This created several cracks in the tank walls. The tank walls with the composites on them were exposed to several environments over the past two years including 1) dry, exposed to UV and freeze-thaw, 2) splash zone, exposed to freeze-thaw, UV, wetting, drying, and aggressive waste water elements, and 3) submerged, exposed to chemical attack of waste water elements. A critical issue for this application is the condition of the concrete-composite bond. The CPAR project, working with SPS and Gannett Fleming evaluated the quality of the bond in July, 1996. In-field tensile bond and shear bond test procedures were applied to the three areas of concern. The bond tests included the popular pull-off test using 2-in. diameter steel plates glued to the 1/4-in. deep in-place drilled core, a pull-off test developed by SPS and verified in the laboratory using 2-in. square plates with perimeter grinding of the laminate, and an SPS developed twist-off shear test. The preliminary results of the tests indicated that deterioration of the composite bond is negligable or non-existent to date.

EXTERNALLY BONDED FRP SYSTEM FOR STRENGTHENING OF BEAMS

The facia beams on a bridge in Butler County, Ohio were upgraded using carbon/epoxy plates fabricated by the Materials Directorate of Wright Laboratory, Dayton, Ohio. The beams were installed following laboratory testing which characterized the beam strength both before and after application of the FRP plates. Following one year of use during which the beams were periodically monitored for performance, one of the beams was removed and tested. A fair amount of FRP debonding was noted by engineers monitoring the performance. Laboratory tests on the removed beam showed a reduction in strength of about 8 percent. Engineers at Wright laboratory are unsure if the reduced strength is due to exposure to weather and traffic, growth of laminate debond areas, or if it is just experimental data scatter. The mode of failure, concrete failing in compression, did not change from previous laboratory tests.

To determine the true nature of the strength reduction, some small scale beams were designed and the carbon/epoxy plates applied to them as on the Butler County bridge. The beams will be shipped to the U.S. Army Corps of Engineers Cold Regions Engineering Laboratory to conducting thermal durability testing. The tests include developing baseline data for standard concrete samples at room temperature and comparing that to data from (a) cyclic freeze thaw tests from -20° F to 120° F, (b) creep tests of a sustained load at 60% of the room temperature bond strength for 60 days at 100° F and 90% relative humidity, (c) flexural tests at -20°F and at 120° F, (d) toughness tests measuring impact energy absorption at room temperature and -20° F, and (e) fatigue tests of one million cycles between a lower limit of 20% ultimate flexural load and a upper limit of 60% at room temperature and -20° F.

EXTERNALLY BONDED FRP SYSTEM FOR STRENGTHENING OF PIER DECKS

Pier 11 is located on the northern waterfront area of Naval Station, Norfolk, Virginia. It is a cast-in-place and precast reinforced concrete structure 1,400 feet (427 m) in length and 150 feet (76 m) wide. Pier 11 is approximately 10 years old and serves nuclear carriers of the Atlantic Fleet. The pier is currently rated (and specifically designed) for 70-ton (620 kN) truck mounted cranes with limited use by 90-ton (800 kN) cranes. However a recent A&E study identified deck slabs in the crane operating lanes of some spans to have design shortfalls that would limit 70-ton (620 kN) crane service. The goal of the project is to upgrade the crane operating area in order that unlimited 70-ton (620 kN) service would be allowed. Upgrade construction using two different CFRP systems, each covering half of the upgrade area, was completed in November, 1996.

The upgraded area on the north side will be instrumented to follow the life cycle of the composite and the composite response to applied patch loads (see impact load tests above). Instrumentation will include strain sensors located at the point of maximum load response and damage sensors along the edges as well as at the point of maximum load response. The former will be traditional wire strain gages embedded in the laminate and aligned with the uniaxial graphite fibers. The latter sensors will be piezoelectric patches or strips embedded in the laminate to sense initiation of edge delaminations and structural damage in the laminate.

FUTURE PLANS

The CPAR team are actively seeking additional demonstration projects. Potential demonstrations which are identified and verified by the team will be described along with the proposed repair/upgrade method.

The final phase of the project will be commercialization and technology transfer of the products that are developed, tested, and demonstrated under this project. The technology transfer will include draft materials specifications, construction practice, and design guidance for using FRP composites for structural repair, leverage for the adoption of this technology with the engineering community (U.S. Army Corps of Engineers, other Federal Agencies, ASTM, codes and standards bodies), and publication of a final report of this project. Information and data collected from this study will enable researchers and engineers to understand the structural behavior of FRP composite materials systems used to repair or upgrade concrete civil engineering structures and more easily use these materials where they are most advantageous.

CONCLUSION

This paper describes several applications of the use of advanced composites for strengthening, repair and/or upgrade of reinforced concrete structures. Properly designed and manufactured composites materials systems can offer superior structural performance and be compatible with existing construction industry practices. Most importantly, the selection and application of composites for repair must be used where the benefits of composites out weigh the use of traditional materials. Composites may not be the answer in some instances but where environmental or structural constraints limit or make repair or upgrade using traditional materials difficult or unfeasible, composites may be perfect for the job.

BIOGRAPHIES

Orange S. Marshall, Jr. is a Materials Engineer and Principal Investigator with the U.S. Army Construction Engineering Research Laboratories. He has conducted R&D at USACERL in the field of composites and polymers for the Army Corps of Engineers.

Pamalee A. Brady is a Structural Engineer and Principal Investigator with U.S. Army Construction Engineering Research Laboratories. She has conducted R&D at USACERL in the field of concrete and masonry seismic and structural analysis and upgrades for the Army Corps of Engineers.

John P. Busel is the Manager of the Market Development of The Composites Institute of The Society of the Plastics Industry, Inc. He is responsible for initiatives for bringing composites into use in the infrastructure.

Advancements in Smart Tagged Composites for Infrastructure Applications

JUSTIN B. BERMAN, ROBERT F. QUATTRONE, JOHN P. VOYLES AND JOHN P. BUSEL

ABSTRACT

Advanced polymer composites can be used to replace steel reinforcing in civil structures and eliminate corrosion as a deterioration mechanism, drastically reducing the cost of maintenance. The lighter weight of many advanced composite structural elements, often coupled with environmental advantages and lower energy consumption during fabrication, offers advantages in erection and shipping costs, even in remote locations. The tagging of advanced composites with micron-sized secondary phase materials can produce "smart" structural systems with enhanced structural health monitoring capabilities. Such systems have the promise of accelerating acceptance and use of advanced composites for infrastructure applications.

Smart material systems have the ability for controlled adaptation of their mechanical properties or dimensions to external stimuli. Tagged smart structural composites exploit the advantages of both smart materials and polymer matrix composites while combining them into one useable structure. In this study, composites were tagged with micron-sized particles and the tags interrogated by passive and active magnetic excitation techniques. The tag response to adverse conditions under hostile loading situations or component cracking can be monitored with conventional NDE equipment.

This paper presents the concepts of passive and active tagging for on-line monitoring of advanced structural polymer composites. A summary of the Corps' research in the tagged composites arena, and proposed applications for tagged composites in infrastructure are discussed.

INTRODUCTION

Civil infrastructure permeates all aspects of our lives. From the highways we drive on

Justin B. Berman, US Army Construction Engineering Research Laboratories, 2902 Newmark Drive, Champaign, IL 61821-1076
Robert F. Quattrone, 2902 Newmark Drive, Champaign, IL 61821-1076
John P. Voyles, 2902 Newmark Drive, Champaign, IL 61821-1076
John P. Busel, Composites Institute, 355 Lexington Avenue, New York, NY 10017

to our wastewater treatments facilities, civil infrastructure accounts for 25% of the US physical capital (Barno and Busel, 1995). This makes construction the largest national industry valued at $2.7 trillion. Of the 575,000 bridges nationwide, over 230,000 are in advanced states of decay requiring over $90 billion in repairs. More than ever, materials need to be developed that facilitate in the repair and maintenance of construction elements at decreased first and life cycle costs. Polymer composite materials offer light weight, high strengths/stiffnesses and are resistant to corrosion. While still not the construction material of choice, construction applications for composites are growing at an average rate of 8.5% per annum. The growth rates can be further increased by the introduction of smart tagged composites for health monitoring of civil structures.

The concept of particle sensing, called "particle tagging", involves embedding micron-size sensory particles into materials, such as composites, concrete, or adhesive layers, to make them an integral part of the host material. When interrogated by suitable instrumentation, the embedded particle sensors interact with their host structures and generate certain types of measurable signatures. The signatures can then be correlated with the material and structural conditions, such as internal material strength, distribution uniformity of critical constituents, state-of-cure, location of voids, etc.

At the US Army Construction Engineering Research Laboratories (CERL) research on various tags and tagging techniques is underway to exploit the advantages tagged composites offer. This research is driven by the fact that the Army maintains an aging inventory of 190,000 facilities requiring $3.8B in backlog of maintenance and repair (BMAR) for which no adequate structural health monitoring capability exists. Smart structural health monitors can more effectively assess conditions and predict maintenance of facilities and systems, including remaining functional life. In conjunction with the Composites Institute, the University of Illinois Urbana-Champaign, and Virginia Polytechnic Institute and State University all three arenas - government, academia and industry - are working to develop standards and experimental work for this promising technology. This paper gives an overview of the progress that has been made with tagged composites research.

INFRASTRUCTURE APPLICATIONS

With the introduction of a new technology into an already well established market, the question "How can infrastructure benefit from smart tagged composites technology?" becomes a key issue. Tagging offers an enabling technology that will enhance current infrastructure materials without necessarily introducing new materials. Some benefits of this added technology were established by Barno and Busel (1995) and are summarized by bullet points in table 1.

Of the potential technologies tagging offers, monitoring of materials processing and in-field determination of structural and physical parameters are the most attractive. Table 2 outlines various processing/in-use structural metrics that could be quantified using either passive or active tagging techniques.

TABLE I. Role of smart tagged composites in infrastructure (Barno and Busel, 1995).

Potential of technology:

- VERIFY PROCESSING OF COMPOSITE
- STRUCTURAL LOADING HISTORY
- ASSESS THE DEGREE AND TYPE OF PROPERTY LOSS
- PROVIDE ADVANCED WARNING OF IMPENDING STRUCTURAL FAILURE
- "LEARNING" AND "HEALING"?

Results of technology:

- LOWER COST (FIRST COST + LIFE CYCLE COST)
- REDUCED MAINTENANCE
- INCREASED PRODUCTIVITY
- FEWER OPERATIONAL PROBLEMS
- GREATER SAFETY AND DURABILITY

TABLE II. Variables of interest for polymer tagging

- CHEMICAL DEGRADATION
- CHEMICAL SHRINKAGE
- CRACKING (BROKEN FIBERS/MATRIX CRACKING)
- CREEP/STRESS RELAXATION
- CURE VARIATIONS
- DEBONDING
- DELAMINATION
- DENSITY VARIATIONS
- FATIGUE DAMAGE

- FIBER MISALIGNMENT
- FIBER WRINKLING
- IMPACT DAMAGE
- INCLUSIONS
- MOISTURE
- RESIN VARIATIONS
- STATE OF STRESS/STRAIN (INTERNAL RESIDUAL AND UNDER LOADING)
- THICKNESS VARIATIONS
- VOIDS/POROSITY

FERROMAGNETIC TAGGED COMPOSITES

Ferromagnetically tagged composites are polymeric composite materials utilizing embedded ferromagnetic particles (referred to as "tags") for sensing. To obtain a uniform distribution of tags in the composite, the largest tag must be much smaller than the smallest reinforcement fiber. Therefore, micron size sensors of sizes 5 microns or less are used. Once embedded, the tags can be interrogated by an external alternating magnetic field at various frequencies. During this interrogation, force/displacement measurement equipment is used to detect the internal vibration of the material created by the tag/polymer interaction. These signals can be used to detect damage in an element. The process for developing active tagging interrogation technology is carried out in multiple steps as illustrated in Figure 1.

Simple laboratory tests confirmed that the technique works using contact sensors (PZT force gage) at the surface of the tagged composite. To improve upon this technique, actual full-scale samples are tagged and evaluated using a non-contact sensor (laser velocimetry). Both of these techniques proved successful. These parts require two-side access via a powerful magnet. The next phase will seek to develop a system that can use this technique with limits to one-sided element access. The final phase will fine tune the arrangement by reducing the size of the interrogation magnet, reducing the magnetic field required to excite the structure, and increasing the non-contact sensor detection range.

Figure 1. Smart Tagged Composites Work Schedule

Five tags of interest were evaluated: Magnetite, NiZn ferrite, MnZn ferrite, Iron Silicide and Lignosite FML. Magnetite and Lignosite FML are the least expensive of the tags and were used in most of the studies. Vinyl ester, polyester and epoxy embedding resins were evaluated for use each with E-glass fiber reinforcement. Three different processing techniques were selected to manufacture the specimens: compression molding, resin transfer molding and pultrusion. The lowest weight fraction was selected so that readings could be made with the interrogation equipment. It was found in the lab studies that 2% weight fraction of a tag in the structural element was sufficient for equipment detection.

LABORATORY STUDIES

These studies demonstrated an active tagging damage evaluation technique for laboratory tagged samples. The ferromagnetic tagged composites (Figure 2) are subjected to an alternating magnetic field. The magnetic field spans many different frequencies which is measured readily with a gaussmeter. A PZT force transducer in contact with the specimen gives the force response. The material characteristics can be determined using both the input (magnetic excitation) and output (PZT transducer) signals. Figure 3 shows the output of the analysis of three specimens: control, large crack, delamination. Variations between the responses verify that a potential damage sensor has been developed using ferromagnetic tagging.

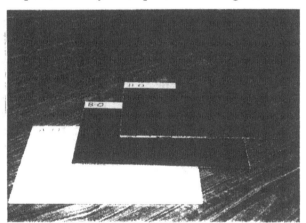

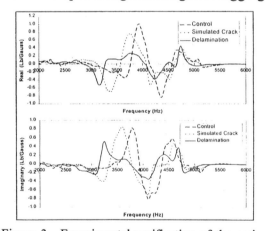

Figure 2. Ferromagnetically tagged composites using glass/polyester and 2% weight fraction magnetite (B-0) and NiZn ferrite (D-0).

Figure 3. Experimental verification of the active tagging technique for laboratory samples using contact sensors.

FULL SCALE STUDIES

Research is currently directed towards scaling up this technique for non-contact NDE of full-scale structural elements. Non-contact sensors such as laser velocimetry, allow for structural element interrogation from a distance. Pultruded channel profiles produced by Creative Pultrusions are being used for this study. Pultrusion was selected since an existing structural profile exists and since the process is automated. Two sets of samples were produced: Both control (yellow or lighter color member) and tagged (olive-green or darker color members) samples were produced (Figure 4). Of the tagged sample, two types were produced: control series and central delamination series. Results of this scale up work show similar results to those seen in the laboratory study. The only difference is that the magnetic field used for interrogation of full size structures had to be increased significantly.

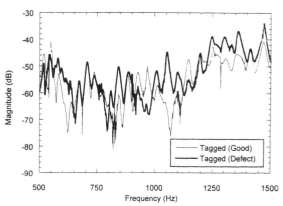

Figure 4. Full scale ferromagnetically tagged composites using glass/polyester and 2% weight fraction magnetite (yellow = control, green = tagged).

Figure 5. Experimental verification of the active tagging technique for full scale members using non-contact sensors.

CONCLUSIONS

While ferromagnetic tagging has been around since the early 1970's, only recently has this technology been sought after for the health monitoring of infrastructure. Studies of recently developed Composites Institute tagged full scale members is underway. Once the proper tagging material for these members has been selected, further improvements in the interrogation technique will be made. Success of the technology will lead to full scale NDE equipment for in service health monitoring of structures.

ACKNOWLEDGEMENTS

Many individuals and research groups contributed to this information found in this paper and each deserve recognition. The Center for Intelligent Material Systems and Structures (CIMSS) of VPI carried out much of this research under contract for the US Army CERL. The Composites Institute and their member companies in the Smart Composites Task Group contributed to the direction, materials, experiments and manufacturing of the tagged composites. Their contributions are highly appreciated.

REFERENCES

Barno, D.S. and J.P. Busel. "Enhanced Infrastructure using Affordable Composites," Composites Institute Report, May 1995.

Quattrone, R.F. and J.B. Berman. "Recent Advancements in Smart Tagged Composites for Infrastructure," Proceedings of the 1996 ASCE Materials Conference, editor: Chong K. P., ASCE, Nov. 10-14, 1996, Washington, D.C., pp. 1045-1054.

Rogers, C.A., Z. Chen and V. Giurgiutiu, "Results of the Application and Experimental Validation of the Active and Passive Tagging Concepts Using Industrial Composite Samples," *CIMSS Technical Report No. 96-003.* February, 1996.

Zhou, S.W., Z. Chaudhry, C.A. Rogers, and R. Quattrone, 1995, "Review of Embedded Particle Tagging Methods for NDE of Composite Materials and Structures," *Proceedings of Smart Sensing, Processing, and Instrumentation, Smart Structures and Materials*, SPIE, San Diego, CA, Feb. 28-Mar. 3, 1995; in press.

Zhou, S.W., Z. Chaudhry, C.A. Rogers, and R. Quattrone, "An Active Particle Tagging Method Using Magnetic Excitation for Material Diagnostics," *Proceedings of the 36th AIAA Structures, Structural Dynamics, and Materials Conference*, April 13-14, 1995, New Orleans, LA; in press.

BIOGRAPHIES

Justin B. Berman works as a Materials Engineer in the Materials Science & Technology division of the US Army Construction Engineering Research Labs. He is currently working on several projects using advanced sensors for infrastructure health assessment.

Robert F. Quattrone is a Metallurgist in the Materials Science & Technology division of US Army CERL. He is the principal investigator on several smart materials projects including Smart Tagged Composites.

John P. Voyles is a student contractor with US Army CERL. He is currently pursuing a Bachelor's degree in Materials Science and Engineering with a special emphasis on polymeric materials.

John P. Busel is Manager of Market Development for the Composites Institute, a division of the Society of the Plastics Industries, Inc. John is largely responsible for transitioning the use of composite materials into new markets such as infrastructure.

Design and Development of FRP Composite Piling Systems

RICHARD LAMPO, ALI MAHER, JOHN BUSEL AND ROBERT ODELLO

ABSTRACT

Under the U.S. Army Corps of Engineers' Construction Productivity Research (CPAR) Program, a three-year cooperative research effort was initiated to develop and demonstrate FRP composite piling systems for marine/waterfront applications. After establishing performance target goals for FRP composite fender, bearing and sheet piling systems, designs to meet these performance goals were developed as part of a design competition among participating manufacturers and design professionals. Specimens were fabricated and tested against the established performance goals. Based on the results of the laboratory tests, piling systems were selected for demonstration in actual waterfront applications.

INTRODUCTION

Traditional piling systems are inherently unsuited for harsh waterfront environments. Pressure-treated timber pilings are subject to attack by marine organisms and pose disposal problems when being replaced. Steel-reinforced concrete piles can fail due to chloride attack on the reinforcing elements and freeze/thaw degradation of the concrete. The problems of corrosion on steel sheet piling are well known and documented. Overall it is estimated that deterioration of wood, concrete and steel piling systems costs the U.S. military and civilian marine and waterfront communities nearly $2 billion annually.

Such traditional practices as pressure treatment of lumber or sandblasting and painting of steel with solvent and/or heavy-metal containing coatings are potentially harmful to the environment and are increasingly being regulated. Properly designed and manufactured composite piling systems should be superior to traditional materials in marine operating

Richard Lampo, U.S. Army Corps of Engineers, Construction Engineering Research Laboratories, P.O. Box 9005, Champaign, IL 61826-9005
Ali Maher, Department of Civil and Environmental Engineering, Rutgers University, Brett & Bowser Roads, Piscataway, NJ 08855
John Busel, Composites Institute, Division of the Society of the Plastics Industry, Inc., 355 Lexington Avenue, New York, NY 10017
Robert Odello, U.S. Naval Facilities Engineering Service Center, 1100 23rd Avenue, Port Hueneme, CA 93043-4370

environments. However, the U.S. construction industry lacks the understanding and sufficient confidence in the design and long-term performance of composite products in civil engineering structures. Composites are already being successfully used in load-bearing civil engineering structures for the chemical processing, oil and gas, and water/wastewater industries. By transferring this technology to wide-spread use in marine/waterfront civil engineering structures, U.S. ports, harbors, and waterways operators can potentially save millions of dollars every year.

Recognizing the opportunities, a proposal was submitted and approved to develop the needed composite piling systems under the U.S. Army Corps of Engineers' Construction Productivity Advancement Research (CPAR) Program. (The CPAR Program was established to promote and assist in the advancement and commercialization of ideas and technologies that can have a direct positive impact on U.S. construction productivity and project costs. Projects under CPAR are shared-cost and shared-effort.) A CPAR Cooperative Research and Development Agreement (CPAR-CRDA) was then developed and signed. The industry/academic Partner on this Project is Rutgers University, New Brunswick, NJ, and the Laboratory Partner on this Project is the U.S. Army Construction Engineering Research Laboratories (USACERL), Champaign, IL. The Composites Institute (CI) also entered into this Agreement as a Partner participant under Rutgers providing significant in-kind support of materials and products for testing and field demonstrations. Another significant Partner participant is the New York /New Jersey Port Authority (NY/NJ PA). Other Laboratory participants include: the U.S. Naval Facilities Engineering Service Center (NFESC), the U.S. Army Cold Regions Research and Engineering Laboratory (CRREL), and the U.S. Army Waterways Experiment Station (WES). An Advisory Board comprising 10 leading U.S. Ports plus key Trade, Technical or Professional Organizations provided "reality checks" for the project.

The objective of this CPAR Project is to develop and demonstrate high-performance polymer composite fender, load-bearing, and sheet pile (bulkheads) systems for marine/waterfront civil engineering applications. Material standards, specifications, and design protocol will be developed for each type piling system.

DEVELOPMENT OF DESIGNS

Existing design guidance, mostly from the Navy and the Corps of Engineers (Naval Facilities Engineering Command, 1987; Corps of Engineers, 1989), for piling and sheet pile systems was compiled, circulated and reviewed. Information on system needs was also collected from the various participant users including Corps of Engineer Districts, the Navy, the NY/NJ Port Authority, New York State DOT and the City of New York, among others. Performance target goals for each type composite piling system were established by the Project Team using the survey results of existing systems, user input relative to system needs and performance expectations, and currently available design guidance for piling systems. Some select performance goals are as follows: The cross-section of the composite fender piles shall not exceed 13x13 inches wide. The fender piles shall not exhibit brittle behavior when subjected to a lateral load at -40°C at a strain rate of 100%/minute. The fender piles shall also have a minimum EI of 6.0×10^8 lb.-in^2. Light-duty, medium-duty, and heavy-duty composite sheet pile systems shall have minimum EIs of 2.48×10^5 kip-in^2/ft, 1.0×10^6 kip-in^2/ft, and 5.5×10^6 kip-in^2/ft, respectively. Composite bearing piles shall, at a minimum, have a loadbearing capacity equal to a 16 inch diameter wood bearing pile--minimum AE of 9.0×10^7 lb. Each of these composite piling systems must be capable of being installed using conventional equipment. At minimum, each of these composite piling systems shall demonstrate cost savings over the life-cycle as compared to the typically used non-composite piling material.

In July 1995, CI's Market Development Alliance announced a Design/Fabrication Competition referencing these Performance Target Goals (Composites Institute, 1995). As a result of the design competition, several innovative designs were submitted from CI member manufacturers and design firms. Five different fender piling designs/products; six different loadbearing pile designs/products; and three different sheet pile designs/products were selected for laboratory testing. Table I shows the participating manufacturers and the products being developed/evaluated for each.

TABLE I -- PARTICIPATING MANUFACTURERS AND PRODUCT SYSTEMS

MANUFACTURER	FENDER	BEARING	SHEET
Creative Pultrusions, Inc.	X	X	X
Hardcore Dupont Composites	X	X	--
International Grating, Inc.	--	--	X
Lancaster Composite	X	X	--
Seaward International, Inc.	X	X	--
Shakespeare Company	--	X	--
Specialty Plastics, Inc.	--	X	--
Trimax Lumber	X	--	X

LABORATORY TESTING

The testing program for each piling type is summarized in Table II.

TABLE II -- SUMMARY OF LABORATORY TESTS

PILE TYPE	TESTS
Fender	1. Bending test to determine EI. 2. Cold bending test to evaluate fracture potential in cold conditions. 3. Cold radial compression test to evaluate behavior in a crushing mode.
Bearing	1. Bending tests to determine EI (for buckling). 2. Compression to determine AE and compressive strength. 3. Creep measurement.
Sheet	1. Bending test to determine EI and bending strength. 2. Coupon testing to determine material properties. 3. Determine potential for built-up structures (to increase moment of inertia, I).

SCREENING TESTS

In each pile category, the test considered the most difficult to meet or the most critical for satisfactory performance was designated as a "screening" test. This approach would allow the manufacturers an opportunity to revise or improve their product (if the results of the screening test were not up to expectations or within the established Target Goals) before additional tests were conducted. The following tests were selected for the screening tests: Fender -- cold radial compression; Bearing -- flexural stiffness; Sheet -- flexural stiffness.

Fender Pile Cold Radial Compression Test

The cold radial compression test was developed based on the conditions and geometries provided by the NY/NJ PA for installations at Port Elizabeth and Port Newark facilities. This test accounted for the performance of wood fender piling and experience from previously installed plastic fender piles that failed in service at those locations. A total of nine different specimen variations from five different manufacturers were tested at Rutgers University. Wood pile specimens were also tested for comparison. Specimens from two of the manufacturers did not meet the minimum requirements as given in the established Performance Target Goals.

The cold radial compression test was developed from specific site conditions at Ports Elizabeth and Newark. After performing the tests and reconsidering the requirements, it became apparent to the Research Team that this crushing condition is not necessarily true for all port facilities. Meeting this cold radial compression requirement is not critical to the function of the fender pile where the fender has the ability to flex in order to absorb the berthing energy. For these specimens, the flexural tests (at ambient and cold temperatures) are considered the critical factor.

Bearing Pile Flexural Test

Specimens of bearing piles from six different manufacturers were received by NFESC for flexural testing. Although two pile designs did not meet the qualitative performance targets, they were considered to be drivable and thus met the intent of the screening test.

Sheet Piling Flexural Test

Specimens of sheet piles from three different manufacturers were received by Rutgers University for flexural bending tests. Because two of the specimens were not flat, a special test fixture had to be fabricated for each corrugated product in order to get a accurate reading of stiffness. A surprising result was that none of the specimens tested met the minimum requirements for EI of even the light duty sheet piling as established in the original Performance Target Goals. A restructuring of the Goals was, therefore, suggested as shown in Table III. All three of the products tested fell into the Light-Duty, Grade 2, category as given above. Since all three products tested represent commercial products already in service in various applications, plans were made to complete the rest of the tests on these products in order to collect additional performance data.

TABLE III -- RESTRUCTURED GOALS FOR SHEET PILING SYSTEMS

CLASSIFICATION	EI (kip-in^2/ft)
Light-Duty, Grade 2	$5 \times 10^3 - 1 \times 10^4$
Light-Duty, Grade 1	$1 \times 10^4 - 5 \times 10^4$
Medium-Duty, Grade 2	$5 \times 10^4 - 1 \times 10^5$
Medium-Duty, Grade 1	$1 \times 10^5 - 5 \times 10^5$
Heavy-Duty, Grade 2	$5 \times 10^5 - 1 \times 10^6$
Heavy-Duty, Grade 1	$1 \times 10^6 - 5.5 \times 10^6$

COMPLETION OF LABORATORY TESTS

At this time, the remaining laboratory tests for the different systems are in various stages of completion. Figure 1 shows the four point bending apparatus being used at Rutgers University to measure flexural properties of the fender piles. The same test fixture will be used by CRREL in Hanover, NH to perform the low temperature flexural property tests. All of the laboratory tests are scheduled for completion by January 1997.

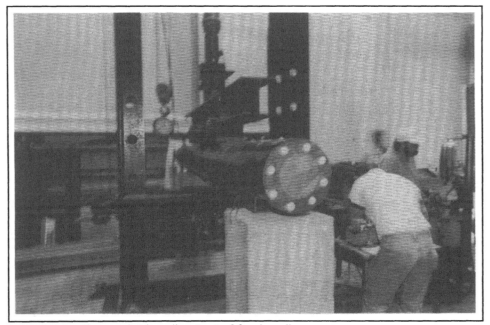

FIGURE 1. Four point bending test of fender piles.

FIELD DEMONSTRATIONS

As the initial screening tests for the fender piles were being completed, the NY/NJ PA approached the Project Team about installing demonstration test piles at their Port Newark facility. The NY/NJ PA wanted to complete the installation of fender piles at Port Newark before winter. Based solely on the cold radial compression test (developed around the conditions at the Port Newark facility), fender pile products from Creative Pultrusions, Seaward International, and Trimax Lumber were recommended for this demonstration test.

Creative Pultrusions' piling, 13 inches in diameter and 60 feet in length, is comprised of pultruded glass/vinyl ester and incorporates a tic-tac-toe profile and a high-density polyethylene (HDPE) bumper. Seaward International's SEAPILE® Composite Marine piling, 13 inches in diameter and 60 feet in length, is comprised of a 100% recycled, HDPE blend, which is reinforced with eight glass/polyester pultruded rebars. Trimax's piling, 10 inches in diameter and 60 feet in length contains 75% recycled HDPE, 20% fiberglass and 5% proprietary additive.

On October 30, 1996, twenty-four fender piles (eight from each manufacturer) were installed at Berth 7, Port Newark, where car carriers are typical ships docked. Figure 2 shows one of the piles being driven with a 9B3 hammer, which exudes 8750 pounds of pressure. With the installation complete, performance monitoring and comparisons to adjacent wood piles will now proceed.

FUTURE ACTIVITIES

After the laboratory tests are completed, additional sites will be selected for installation and monitoring of the piling systems not installed at the Port Newark site. With the CPAR Project officially ending in December 1997, CI will work to continue the performance monitoring of the installed demonstration systems. Participation in the newly formed ASTM Task Group on Systems for Marine/Waterfront Applications is encouraged. The development of industry consensus specifications for these products will be of equal benefit to both Government and private sector users.

FIGURE 2. Driving of demonstration fender piles at Port Newark.

REFERENCES

Composites Institute, *Market Development Announcement--Next Generation Composite Pile Design/Fabrication Competition*, July 1995.

Corps of Engineers Guide Specification, CEGS 02361, *Round Timber Piles*, July 1989.

Naval Facilities Engineering Command, Military Handbook, MIL-HDBK 1025/1, *Piers and Wharfs*, October 1987.

Development and Demonstration of a Modular FRP Deck for Bridge Construction and Replacement

ROBERTO LOPEZ-ANIDO,[1] HOTA V. S. GANGARAO,[2]
JONATHAN TROVILLION[3] AND JOHN BUSEL[4]

ABSTRACT

A partnership of Academia, State and Federal agencies, and the composites industry has developed a system intended to replace deteriorated concrete bridge decks with modules made of non-corroding composite material. Deck modules were designed with enhanced cross-section and fiber architecture. Testing prototypes and field demonstration decks were fabricated by VARTM and by pultrusion. Two bridge demonstration projects are planned to be constructed. The composite deck system is expected to be utilized to replace aged concrete decks, to construct new bridge decks, and to build all-composite short-span bridges.

INTRODUCTION

A properly designed, fabricated, and installed fiber reinforced polymer (FRP) composite bridge deck can be used very efficiently to replace deteriorated concrete decks or to build new decks. In addition, FRP decks can be assembled with optimized FRP structural components to develop an all-composite short-span bridge system. From the structural reliability viewpoint, decks provide adequate redundant capacity when combined with the stiffening system underneath the deck. From the durability viewpoint, conventional bridge decks typically need repair and eventual replacement in about 10 to 14 years. Therefore, bridge decks have to be highly compatible with composite applications: we can apply a durable composite material without compromising the integrity of main structural elements.

In order to upgrade the load capacity or rating of highway bridges, deck construction and replacement techniques should yield a light weight (approximately 16 to 20 lb./sq.ft.) deck system. FRP composite materials can be used to develop a modular bridge deck and superstructure systems that have high strength and stiffness to weight ratio, that are non-corrosive, and that have good fatigue resistance.

---

[1] Research Assistant Professor, Department of Civil and Environmental Engineering (CEE), and Constructed Facilities Center (CFC), West Virginia University (WVU), Morgantown, WV 26506-6103.
[2] Professor, CEE, and Director of CFC, WVU.
[3] Materials Engineer, US Army Corps of Engineers, Construction Engineering Research Laboratory (CERL), Champaign, Illinois.
[4] Manager, Market Development, Composites Institute, Society of Plastic Industries, New York, NY.

The objective of this work is to present a new technology for composite bridge decks. The composite deck product was designed to meet highway bridge performance requirements and be cost-competitive. The deck product was manufactured using two different processes that have distinctive advantages in the phases of prototyping and mass-production. Two demonstration bridges have been selected to apply the new composite deck system: (1) A bridge with a composite deck on steel beams, and (2) An all-composite short-span bridge. The potential of the composite deck to compete with existing bridge deck technologies is discussed.

THE TEAM

The deck product was designed and tested under the CPAR program of the US Army Corps of Engineers. The CPAR project is called "Development and Demonstration of Hybrid Advanced Design Composite Structural Elements." The main participants in this project are the Construction Engineering Research Laboratory, West Virginia University-Constructed Facilities Center, and the Composites Institute. Several composite industries have participated during all the phases of this program and provided not only in-kind contributions, but also valuable technical expertise. Federal Highway Administration and West Virginia Department of Transportation, Division of Highways (WVDOH) are sponsoring the development of the composite deck system for concrete bridge deck replacement through the Priority Technologies Program (PTP). Finally, District Four of WVDOH has been actively involved in the preparation of bridge plans and will construct Wick Wire Run Bridge in Spring 1997. Personnel from District Seven of WVDOH have proposed to construct an all composite short-span bridge, Laurel Lick Bridge, including the deck, the beams and the abutments.

PRODUCT DESCRIPTION

The composite deck cross-sectional shape was designed with efficient fiber architecture to withstand highway bridge loads. After several iterations, and based on previous experience with other composite deck configurations, it was found that a cross-section made of full-depth hexagons and half-depth trapezoids has enhanced performance for bridge decks. The resulting product is called H-deck. The H-deck modules are placed transversely to the traffic direction and are supported by longitudinal steel or FRP beams (See Figure 1). The supporting beams can be spaced up to 9 ft. (2.74 m) apart. The H-deck system has been designed to comply with AASHTO highway bridge design requirements. The fiber architecture incorporates E-glass fibers in the form of multi-axial stitched fabrics, rovings, and chopped strand mats. The matrix is a vinylester resin with good weatherability and resistance to harsh environments.

This product was designed to replace deteriorated concrete decks. In consultation with Federal and State highway officials, the depth of the H-deck was constrained to 8 in (200 mm) to match typical in-service concrete deck systems. The H-deck modules are joined through a shear key that provides mechanical interlock and an extensive bonding surface. At the manufacturing plant, and under controlled conditions, high performance adhesive bonding is used to assemble the component into large modules. During field installation, the large modules are connected with shear keys. In the shear keys that are connected in the field, mechanical connectors (blind fasteners) are used in addition to adhesive bonding to guarantee the required shear transfer capacity. It is likely that good bending transfer capacity may be attained, but this issue is being evaluated.

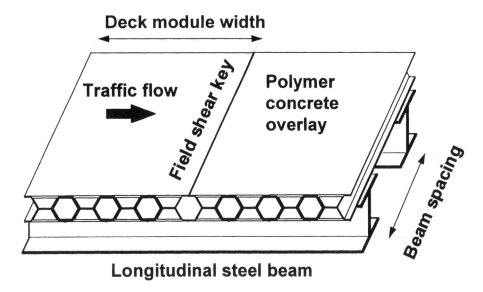

Figure 1. PORTION OF FRP DECK SUPPORTED ON STEEL BEAMS

ROAD TO IMPLEMENTATION

The two demonstration projects selected have distinctive characteristics with regard to the use of composites. Both applications must meet all serviceability and strength limit states.

WICK WIRE RUN BRIDGE

Construction of Wick Wire Run Bridge is planned during January-March 1997. This bridge is located off US Route 119 in Taylor County, WV. The bridge will be constructed with a composite H-deck supported by four longitudinal steel beams spaced 6 ft. (1.83 m) apart. The length of the bridge is 30 ft. (9.14 m) and the width is 21.7ft. (6.60 m).

This concept can also be used for concrete bridge deck replacement. The advantage of the composite deck for replacing deteriorated concrete decks is accelerated erection time leading to reduction in user inconvenience (minimum disruption to vehicular traffic). In addition, the lightweight composite deck (approximately 1/5 to 1/6 of the weight of a concrete deck) can result in an increase in load rating due to reduction in dead load. Nevertheless, it is necessary to evaluate the reduction in the moment capacity of a steel beam that was designed for composite-action (here we use the term in the structural sense) with a concrete slab, when the concrete slab is replaced by an FRP composite deck.

LAUREL LICK BRIDGE

The second bridge demonstration project is a short-span all-composite bridge. Laurel Lick Bridge is scheduled to be constructed under the WVDOH program for bridges with less than 20 ft. (6.10 m) span length. This bridge is located on county Road 26/6 in Lewis County, WV. The bridge will be constructed with an FRP H-deck supported by WF pultruded beams. This superstructures will be supported by WF pultruded columns acting as abutments. Pultruded multi-cellular panels will be placed in-between the columns and will support the lateral soil pressure. The length of the bridge is approximately 20 ft. (6.1 m) and the width is 16 ft. (4.9 m).

The WF structural elements were designed with optimized fiber architecture and manufactured by pultrusion (See Lopez-Anido et al. 1996).

The composite short-span bridge concept is based on pre-fabrication and modular construction. The advantages of the all-composite short-span bridge are the reduction in use of heavy equipment, and short time field installation. The short-span composite bridge system can be developed as a modular construction kit.

FABRICATION

The H-deck product was designed to be manufacturing transparent. Two different manufacturing processes have been used to produce the H-deck during the phases of laboratory prototyping and field demonstration. In both processes, an efficient fiber architecture was designed with several layers of multi-axial fabrics in vinylester matrix.

VARTM

H-deck prototypes were manufactured by Vacuum Assisted Resin Transfer Molding (VARTM). Some aspects of this manufacturing process are patented as SCRIMP™ (Seeman Composites Resin Infusion Molding Process). This manufacturing process requires minimum tooling, and can be used to fabricate large deck modules. Heavy weight multi-axial stitched fabrics (up to 64 oz./sq.yd.) with binderless chopped strand mats manufactured by Brunswick Technologies Inc. (BTI), were used as reinforcements. Foam was used to mold the cells in the VARTM prototypes. In this VARTM process, the fabrics are laid up dry by hand, then vacuum is applied, and finally resin is infused. The disadvantages of this fabrication process for the proposed FRP deck are that is labor intensive, and that product uniformity is difficult to maintain resulting in higher dimensional tolerances. These disadvantages can be partially overcome by using more expensive tooling, and more efficient production techniques such as fabric pre-forms. Fiber volume fractions ranging from 50 to 55 % were attained resulting in high stiffness. VARTM was the process of choice to fabricate H-deck testing prototypes. The test prototypes were fabricated by HardCore-DuPont, L.L.C. Deck prototypes were satisfactorily tested under static and fatigue loads. VARTM proved to be a convenient fabrication process for prototyping the FRP deck.

PULTRUSION

H-deck modules for laboratory testing and field demonstration have been fabricated by pultrusion. The main disadvantage of pultrusion is the initial high cost of tooling, and therefore it was not the choice for prototyping. Standard pultruded sections are mainly reinforced with unidirectional rovings (a group of untwisted parallel strands). Although, continuous or chopped strand mats are used in standard pultruded sections to provide a minimum transverse reinforcement, the sections typically lack in transverse fiber reinforcement to provide continuity between members, e.g., flange and web. Recently, it was shown by the authors that pultruded structural shapes, e.g., wide-flange beam, with efficient bi-directional fabric architecture exhibit enhanced strength and stiffness response, and ensure fiber continuity between flange and web elements (See Lopez-Anido et al., 1996). Based on this new fiber architecture concept and the pultrusion expertise that was developed, pultruded sections equivalent to the VARTM prototype were designed. Creative Pultrusion, Inc. fabricated two dyes and pultruded double-trapezoid and hexagonal components (See Figure 2). Deck components are pre-assembled in a shop under controlled conditions, ensuring quality and taking advantage of their lightweight.

The efficient fiber architecture of the pultruded components incorporates rovings and triaxial stitched fabrics with binderless CSM that were custom designed and fabricated by BTI. The triaxial fabric weighs 40 oz./sq.yd. The advantages of pultrusion are: (1) low labor cost (the process requires little operators input other than to maintain materials supply), (2) low operating costs; (3) minimal material wastage; and (4) high production rate. Assuming a unit weight for the pultruded composite of 110 lb./cu.ft., the weight of the deck results in 20 lb./sq.ft. However, it is anticipated that the deck weight can be reduced approximately 20% when long-term experimental data and field evaluations become available allowing to have a more precise calibration of safety factors.

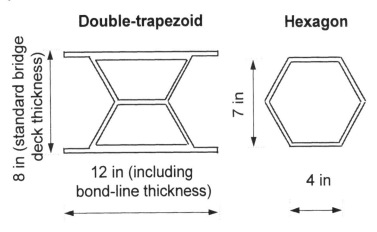

Figure 2. PULTRUDED COMPONENTS

Deck components for laboratory testing were fabricated by Creative Pultrusions, Inc. Pultrusion was also the process of choice to fabricate the H-deck for Wick Wire Run Bridge. Provided there is a sustained growth of the composite bridge market, the pultrusion industry may consider investing on fabricating larger dyes leading to consolidation of deck parts.

The pultruded components are connected in a shop with an adhesive system that is designed to carry the full shear force. The adhesive to bond the deck components needs to satisfy the following properties: good elongation, high peel and energy absorbing properties, fatigue resistance, environmental resistance (humidity, salt spray, cold and hot environments), working time of at least 30 minutes, minimum surface preparation, acceptance of variable bond line thickness (10-250 mils), and good gap filling capabilities (to compensate fabrication tolerances).

FIELD INSTALLATION

In the field, deck modules up to 8 ft (2.45 m) wide are connected with shear keys (See Figure 1). The same adhesive used to bond the deck components in the shop can be used in the field provided gap tolerances and ambient temperature are within acceptable values. Blind fasteners (FLOORTIGHT™) from Huck International, Inc., are also used to facilitate the connection of modules with shear keys, and to provide a reserve shear strength. Then, the composite deck modules are tied down to the main supporting beams (steel or composite) with blind bolts (BOM™) from Huck International, Inc. Finally, a thin overlay or wearing surface is placed on top of the H-deck. This overlay is made of polymer concrete that is bonded to the FRP deck. The polymer concrete system shall provide a satisfactory riding surface (e.g., abrasion resistance, skid resistance).

DISCUSSION AND CONCLUSION

Deck deterioration due to harsh environments and traffic loads is a main concern for users and owners of highway bridges. Composite decks can overcome these problems. While the initial costs associated with using composites are higher than those for conventional bridge materials, the following considerations "equalize" composites' deck cost: (1) reduction in erection time due to modular construction, (2) high load ratings due to the high-strength to weight ratio of the composite deck, (3) long service life and reduction in maintenance cost due to fatigue and corrosion-resistant properties, and (4) less use of heavy equipment due to the modules' lightweight.

The composite H-deck system was developed for bridge deck replacement and short-span bridge applications because they represent a very important volume of construction materials and they will have easier market acceptance than long-span bridges within the civil engineering community.

The success of FRP bridge decks requires a systems approach leading to design guidelines and specifications, long-term performance evaluations, and the development of construction and replacement technologies.

ACKNOWLEDGMENT

Part of the work presented in this article was sponsored by the CPAR Program of the US Army Corps of Engineers in collaboration with the Construction Engineering Research Laboratory, and the Composite's Institute of the Society of Plastic Industries. Another part of the work presented herein was sponsored by the PTP Program of Federal Highway Administration and WVDOH. The financial support is gratefully acknowledged.

REFERENCES

Lopez-Anido R., GangaRao, H.V.S., Al-Megdad, M. & Bendidi, R. (1996), "Optimized Design of Fiber Architecture For Pultruded Beams," Composites Institute's 51st Annual Conf., 20-D.

BIOGRAPHIES

Roberto Lopez-Anido, Ph.D., is Research Assistant Professor with the Department of Civil & Environmental Engineering, and the Constructed Facilities Center at West Virginia University.

Hota V. S. GangaRao, Ph.D., P.E., is Professor with the Department of Civil & Environmental Engineering, and Director of the Constructed Facilities Center at West Virginia University.

Jonathan Trovillion, is Materials Engineer with the Construction Engineering Research Laboratory (CERL) of the U.S. Army Corps of Engineers.

John Busel, is Market Development Manager with the Composites Institute, of the Society of Plastic Industries.

Composite Elements for Navy Waterfront Facilities

ROBERT J. ODELLO, GEORGE E. WARREN, DANIEL HOY AND JOHN P. BUSEL

ABSTRACT

The U. S. Navy's Naval Facilities Engineering Service Center (NFESC) and the Composites Institute of the Society of the Plastics Industry (CI/SPI) have entered into a Cooperative Research and Development Agreement (CRADA) to develop and demonstrate composites to repair and upgrade pier structures. The objective of this project is to establish performance requirements and develop products for the Navy that provide a cost-effective and durable infrastructure to support the Fleet. This project will identify products that will have significant applications to commercial waterfront facilities.

This paper will discuss how composites were used to assist the Navy in constructing a waterfront test site. It will also include examples of the Navy's effort in the development of a performance specification for composite fender piles. The paper will describe demonstration projects that showcase the use of composites such as pier upgrades at Norfolk, VA, San Diego CA, and Pearl Harbor HI.

The authors will discuss recommendations from the Navy on composites development for the marine environment. In addition, development of standard design methodologies and product specifications for composite materials for Navy Facilities will be demonstrated.

INTRODUCTION

In 1994, the Naval Facilities Engineering Service Center (NFESC) signed a Cooperative Research and Development Agreement (CRADA) with the Composites Institute of the Society of the Plastics Industry (CI/SPI). This CRADA was established to combine the joint resources of the CI/SPI and the NFESC to develop test and demonstrate structural components, subcomponents, materials and methodologies for the construction of waterfront facilities. This agreement seeks to find economical substitutes for known structural and utility system components that are cost effective, durable and economical to maintain. The CI/SPI will provide materials to be tested, data, comparisons with alternative materials and technical/structural design support. NFESC will provide test sites, equipment and the Navy's

Robert J. Odello, George E. Warren and Daniel Hoy, U.S. Naval Facilities Engineering Service Center. John P. Busel, Composites Institute.

unique technical expertise and knowledge for conducting the testing and evaluating the results of the testing.

Both parties will benefit from the agreement. The Navy will obtain first hand knowledge of new structural materials and components and will be able to perform tests in environments appropriate to Navy mission requirements. The Naval Facilities Engineering Command, the Navy's construction agent, will be able to develop design methodologies and construction specifications that will allow the use of such materials in Navy contracts. Industry members of the CI/SPI will also gain this knowledge and will be able to compete equitably with other materials in this market. The Navy will thus be able to build and maintain a more cost effective infrastructure and the composites industry will be able to expand into the marine waterfront market for both the Navy and other owners of the waterfront infrastructure.

ADVANCED WATERFRONT TECHNOLOGY TEST SITE

The first major effort relating to the CRADA was the fabrication of the Advanced Waterfront Technology Test Site (AWTTS). The AWTTS is a 46 m (150 ft) long wharf located in the entrance to the harbor of Port Hueneme, California. The deck is 5.5 m (18 ft) wide and typical spans are 3 m (10 ft) These spans were designed to simulate one-half scale equivalents of typical Navy piers. NFESC designed and constructed the test site with assistance from the U.S. Army Corps of Engineers Construction Engineering Research Laboratory in support of the Army's Construction Productivity Advancement Program (CPAR).

Two of the spans were 6 m (20 ft) long. A prestressed concrete deck constructed with carbon fiber prestressing tendons sits on one of the 6 m (20 ft) spans. It was constructed under the direction of the South Dakota School on Mining and Technology as part of a related CPAR project. The second 6 m (20 ft) span has a deck made entirely of fiber reinforced plastic materials. It is a demonstration of how such a system could be fabricated with the technology available at the time the AWTTS was being built. The all composite span uses pultruded box beams to span between concrete pile caps. The pile cape use glass fiber reinforced polymers for reinforcement. Glass fiber reinforced plastic pipes were used to clamp the box beams together laterally in an attempt to produce composite structural action. Tests conducted by NFESC indicate that this clamping action was not effective. The deck consisted of glass fiber reinforced polymer flat panels, mechanically connected to the box beams.

In addition to the access and working deck the AWTTS has one 3 m (10 ft) span for a control specimen and five spans devoted to structural testing. NFESC fabricated the test slabs and installed them in March of 1995. Four additional spans of the AWTTS are dedicated to materials testing and long term exposure. current tests involve epoxy coated steel reinforcing bars and dual phase steel bars in concrete slabs and pipe and utility conduit supporting devices made of composite materials.

A catwalk or working platform under the structural test sections has been a major contribution to the AWTTS by CI/SPI member companies. The catwalk consists of all fiber reinforced plastic gratings, structural members and hand rails to provide a safe working and access area for NFESC structural engineers and technicians to conduct structural tests on the decks. Five companies contributed to the design and fabrication of the catwalk components.

These included Fibergrate, MMFG, International Grating, IKG Fiberglass Systems and Creative Pultrusions. The effort was coordinated by Ashland Chemicals. The companies fabricated the components and shipped them to Port Hueneme as part of their contribution to the CRADA, and NFESC engineers with the help of U.S. Navy divers assembled the catwalk. Although no underwater work was involved, the divers were able and willing helpers. To date, NFESC engineers and technicians have used the catwalk for extensive tests on concrete pier deck repair and upgrade using FRP methodologies.

COMPOSITE FENDER PILES

The primary energy absorbing component of U.S. Navy fendering systems is timber fender piles. Ship berthing forces are distributed to the piles through camels, long, horizontal floating objects which can also be make of wood. The fender piles must absorb energy from berthing ships but must possess sufficient stiffness so that excessive deflections do not threaten bearing piles or the ship itself. However, the Navy is faced with the problem of scarcity and increased cost of procuring and utilizing quality lumber in waterfront applications as well as the eventual disposal of treated timber at the end of its useful life. The development of high quality composite products offers an attractive option for meeting the Navy's construction needs as well as meeting new, more stringent environmental restrictions on the use of treated timber in the waterfront.

The first step in the development process was to obtain products that perform the same functions as wood and worked in the same configurations. The represents the substitution phase of the classical technology development process. The ultimate goal is to develop general performance specifications for a fendering system. However the first step was to develop performance target goals.

The Construction Productivity Advancement Research (CPAR) program with the Army Corps of Engineers Construction Engineering Research Laboratory, Rutgers University, and CI/SPI was the vehicle for the development of performance target goals. The Army Cold Regions Research and Engineering Laboratory and the NFESC were also participants in the CPAR. The overall goal of the CPAR was to develop a family of composite pile systems for marine waterfront applications. It included bearing piles and sheet piles as well as fender piles. We will discuss the fender pile aspects as an example. The performance target goals for the fender pile incorporate both the physical and structural properties of timber piles and the environmental constraints imposed on any material in the marine waterfront. The following is a summary of the performance target goals for composite fender piles:

I. Dimensions/Appearance

       1. Cross-section may be any shape but not to exceed 0.33m x 0.33m (nominal 13 in. x 13 in.)

       2. Must have continuous length of at least 21.34m (70 ft.); longer piles may use splices but spliced piles must meet all other mechanical and performance properties

II. Performance Requirements

       1. Under normal service conditions, mechanical properties shall not degrade more than 10% over the design life

       2. Less than 5% weight increase due to water absorption

3. Shall not pose a hazard to the environment and shall meet standard regulatory requirements for leaching, flame spread and potential ignition.

4. Shall exhibit ductile behavior even at air temperature extremes

5. Shall resist loads from various angles of approach

III. Mechanical Properties

1. Minimum EI = 1,735 kN-m^2 (6.0 x 10^8 lb-in^2)

2. Minimum outer fiber stain at fracture of 2% in bending

3. Minimum energy absorption capacity of 6,780 joules (5.0 ft-kips)

IV. Installation/Fabrication

1. Drivable with standard pile driving equipment

2. System connections with standard mechanical methods such as bolting, collar or wire wrap with staples; drillable or can be hot-lanced.

V. Cost

1. Life cycle cost shall be equivalent to treated/untreated wood

In a separate effort supported by the Naval Facilities Engineering Command Criteria Office NFESC is obtaining data to develop Navy specifications for composite fender piles. Structural tests will evaluate several products for stiffness, load-deflection characteristics, cyclic loading effects and failure mode. In addition NFESC will conduct several tests to determine durability parameters. These will include ultraviolet, chemical and abrasion resistance, brittleness and water adsorption. The testing and analytical efforts will determine the range of properties available with present products and compare those results with the requirements of Navy fendering systems. The ultimate goal is to develop specifications which meet the performance requirements and are within the capabilities of the industry to manufacture appropriate products.

DEMONSTRATION PROJECTS FOR PIER UPGRADES

NFESC is also investigating concepts which use composite materials for strengthening existing piers. Within the current environment of decreasing funding for military construction and changing operational requirements the Navy has several piers which must be strengthened to accommodate crane loadings and ship operations for which they were not originally designed. Based on the successful work of others and subscale tests conducted at the AWTTS, NFESC is conducting demonstrations at the Naval Stations at Norfolk, VA, San Diego CA and Pearl Harbor, HI. These sites represent the three largest U.S. Navy activities world-wide and also represent three significantly different exposure environments. Costs have been shared between the activity operating funds and NFESC research and development funds..

NAVAL STATION NORFOLK VIRGINIA

The demonstration at Norfolk involves a single 6.7 m (22 ft) span of a 0.48m- (19 in.) thick reinforced concrete deck. The pier is currently rated (and specifically designed) for 620 kN (70-ton) truck mounted cranes with limited use by 800 kN (90-ton) cranes. Recent analyses have identified design shortfalls that would limit 620 kN crane service in the area of the 6.7 m spans. Our goal is to upgrade the crane operating area of the span between pile caps 50 and 51 in order that unlimited 620 kN service would be allowed. This area includes lanes adjacent to both curbs. This area includes lanes adjacent to both curbs. This is a relatively small deck area of less than 100 m² (100 yd²).. The maximum outrigger float load is 590 kN (155,000 lbs) applied on a 0.61 m (24-inch) pad. The uniform live load requirement is 50 kPa (1000 psf). The piles are more than adequate to support the required cranes plus lateral loads from wind and berthing. Therefore, the piles will not be strengthened.

NAVAL STATION SAN DIEGO, CALIFORNIA

Pier 12 at the Naval Station, San Diego, California is a cast-in-place, reinforced concrete structure 444 m (1,458 feet) long and 9.1 m (30 feet) wide. 270 kN (30-Ton) truck mounted cranes are currently limited to the pile caps and the central deck for setting outriggers. An elaborate colored marking system has been painted on the decks to locate safe areas for crane operations. Our objective is to upgrade 20 spans to serve 450 kN (50-ton) cranes or a uniform load of 420 kPa (750 psf). The maximum crane loads are 490 kN (110,000 lbs) on 0.6 m (24-inch) square outrigger floats. The upgrade includes adding composite reinforcing equivalent of 12 kN/cm (7 kips/ inch) transversely and 17.5 kN/cm (10 kips/inch) longitudinally to the bottom of the 0.2 m (8 inch) deck section over a total area of 190 sq. m (225 sq. yds). We will add a composite reinforcing equivalent of 12 kN/cm (7 kips/inch) longitudinally to the bottom of the 0.6 m (24 inch) deck section. On the top surface of the deck we are reinforcing with an equivalent of 12 kN/cm (7 kips/inch) over full width and 1 m (3 feet) on either side of each of 22 pile caps an area of 500 sq. m (600 sq. yds). 132 piles will be reinforced with polymer reinforced composites between the waterline and the pile caps to increase strength and ductility.

NAVAL STATION PEARL HARBOR HAWAII

Bravo 25 is located near the eastern end of berthing wharves Bravo 22 through 26 at Naval Station Pearl Harbor. It is a cast in place reinforced concrete deck and superstructure supported by precast concrete piles. The Bravo wharves are over 50 years old. Bravo wharves were originally designed to support 445 kN (50-ton), rail mounted, portal cranes as well as 42 kPa (900 lbs per sq. ft). Over time the concrete and steel reinforcing have deteriorated and track mounted cranes have been replaced by truck mounted, mobile cranes. Mobile crane load limitations currently placed on Bravo wharves are very restrictive. Maximum uniform live load is limited to 23 kPa (490 psf).

Our objective is provide access on the deck for a 450 kN (50-ton) mobile truck crane with a maximum outrigger load of 980 kN (110 kips) on a 0.6 m (24-inch) square outrigger pad. We also want to increase the maximum uniform load to 36 kPa (750 psf). To accomplish this we intend to increase the capacity (upgrade) the deck, the rail stringers and the transverse girders. We expect to be upgrading approximately 73 m (240 feet) of deck (450 sq. m (4800 square feet)) and 80 piles. We plan to add reinforcement to the bottom of the 0.22 m (8 1/2-inch) deck between the rail slabs, to the bottom of the 0.34 m (13 1/2-inch) deck between the continuous outboard girder and the curb, and to the bottom of the rail slabs. We also intend to reinforce the top of the deck over each of the transverse girders. About half of the piles (over 100) will be strengthened with composite wrap and mortar between the girder intersection and the waterline.

FUTURE EFFORTS

In our work to date we have focused on composite fender piles and fendering systems and on the development of technologies for the repair and upgrade of piers and wharves. The objective of these efforts is to develop standardized design methods and guide specifications to permit successful use of these products. the laboratory and field demonstration tests are designed to validate the analytical methods and define the tolerable ranges for critical parameters. Without this knowledge, the products may not be economically or operationally successful. Similar efforts are necessary to implement other composite products.

The Navy's waterfront environment offers many opportunities for the application on composite products. Navy piers and wharves have extensive utility systems including water, steam, sewerage, electrical power and lighting systems. Corrosion resistant elements for these systems may significantly reduce facilities costs. Composite pipes, pipe hangers, conduits and light poles are possible items to evaluate and demonstrate in the future. Other possibilities involve the use of composite members for brows or shore-to-ship access structures. To insure safe and cost effective use of composites in these situations, the Navy will want some level of testing and evaluation in order to identify or develop appropriate specifications to allow competitive acquisition of these products.

New Developments in SMC/BMC Materials: From Low Density to Low Pressure, with a Few Stops in Between

New Advances in Low Pressure Sheet Molding Compound

LOUIS DODYK, BURR LEACH AND ANDREW RATERMANN

ABSTRACT

There continues to be a growing interest in low pressure sheet molding compound for a vast array of products ranging from automotive to heavy truck to industrial applications. Much progress has been made in recent years developing newer materials to meet these applications.

By formulating new systems using low pressure SMC, the molder can reduce tooling costs without the need to install large expensive presses. Materials are being formulated to mold at 100 psi, although the more typical low pressure SMC applications are at 250-300 psi compared to conventional SMC molded at 800-1200 psi molding pressure.

As more open molders and resin transfer molders switch to closed molding due to higher production rates, reduction of VOC's, and improved housekeeping, the demand for specialty low pressure compounds will increase. This paper will try to capture some of those new opportunities that are being generated using low pressure sheet molding compound.

INTRODUCTION

The SMC Automotive Alliance (SMCAA), a trade group that monitors industry trends, projects a 20 percent jump in SMC usage for automotive use in 1996 - from 200 million pounds in 1995 to 240 million pounds. In 1997, that usage is expected to grow to 250 million pounds, or an increase of 4.2 percent. Most of the growth over the past few years is for body panels and more recently, structural applications where SMC can utilize its inherent advantages over steel, aluminum and thermoplastics.

The SMCAA projects a 63% increase in SMC usage through the year 2000. Much of this growth will take place in both the light and heavy truck segments. One material which could make major inroads in the heavy truck segment is the use of low pressure SMC for Class A exterior parts. In addition, other applications utilizing low pressure SMC technology could increase that projection immensely.

Louis Dodyk, Cambridge Industries, 1700 Factory Avenue, Marion, Indiana,46952
Burr L. Leach, 1700 Factory Avenue, Marion, Indiana 46952
Andrew Ratermann, 1700 Factory Avenue, Marion, Indiana 46952

This paper will discuss some of those applications from Class A automotive and heavy truck to pigmentable grades to lower temperature (180°F) and even lower pressures (100 psi) using new low pressure compounds developed by Cambridge Industries with the help of its supplier base.

CLASS A LOW PRESSURE

Over the past two years, Cambridge Industries, in conjunction with its suppliers, has developed a Low Pressure Class A SMC that was capable of being produced on its current SMC machine. This accomplishment allowed us to get into production quicker without the need for adding expensive equipment to the current system, and concentrate our efforts on the chemistry and formulation of the material. We, therefore, were able to make a direct comparison of our standard Class A SMC, molded at 1000 psi, to our new low pressure SMC molded at 300 psi.

Table I shows the average mechanical properties of our standard Class A 7160 SMC molded at 1000 psi and 300 psi compared to our low pressure Class A 7525 SMC molded at the same pressures:

TABLE I
MECHANICAL PROPERTIES OF 7525 LOW PRESSURE SMC VS. 7160 STANDARD SMC AT VARIOUS PRESSURES

MATERIAL	7160	7160	7525	7525
Molding Pressure (Psi)	1000	300	1000	300
Tensile Strength (MPa)	84	79	85	88
Tensile Modulus (MPA)	11,000	11,000	10,500	10,900
Flexural Strength (MPa)	169	165	195	198
Flexural Modulus (MPa) Secant @ .5 mm defl. Secant @ 2.5 mm defl. Tangent	10,468 7,789 11,000	10,520 7,800 11,000	8,500 7,400 9,000	8,734 7,432 8,701
Izod Impact (J/m) Notched Unnotched	1000 1149	1013 1141	800 1116	1022 1182
Water Absorption ((%)	.50	.70	.57	.59
Glass Content (%)	26.2	26.0	28.0	27.4
Specific Gravity	1.9	1.9	1.8	1.84
Shrinkage in/in. (Cold part/Cold mold)	+.0006	+.0006	+.0009	+.0009

The average mechanical properties are very comparable between these two materials. The only exception being that the Flexural Modulus values are slightly lower for the low pressure compounds.

One area that was a major concern when molding at lower pressures is the tendency of the SMC to absorb water.

Table II compares the effect of molding pressure on water absorption for these two materials:

TABLE II - WATER ABSORPTION COMPARISON

MOLDING PRESSURE	WATER ABSORPTION	SMC TYPE
300 PSI	.70	7160
300 PSI	.59	7525
500 PSI	.64	7160
500 PSI	.61	7525
1000 PSI	.53	7160
1000 PSI	.57	7525

The 7525 SMC performs as well regardless of molding pressure, however, as the pressure decreases, the 7160 SMC increases in water absorption.

MATERIAL FLOW

The flow of 3 different low pressure SMC materials (7421, 7425 and 7490) was measured at 300 and 1000 psi and compared to typical Class A SMC using a spiral flow mold. Measurements were made at 7 and 10 days (Table III). Low pressure SMC has greater flow compared to typical Class A SMC, both at 300 and 1000 psi. However, the important advantage of LPSMC is that at 300 psi, the flow is greater than standard SMC at 1000 psi. All of these materials are formulated for fast cure applications at 300°F and contain 25-27% fiberglass. Therefore, the increased flow is a direct result of the LPSMC flow characteristics at decreased pressure. The low pressure SMC has molded well at 300 psi even after being one month old and stored at 80°F.

TABLE III - SPIRAL FLOW AT 295°F

SMC	7 Days 300 psi	7 Days 1000 psi	10 Days 300 psi	10 Days 1000 psi
7421	35	46	35	44
7490	23	44	23	42
7425	34	39	27	38
7160	16	32	12	31

ADHESIVE BONDING

Table IV compares various adhesives tested for the low pressure SMC. As the results indicate, the 7525 SMC passes the GM3629 Specification with all the adhesives tested.

TABLE IV - SMC 7525 BOND ADHESION

AVERAGE VALUES

TESTING CONDITIONS	GM-3629M SPECIFICATION	FUSOR 320/322	GENTAC 302/4001	ASHLAND 7000/7002
Room Temp Load	3400 (MPa)	3900	3452	4008
Room Temp Failure Mode	--	100% FT.	100% FT.	100% FT.
82°C Load	1400 (MPa)	1693	2505	3472
82°C Failure Mode	--	100% CF	100% FT.	96.3% FT. 3.7% AF

KEY: FT is Fiber Tear; AF is Adhesive Failure; CF is Cohesive Failure

IN-MOLD COATING AND PRIMER ADHESION

Tables V and VI show the results of testing 7525 SMC with GenGlaze EC610 IMC, EC800 IMC and Siebert-Oxidermo BP-9471 primer. The material passes all tests performed, comparable to standard SMC.

TABLE V
IMC ADHESION USING GENGLAZE IMC WITH 7525 LOW PRESSURE SMC

CONDITION	TAPE ADHESION		CROSSHATCH		BRITTLENESS		GRAVELOMETER (0 F)	
	EC-610	EC-800	EC-610	EC-800	EC-610	EC-800	EC-610	EC-800
Initial	100%	100%	0	0	Pass	Pass	9	9
E-Coat Bake 1 hr. @ 400 F	100%	100%	0	0	Pass	Pass	9	9
Heat Aged (7 days @ 70 C)	100%	100%	0	0	Pass	Pass	9	9
Humid Aged (96 hrs. @ 100 F and 100% RH)	100%	100%	0	0	Pass	Pass	9	9
Water Immersion (10 days @ 32 C)	100%	100%	0	0	Pass	Pass	9	9

TABLE VI

Primer Adhesion Crosshatch*	Average Values
Primer adhesion on bare substrate	0.5
Primer adhesion on bare substrate after ELPO bake	1
Primer adhesion on bare substrate after 96 hour heat (38°C) and humidity (100% RH)	1
Primer Adhesion Gravelometer **	**Average Values**
Primer adhesion on bare substrate	8
Primer adhesion on bare substrate after ELPO bake	8.5
Primer adhesion on bare substrate after 96 hour heat (38°C) and humidity (100% RH)	8.5

*Crosshatch Key	**Gravelometer Key
0 is best, 10 is worst	10 is best, 0 is worst
Less than 3 is acceptable	Greater than 7 is acceptable
Test Method: Ford Lab Test Method BI6-1(MJ)	Test Method: GM 9508P

LOWER COST MOLD

Lower pressure molding causes less stress and side thrust on a mold. Using standard SMC mold cutting techniques, a 15% minimum decrease in mold cost can be expected resulting from a smaller block of steel being required. Small run-off and less steel under the mold is required. However, the size of the steel block must be large enough to prevent thermal stresses during cutting of the mold.

Although epoxy tools can reduce cost of molds, current methods of making epoxy tools do not provide the uniform heat or lack of distortion for a Class A surface that LPSMC can provide. Aluminum molds also have the potential to reduce tool costs. However, the surface of aluminum molds is too porous for use with LPSMC. Tight, long wearing shear edges are required for LPSMC.

New methods of mold production and different types of molds offer further potential for cost reductions. These methods include reinforced castings, aluminum molds with a hardened metal coating, or light weight mold inserts that fit in a reusable container that remains in the mold.

NEW APPLICATIONS IN LOW PRESSURE SMC

With the development of Low Pressure SMC using a standard SMC machine and thickened with conventional earth oxide thickeners, Cambridge Industries, with the help of its resin suppliers, has been able to formulate various materials for specific applications. In addition, because of the lower most cost, applications that were in RTM or open mold are now finding themselves being produced in Low Pressure SMC.

Since most of these applications are in the industrial division or non-automotive, the use of Class A materials has diminished. This opens up the ability to provide the molder with a low pressure SMC has molded-in color and either eliminates prime coating or at least allows the molder to prime or topcoat more effectively. Also, by switching to a color molded-in system, the molder can specify that the material be UV stable which again could possibly eliminate the high cost of painting depending on the application. We will not go into specific parts which are utilizing this technology to protect our customers' confidentiality; however, based on what we have seen over the past year, almost any application can be formulated to mold at low pressures -- whether it is automotive, heavy truck, industrial, pigmentable, UV stable, etc. -- whatever the customer wants.

LOW TEMPERATURE/LOW PRESSURE SMC

For the past year, Cambridge has been working with its supply base to develop an SMC material capable of being molded at 180°F and 100 psi. This would allow a molder to utilize "soft" molds, such as epoxy, kirksite or cast aluminum with hot water as the heat source. These molds could be considered to provide economical tooling for low volume production runs with physical properties equivalent to standard SMC production parts.

Table VII is a comparison of a 25% glass reinforced SMC molded at 80 psi and 180°F and a 35% glass reinforced SMC molded at 100 psi and 180°F from epoxy tooling compared to a 26% glass reinforced SMC molded at 1000 psi and 300°F off steel tooling. The physical properties of the low temperature/low pressure molding compare very well to the conventional SMC parts that this could open the door to as the material of choice for prototype molds or, even greater expectations is to use this material for low volume production runs.

TABLE VI I
MECHANICAL PROPERTY COMPARISON OF MOLDED PARTS USING
LOW PRESSURE/LOW TEMPERATURE SMC VS. STANDARD SMC

	25% Glass/ 80 psi	35% Glass/ 100 psi	Production R25-1013	Production R25-1013R
Tensile Strength (Mpa) AVE RANGE	50.3 38.43-64.42	90.8 69.64-123.4	85.6 81.93-89.12	74.6 68.72-81.66
Tensile Modulus (Mpa) AVE RANGE	10,440 9505-12,200	14,620 12,780-17,190	12,480 11,990-13,180	11,780 11,030-12,270
Flexural Strength (Mpa) AVE RANGE	140.9 110.5-182.1	255.4 219.4-281.3	176 152.5-189.8	160 150.7-178.2
Secant Flex Modulus @ 2.5min AVE RANGE	5602 5158-6360	9264 9032-9678	6659 6092-7064	6111 5684-6852
Secant Flex Modulus @ .5 min. AVE RANGE	8428 7411-9128	11,210 10,950-11,620	9216 8470-10,240	9968 9778-10,230
Tangent Flex Modulus AVE RANGE	10,330 9563-10,820	13,160 13,040-13,270	12,170 11,880-12,630	11,810 11,130-12,230
Izod Impact Notched J/m AVE RANGE	1088 844-1294	1439 1148-1778	1072 911-1244	1192 1117-1241
Izod Impact Unnotched J/m AVE RANGE	1159 1052-1373	1414 1192-1658	1413 1196-1636	1154 1067-1248
Water Absorption AVE RANGE	.20 .19-.21	.17 .16-.18	.54 .5-.58	.45 .42-.49
Specific Gravity AVE RANGE	1.777 1.75-1.8	1.86 1.84-1.87	1.913 1.88-1.95	1.923 1.9-1.95
Glass % AVE	23.2	35.5	26.1	26.1

TABLE VIII is a comparison of the mechanical properties of the low pressure, low temperature SMC flat panels molded at 180°F and 300 psi compared to a standard SMC molded at 300°F and 1000 psi. Again, the properties are very comparable to standard SMC.

TABLE - VIII
MECHANICAL PROPERTIES OF FLAT PANELS OF
LOW TEMPERATURE/LOW PRESSURE SMC VS. STANDARD SMC

	LOW PRESSURE SMC A	LOW PRESSURE SMC B	STANDARD SMC
Tensile Strength (MPA)	79	143	84
Tensile Modulus (MPa)	8990	12075	10954
% Elongation	1.89	1.95	1.3
Flexural Strength (MPa)	165	242	178
Tangent Flex Modulus (MPa)	8563	11275	10200
Water Absorption (%)	0.25	0.30	0.60
Percent Glass	24.4	38.4	27

CONCLUSION

With the current technology described in this paper, the potential for low pressure SMC applications appears to be unlimited. The challenge we face today is to get the design engineers to think SMC first, then we can begin to really grow this business. Not only can low pressure SMC be used in the automotive and heavy truck industry, but also in any other applications where color, weatherability, dimensional stability, corrosion resistance, light weight and all the rest of the attributes of reinforced plastics can be used to its fullest potential.

BIOGRAPHIES

LOUIS DODYK

Louis Dodyk is the SMC Sales and Service Manager for Cambridge Industries in Marion, Indiana. In addition to his sales and service responsibilities, he is responsible for the daily operations of the SMC operation. Dodyk has spent most of his almost 20 years in the reinforced plastics industry working for GenCorp until that division was sold to Cambridge Industries. Dodyk received his B.S. in Plastics Engineering from the University of Massachusetts - Lowell in 1975.

BURR LEACH

Burr ((Bud) Leach is a Technical Specialist at Cambridge Industries, formerly GenCorp Reinforced Plastics. Bud has spent 25 years working in the areas of material formulation, manufacturing and molding process development for Reinforced Plastics. Leach received his B.S. in Chemistry from Manchester College in 1962, and an M.A. from Ball State University in 1970.

ANDREW RATERMANN

Andrew Ratermann is a Senior Research Chemist at Cambridge Industries. Previously, he was employed by GenCorp, Inc. in both the Reinforced Plastics Division and the Corporate Research Center. He received a B.S. in Chemistry from the University of Missouri in Saint Louis in 1976 and a Ph.D in Inorganic Chemistry from Indiana University in 1983.

The Use of Divinylbenzene Monomer for Improved Fiber-Reinforced Composites

K. E. ATKINS, R. C. GANDY, C. G. REID AND R. L. SEATS

ABSTRACT

Thermosetting resins are widely known to shrink on curing and this attribute introduces many property deficiencies in the resultant molded product. Low profile additive contained systems function to reduce shrinkage inherent in a resin system during the curing reaction and to thereby improve dimensional stability and surface smoothness.

A major advance in commercial thermosetting molding technology was the introduction of chemically thickened systems utilizing low profile additives. These chemically thickened systems help carry the glass fiber to the extremities of the mold during the crosslinking of the system.

While low profile unsaturated polyester fiber glass reinforced molding systems have gained wide acceptance in the molding industry because of good surface appearance, dimensional stability, physical properties, production and assembly costs, and weight savings, there. remains a need for further improvements.

In compression molding there is now provided a means for making low profile, faster curing, fiber-reinforced molding compositions that can be used to generate molded parts at high temperature molding conditions which possess exceptional surface smoothness with increased physical properties, particularly interlaminar tensile shear strength at elevated temperature, and also exhibit a greater resistance to cracking from the molding operation.

With respect to injection molding there is also now provided a means to generate molded parts with higher flexural properties at elevated temperatures which should reduce cracks during the handling of the hot part immediately after molding and other elevated temperature processes. Articles from both types of molding exhibit higher gloss characteristics as well.

These enhanced physical characteristics of faster cure, improved interlaminar strength and higher flexural characteristics at elevated temperatures, along with higher gloss are accomplished by the addition of a highly reactive olefinically unsaturated compound, divinylbenzene monomer, to the resin system. When used with styrene, it becomes the monomer matrix of choice for fast-cure systems.

K. E. Atkins, R. C. Gandy, C. G. Reid, R. L. Seats, Union Carbide Corp., P. O. Box 8361, South Charleston, WV 25303-0361

BACKGROUND

Unsaturated polyester resins are widely employed commercially in a variety of reinforced fabrication systems including among others matched metal-die compression, transfer, pultrusion and injection molding. These systems involve curing a formulated compound at high temperatures and pressures in hardened and chrome plated molds. These methods provide the highest volume and highest part uniformity of any thermoset molding technique.

Thermosetting resins are widely known to shrink on curing and this attribute introduces many property deficiencies in the resultant molded product. Low profile additive resin systems have made a significant contribution to commercial thermosetting molding resin systems of all varieties. The low profile additives function to reduce shrinkage inherent in a resin system during the curing reaction and to thereby improve dimensional stability and surface smoothness.

A major advance in commercial thermosetting molding technology was the introduction of chemically thickened systems utilizing low profile additives. Chemically thickening is always employed in sheet molding compounds ("SMC"), and is increasingly being used in bulk molding compounds ("BMC"). Thus, the combination of thickened systems in low profile additive resin systems has made a major contribution to the commercial expansion of polyester molding.

Low profile unsaturated polyester fiber glass reinforced molding systems (the combination of fiber reinforced resin system and LPA resin system) have gained wide acceptance in the transportation industry because of good surface appearance, dimensional stability, physical properties, production and assembly costs and weight savings versus the use of metal. However, as applications for resin systems have grown, more demanding standards are being developed and imposed by the users of them.

Some answers to these increased demands have previously been addressed by improvements in the low profile resin system, see e. g. U. S. Pat. Nos. 4,374,215 and 4,525,498 and 4,673,706 and 4,755,557, etc. issued to Atkins et. al.

INTRODUCTION

As standards rise, subtle issues previously not addressed in a critical manner are now being recognized. For example, there has developed a recognition that low profile additives tend to reduce the physical properties of the molded laminate. This is particularly pronounced at the elevated temperatures at which the parts are formed under pressure; i., e., 130°C-170°C. This reduction in properties is seen to be the cause for defects such as cracking, delamination (blistering) and others.

Blistering is the most severe defect since it almost always results in scrapping of the part. Considerable attention has been given to this defect in the past. An excellent publication discussing causes of and methods to reduce blistering is by R. M. Griffith and H. Shanoski, "Reducing Blistering In SMC Molding," *Plastics Design and Processing*, pp. 10-12 (February 1977).

The most likely mechanism for blister formation stems from pressurization of gas within the part during molding, followed by crack initiation and extension of the cracks when the press opens. The pressure results from expansion of air or other volatiles trapped or dissolved in the part. The gas pressure causes the delaminated area to deform into a blister. This deformation is usually permanent although reduced in size, even when the pressure is relieved.

Trapping of gas during molding is a probable consequence of a practical molding operation. In addition, molded parts may again become pressurized internally when later heated either during painting or bonding operations. Therefore, an adequate level of interlaminar strength is required in the finished moldings to prevent blister formation.

In addition to the chemically thickened molding compositions mentioned above there is still a major and ever growing interest in compositions that are not chemically thickened.

These are widely used and are of particular interest in the high speed process of injection molding. Because of its inherent faster cycles and ease of automation as compared to compression molding, injection molding of fiber reinforced systems is undergoing great growth for the production of automotive body panels.

Frequently, unthickened compounds are desired in this application because they allow better control of the compound's viscosity and provide greater shelf life of the material without requiring machine and molding parameter changes. However, with this type of material there remains the need to obtain better surface smoothness and shrinkage control as contrasted with the systems used in compression molding.

A problem with these injection molded compositions is that they tend to crack in areas of reinforcing ribs and bosses in the molded part. These cracks can reduce the overall strength/integrity of the molded part (such as a head lamp reflector or door) and also serve as a source of gas release that causes paint film disruption ("popping") during finishing of the part. Reduction or elimination of this defect would lead to greater utility of injection moldable compositions and expand the design latitude for parts produced from them.

There exists a demand for a superior resin system which provides under conventional high temperature molding conditions, such as practiced at temperatures of from about 140° to about 200°C., molded parts possessing exceptional surface smoothness with increased physical properties, particularly interlaminar tensile shear strength at elevated temperature.

In addition, there exists a demand for a superior resin system which provides under conventional high temperature molding conditions, such as practiced at temperatures of from about 140° to about 200°C., molded parts possessing minimum cracking at the surface and in the interior of the parts, resulting in greater serviceability of the part.

This paper will describe how to meet these demands and others by the use of a novel molding composition that utilizes known materials in an uniquely beneficial manner. Basically this is accomplished by the addition of a small but meaningful amount of a highly reactive olefinically unsaturated compound (divinylbenzene monomer) to a resin system.

EXPERIMENTAL

GENERAL PROCEDURES FOR PREPARATION OF SHEET MOLDING COMPOUND

All the liquid components were weighed individually into a 5 gallon open top container placed on a Toledo balance. The contents of the container were then mixed (in a hood) with a high speed Cowles type dissolver. The agitator was started at a slow speed, then increased to a medium speed to completely mix the liquids over a period of 2-3 minutes. The mold release agent was next added to the liquids from an ice cream carton and mixed until completely dispersed. The filler was next added gradually from a tarred container until a consistent paste was obtained and the contents were then further mixed to a minimum temperature of 32°C. The thickener was next mixed into the paste over a period of 2-3 minutes, the mixer was stopped and 175 grams of the paste was removed from the container and transferred to a wide-mouthed 4 oz. bottle. The paste sample was stored in the capped bottle at room temperature and the viscosity measured periodically using a Model HBT 5X Brookfield Synchro-Lectric Viscometer on a Helipath Stand.

The balance of the paste is next added to the doctor boxes on the sheet molding compound (SMC) machine where it is further combined with glass fibers (~1" fibers). The sheet molding compound is then allowed to mature to molding viscosity and then molded into the desired article.

GENERAL PROCEDURES FOR PREPARATION OF BULK MOLDING COMPOUND

All the liquid components were weighed individually into a Hobart mixing pan placed on a Toledo balance. The pan was attached to a Model C-100 Hobart mixer (in a hood). The agitator was started at a slow speed, then increased to medium speed to completely mix the liquids over a period of 3-5 minutes. The agitator was then stopped and the internal mold release agent was next added to the liquid from an ice cream carton. The Hobart mixer was restarted and the mold release agent mixed with the liquid until it was completely wet out. The filler was next added to the pan ingredients (agitator off) then mixed, using medium to high speed, until a consistent paste was obtained. The mixer was again stopped and the weighed amount of thickening agent, if desire, was mixed into the paste over a period of 2-3 minutes, the mixer was again stopped and ~175 grams of the paste were removed from the pan (using a large spatula) and transferred to a wide-mouthed 4 oz. bottle. The paste sample was stored in the capped bottle at room temperature and the viscosity measured periodically using a Model HBT 5X Brookfield Synchro-Lectric Viscometer on a Helipath Stand.

After removal of the paste sample, the contents were reweighed and styrene loss made up, the chopped glass fibers were added slowly (from an ice cream carton) to the paste with the mixes running on slow speed. The mixer was run for 30 seconds after all the glass was in the paste. This short time gave glass wet out without glass degradation. The pan was then removed from the mixer and separate portions of the BMC mix of appropriate amount were removed using spatulas and transferred to aluminum foil lying on a balance pan (balance in the hood). The mix was tightly wrapped in the aluminum foil (to prevent loss of styrene via evaporation) and stored at room temperature until the viscosity of the retained paste sample reached molding viscosity as chemically thickened. The weight of the BMC added to the foil varies with the molding application.

COMPRESSION MOLDING

Flat panels 18 inches X 18 inches of varying thickness were molded using an extremely smooth, highly polished matched metal die set of chrome plated molds. Both platens are oil heated separately to allow for varying temperature. The molding was conducted in a 200 ton press and the panel removed with the aid of ejector pins. A standard molding temperature is 290-300°F. at pressures of 500 to 1000 psi. Laminate thicknesses are typically 0.10 and 0.125 inches.

INJECTION MOLDING

The BMC paste for these moldings was compounded using a high speed dissolver (Jaygo). The fiberglass was mixed with the paste in a Draiswerke T50FM1G low intensity mixer. This BMC was injection molded on a 140 ton Billion ZMC machine. A 18 cm by 36 cm flat panel mold was used. The material was injected along the long axis of the mold.

METHOD OF SHRINKAGE MEASUREMENT

A 18 X 18 inch X 0.125 inch flat panel is molded in a highly polished chrome plated matched metal die mold in a 100 ton press. The exact dimensions of the four sides of this mold are measured to the ten-thousandths of an inch at room temperature. The exact length of the four sides of the flat molded panel is also determined to the ten-thousandths of an inch. These measurements are substituted into the equation below:

(a-b)/a=inch/inch shrinkage whereby inch/inch shrinkage X 1000=mils/inch shrinkage

a=the sum of the lengths of the four sides of the mold at room temperature.
b=the sum of the lengths of the four sides of the molded panel at room temperature.

A positive (+) number recorded indicates an expansion of the molded part as measured at room temperature compared to the dimension of the mold measured at room temperature. A negative (-) number indicates shrinkage by the same comparison. The larger the positive number, the better the performance.

INTERLAMINAR SHEAR TEST

This test is to measure the interlaminar strength of a molded fiberglass specimen by perpendicular tensile forces. The test is restricted to flat or nearly flat sections of the molding. Equipment used in the test are set forth immediately below.

1. Instron tensile testing machine; 10,000 pound cell. Universal Testing Machine Jaws.
2. Instron oven with temperature control for 300°F.
3. Metal mounts for bonding to specimens with diameters at bonding area of 1.59 inches (2 square inch area). Two (2) mounts requires per specimen, each pair numbered. Mounts threaded for attachment to Universal Testing Machine Jaws.
4. Alignment fixture for holding metal mounts, and specimen during oven curing of the bond.

Flat panels were molded from standard type SMC's (resin, filler, fiberglass, in normal amounts). A standard mold temperature of 290-300°F. along with molding pressures of 1000±200 psi. were used to mold panels of a nominal 0.100 inch thickness. The Bonding Adhesive used in the test was Hysol Adhesive EA 934.

PROCEDURE
A. Specimen Preparation
1. Cut (3) specimens approximately 17/8 inch square from the flat area of the test specimen, at least 3 inches away from any visible defects such as blisters, molded in flash, non-fills etc.; and 1 inch from the external edge of the part (if possible).
2. Lightly sand (240 grit) both bonding surfaces of the specimen and wipe clean with alcohol.
3. Mix a small amount of the (2) component bonding adhesive together (ten specimens require approximately 33 grams).
4. Apply 20-40 mil film of adhesive on opposing faces of the metal mounts.
5. hand squeeze the FRP specimen between the two metal mounts to insure good contact.
6. Place the specimen with mounts into the alignment fixture. Observe the alignment to be sure that sufficient FRP is overhanging all edges of the mount.
7. After all the specimens are in the alignment fixture, place into a preheated oven.
8. Cure the bonding adhesive for 30 minutes at 200°F.
9. Sand or grind off the excess edges of the FRP square to the diameter of the metal mounts.
B. Testing
1. Preheat the Instron over (with all the specimens inside) for 2 hours at 300°F.
2. Each specimen is tensile pulled in the oven at 300°F at a cross-head speed of 0.05 inches per minute.
3. Record the failure load (pounds).
4. Determine the interlaminar strength (psi) by dividing the failure load by 2 square inches.
5. report the interlaminar strength as the average of the specimens tested for the subject area.

SURFACE SMOOTHNESS
As the use of low profile reinforced plastics further expands into more body panel applications the criteria and standards for surface smoothness has dramatically increased. An instrument developed by Diffracto was adopted by GM for evaluation of their U-van body parts and has fast become an industry standard for comparison. Surface smoothness characteristics of molded panels were determined using Diffracto's D-Sight Audit Station™ with their surface waviness analysis algorithm program.

CURE CHARACTERISTICS
Cure time results were determined by the use and operation of the ICAM-2000 dielectrometer from Micromet Instruments, Inc. Sensors for dielectric properties measure ion viscosity (resistivity). The ion viscosity and the slope of the ion viscosity are used to follow the cure state of the material from the beginning to the end of the reaction.

MECHANICAL PROPERTIES
Standard ASTM methods were used to determine various flexural properties.

RESULTS AND DISCUSSION

BLISTERING

The most likely mechanism for blister formation stems from pressurization of gas within the part during molding, followed by crack initiation and extension of the cracks when the press opens. Pressure results from expansion of air or other volatiles trapped or dissolved in the part. The gas pressure causes the delaminated area to deform into a blister. Although reduced in size when the pressure is relieved, this deformation is usually permanent.

Trapping of gas during molding is a probable consequence of practical molding operations. In addition, molded parts may again become pressurized internally when later heated either during bonding or painting operations. Therefore, an adequate level of interlaminar strength is required in the finished moldings to prevent blister formation. The addition of divinylbenzene to a typical automotive SMC formulation substantially increases the interlaminar strength of the resultant substrate molding, as shown in Table 1, thus decreasing the blister formation problem.

Frequently, these systems are not chemically thickened, allowing better control; of the compound viscosity and providing longer shelf life. However, with these types of materials, there remains the need to obtain better surface smoothness and better dimensional stability, which may eliminate painting and priming steps.

Microcracks in areas of reinforcing ribs and bosses can reduce the overall strength and integrity of the molded part and also serve as a source of gas release that causes paint film disruption ("popping") during finishing operations. The use of divinylbenzene can reduce or eliminate this defect by improving the physical properties of the molded part. This leads to expanded latitude for the molder. Table 2 and Figures 1 through 6 illustrate the kinds of improved physical properties that can be obtained by using divinylbenzene.

Injection molding formulations utilizing divinylbenzene (Table 2) have also shown a higher degree of flexibility and higher impact strengths. This results from the improved resin and fiberglass reinforcement interface and is of particular importance for providing potential dent resistance to the manufactured part. The use of divinylbenzene in formulas, such as those shown in Table 2, also yields parts that have outstanding gloss and surface properties.

CURE TIME AND SURFACE SMOOTHNESS

The effect of divinylbenzene on cure time and part surface smoothness was studied in Class A sheet molding compound (SMC) formulations. A typical formulation as shown in Table 3 exhibited a 10 percent reduction in cure time with no degradation of surface smoothness, as measured on Diffracto's D-sight surface instrument.

SUMMARY

The use of divinylbenzene has been shown to generate molded parts at higher production rates and with exceptional surface smoothness and increased physical properties. Small additions of divinylbenzene provide improvements in interlaminar tensile shear strength, improved flexural strength, and a greater resistance to cracking from the molding operation. When used with styrene, it becomes the monomer matrix of choice for fast-cure systems.

BIOGRAPHIES

K. E. Atkins

K. E. Atkins received a B. S. degree in Chemistry from West Virginia Institute of Technology in 1960 and an M. S. (Chemistry) from West Virginia University in 1965. He has been employed at Union Carbide's South Charleston, West Virginia Technical Center since 1960 and was appointed a Corporate Fellow in 1987. He is Technology manager of the PVA Performance Polymers group and former Chairman of the Union Carbide Materials

Science Corporate Technology Team. He was awarded the American Chemical Society's Thomas Midgley Award in 1989 for chemical contributions to the automotive industry.

R. C. Gandy

R. C. Gandy received a B. S. degree in Chemistry from Morris Harvey College in 1963. He is presently a Senior Development Scientist in the PVA Performance Polymers area of the Specialty Polymers & Products Division of Union Carbide. He has been employed at Union Carbide's, South Charleston, West Virginia Technical Center since 1956; the last 20 years in the PVA Performance Polymers area.

C. G. Reid

C. G. Reid received a Ph. D. from Case Western Reserve University in 1969. With more than 10 years' experience in the research and development of specialty resins with Celanese Corp., he joined Union Carbide in 1984 and is a Development Scientist at the South Charleston, West Virginia Technical Center. He is currently working with the continuing commercialization and new product development for polyvinyl acetate polymers.

R. L. Seats

R. L. Seats received a B. S. degree in Chemical Engineering from the University of Illinois. He joined Union carbide after three years with Rockwell International in the Materials Engineering Department. Currently he is a Development Scientist in the PVA Performance Polymers group at Union Carbide's, South Charleston, West Virginia Technical Center.

TABLE 1: Effect of Divinylbenzene On The Interlaminar Shear Strength Of A Typical Automotive SMC Formulation		
	Parts By Weight	
Ingredients	**Control**	**With DVB**
Polyester Resin	55	55
NEULON® Polyester Modifier TPlus	37	37
Styrene	7	7
Divinylbenzene	–	6.24
Inhibitor	0.4	0.4
Water,% adjusted to in A-Side	0.18	0.18
Viscosity Reducer	1.7	1.7
Zinc Stearate	4	4
Calcium Carbonate, 5 micron	185	185
Catalyst 1	1.3	1.3
Catalyst 2	0.5	0.5
Calcium Oxide Thickener System	12.5	12.5
Fiberglass, 1-in chopped	28 percent by weight	
Composite Properties		
Interlaminar Shear at 149°C, psi	491	787
Flexural Strength, MPa	140.6	144.8
Flexural Modulus, MPa	7440	7785

TABLE II: Effect Of Divinylbenzene On Flexural Properties Of Injection Molded BMC

Ingredients	A	Parts by Weight B	C	D
Polyester Resin	60	60	57	57
UNION CARBIDE®Low Profile Additive LP-40A	40	—	—	—
UNION CARBIDE®Low Profile Additive LP-4022	—	40	37	37
Divinylbenzene	—	3.12	6.24	9.36
Inhibitor	0.4	0.4	0.4	0.4
Catalyst 1	1.5	1.5	1.5	1.5
Catalyst 2	0.4	0.4	0.4	0.4
Zinc Stearate	2	2	2	2
Calcium Stearate	2	2	2	2
Calcium Carbonate, 3 microns	220	250	220	220
Fiberglass, 1/2-in chopped		20.0 percent by weight		
25°C Flexural Properties				
Strength, psi (Mpa)	14997(103.4)	15680(108.1)	18830(126.3)	18145(125.1)
Modulus, MM psi (Mpa)	2.00 (13789)	2.02 (13927)	1.97 (13582)	1.96 (13514)
150°C Flexural Properties				
Strength, psi (Mpa)	5804 (40.0)	6325 (43.6)	7373 (50.8)	9513 (65.6)
Modulus, MM psi (Mpa)	0.681(4695)	1.00 (6895)	0.971 (6634)	1.05 (7239)
60°Gloss of Molded Panels	50			80

TABLE III: Effect Of Divinylbenzene On Cure Time and Surface Smoothness		
	Parts by Weight	
Ingredients	Control	With DVB
Polyester Resin	57	57
NEULON® Polyester Modifier TPlus	38	38
Styrene	5	5
Divinylbenzene	–	3
Inhibitor	0.5	0.5
Water,% adjusted to in A-Side	0.18	0.18
UNION CARBIDE® Viscosity Reducer VR-3	2	2
Zinc Stearate	2	2
Calcium Carbonate, 5 micron	200	200
Catalyst	1.5	1.5
Calcium Oxide Thickener System	12.5	12.5
Fiberglass, 1-in. chopped	28 percent by weight	
Composite Properties		
Cure Time (1)	49	44.9
Surface Smoothness (2)	140.6	144.8
(1) Determined by Micromet Cure Analyzer		
(2) Diffracto's D-Sight Number		

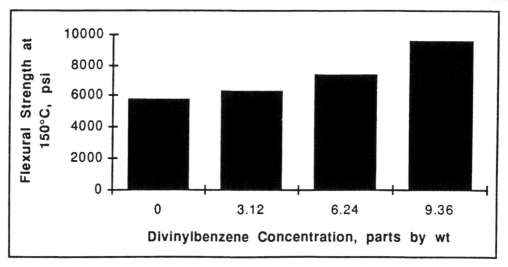

Figure 1: Effect of Divinylbenzene on Flexural Strength at 150° C in a Typical Automotive SMC Formulation

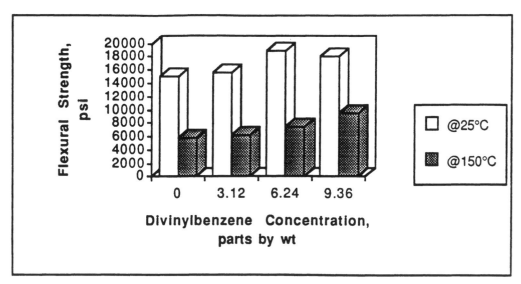

Figure 2: Effect of Divinylbenzene on Flexural Strength in a Typical Injection Molded BMC Formulation

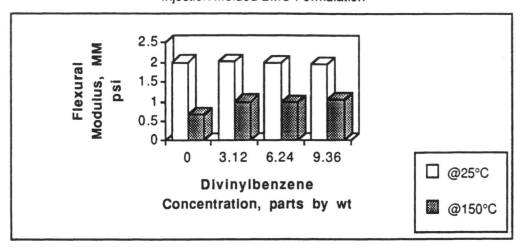

Figure 3: Effect of Divinylbenzene on Flexural Modulus in a Typical Injection molded BMC Formulation

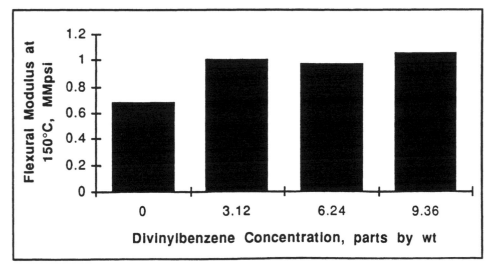

Figure 4: Effect of Divinylbenzene on Flexural Modulus at 150°C in a Typical Automotive SMC Formulation

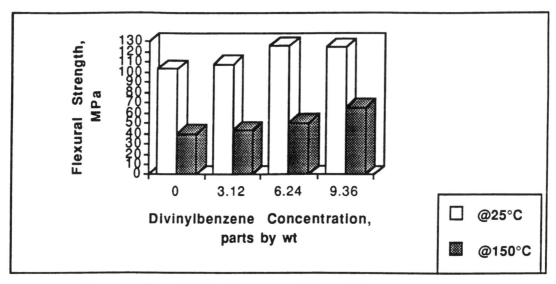

Figure 5: Effect of Divinylbenzene on Flexural Strength in a Typical
Injection Molded BMC Formulation

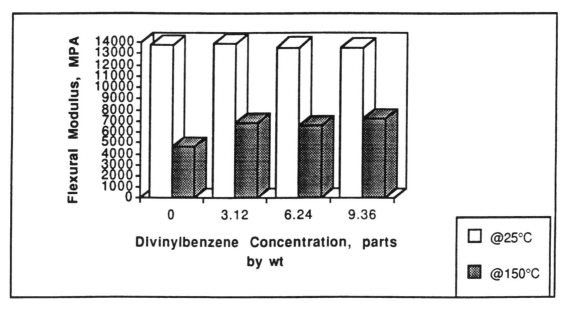

Figure 6: Effect of Divinylbenzene on Flexural Modulus in a Typical
Injection Molded BMC Formulation

A New Structural SMC Resin

STEVEN P. HARDEBECK AND CLARK B. WADE

ABSTRACT

Composites have been engineered into automotive applications for many years because they are light weight, tough, corrosion resistant and easily designed into the vehicle. Recently automotive engineers have expanded the use of FRP composites to include under the hood applications where all of these attributes can be realized. Structural FRP formulations fall into three major categories; corrosion resistance, heat resistance, and high strength.

Vinyl esters have been traditionally specified for these jobs to provide a strong heat resistant matrix. The major drawback with vinyl esters is that they are difficult to thicken consistently in SMC formulations. Blends of polyester and vinyl esters appear to solve the thickening problem at the expense of a weaker, less heat resistant matrix.

This paper will present our development efforts towards a new structural SMC resin that can be thickened without sacrificing mechanical strength and heat resistance.

INTRODUCTION

The use of Sheet Molding Compound (SMC) in automotive applications has grown from a mere 40 million pounds in 1970 to an estimated 240 million pounds in 1996. By the year 2000, nearly 400 million pounds of SMC will be used in wide variety of automotive applications [1]. It is interesting to note how this growth has evolved. In the early 1970's the largest application for automotive SMC was in Grille Opening Panels (GOP). The reason for this was that designers discovered that a fiberglass reinforced plastic (FRP) GOP offered parts consolidation, styling freedom, weight savings, and ease of assembly [2]. In addition, SMC formulations were being modified to improve the surface of the finished part. The number of GOP applications continued to grow into the 1980's. Styling

Steven P. Hardebeck, Alpha/Owens Corning, 2552 Industrial Dr., Valparaiso, Indiana 46383
Clark B. Wade, Alpha/Owens Corning, 2552 Industrial Dr., Valparaiso, Indiana 46383

changes in the late 1980's and early 1990's replaced the GOP with the Grille Opening Reinforcement (GOR). Today GOP's account for about 8% (19 million pounds) of the total SMC automotive market [3].

With the success of the GOP automotive designers began to choose SMC for larger body panels. In the mid 1980's two vehicles, the Pontiac Fiero, and GM APV Van, were launched. These vehicles used Class A SMC body panels as the major exterior surface material. Today SMC is used on 25 passenger car and truck hoods. Body panel applications account for over 75% of the automotive SMC produced. The reason for this wide spread acceptance is the same as in the 1970's plus the many proven success stories throughout the last twenty years. SMC is no longer considered as a material for only low-volume production jobs.

The newest growth area for SMC is in structural parts. Structural panels have grown from 14 million pounds in 1993 (9%) to 51 million pounds in 1996 (21%). In 1997, one half of the new pounds in the automotive area will come from structural components [4]. Most of this growth is attributed to applications on high volume production vehicles.

Besides SMC's light weight, design flexibility and lower total cost, there are other reasons for this growth. SMC has a strong well-established supplier base who are accustomed to mass production requirements. SMC is very heat resistant. It has lower thermal conductivity and has better noise damping properties than steel. It adapts easily into current manufacturing and assembly plants. An SMC part can consolidate several components into one tool, allowing designers to develop modular assemblies. Finally, SMC properties and capabilities have been successfully demonstrated in the industry for over twenty years [5].

Structural SMC formulations fall into three major categories; corrosion resistance, heat resistant, and high strength. Often the attributes of one formulation will overlap into the other two areas. Examples of current production jobs requiring structural formulations are:

-Radiator Supports
-Fuel Tank Heat Shields
-Cowl Plenum
-Cross Vehicle Beams.
-Valve Covers

Vinyl esters have traditionally specified into structural formulations because they provide a strong heat and corrosion resistant matrix. The major draw back with vinyl esters is that they are more expensive and are difficult to thicken consistently in SMC formulations. Blends of polyesters and vinyl esters appear to solve these problems at the expense of a weaker, less heat resistant matrix.

The following study was designed to look at the individual resin components of a structural formulation and to develop a structural formulation that would be strong, heat resistant, and provide a consistent thickening profile.

EXPERIMENTAL

This study was divided into two parts. The first part evaluated various low shrink additives with a given unsaturated polyester resin (UPR). The SMC formulation is shown in Table I. The four low profile additives (LPA) types were polystyrene, polyvinylacetate, saturated polyester, and polyacrylic. Also included was a formulation that contained no LPA. The formulations were compounded on a laboratory SMC machine, and matured for three days. Test panels were molded on a 30.5 x 30.5 cm. flat mold at 152°/146°C for 2 minutes at 6.9 MPa. The charge pattern covered approximately 70 % of the tool surface. A 2 ply charge, with one ply rotated 90°, was used to produce 2.5 mm thick panels.

The test panels were allowed to cool for at least 24 hours, then measured for shrinkage. Shrinkage values were calculated cold part verses cold tool. The heat aged panels were placed vertically in a pre-heated, forced air oven controlling at 180°C. At the end of 1000 hours the panels were removed and again allowed to cool for 24 hours. Tensile and flexural specimens were cut from both the "as made" and "conditioned" panels according to ASTM test procedures. Tensile testing was performed per ASTM D-638, and flexural testing per ASTM D-790.

The second part of the experiment used the LPA from the first part that gave the best shrink control and the best retention of physical properties after heat aging. This LPA was then used to compound additional SMC formulations that contained various unsaturated polyesters. The unsaturated resins were a high reactive isophthalic (ISO), high reactive dicyclopentadiene (DCPD), a high reactive resin made from recycled polyethylene-teraphthalic (PET), and a polyester/vinylester hybrid (PE/VE). Also included was a commercial vinyl ester (VE). These compounds were made and tested as outlined above.

The optimized formulation was then characterized for liquid properties, clear casting properties, chemical thickening, and mechanical properties per Ford Engineering Material Specifications WSA-M3D164 and the Automotive Composites Consortium Automotive Structural Composites Test Procedures.

RESULTS

The first part of the study was designed to determine the LPA that gave the best shrink control and the best retention of physical properties. Table II compares the shrink control of four LPA types and a control formulation without LPA. In this resin system the saturated polyester LPA exhibited the best shrink control.

Table III lists the tensile properties of the same systems "as made" and after heat aging. All the systems had similar tensile strength values "as made". After heat aging the saturated polyester LPA had the highest retention and the system without LPA had the lowest retention. A similar trend was observed for the tensile modulus values. The tensile elongation values showed mixed results. While the saturated polyester LPA and no LPA systems were initially high when tested "as made", they both showed the lowest values after heat aging.

Table IV lists the flexural properties "as made" and after heat aging. Again the saturated polyester LPA showed the highest flexural modulus retention after heat aging, and higher flexural strength retention than the system with no LPA. From these two tables we concluded that the saturated polyester LPA exhibited the best overall retention of properties.

The second part of this study used the saturated polyester LPA from the first study, in combination with various unsaturated polyester resins to determine the best combination. A commercially available vinylester was included in the study. Physical properties were tested "as made" and after heat aging. Table V lists the tensile properties and Table VI lists the flexural properties for these compounds.

The polyester/vinylester hybrid system had the highest tensile strength both before and after heat aging. It also had the highest flexural strength and modulus. The PE/VE hybrid system retained 100% of its original flexural modulus after heat aging. From this data it was decided that the PE/VE hybrid resin system offered the best mechanical properties in terms of overall strength and retention of strength after heat aging.

Table VII lists the typical liquid properties of the PE/VE hybrid resin system. Table VIII contains the resin casting mechanical properties for this system. The chemical thickening profile is outlined in Table IX for the hybrid resin using the SMC formulation from Table I.

Tables X and XI lists the typical SMC mechanical properties for the VE/PE hybrid system for Ford's Engineering Material Specification WSA-M3D164 and following the Automotive Composites Consortium Structural Composites Test Procedures.

CONCLUSIONS

This new structural SMC resin is a polyester/vinylester hybrid with a saturated polyester thermoplastic additive. It can be chemically thickened to a consistent profile. Also it has high initial mechanical properties and retains these properties after heat aging under extreme conditions. This new resin system is suitable for use in a variety of applications including radiator supports, fuel tank heat shields, and cross car beams.

REFERENCES

1) SMC Automotive Alliance (SMCAA) "1996 Model Year Passenger Car and Truck SMC Components".

2) SMC Automotive Alliance (SMCAA) "1992 Model Year Passenger Car and Truck SMC Components".

3) SMC Automotive Alliance (SMCAA) "1996 Model Year Passenger Car and Truck SMC Components".

4) SMC Automotive Alliance (SMCAA) September 26, 1996 "SMC in the Automotive Industry - Growth and Applications' Trends" .

5) SMC Automotive Alliance (SMCAA) "1996: It's Leap Year".

BIOGRAPHIES

Steven P. Hardebeck
    Steven P. Hardebeck received his B.S. in Chemistry from Valparaiso University in Valparaiso, Indiana, and his M.S. in Chemistry from Case Western Reserve University in Cleveland, Ohio. Mr. Hardebeck joined Alpha/Owens-Corning (Owens Corning) in 1979, and is a Product Leader located in Valparaiso, Indiana.

Clark B. Wade
    Clark B. Wade received his B.S. in Chemical Engineering from The Ohio State University in Columbus, Ohio. Mr. Wade joined Alpha/Owens-Corning (Owens Corning) in 1989, and is a Technical Service Specialist located in Valparaiso, Indiana. Prior to joining AOC Mr. Wade was employed with The Goodyear Tire and Rubber Company in Jackson, Ohio.

TABLE I - SMC FORMULATION

Ingredient	PHR
UPR	50.0 (solids)
LPA	6.0 (solids)
Styrene	44
tBPB (initiator)	1.5
SP-9139 (inhibitor)	0.1
CM-2015 (pigment)	0.25
Zinc Stearate (internal release agent)	4.0
Omyacarb 5 (calcium carbonate filler)	40
PG-9033 (chemical thickener)	2.5
973 1" (glass roving)	50%

SP-9139, CM-2015 and PG-9033 are manufactured by Plasticolors, Astabula, Ohio

Omyacarb 5 is manufactured by Omya Inc., Proctor, Vermont

973 Roving is manufactured by Owens Corning, Toledo, Ohio

TABLE II - SHRINKAGE OF VARIOUS LPA'S
@ 55% Glass Content

LPA	POLYSTYRENE	POLYVINYL-ACETATE	SATURATED POLYESTER	ACRYLIC	NONE
SHRINKAGE (%)	0.067	0.048	0.035	0.072	0.074

TABLE III - TENSILE PROPERTIES OF VARIOUS THERMOPLASTIC ADDITIVES @ 50% GLASS CONTENT

LPA Type	POLY-STYRENE	POLYVINYL ACETATE	SATURATED POLYESTER	ACRYLIC	NONE
Strength (MPa) as made	150	151	160	159	158
Strength (MPa) heat aged*	132	144	153	150	112
Strength Retention (%)	88	95.4	95.6	94.3	70.9
Modulus (GPa) as made	14.6	15.6	13.5	15.1	13.7
Modulus (GPa) heat aged*	13.6	14.6	13.1	14.4	10.9
Modulus Retention (%)	93.2	93.6	97	95.4	79.6
Elongation (%) as made	1.61	1.63	1.80	1.70	1.80
Elongation (%) heat aged	1.34	1.36	1.39	1.51	1.27
Elongation Retention(%)	83.2	83.4	77.2	88.8	70.6

*1000 hours @ 180°C

TABLE IV - FLEXURAL PROPERTIES OF VARIOUS THERMOPLASTIC ADDITIVES @ 50% GLASS CONTENT

LPA Type	POLY-STYRENE	POLYVINYL ACETATE	SATURATED POLYESTER	ACRYLIC	NONE
Strength (MPa) as made	323	327	314	331	299
Strength (MPa) heat aged*	236	253	258	292	157
Strength Retention (%)	73	77.4	82.2	88.2	52.5
Modulus (GPa) as made	15.5	15.0	13.7	15.5	13.4
Modulus (GPa) heat aged*	13.0	12.8	12.9	14.1	10.3
Modulus Retention (%)	83.9	85.3	94.2	91	76.9

*1000 hours @ 180°C

TABLE V – TENSILE PROPERTIES OF VARIOUS UNSATURATED RESINS
@ 50% GLASS CONTENT

Resin Type	ISO	DCPD	PET	PE/VE Hybrid	VE
Strength (MPa) as made	161	159	172	177	163
Strength (MPa) heat aged*	153	159	165	169	150
Strength Retention (%)	95	100	95.9	95.5	92
Modulus (GPa) as made	13.5	14.0	13.1	14.5	15.0
Modulus (GPa) heat aged*	13.4	14.9	13.4	14.7	15.9
Modulus Retention (%)	99.3	106.4	102.3	101.4	106
Elongation (%) as made	1.80	1.68	2.19	1.96	1.69
Elongation (%) heat aged	1.72	1.55	1.75	1.83	1.46
Elongation Retention(%)	95.6	96.8	79.9	93.5	86.4

*1000 hours @ 180°C

TABLE VI – FLEXURAL PROPERTIES OF VARIOUS UNSATURATED RESINS
@ 50% GLASS CONTENT

Resin Type	ISO	DCPD	PET	PE/VE Hybrid	VE
Strength (MPa) as made	308	319	333	352	278
Strength (MPa) heat aged*	286	286	304	288	259
Strength Retention (%)	92.9	89.7	91.3	81.9	93.2
Modulus (GPa) as made	14.8	14.6	14.2	15.6	12.7
Modulus (GPa) heat aged*	14.6	15.1	14.9	15.7	15.2
Modulus Retention (%)	98.6	103.4	104.9	100.8	120

*1000 hours @ 180°C

TABLE VII - LIQUID PROPERTIES

Property	Value
Viscosity (cPs) RVT #2 @ 20 rpm	760
Non-volatile (%)	58.5
82°C SPI Gel (min.)	8.1
Gel-Peak (min.)	10.5
Peak Exotherm (°C)	230
Acid Value	11.5
Weight/gallon (lbs/gal)	9.01

TABLE VIII - CASTING PROPERTIES

Property	Value
Tensile Strength (MPa)	73.1
Tensile Modulus (GPa)	3.42
Elongation (%)	2.87
Flexural Strength (MPa)	99.1
Flexural Modulus (GPa)	3.7
HDTUL (°C)	122
Hardness (HB)	43

TABLE IX - CHEMICAL THICKENING DATA

Time	2.0 phr PG-9033	2.5 phr PG-9033
1 Hour RVT # 7 @ 1rpm (McPs)	12.5	21.7
1 Day HBT-F @ 1 rpm(MMcPs)	5.7	10.3
2 Day	9.2	16.3
3 Day	12.5	22.4
7 Day	15.5	28.1
14 Day	24.9	50.2

TABLE X - MECHANICAL PROPERTIES
PER FORD ENGINEERING MATERIAL SPECIFICATION
WSA-M3D164

Property	Specification	Value
Density	1.78-1.86	1.81
Glass Content	50% minimum	56.5%
Tensile Strength	190 MPa minimum	193 MPa
Flexural Strength	345 MPa minimum	345 MPa
Flexural Modulus	14.5 GPa minimum	15.6 GPa
Shear Modulus	3.92 GPa minimum	5.71 GPa
Izod Impact Strength	1.3 KJ/m minimum	1.3 KJ/m
Water Absorption	0.35% maximum	0.22%
Heat Aging Performance*		
Tensile Strength Change	-25% maximum	-4.3%
Flexural Strength Change	-25% maximum	-18.1%
Izod Impact Change	-25% maximum	-14.9%
Flammability Burn Rate	100mm/min Self Extinguishing SE	0 mm/min SE

* 1000 hours @ 180°C

TABLE XI – MATERIAL PROPERTIES SUMMARY
PER AUTOMOTIVE COMPOSITES CONSORTIUM TEST PROCEDURES

Test Temperature	-40°C	23°C	120°C
Tensile			
Strength (MPa)	187	177	167
Modulus (GPa)	17.15	15.84	12.61
Poisson's Ratio	NT	0.44	NT
Failure Strain (%)	1.9	1.66	1.61
Failure Energy (J)	8.65	7.01	6.05
Compression			
Strength (MPa)	262.7	231.9	130.8
Shear			
Strength (MPa)	163	150	62
Modulus (GPa)	5.02	3.24	2.62

Coefficient of Linear Thermal Expansion $(1/°C*10^{-6})$	
-30 to +30 °C	17.6
+30 to +80°C	19.4
+80 to +125°C	15.8

Glass Content (%)	51.1

Specific Gravity	1.81

Glass Transition Temperature Tg (°C)	110

NT= not tested

High Performance Vinyl Ester Molding Compound

KOICHI AKIYAMA, KENICHI MORITA, HIDEKI TERADA,
HIROYA OKUMURA AND TAKASHI SHIBATA

Abstract

High performance vinyl ester molding compound has been developed. It is based on newly developed high performance vinyl ester resin system. Typical two types of vinyl esters, bisphenol A type and phenol novolac type have different properties. Bisphenol A type vinyl ester tend to be tough, but performance at high temperature is inferior than novolac type. On the other hand, novolac type vinyl ester shows good performance at high temperature, but it tend to be brittle. Novolac vinyl ester resin is preferred to be used for high temperature applications. But it is difficult to be molded in some cases, because its crosslinking density is too high. The new specially modified vinyl ester resin shows both excellent performance at high temperature like phenol novolac type vinyl ester and good moldability like bisphenol A type vinyl ester. This vinyl ester resin is also modified with carboxylic anhydride to be matured with typical metal oxide or hydroxide like MgO, $Mg(OH)_2$. Its price is comparable to bis phenol A type vinyl ester. Impregnation, maturation and moldability of this high performance vinyl ester SMC, BMC and TMC is similar to unsaturated polyester molding compound. The molding compounds show excellent mechanical properties at high temperature, toughness and creep properties. They are suitable for high performance automotive, corrosion and construction applications. In addition, the resin is colorless clear liquid although almost all vinyl ester resin is yellow or brown liquid. It means this system can be used for applications that is necessary to be pigmented.

Introduction

Unsaturated polyester resin has been mainly used for fiber reinforced plastics. Resin properties have been improved year by year to satisfy criteria of new applications, thus resin performance has been getting higher. Some of high performance unsaturated polyester resin are used for severe applications like high temperature or corrosion resistance. But properties expected for some advanced fiber reinforced plastics have been getting too high for unsaturated polyester resin, recently. Vinyl ester is a realistic alternative in such case, because some properties of vinyl ester is much better than these of unsaturated polyester and molding of it is similar like that of unsaturated polyester. High temperature properties, fatigue properties, toughness and corrosion resistance are typical advantages of vinyl ester resin comparing to unsaturated polyester. And it can be molded with similar molding process as unsaturated polyester resins. Especially vinyl ester can be easily used for molding process that doesn't require chemical thicken-

Koichi Akiyama, Kenichi Morita, Hideki Terada, Hiroya Okumura and Takashi Shibata, Takeda Chemical Industries, Ltd.

ing such as hand lay up, spray up and casting.

In order to use vinyl ester in molding compound, there are two procedure for vinyl ester thickening, isocyanate thickening and metal oxide thickening. Isocyanate thickening will not show good stability in production. Controlling isocyanate thickening is very difficult, because isocyanate thickening is affected by water content of compound very sensitively and isocyanate thickening produce covalent bond that isn't dissociated at molding temperature. As well known, isocyanate reacts with water and generate carbon dioxide. The gas remained in molding compound results in pinholes. Viscosity of thickened compound is also varied by water content. Water content is changed by all raw material moisture, so it is difficult to be controlled. In isocyanate thickening, viscosity is rise up by urethane bond generation. The urethane bond isn't dissociated at molding temperature. In case of viscosity built up too high, compound can't be molded even with high pressure. The other hand, too low viscosity gives very poor handling. Proper thickening viscosity target is very narrow.

For metal oxide thickening, some resin modifications is necessary. Terminal carboxylic group was essential for metal oxide thickening. Molecular weight of vinyl ester resin also very important for thickening ability. It should be relatively high to get good thickening ability. In case of bisphenol A type vinyl ester, large number of bisphenol A unit should be used to get good thickening. But large number of bisphenol A unit gives low heat distortion temperature, or HDT because of too low crosslinking density. Novolac type vinyl ester tend to give good maturation properties, but HDT tend to be too high. It results in poor moldability. Novolac type vinyl ester also tend to be expensive. Developed thickenable vinylester resin is based on bisphenol A unit, but molecular weight is raised by urethane modification and also has terminal carboxylic group. Degree of crosslinking is controlled to give good thickening properties, excellent mechanical properties and moldability. Any style of molding compound such as SMC, BMC and TMC®, can be prepared by this resin system with conventional maturation procedure similar as that for unsaturated polyester resin system.

The high performance TMC® were developed for advanced sanitary application. The TMC showed excellent moldability, toughness, corrosion resistance and hot water resistance. But this excellent properties also will be exhibited in the other advanced application like automotive, corrosion resistance and high temperature.

Modified thickenable vinyl ester resin

The present carboxylic acid-pendant urethane-modified vinylester resin attained high crosslinking-density and excellent heat-resistance by allowing to react the hitherto-known vinylester resin with isocyanate components to give higher molecular weight compounds. Additionally, this type of the resin achieved good thickening property by subjecting hydroxyl groups in the highly molecularized urethane modified vinylester resin to the addition reaction with polybasicacid anhydride.

Fundamental vinylester resin was prepared by following the conventional procedure, namely molar equivalent amounts of bisphenol A type epoxy resin and methacrylicacid were employed and both conversions of epoxide and methacrylicacid were traced by titration. The reaction between epoxide and methacrylicacid was proceeded almost quantitatively, and that more than 99% epoxide was reacted with methacrylicacid before isocyanate compounds were introduced.

Urethanation was carried out to react hydroxyl groups on the fundamental vinylester resin with isocyanates until the typical isocyanate peaks in the IR spectra were completely disappeared.

In the present urethane-modified vinylester resin, totally 25% hydroxyl groups of the fundamental vinylester resin was reacted with isocyanate groups of the hydrogenated xylylene diisocyanate.

A certain kind of isocyanate compounds, such as tolylene diisocyanate (TDI) or 4,4'-diphenylmethane diisocyanate (MDI), are cost-effective, however, so-called non-yellowing isocyanate compounds, exemplifying isophorone diisocyanate (IPDI), xylylene diisocyanate (XDI) or hydrogenated xylylene diisocyanate, are desirably employed when the appearance of the cured products is regarded as important.

Finally, remained hydroxyl groups on the urethane-modified vinylester resin were subjected to the addition reaction with the polybasicacid anhydride, more specifically, with tetrahydropthalic anhydride.The prepared resin contained 30% styrene monomer ,and indicated.4500cps viscosity and 15mgKOH/g total acid value.

High Performance vinyl ester TMC

The vinylester TMC can be produced by conventional TMC process[1] with magnesium oxide, or MgO as thickener. Thickening behavior are shown in Figure 1. It shows very good viscosity build up. Handling properties of thickened TMC is similar as typical SMC based on unsaturated polyester. SMC also can be formulated with this vinyl ester resin system. The vinyl ester TMC can be molded with quite the same molding condition as that for unsaturated polyester molding compounds. Molding parameter for it is typically 8-13 MPa of molding pressure and 125-150 °C of molding temperature.

Bath tub and sanitary counter top molded with the TMC are evaluated. Appearance is very good. When combination of this resin system and proper grade of aluminum trihydrate or glass powder is selected, molded parts having translucency that is often used for artificial marble products also can be formulated with the system. The resin system is colorless, although general vinyl ester system tend to have deep color. The largest FRP market in Japan is sanitary that is necessary to be pigmented. Deep colored resin can't be used for this application. This colorless resin system is achieved by the improved cooking procedure shown above.

Bath tubs molded with developed vinyl ester TMC and unsaturated polyester TMC are compared. Mechanical properties of developed TMC is shown in Table 1. Test pieces were taken from side wall of bath tub along both vertical direction and horizontal direction against material flow direction to evaluate glassfiber formation. Although there are difference between vertical and horizontal strength, mechanical properties of VE-TMC are much higher than these of UP-TMC. It is said that Dupont impact test can be simulate well resistance for impact in real use. Toughness was also evaluated by dropping steel ball or sand bag on the surface of the bathtub. The results showed that VE-TMC is tougher than UP-TMC.

Hot water resistance is very important characteristics for sanitary applications, especially for bath tub. Vinyl ester TMC showed excellent hot water resistance. Artificial marble bath tub is remarkably growing market in Japan[2]. 4000 ton/year of molding compounds is consumed in the market. Bath tub should have excellent hot water resistance. In general use, hot water was filled in a bath tub once a day. Bath tub should be used for ten years with this condition. Recently hot water circulation filtering system called 24 hours bath system, was developed in Japan. With this circulation system, bath tub is filled 40 ˚C water all the day. People can take a bath any time when they want to be in with almost the same energy cost. Bath tub don't have to be cleaned, because water is filtered all the time. This application is getting popular day by day. For this application, extremely high water resistance will be needed for bath tub. Hot water resistance of the resin system is shown in Figure 1. This vinyl ester TMC is said to met the criteria of it.

Possibility for the other applications

This vinyl ester resin system was originally developed for sanitary applications, because it is the largest in Japan. Therefore superior properties of the resin system is not only for sanitary, but also for the other advanced applications, such as automotive, high temperature application and corrosion resistance and so on. SMC based on the vinyl ester system was evaluated. SMC was made by conventional procedure and then molded at 140 ˚C, 10MPa. Flexural strength both at 25 ˚C and at 140 ˚C of vinyl ester is higher than that of unsaturated polyester SMC. Adhesion between resin and glass fiber will be stronger in vinyl ester SMC than in unsaturated polyester SMC. Vinyl ester SMC also showed better tensile and impact properties than unsaturated polyester SMC. Thermal creep properties of it was investigated. SMC was treated at 130 and 150 ˚C. Flexural strength of both vinyl ester and unsaturated polyester system decrease at early period of the test. Strength of vinyl ester system decrease slightly after 1000 hours treatment, but decreasing tendency is still continued in unsaturated polyester system. Reliability of long term high temperature exposure will be much better in vinyl ester system. High performance can be obtained by conventional impregnation process with the vinyl ester system. It is expected to be applied for many advanced applications.

Note

TMC® is registered as trade name of Takeda's thick molding compound.

References

1)Iwai,T., et al. 1978. 33rd SPI Session 4-B
2)Akiyama,K., et al. 1995. 50th SPI Session 20-A

Biographies

Koichi Akiyama

Koichi Akiyama received his M.S. degree in applied chemistry from Osaka University and joined Osaka Research Laboratories, Chemical Products Division, Takeda Chemical Ind., Ltd., in 1988. He has been working in Tokyo Research Laboratories from 1992. He is responsible for material and molding development of thermoset molding compound.

Kenichi Morita

Kenichi Morita received his B.S. degree in chemistry from Tokyo Met. University. He has been in Tokyo Research Laboratories, Chemical Products Division, Takeda Chemical Ind., Ltd., since 1992. He is responsible for material and molding development of thermoset molding compound.

Hideki Terada

Hideki Terada received his B.S. degree in applied chemistry from Kobe University and joined Osaka reserch laboratories, Chemical Products Division, Takeda Chemical Ind., Ltd., in 1992. He has been working in Tokyo Research Laboratories from 1995. He is responsible for material and molding development of thermoset molding compound.

Hiroya Okumura

Hiroya Okumura received a Master of Science degree in polymer science from Kyoto University. He joined Osaka Research Laboratories, Chemical Products Division, Takeda Chemical Ind., Ltd., in 1988. He is currently a member of a research team focused on vinylester resin in Tokyo Research Laboratories.

Takashi Shibata

Takashi Shibata is a senior reseach manager of research Laboratories, Chemical Products Division, Takeda Chemical Ind., Ltd. He earned his B.S. degree in chemistry from Osaka University in 1970. He is responsible for R & D of vinyl ester resins.

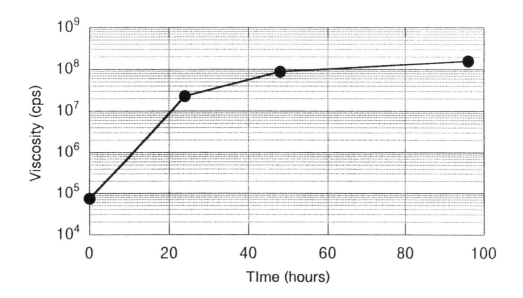

Figure 1 Thickening Properties of Vinyl Ester Compound

Compound contains 150 phr of aluminum trihydrate and 1 phr of magnesium
Viscosity was measured at 40 °C

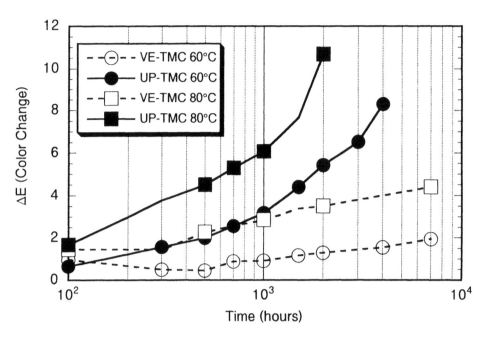

Figure 2 Hot Water Resistance of Vinyl Ester TMC

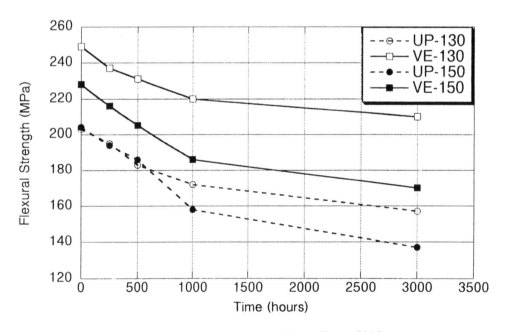

Figure 3 Heat Resistance of Vinyl Ester SMC

Table 1 Mechanical Properties of Vinyl Ester TMC

		UP-TMC	VE-TMC
Glass Content (%)		7	7
Dupont Impact (J)	Front	1.77	3.14
	Back	2.50	3.29
Flexural Strength (MPa)	Vertical	49.28	74.05
	Horizontal	34.82	57.69
Flexural Modulus (GPa)	Vertical	9.16	10.35
	Horizontal	8.58	9.61
Tensile Strength (MPa)	Vertical	11.53	29.57
	Horizontal	9.23	20.74
Tensile Modulus (GPa)	Vertical	1.60	1.98
	Horizontal	1.63	1.63

Data was measured with test pieces taken from bath tub

Table 2 Mechanical Properties of Vinyl Ester SMC

	Temperature	VE-SMC	UP-SMC
Glass Content (%)		30.0	30.0
Flexural Strength (MPa)	23°C	231.5	194.8
	140°C	121.4	96.9
Flexural Modulus (GPa)	23°C	10.7	10.0
	140°C	7.6	7.4
Tensile Strength (MPa)	23°C	105.3	88.7
Tensile Modulus (GPa)	23°C	13.6	11.3
Izod Impact (kJ/m^2) (flatwidth, unnotched)	23°C	73.4	55.4

Data was measured with test pieces taken from flat panel

Development of Low Density SMC Formulations in the Transportation Market

JAMES R. LEMKIE AND MICHAEL SOMMER

The objective of this paper was to develop new compounds for transportation applications with lower density but with similar properties as standard SMC formulations. Automotive parts need to weigh less for cars and trucks to be able to use the same platform but add more options to the vehicle, especially in the area of safety such as air bags, side air bags, side impact protection etc. or to reduce the total weight of the car to achieve better fuel consumption. Exterior automotive parts require excellent surface quality and are usually painted. Also functional or structural parts can be made from SMC with high demands in physical strength and less demand for surface quality. In mass transportation, weight reduction becomes important with the new generation of high speed trains and metros for increased loads or reduce energy consumption. In mass transportation, safety plays a much bigger role in the formulations for flame retardant applications. Therefore our focus was then put into three areas:

1. Low profile Class A SMC
2. Low Profile SMC with good mechanical properties
3. Flame retardant LS/LP SMC

The development goals are listed in table 1:

Table 1: Objectives of the development

Requirement/ Type	LP Class A	LP	LS Flame retardant
Density	1.4 kg/dm^3	1.3 kg/dm^3	1.5 kg/dm^3
Surface	excellent	standard	standard
Mechan. properties	standard	high	standard
Flame retardancy	none	none	UL 94 V 0
Applications	Visible parts on cars and trucks with painting	Functional parts on cars and trucks such as noise incapsulations, dashboard beams, radiator supports etc.	In- and Outside parts as well as functional parts sometimes pigmented in trains or trams such as side panels, ceilings etc.

James R. Lemkie and Michael Sommer, BYK-Chemie USA.

We put our effort into developing processible SMC formulations in these three areas. BYK-Chemie wetting and dispersing additives were used to control paste viscosity and homogeneity, to prevent separations and to maintain or improve mechanical properties. This will be explained in detail at the corresponding formulations for each area.

Some application areas are already known such as radiator supports and noise incapsulation shields where the weight reduction achieved of 15 to 20% can be made over standard SMC. Visible parts such as hoods or front end panels are molded in standard density SMC and are painted, examples are the hoods of Alfa Spider/GTV, Ford Mustang and Chrysler Sebring. Low density SMC formulation could reduce the total weight of 25 to 30 kg down to 20 to 25 kg. Low density formulations using light weight fillers are known but faced several problems. One problem is the drop in physical properties which can be overcome by not only adjusting the volume of the glass content but also using BYK®-W 972 to help maintain the level of mechanical properties. Another problem is the sanding of light filler containing parts which are torn out of the surface during sanding to form fine holes that entrap paint and solvents which may causes to bubble in the coating after curing in the oven. All these problems were taken into consideration when we developed the new low density formulations for this paper.

Introduction

Our program is broken into two major areas of study of low density.   Low profile application are better dimensional stablity and physical properties. Low shrink applications will address the needs of flame retardant requirements.

Low profile development required three different formulations to meet the requirements of the study. All formulations will use 3M Scotchlite microbubbles to achieve lower density in the formulation. Low profile formulations will require the use fine grade of calcium carbonate to achieve good flow behaviour and surface properties.

To achieve the flame retardancy, ATH was used as a more environmentally acceptable than using halogenated flame retardants. ATH levels were varied from 200 phr to 280 phr depending on the thickenness of the part to obtain a UL 94 V-0 rating. For thicker parts, the ATH level can be reduced to meet the flame rating and also decreases density again. The ATH filled Low Shrink systems were pigmented. BYK®-W 972 was used to stabilize the LS additive in the systems and help to improve the pigmentation. To homogenize the paste, BYK wetting and dispersing additives BYK®-W 996 and BYK®-W 9010 were used. When using low density fillers viscosity measurement do not accurately depict the viscosity of standard density SMC, our test in low density SMC give the appearance of some kind of thixotropic behaviour. Of course, it is important to keep the volume of the fiber glass content constant when comparing physcial porperties to standard density SMC. To control the drop in physical properties, 1 to 2 phr of BYK®-W 972 help to maintain the physical properties levels.

Details of the development

1. Development of a CLASS A Low Profile Formulation

3M Scotchlite micro bubbles K 37 were used as a light filler because of their high crunch strength which allows the microbubbles to withstand the shear of mixing and standard pressures of compression molding. First a comparison in the same low density low profile formulation was made to find out the influence of BYK-W 972 on the mechanical properties. The formulations contain a high load of microbubbles (55phr) and 100 phr of calcium carbonate. 6 phr of BYK-W 996 together with 2 phr of BYK W 972 in the second formulation for better wetting. The W 996 wets very well the CaCO3 and ATH but the bubbles are better wetted by the W 972 which results in a viscosity reduction and slightly better properties as shown in table 2:

Table 2: Mechanical properties comparsion

Property / Formulation	Low Density LP 1	Low Density LP with W 972	
Density	1.4	1.4	g/cm^3
Flexural strength	89	97	N/mm^2
E-modulus	6.1	6.2	kN/mm^2
Tensile strength	46	51	N/mm^2
E-modulus	7.9	7.7	kN/mm^2
Shrinkage	0	0	%

A formulation based on a PG-Maleate resin and a PVAc solution with filler load of up to 100 parts of calcium carbonate and 40 to 65 parts of microspheres were produced. It is shown in table 3:

Table 3: Formulations for Class A Low Density SMC

	phr		
Raw materials	Standard Class A	LD Class A a-side	LD Class A b-side
PG-maleate resin	60	60	60
Neulon T+	40	40	40
BYK-W 996	**1.8**		**5**
BYK-W 972		**2**	**2**
Styrene			10
TBPB	1.5	1.5	1.5
Zinc stearate	5	2.5	2.5
Calcium stearate		2.5	2.5
Scotchlite K37		5	50
Calcium carbonate 3μ	180	100	20
Luvatol MK 35 NV	3	3.5	6
Roving 25mm (w/w)	30	38	
	18	19	
Density molded part	**1.85** g/cm^3	**1.4** g/cm^3	

The standard SMC contains 180 phr of calcum carbonate and 30 weight-% of fiberglass roving. The low density formulation was prepared with an a and a b-side that use different formulations with different filler load as shown in table 2.

These low density formulations will enhances the impregnation of the glass roving bed. We created sandwich of the pastes with the a-side and b-side being put in different doctors during compounding. During molding the a-side is placed next to both mold surfaces and b-sides should always be put towards each other in the inner side of the charge. The a-side is used to give a smooth surface during molding as the b-sides with the high amount of bubbles are on the inside to control the density. The a-side formulation should avoids the pulling out of bubbles during sanding. This example has worked well for molding panels. We are continuing our testing of this formulations on a larger part with more flow to see how uniform the surface will be and the effect of sanding the part on paintability. To maintain the physical properties, 2 parts of **BYK®-W 972** were used in the both formulations. The b-side requires 5 parts of **BYK®-W996** and 10 parts styrene to achieve a viscosity level that is feasible on the SMC line. The SMC molds very well and has an excellent surface.

The density is 1.3 g/cm³. Table 4 shows the mechanical properties of the formulation in comparison to the standard Class A SMC.

Table 4: Results

Property / Formulation	Std-LP	LD-Class A	
Density	1.85	1.3	g/cm³
Flexural strength	160	157	N/mm²
E-modulus	10.5	8.8	kN/mm²
Tensile strength	75	67	N/mm²
E-modulus	9.0	7.3	kN/mm²
Shrinkage	0	-0.04	%
Surface	Class A	Class A	

2. Development of Low Profile Formulations for Functional Parts

The functional formulation is based on a standard resin and a PVAc solution formulations with filler load of up to 100 parts of calcium carbonate and 40 to 65 parts of microspheres were produced.. We again used the 3M Scotchlite micro bubbles K 37 but increased the fiberglass content of LD LP formulations.

Table 5: Formulations for Low Density SMC for functional parts

Raw materials	phr			
	Standard Class A	Ultralight	LD LP a-side	LD LP b-side
PG-maleate resin	60	80	80	70
Neulon T+	40	20	20	30
BYK-W 996	**1,8**	**6**		**5**
BYK-W 972		**2**	**2**	**2**
Styrene		10		10
TBPB	1.5	1.5	1.5	1.5

BYK-W 996	**1,8**	**6**		**5**
BYK-W 972		**2**	**2**	**2**
Styrene		10		10
TBPB	1.5	1.5	1.5	1.5
Zink stearate	5	2.0	2.5	2.5
Calcium stearate		2.5	2.5	2.5
Scotchlite K37		65	5	50
Calcium carbonate 3µ	180	30	100	20
Luvatol MK 35 NV	3	4	4,5	
MgO				1,4
Roving 25mm (w/w)	30	38	50	
(v/v)	19	15	30	
Density molded part	**1.85** g/cm^3	**1.1** g/cm^3	**1.5** g/cm^3	

The Ultralight SMC contains 65 phr of low density filler and 30 parts of carbonate which gives an excellent density for parts where mechanical properties are less important. The low density LP formulation was prepared with an a-side and a b-side that use different formulations with different filler load as already described in part 1 and shown in table 5.

Molding will be done as described for Class A SMC for the LD-LP formulation and as a standard SMC for the Ultralight material. The physical properties are listed below:
To maintain the physical properties, 2 parts of **BYK®-W 972** were used in all formulations. Ultralight requires 6 phr of **BYK®-W996** to get better wetting and the b-side of the LD-LP requires 5 parts of **BYK®-W996** and 10 parts styrene to achieve a viscosity level that is feasible on the SMC line. The SMC molds very well and gives the require physical properties. **The density is 1.1 g/cm^3** for the Ultralight and **1.5 g/cm^3** for the high strength material.

Table 6: Results

Property / Formulation	Ultralight	LD-LP	
Density	1.1	1.5	g/cm^3
Flexural strength	120	202	N/mm^2
E-modulus	6.0	11.5	kN/mm^2
Tensile strength	52	97	N/mm^2
E-modulus	6.2	9.7	kN/mm^2
Shrinkage	0.04	0.0	%

3. Development of a flame retardant Low Density Formulation

The flame retardant low density formulation based on a standard resin and a polystyrene solution with filler load of up to 280 parts of ATH and 20 to 40 parts of low density fillers were produced.

Table 7: Formulations for flame retardant low density SMC

Raw materials	phr		
	Standard ATH	LD ATH 1	LD ATH 2
Std-ortho resin	80	80	80
PS-solution	20	20	20
BYK-W 996	**6**	**6**	**6**
BYK-W 972		**2**	**2**
Styrene	5	12	12
TBPB	1.5	1.5	1.5
Zinc stearate	2	2	2
Calcium stearate	3	3	3
Scotchlite K37		20	
Fillite 100			40
ATH 21 μm	200	200	200
ATH 4μm	80	80	80
Yellow pigment paste	7	7	7
Luvatol EH 35	6	6	6
Roving 25mm (w/w)	24	27	28
(V/V)	18	17	18
Density molded part	**2.0 g/cm³**	**1.6 g/cm³**	**1.7 g/cm³**

We used the 3M Scotchlite microbubbles K 37 and the Fillite 100 powder were used as the low density filler. Again it was possible to achieve this high filler load of ATH plus microbubbles by using **BYK®-W996** together with VR 3 and styrene. The following table shows the mechanical data which were achieved by adding 2 phr of **BYK®-W 972** vs the standard formulation.

Table 8: Results

Property / Formulation	Std-ATH	LD-ATH 1	LD ATH 2	
Density	2.0	1.6	1.7	g/cm³
Flexural strength	135	127	123	N/mm²
E-modulus	9.3	8.7	8.5	kN/mm²
Tensile strength	56	55	54	N/mm²
E-modulus	8.9	8.5	7.4	kN/mm²
LOI (1.1")	66	64	65	%
Shrinkage	0.14	0.14	0.11	%

Both formulations can be molded with normal charge pattern under standard conditions (140 bis 155 °C, 80 bar). The parts made from LD-ATH 1 are homogenously yellow pigmented whereas the parts made from LD-ATH 2 show a darker color due to the grey color of the fillite. Due to the higher density of the Fillite 100 it is also required top use twice the amount to achieve the same weight reduction compared to Scotchlite K 37.

Summary of the results

Low density-SMC formulations can be produced with low density fillers such as 3M Scotchlite K 37 microbubbles or Fillite 100 bubbles. **BYK®-W 996 or BYK®-W 9010** are used as wetting and dispersing additives and **BYK®-W 972** is used to maintain mechanical properties under maintaining the glass volume content. The density of the LD fillers are:

Scotchlite K 37:   0.35 - 0.39
Fillite 100:   0.60 - 0.80

Limitations are the fact that LD fillers cannot be used in high quantities due to the viscosity increase of the paste. For Fillite 100 also the grey color is a limitation for pigmentation of the compound.

Class A SMC parts are feasible with a density of 1.3.

Ultralight LP SMC can be made with a density of 1.1 and high strength LPSMC with 1.5.

Flame retardant LD SMC was produced at densities of 1.6 - 1.7 with high load of ATH and good flame retardancy levels.

Compounding and Molding Parameters that Lead to Edge Paint Pops in SMC Composite Panels

MICHAEL W. KLETT AND STEVEN J. MORRIS

ABSTRACT

Edge paint pops are a major problem in Class 'A' automotive SMC body panels because of additional costs incurred with rework. These defects appear as small volcano-like craters in the overlying paint films at or near the molded panel edges. The pops originate in the composite matrix and are independent of any surface preparation such as filing or sanding during "deflashing." A single explanation for the cause of paint edge popping has eluded molders and material suppliers primarily because of the indiscriminate nature of the problem and the numerous variables associated with compounding and molding. The objective of the present study was to characterize the effect of raw materials (specifically the resin paste and fiber glass reinforcement) and molding parameters on edge paint pops. A test procedure for analyzing numerous experimental variables with reproducible results was established. This test ultimately led to the identification of the major factors and interactions that influence edge paint popping. The factors which we found to have the largest impact on edge paint popping include molding closure speed, resin paste formulation, and fiber orientation. The specific effect from the fiber glass sizing was found to be small. The magnitude of edge paint popping due to fiber glass sizing was primarily dependent on the resin paste. Our results indicate the potential to significantly reduce and perhaps eliminate edge paint pops through optimization of the molding parameters, resin paste, and fiber glass reinforcement.

INTRODUCTION

Automotive SMC applications have grown from 40MM lbs. in 1970 to a 250MM lb. business today (SMC Car and Truck Component List, 1997). Approximately 200MM lbs. are for exterior body panels. SMC's success is due to the efforts of molders and material suppliers to evolve the technology to a point where, on an appearance basis, painted SMC parts are virtually indistinguishable from steel. Most of the original cosmetic issues with SMC, such as dimensional distortion, long-term waviness, fiber print, and porosity, have been addressed and, to a large degree, have been resolved. However, paint defects

Michael W. Klett and Steven J. Morris, PPG Industries, Inc., P.O. Box 2844, Pittsburgh, PA 15230

(specifically edge paint pops) continue to plague the industry. The current perception in the SMC community is that the paint edge pop problem is perhaps the major technical issue that limits growth of SMC into exterior automotive applications.

Paint pops have been reported to arise from the rapid expulsion of gases generated from absorbed paint solvent from either the multilayer paint systems or from the substrate (Cheevers, 1986; Edwards, 1985). Substrate-related pops appear as volcano-like eruptions in the paint film and can occur at various locations on the finished part. The edge paint pops examined in this study occurred exclusively at or near the molded panel edges and were independent of any surface preparation such as filing or sanding during "deflashing." To our knowledge, there have been no published studies that address the cause(s) of edge paint pops. Part of the reason is perhaps the indiscriminate nature of the problem and the numerous variables associated with manufacturing a finished SMC part. The objective of this study was to identify the major factors relating to material selection and molding processes that contribute to edge paint pops.

EXPERIMENTAL

MATERIAL SELECTION, COMPOUNDING, AND MOLDING PROCEDURES

Two SMC resin paste formulations capable of affording a Class 'A' surface appearance (Loria values of 80-90 and 100-110) were used to compound the four commercial fiber glass rovings described below. Letter designations have been used for the resin formulations (i.e., R1 and R2) and the fiber glass rovings (i.e., G1, G2, G3, and G4) in order to maintain objectivity. The SMC was produced on a 70 cm (24") wide Finn and Fram machine with chain mesh compaction. Sheet weights were held constant at 5.5 kg/m^2 (18 oz/ft^2) with glass contents at 28-30%. Panels were molded from either a 40.6 cm by 40.6 cm (16" by 16") or a 45.7 cm by 45.7 cm (18" by 18") flat panel mold. The latter tooling had vacuum-assist capability. Unless otherwise noted, the charges were cut for 25-30% mold coverage and molded at 150°C (300°F) under 68 atm. (1000 psi) and 6-7 sec. closure times. Commercial base coat and clear coat systems were used to paint the molded SMC panels.

PANEL PREPARATION, PAINTING, AND POP RECORDING PROCEDURES

Flat panels were molded to a thickness of 2.9-3.0 mm (0.115-0.120"), and only the molded edges (i.e., the 3-mm wide area) were painted and examined for edge paint pops. Preparation of the panels for painting involved first removing the edge flash by filing with a 3-square second cut file. In order to minimize fiber pulls, filing was performed at a 45° angle towards the panel. The panels were sheared into equal quarters and rotated until the sheared edges for all four quarters faced the same direction. These quarter panels were stacked on top of each other 40-48 high with 0.3-mm thick paper spacers between each quarter panel. Each block of quarter panels represented one complete iteration of a designed experiment (DOE), with each successive block a replicate of the first block.

For ease of handling and minimizing the movement of the panels, two 10-mm (3/8") holes were drilled through each block of quarter panels through which threaded bolts were inserted. Wing nuts were then applied to the bolts. Before the blocks were painted, the molded panel edges were cleaned with a lint-free cloth and isopropanol and then dried with high-pressure air. The blocks were positioned such that both molded panel edges could be painted in one application. The paint cycle consisted of two applications of first top coat and then clear coat with a one-minute flash between each application. After an additional ten-minute flash, the panels were baked in the oven at 140°C (285°F) for 30-60 minutes until the paint was dry. The final film build was 20-30 microns (8-11 mil) for the top coat and 40-50 microns (16-20 mil) for the clear coat. A primer application step was omitted in this study because our initial experiments with primer showed only a negligible reduction in edge paint popping.

Counting of the pops was restricted to one observer, and random checks were made by one additional person to verify the count. The pops that were recorded were those that were visually distinguishable as having the volcano-like morphology described below. In the present study, these pops measured 0.5-1.0 mm (20-39 mil) in diameter at the base. Pops below 0.5 mm could not be unambiguously characterized without the use of a microscope and, therefore, were excluded from the analysis.

EXPERIMENTAL DESIGNS AND STATISTICAL ANALYSIS OF DATA

Full-factorial DOEs with 8-10 replicates per condition were used throughout the study unless otherwise noted. Classical (ANOVA) and nonparametric techniques (Kruskal-Wallis) were used for data analysis. Nonparametric techniques were used because of the inability to validate the assumptions underlying the ANOVA techniques. Conclusions from both analysis techniques were consistent, and the observed trends were reproducible.

RESULTS AND DISCUSSION

MICROSCOPY

An example of an edge paint pop is given in Figures 1 and 2. These pops exhibit "volcano-like" morphology, i.e., steep-walled craters typically containing one small "vent hole" at the bottom, identical to those described by Cheevers (1988). The pops ranged in size from 0.2-1.0 mm (8-39 mil) across the outside base of the pop and 0.1-0.7 mm (4-28 mil) across the inside width of the crater. The "vent holes" measured 0.01-0.03 mm (0.4-1.2 mil) in diameter. Several pops were chosen at random, and the paint was removed by an aggressive solvent mixture until the surface of the SMC was exposed. Microscopic examination of this area revealed a small hole (0.01-0.03 mm in diameter) in the SMC substrate corresponding to the "vent hole" in the original pop. Comparable holes were also observed on the unpainted panel edges.

In order to observe the origin of the "vent hole" within the substrate, several of the cratered sections were successively polished on edge and examined by scanning electron microscopy after each polishing step. This was continued until a cross section of the "vent hole" was observed. Beneath the pop, a capillary-like void connected the vent hole to a small delamination within the substrate. The size of the delaminations from the polished side varied from 250 microns (98 mil) to up to 750 microns (295 mil) in length. The same polishing procedure was used on several "holes" found on the unpainted panel with identical results. Our observations indicate that the delaminations or "microcracks" are originally present in the SMC substrate, and the painting process only magnifies the defect.

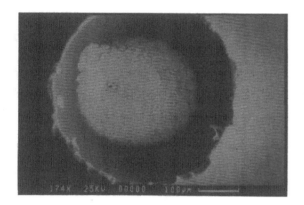

Figure 1. Top View of an Edge Paint Pop

Figure 2. Cross Section of an Edge Paint Pop Showing "Vent" And Delamination

MOLDING COMPOUND

The three aspects of the SMC compound investigated in this study included resin formulation, glass fiber reinforcement type, and the orientational bias of the fibers in the SMC. The contribution of resin formulation and the glass fibers towards edge paint pops is summarized in Tables I, II, and III.

TABLE I - STATISTICAL ANALYSIS OF THE EFFECT OF RESIN PASTE FORMULATION
ON EDGE PAINT POPS

DOE No.	Resin Paste	Panels Molded	Pop Mean	Lower Confidence Level (95%)	Upper Confidence Level (95%)	Kruskal-Wallis Rank
1	R1	48	0.31	0.06	0.56	40.5*
1	R2	48	1.10	0.85	1.30	56.5**
2	R1	60	1.88	1.30	2.46	41.0*
2	R2	60	4.77	4.18	5.35	80.0**

Note: Statistically significant (nonparametric) differences exist between * and **.

Resin paste formulation has a strong influence on the level of edge paint pops. Because of the nonnormal distribution of the data, classical ANOVA techniques were not valid. However, the consistent trends observed from experiment to experiment give credence to the analyses. Analogous conclusions for the effect of resin paste formulation were reported for paint pops on both panel surfaces (Edwards, 1985; Cheevers, 1988) and water-jet cut edges (Mitani, et al., 1990).

Our data also indicates that the glass fiber type impacts the incidence of pops, but to a smaller extent than that due to the resin paste. The magnitude of the effect of edge paint popping due to glass appears to be dictated by the resin paste (Tables II and III). In other words, some resin paste formulations are more robust or less sensitive to the effects of glass type. Mitani et al., reported a similar observation and attributed the level of pops directly to the relative solvent absorption power of the glass sizing (Mitani et al., 1990). Specifically, "very hard" (low styrene solubility, e.g., ~30%) glass rovings afford fewer pops than "semi-hard" (mid-range styrene solubility, e.g., ~50%) glass rovings. Our results (Tables II and III), however, do not support any correlation of roving "hardness" (styrene solubility) with edge pops.

TABLE II - STATISTICAL ANALYSIS OF THE EFFECT OF GLASS ROVING IN RESIN PASTE R1

Glass Roving	Roving Classification	Count	Pop Mean	Lower Confidence Level (95%)	Upper Confidence Level (95%)	Kruskal-Wallis Rank
G1	"Medium-Hard"	180	0.53	0.44	0.63	289*
G2	"Hard"	176	0.41	0.33	0.51	270*
G3	"Medium-Hard"	116	0.75	0.64	0.86	324**
G4	"Hard"	116	0.64	0.53	0.75	309**

Note: Statistically significant (nonparametric) differences exist between * and **.

TABLE III - STATISTICAL ANALYSIS OF THE EFFECT OF GLASS ROVING IN RESIN PASTE R2

Glass Roving	Roving Classification	Count	Pop Mean	Lower Confidence Level (95%)	Upper Confidence Level (95%)	Kruskal-Wallis Rank
G1	"Medium-hard"	104	1.66	1.47	1.86	166**
G2	"Hard"	104	0.74	0.54	0.94	121*
G3	"Medium-hard"	40	0.95	0.63	1.27	134*
G4	"Hard"	40	1.42	1.11	1.74	161**

Note: Statistically significant (nonparametric) differences exist between * and **.

An inherent characteristic of SMC is the bias of glass fibers along the compounding or machine direction. This fiber alignment bias is apparent in the mechanical properties of molded panels. For example, compound produced from our equipment consistently showed a 5-10% difference in tensile strengths between the machine and cross machine directions. The contribution of this fiber bias towards edge paint pops was evaluated for each molding from two independent experiments. Preparation of the charges involved cutting the SMC and stacking the three plies in the same orientation as the compound was produced (machine direction). The charges were oriented in the mold either along a top-to-bottom or left-to-right fashion to minimize any effects potentially due to mold inconsistencies.

Table IV summarizes the edge pop means and machine/cross machine ratios for the two orientations (i.e., top-to-bottom and left-to-right). The consistent machine/cross machine ratio of 1.6-1.7 from two independent DOEs indicates that glass fiber orientation has a significant effect on edge paint pops.

One hypothesis to explain this phenomenon is fiber realignment within the SMC during flow. For instance, when a lateral force (such as that which occurs during flow) is applied to fibers, the fibers realign to an orientation that resists the initial force. In other words, fibers oriented parallel to the lateral force will rotate 90°, whereas no realignment is necessary for fibers aligned perpendicular to the lateral force. This rationale follows the observation that molded SMC panel porosity increases with increasing fiber length (Das et al., 1980). Relating this to the results in Table IV, the level of turbulence for the machine direction should be higher than that for the cross machine direction.

TABLE IV - ANALYSIS OF EDGE POPS AS A FUNCTION OF FIBER ORIENTATION

DOE No.	Panels Molded	Top-to-Bottom Machine/ Cross-Machine	Left-to-Right Machine/ Cross-Machine	Grand Mean Machine/ Cross-Machine
1	120	1.78/1.07	2.33/1.46	1.67
2	96	1.27/0.95	1.66/0.85	1.61

MOLDING PARAMETERS

The relative magnitude of edge paint popping for the glass roving and molding parameters (i.e., pressure, temperature, and closure speed) was assessed through the DOE shown in Table V. This design additionally provided information on interactions between the glass roving and each of the molding parameters examined. Only one resin paste formulation (R1), which afforded the lowest level of pops, was used. Rovings G1 and G2 were used because they had the smallest and largest effect, respectively, on edge paint pops in the experiments described above. The range of molding conditions used was evaluated prior to starting the DOE in order to ensure that no surface defects (such as porosity or blisters) were visible. Differences due to the level of cure resulting from the two extremes in temperature (i.e., 138°C and 160°C [280°F and 320°F]) were eliminated in this case by adjusting the molding times to 4.5 minutes and 2.0 minutes, respectively. DSC analysis of the panels confirmed that identical levels of cure were obtained.

TABLE V - DESIGNED EXPERIMENT FOR DETERMINING THE EFFECT OF ROVING AND MOLDING PARAMETERS ON PAINT EDGE POPS

Exp. No.	Roving	Pressure (atm.)	R*Press.	Temp. (°C)	R*Temp.	R*Clos.	Closure (sec.)	Total Pops	No. of Panels
1	G1	34	1	138	1	1	1	40	8
2	G1	34	1	160	2	2	10	2	8
3	G1	68	2	138	1	2	10	5	8
4	G1	68	2	160	2	1	1	47	8
5	G2	34	2	138	2	1	10	14	8
6	G2	34	2	160	1	2	1	24	8
7	G2	68	1	138	2	2	1	33	8
8	G2	68	1	160	1	1	10	9	8

Responses for the variables are summarized in Table VI, and interactions are given in Figures 3, 4, and 5, respectively. Closure speed showed the strongest effect. Slow closure (10 seconds) decreased the level of edge pops by a factor of 5 relative to fast closure (1 second). Glass type, pressure, and temperature showed a small impact on edge paint pops. The small effect due to glass roving is consistent with the numerous panels molded for the full-factorial experiments described above.

TABLE VI - STATISTICAL ANALYSIS OF THE EFFECT OF ROVING AND MOLDING PARAMETERS ON EDGE PAINT POPS

Variable	Variable Level	Panels Molded	Pop Mean	Lower Confidence Level (95%)	Upper Confidence Level (95%)	Kruskal-Wallis Rank
Roving	G1	32	2.94	2.11	3.77	32.1
	G2	32	2.50	1.67	3.33	32.9
Closure	1	32	4.50	3.67	5.33	43.6*
(sec.)	10	32	0.94	0.11	1.77	21.4**
Pressure	34	32	2.50	1.67	3.33	31.4
(atm.)	68	32	2.94	2.11	3.77	33.6
Temperature	138	32	2.88	1.04	3.71	34.7
(°C)	160	32	2.56	1.73	3.39	30.3

Note: Statistically significant (nonparametric) differences exist between * and **.

Figures 3, 4, and 5 show the analysis of interactions between glass roving and the molding parameters. Interactions noted for glass-temperature and glass-closure are signified by the intersecting lines. This data indicates that the optimal performance of a glass roving is dependent on the molding conditions in much the same way that the resin paste determines the limitations of the roving.

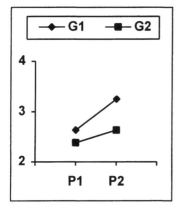

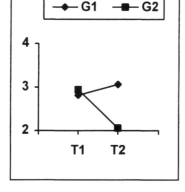

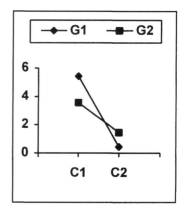

Figure 3. Interaction Response Plot for Glass Roving and Molding Pressure

Figure 4. Interaction Response Plot for Glass Roving and Molding Temperature

Figure 5. Interaction Response Plot for Glass Roving and Closure Speed

Vacuum-assisted molding has been reported to substantially reduce the level of porosity and pinholes in molded SMC parts (Tomiyama et al., 1991; Yamada et al., 1992). The potential of this technique to reduce edge paint pops led us to investigate vacuum as a variable in two independent DOEs. Standard molding conditions of 68 atm. (1000 psi) molding pressure, 150°C (300°F) platen temperature, 6-7 second closure, and 25% charge coverage were used. Under these conditions, vacuum was either off or on. The vacuum was activated when the platen was 3 mm (0.12") above the charge. The level of vacuum was 635 mm (25") Hg and was applied before full tonnage was attained. The results (Table VII) from numerous moldings indicated that vacuum provided no advantage in reducing the edge paint pops.

TABLE VII - SUMMARY OF THE EFFECT OF VACUUM ASSISTED MOLDING
ON PAINT EDGE POPS

DOE No.	No. of Panels Molded	Total No. of Pops (Mean)	
		Vacuum Off	Vacuum On
2	40	99 (0.67)	---
2	40	---	121 (0.79)
3	48	117 (0.61)	---
3	48	---	123 (0.64)

CONCLUSION

From our studies, it is clear that the origin of the edge pops resides in "microcracks" in the substrate and these "microcracks" are present before the painting process. The SMC factors which we found to have the largest impact on edge paint popping include molding closure speed, resin paste formulation, and fiber orientation. The specific effect from the fiber glass sizing was found to be small. The magnitude of edge paint popping due to fiber glass sizing was primarily dependent on the resin paste. Our results indicate the potential to significantly reduce and perhaps eliminate edge paint pops through optimization of the molding parameters, resin paste, and fiber glass reinforcement.

ACKNOWLEDGMENTS

The authors wish to acknowledge many individuals for their part in compounding, molding, painting, pop counting, microscopic analysis, data analysis, and typing of this manuscript. Included among these helpful people are Thomas Vorp, Michael Baxter, Bill Pish, Beth Docherty, Bob Cooper, Michael Farnish, Jim Terrell, Lisa Li, Karen Copeland, and Sandy Nalesnik. The authors also wish to thank Ken Atkins, Rob Seats, and Ray Gandy at Union Carbide for the use of their molding equipment and David Dana for valuable discussions and insights.

REFERENCES

1997 SMC Car and Truck Component List published by the SMC Automotive Alliance, Composites Institute, New York, NY.

Cheevers, G. D. 1986. *J. Coating Technology* 58, 742, 25.

Cheevers, G. D. 1988. *J. Coating Technology* 60, 762, 61.

Das, B., et al. 1980. 36th Tech Conf Reinforced Plastics/Composite Inst. SPI, Session 2-B.

Edwards, H.R. 1985. 40th Tech Conf Reinforced Plastics/Composite Inst. SPI, Session 16-A.

Mitani, T., et al. 1990. 45th Tech Conf Reinforced Plastics/Composite Inst. SPI, Session 1-D.

Tomiyama, T, et al. 1991. 46th Tech Conf Reinforced Plastics/Composite Inst. SPI, Session 5-D.

Yamada, H., et al. 1992. 47th Tech Conf Reinforced Plastics/Composite Inst. SPI, Session 1-B.

BIOGRAPHIES

Michael W. Klett is a Research Associate at the PPG Fiber Glass Research Center in Pittsburgh, PA. He received his B.S. degree in Chemistry from California University of Pennsylvania in 1979 and his Ph.D. degree in Physical Organic Chemistry at Iowa State University in 1985. Since joining PPG in 1985, Dr. Klett has worked in a variety of areas in fiber glass product development for RP applications and has authored numerous publications on pultrusion. He is currently project leader for the SMC program.

Steven J. Morris is currently a Senior Molding Specialist at the PPG Fiber Glass Research Center in Pittsburgh, PA. He received his B.A. degree from Grove City College, Grove City, PA, in 1972. He joined PPG in 1974 and has worked since that time with both the Fiber Glass and the Coatings and Resins Divisions in composite process and materials characterization. His work has included contributions to the development of various new fiber glass reinforcements, resin systems, and high-strength molding compounds.

SESSION 18

Composites in Action

The format of this session was a hands-on demonstration; therefore, there were no formal papers.

SMART Material Systems: New Developments in the Monitoring of Structural Composites

Damage Detection in Composite Rotorcraft Flexbeams Using the De-Reverberated Response

K. A. LAKSHMANAN AND DARRYLL J. PINES

ABSTRACT

This paper discusses the application of wave propagation analysis to detect damage in rotorcraft flexbeams in the form of cracks and delaminations. The approach used involves examining the local scattering properties of structural discontinuities in the frequency domains to assess the effect of damage on the modal response of the structure. A model using a simulated cantilevered beam is used to illustrate the performance of the wave model approach for dynamic detection of damage in the form of transverse cracks imparted to composite rotorcraft flexbeams. Experimental validation is carried out on a graphite/epoxy cantilever beam test specimen with $(0/90)_S$ ply orientation in vacuum. Damage in the form of a transverse crack, which extends across the entire width of the specimen is imparted at depths of 25, 50 and 75 % of the beam thickness to the test specimen. A piezo-electric actuator mounted near the root is used to excite the structure dynamically to help locate and determine the extent of damage. The resonant frequencies and scattering properties are used as performance metrics to compare the experimental behavior against the analytical predictions from the wave modeling.

1 INTRODUCTION

The rotorcraft system is a highly complex system. Under normal operation, the rotorcraft system is subjected to a variety of loading conditions which can lead to reduced performance of various mechanical parts including the rotor heads, bearings, pitchlinks and individual blades over time. To reduce maintenance costs and improve overall system reliability, there has been an increased focus on developing bearingless and hingeless rotor systems (See figure 1). Such systems attempt to eliminate the need for bearings by introducing composite flexbeams which permit motion in the flap and lag direction without the need for moving parts. A torque tube is then used to provide torsional rigidity in the pitch axis. Under normal operation these composite flexbeams are also subjected to a variety of loading conditions which can lead to development of damage in the form of small cracks or ply delaminations. In time, if these defects go undetected, they can lead to reduced performance and ultimately catastrophic failure of the rotor system.

K. A. Lakshmanan and D. J. Pines, University of Maryland, College Park, MD 20742.

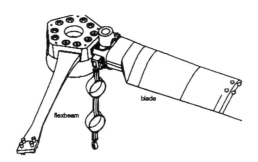

Figure 1: A flexbeam in a rotor system

1.1 MOTIVATION

Recently, there has been a considerable amount of interest in monitoring the health of rotorcraft structures. However, very little effort has been focussed on detecting damage in rotating structures. Most of the previous methods have focussed on loads monitoring and fault simulation. This paper attempts to develop a strategy using wave models to detect damage in rotorcraft flexbeams. Considerable effort has been expended in recent years to develop damage detection and health monitoring methods for structures based on traditional modal analysis techniques. These techniques are often well suited for discrete structures such as trusses used in large space structures or large civil engineering applications (eg. bridges) where catastrophic failures are usually caused by some low frequency change in the global behavior of the system. The global changes are monitored by observing changes in the modal mass, stiffness or damping parameters. However in case of small defects such as cracks, the modal approach is not very efficient. This is because small defects such as cracks exhibit high frequency effects and cannot be discovered easily by examining changes in the modal mass, stiffness or damping matrices. At high frequencies, the modal structural models are subject to uncertainty. Wave propagation models on the other hand are well suited for these small defects since one simply examines the local impedance change across the structural discontinuity. An impedance change is revealed in the frequency dependent scattering matrix that relates incoming and outgoing waves as they scatter across the discontinuity. Such scattering dynamics have magnitude and phase properties that help determine the location, size and nature of the discontinuity.

1.2 STATE-OF-THE-ART

Previous attempts have been made to use wave propagation models to study damage. Keilers et al [1, 2] performed an investigation on using piezoelectrics built into laminated composite structures to detect and characterise a delamination. They used wave response and a frequency domain model to succesfully predict delamination in composite beams. Doyle [3] has studied the use of wave models to characterise cracks in beams but he does not discuss the effect of boundary conditions on the wave response. Methods have been developed by Pines et al. [4, 5] to sense the directional waves and filter the wave amplitudes from the measured transfer function response. The approach taken in this paper follows the analysis developed by Lakshmanan and Pines [6].

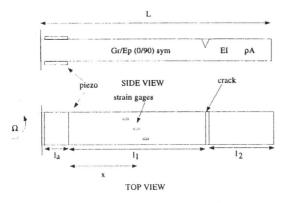
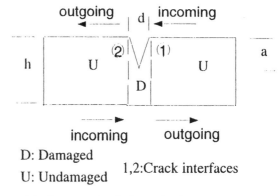

(a) Schematic representation of a flexbeam with a transverse crack

(b) Closeup of a B-E model for a transverse crack

D: Damaged   1,2:Crack interfaces
U: Undamaged

Figure 2: Schematic model for a transverse crack

1.3   OUTLINE OF THE PAPER

This paper develops the first part of the damage detection analysis, i.e. characterising the effect of a particular kind of damage on the global and local system response. The first part of the paper develops the analysis of structures in terms of travelling waves. Following this development, impedance models characterising structural discontinuities are developed for a transverse crack in a cantilever beam. The effect of this model is simulated for a moment acting on a cantilever beam. Experimental results were obtained on a beam in vacuum. The beam was tested at various rotational speeds and the effect of damage and rotational speed on the transfer function response was observed.

2   TRANSVERSE CRACK MODELING

Modeling chordwise through crack phenomena in rotorcraft flexbeams involves relating the depth and width of the crack to changes in the macroscopic physical properties of the structure in the region where the damage exists. Typically with finite elements this involves meeting the compatibility and force constraints from the undamaged region to the damaged region. This tends to be difficult with typical beam finite elements. Often the local effects become obscured in the global response predictions of the structure with damage.

2.1   CONTINUUM MECHANICS FORMULATION

A simplified distributed parameter approach is to neglect the relative rotation of the undamaged and damaged sections of the beam and treat the structure as a Bernoulli-Euler beam element with different physical properties in the damaged section. See fig 2. By examining how the incoming and outgoing waves scatter, one can connect the dynamics of the damaged region to those of the undamaged region through frequency dependent scattering matrices at the right and left interfaces of the damaged region. The dispersion relation for a B-E beam element suggests that transverse bending will support four wave

Parameter	Description	Units
EI	Bending stiffness	1.26 N-m^2
ρA	Mass per unit length	0.0797 kg/m
L	Length	13.5625 in
l_1	piezo to crack distance	12.3125 in
l_2	crack to tip distance	1.21875 in
d	crack width	0.03125 in
h	beam thickness	0.082 in
a/h	crack depth/thickness	25,50 or 75%
x	piezo-to middle straingage	8.000 in

Table 1: Properties of the experimental Graphite/Epoxy $(0/90)_S$ beam specimen

modes, two pairs of wave modes travelling in each direction. This implies that at the crack interface the input-output relation is given by the following equations.

$$\bar{w}_o = S_{DR}\bar{w}_i = \begin{bmatrix} S_{DR1} & S_{DR2} \\ S_{DR3} & S_{DR4} \end{bmatrix}_{4x4} \quad \bar{w}_o = \begin{bmatrix} T & R \\ R & T \end{bmatrix}_{4x4} \bar{w}_i$$

$$\bar{w}_o = S_{DL}\bar{w}_i = \begin{bmatrix} S_{DL1} & S_{DL2} \\ S_{DL3} & S_{DL4} \end{bmatrix}_{4x4} \quad \bar{w}_o = \begin{bmatrix} T & R \\ R & T \end{bmatrix}_{4x4} \bar{w}_i$$

(1)

where the terms R and T correspond to 2x2 sub matrices indicating reflection and transmission of the incident dynamics. These input/output equations relate how incoming waves interact at the left and right interfaces of the damaged region to generate outgoing waves. Information about how the crack affects the local impedance of the structure is contained in these elemental scattering matrices. Traversing the crack region is accomplished through the frequency dependent transition matrix of equation which connects wave modes spatially from one location to another location in the structure. In this manner, the scattering dynamics of the left and right crack interfaces are coupled.

3 EXPERIMENTAL VALIDATION

To validate the analytical predictions developed by Lakshmanan and Pines [6], several FRP composite flexbeams were fabricated (See Figure 3)and tested on a rotating test stand in vacuum. The properties of these experimental specimens are listed in Table 1. Damage in the form of a transverse crack was artificially imparted to the test specimens at a distance of $\approx$ 1.5 inches from the tip of the flexbeam. Crack depths of 25, 50 and 75 % were studied to determine their effect on the global/local system response. Transfer function response data was taken for each specimen at RPM rates from 0 to 900 at increments of 250 RPM. Figure 4 compares the analytical and experimental transfer function response for the middle strain gage sensor in the case of damage at three RPM speeds (0,500,900). For the assumed model structural damping of 0.1%, the analytical prediction with damage appears to match the experimental transfer function at the three different rotational rates. The discrepancy betweeen the measured and analytical results is

RPM	Mode number	Frequency(Hz)			
		0%damage	25%damage	50% damage	75%damage
0	one	15.64	15.64	15.64	15.64
	six	1120.48	1119.02	1117.55	1110.22
250	one	17.10	17.10	17.10	17.10
	six	1120.48	1119.02	1116.08	1111.69
500	one	20.05	20.05	20.05	20.05
	six	1123.41	1123.41	1119.02	1113.16
750	one	22.97	22.97	22.97	22.97
	six	1124.88	1124.88	1120.48	1116.08
900	one	24.43	24.43	24.43	24.43
	six	1127.81	1127.81	1123.41	1119.02

Table 2: Experimentally determined damped natural frequencies of the beam(Hz)

profound away from resonance where sensor signal to noise ratio is low. Figure 5 compares analytical and experimental transfer function responses for the case of 75% damage at the three respective RPM models. As expected the loss of cross-sectional area in the damaged region reduces the effective stiffness locally. At low frequencies this effect is not observed in the transfer function response. However at higher frequencies, the softening in the damaged region causes a leftward shift in the higher frequency modes (modes 5 and 6).

This progresssive leftward shift is examined in more detail in Figure 6 where the analytical and experimental undamaged and 75% damaged cases are compared for the sixth mode of the rotating cantilever flexbeam at 500 RPM. In this figure the analytical model predicts a greater shift than that observed experimentally. This difference between experiment and theory is believed to be partially due to poor modeling of the beam boundary condition and the assumed structural damping of 0.1%. Nevertheless, the observed trends between theory and experiment are similar. This effectively highlights one of the advantages of using a wave model as compared to a modal method to detect damage. In this case, investigating the behavior of the first mode alone would not have indicated the onset of damage. We need to look at the higher frequency modes and we know that at the higher frequencies the accuracy of the modal analysis is poor. The elements of the scattering matrix are changed as a result of the damage. The change in these elements has to be studied and quantified in order to assess the nature and location of damage. We find that the scattering matrix can then be used as a performance metric. The elements of the matrix depend on the wave amplitudes and the phase lag of the waves as they are scattered at the crack interface. As the rotation rate is increased, the centrifugal loading adds stiffening to the experimental test specimen. However this is a low frequency effect (as blade RPM is much lower than the frequency of interest) and so we expect that the higher modes would be less affected by the increase in RPM. A decreasing shift was observed in the case of the higher modes. Thus while the first mode shifted by 10 Hz (15 to 25 Hz), the sixth mode shifted only by about 7 Hz (1120 to 1127 Hz). This seemed to validate our assumption that as we increase the RPM, the effect on the higher modes is progressively lesser. This leads us to believe that this model which was developed for a stationary beam can be applied to a rotating beam in vacuum. The effect of centrifugal loading on the undamaged and damaged damped natural frequencies are illustrated in Table 2.

4 SUMMARY AND CONCLUSIONS

This paper has investigated the use of local wave models to characterise damage in composite rotorcraft flexbeams. Experimental study on a FRP beam has demonstrated that there is a leftward shift in the transfer function at the higher frequencies due to a transverse crack in the beam. It was observed that this shift was more in case of the higher modes; the first mode actually did not show any noticeable effect even at 50% damage. The analytical model for the damaged and undamaged beam has been developed and it is hoped that the scattering matix can be used as a performance metric. For study of damage assesment the transfer functions with RPM indicated that the model can be extended to a rotating beam at the higher frequencies as the rotations primarily resulted in a stiffening effect which was less pronounced in the higher modes. Thus the paper has demonstrated the applicability of local wave models in the investigation of incipient damage which is not easily detectable using modal analysis techniques.

5 ACKNOWLEDGEMENTS

This work was supported by the U.S. Army Research Office under the Smart Structures University Research Initiative, contract no. DAAL03-92-G-0121, with Dr. Gary Anderson serving as contract monitor.

References

[1] Keilers, C.H. JR., and Chang, F. "Identifying Delamination in Composite Beams Using Built-In Piezoelectrics: Part 1- Experiments and Analysis", *Journal of Intelligent Material Systems and Structures*, Vol. 6 September 1995, pp. 649-663.

[2] Keilers, C.H. JR., and Chang, F. "Identifying Delamination in Composite Beams Using Built-In Piezoelectrics: Part 2- An Identification Method", *Journal of Intelligent Material Systems and Structures*, Vol. 6 September 1995, pp. 664-672.

[3] Doyle, J. "Determining the Size and Location of Transverse Cracks in Beams", *Experimental Mechanics*, September 1995, pp. 272-280.

[4] Pines, D.J., Miller, D.W. and A.H. von Flotow, "Directional Filters for Sensing One-Dimensional Structural Dynamics", AIAA Paper 92-2333, *AIAA SDM Conference*, Dallas, Tx, April 1992.

[5] Pines, D.J. "Distributed Wave Sensors and Actuators for Structural Control", *2nd Workshop on Smart Structures held at the University of Maryland*, College Park, September 1995.

[6] Lakshmanan, K.A., and Pines, D.J. "Damage Detection In Rotorcraft Flexbeams Using Wave Models", *to be presented at the SPIE Far East Conference*, Bangalore, India, December 11-14, 1996.

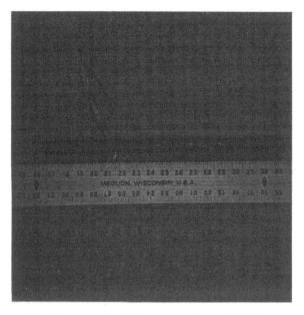

(a) photograph of the experimental Gr/Ep beam specimen

(b) close-up of the crack

Figure 3: Photographs of the experimental beam specimen

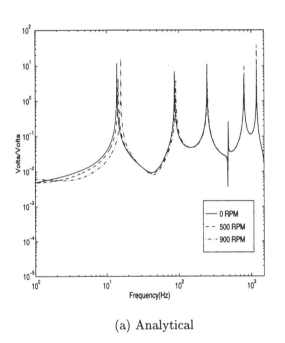

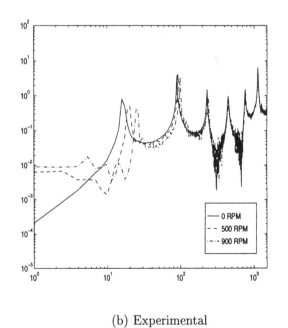

(a) Analytical

(b) Experimental

Figure 4: Comparison of analytical and experimental shift with RPM speeds

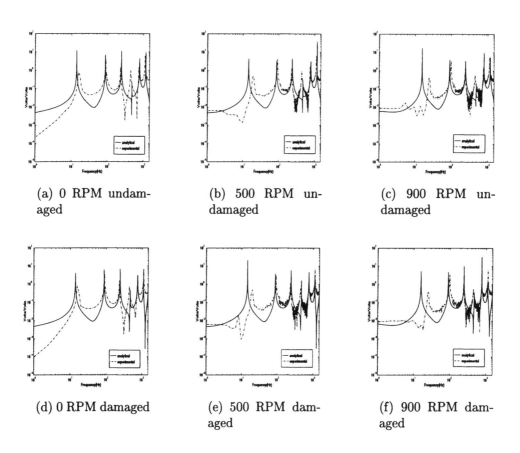

(a) 0 RPM undamaged

(b) 500 RPM undamaged

(c) 900 RPM undamaged

(d) 0 RPM damaged

(e) 500 RPM damaged

(f) 900 RPM damaged

Figure 5: Comparison of undamaged and damaged transfer functions at various RPM speeds

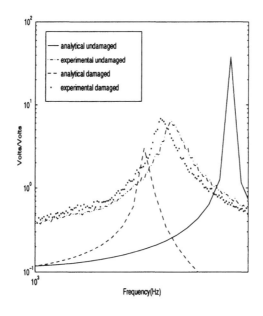

Figure 6: Closeup of the sixth mode(75%damaged and undamaged) at 500 RPM

Embedded Fiber Optic Sensor for Filament Wound Composite Structures

C. C. CHANG, RICH FOEDINGER, JIM SIRKIS AND TERRY VANDIVER

ABSTRACT

This paper describes the use of embedded in-fiber Bragg grating sensors for monitoring the condition of filament wound composite pressure vessels. The operational characteristics of Bragg grating sensors are covered first, followed by an explanation of the burst testing program conducted. The paper concludes with a summary of the results and of the lessons learn from this effort.

INTRODUCTION

After twenty years of research and development, fiber optic strain sensors are becoming accepted by the traditional stress analysis community as viable sensor options for several niche application areas. The primary agents responsible for the impressive level of development in fiber optic sensors that has occurred over the past decade are: (1) ever lower costs and ever better performance of optical fiber components, (2) development of practical optical fiber strain sensor configurations that in many ways mimic the functionality of standard resistance strain gages, (3) further development of commercial products, and (4) greater exposure of fiber optic sensors to traditional strain measurement communities. Though fiber optic strain sensor technology is maturing, it will most likely never rival the cost and simplicity of resistance strain gage technology. As a result, fiber optic strain sensors will be relegated to niche applications not well suited for resistance strain gages. The applications that are attracting the most attention are those involving: (1) high temperatures, (2) electromagnetic noise, (3) electromagnetic pulses, (4) ground loops, (5) volatile chemicals, (6) optical multiplexing, and (7) embedded applications involving composite and other structural materials. Other interesting characteristics of optical fiber strain sensors include a strain-to-failures approaching 100,000 $\mu\varepsilon$ and gage lengths ranging from ~10 microns to several meters.

C. C. Chang and Jim Sirkis, Smart Materials and Structures Research Center, University of Maryland, College Park, Maryland.
Rich Foedinger, Technology Development Associates, Inc., Wayne, Pennsylvania.
Terry Vandiver, Army Missile Command, Redstone Arsenal, Alabama.

One advantage offered by optical fiber sensors their geometrical similarity to reinforcing fibers used in composite materials. This similarity suggests that embedded optical fibers would be less disruptive to the local microarchitecture of composites than traditional sensors. This heuristic argument holds to a certain degree, particularly when Kevlar or boron reinforcing fibers, which have roughly the same diameter as optical fibers, are used. However, optical fibers are roughly a factor of ten larger than the more common glass or graphite reinforcing fibers. Nevertheless, embedding optical fibers so that they are oriented parallel to the local reinforcing fibers has proven to be an effective means of minimizing the disruptive nature of embedding optical fibers in fiber reinforced composite structures. The fact that optical fibers can with stand typical thermoset and thermoplastic curing cycles is another compelling reason to pursue embedded fiber optical strain sensor technology. These factors have lead to embedded optical fiber sensor applications involving the X-33 Reusable Launch Vehicle [Melvin (1996)], composite power distribution poles [Udd et al (1996)], airframe repair patches [Chang and Sirkis (1996)], filament wound composite pressure vessels [Foedinger and Repper (1996); Sirkis and Chang (1996)], as well as others. In the case of filament wound composite structures, there is a rather unique incentive to explore embedded fiber optic sensors. Filament wound structures are one of the few types of components that allows optical fibers to be embedded with minimal impact on the manufacturing processes.

FILAMENT WOUND COMPOSITE PRESSURE VESSELS

Successful development of filament wound structures with embedded fiber optic strain sensors involves several key elements, including (1) compatibility between sensors and the host structure, (2) methods of embedding optical fibers into the composite laminate without damaging the sensor, (3) connectorization to sensor instrumentation, (4) sensor data interpretation, and (5) characterizing pressure vessel strength degradation effects. A recent collaborative research effort between Technology Development Associates, Inc., the University of Maryland and the Army Missle Command addressed many of these issues. In this research project, Bragg grating sensors were embedded in 5.75 inch diameter T1000G/UF-3325 carbon/epoxy pressure bottles. The specimens adhered to ASTM D 2585 Standard Test and Evaluation Bottle (STEB) fabricated on a McClean Anderson W60 4 axis computer controlled winding machine. Several different composite stacking sequences were investigated, and the sensors were embedded at several orientations relative to the reinforcing fibers. These pressure vessels were then burst tested using pressurized water while at the same time recording fiber sensor and collocated resistance strain gage data.

Figure 1 shows a photograph of one of the [(+/-)12°/90°/90°/(+/-)12°/90°/90°] composite pressure vessels used during the testing program. In this particular vessel, the optical fiber was embedded parallel to the +12° ply near the mid-plane. The optical fiber exiting the pressure vessel near its apex is protected using a combination of Teflon tubing and high temperature RTV, as indicate in the blow-up photograph provided in Figure 2. Analysis of the strain data produced by the fiber optic sensor and traditional strain sensors proved very interesting, and allowed for the transverse strain sensitivity of the embedded sensor to be analyzed. The burst tests, augmented with extensive two dimensional finite element analysis, indicated that while the embedded sensors produced highly localized stress concentrations, they did not change the overall burst strength of the pressure vessels.

Figure 1. Pretest Photograph of a 5.75" Diameter Pressure Vessel With a Bragg Grating Strain Sensor Embedded in the Helical Direction.

Figure 2. Enlargement of the Optical Fiber Egress From the Pressure Vessel in Figure 1.

BRAGG GRATING SENSORS

In-fiber Bragg grating sensors are relative newcomers to the fiber sensors world that offers unparalleled optical multiplexing potential. By multiplexing, we mean that many sensors can be located at different spatial locations along the fiber length, yet the signals from the individual sensors can still be addressed. In-fiber Bragg grating sensors are formed by creating a periodic refractive index variation in the core of an fiber optic, as illustrated in Figure 3. This periodicity can be formed in the fiber core by one of many different fabrication techniques [Morey et al. (1989); Morey et al. (1991); Dunphy et al. (1990)], all of which take advantage of the photorefractive effect. Broadband light is launched into the fiber interacts with the periodic index variation so as to reflect light at a single wavelength, known as the Bragg wavelength, and produces a dip in the transmitted spectrum at the same wavelength. It has been demonstrated that the Bragg wavelength, λ_b, is related to the grating pitch, Λ, and the average refractive index n_o, by

$$\lambda_b = 2n_o\Lambda. \tag{1}$$

It has been further shown that this wavelength will shift by an amount $\Delta\lambda_b$, as a function of the

$$\frac{\Delta\lambda_b}{\lambda_b} = (1 - \frac{1}{2}n_o^2 P_{12})\epsilon_1 - \frac{1}{4}n_o^2(P_{11} + P_{12})(\epsilon_2 + \epsilon_3) + \xi\Delta T, \tag{2}$$

strain and temperature as where the symbols used in this equation have the usual meanings. While applying this equation, one should note that the strains represented by ϵ_i are the total strain, i.e., the combination of thermal and mechanical strains in the fiber. When a Bragg grating is

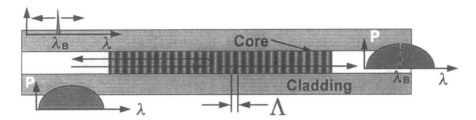

Figure 3. Schematic Illustrating the Operational Principles of Bragg Gratings Sensors.

bonded to the surface of the structure being monitored, the strain state in the fiber is uniaxial, and therefore the wavelength-strain model simplifies to

$$\frac{\Delta\lambda_b}{\lambda_b} = (1 - \frac{1}{2}n_o^2[P_{12} - \nu(P_{11} + P_{12})])\epsilon_1 + \xi\Delta T. \tag{3}$$

The features of Eqs. (2) and (3) that are most relevant to the present discussion are that there are three strain components in Eq. (3), and that thermal contributions to the wavelength change are on the same order of the strain contribution. For this reason, intrinsic sensors have much higher thermal apparent and transverse strain sensitivities strain than extrinsic sensors. The fiber sensor community is presently grappling with these issues and will no doubt devise sensor designs capable of reducing these apparent strain sensitivities.

RESULTS

Two sensor configurations were investigated.The first configuration had the Bragg grating embedded locally parallel to the mid-plane $+12°$ ply so that the reinforcing fibers surround the optical fiber as illustrated in Figure 4. The second configuration had the Bragg grating located

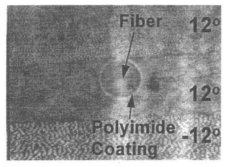

Figure 4. Micrograph of an Optical fiber Embedded Parallel to Local Reinforcing Fibers.

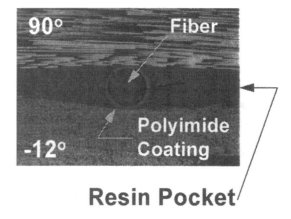

Figure 5. Micrograph of an Optical Fiber Embedded so that it is Not Parallel to the Local Reinforcing Fibers.

next to the mid-plane +12° ply, but oriented in the 0° direction. The reinforcing fibers in this case do not mold around the optical fiber. Instead, they form a lenticular resin pocket similar to the one shown in Figure 5. The sensor data will show that the optical fiber orientation relative to the local reinforcing fibers has an impact on the embedded sensor data interpretation.

The pressure vessels were pressurized using water in 200psi increments until the vessels burst. Internal pressure, strain from the surface mounted resistance strain gage, and from the embedded Bragg grating sensor were all recorded using a data acquisition system. The Bragg grating sensors were read-out using an optical spectrum analyzer. While expensive as a sensor read-out system, the optical spectrum analyzer provided a wealth of information which ultimately proved very useful in interpreting the embedded sensor results. Figure 6 compares the strain measured by the Bragg grating embedded parallel to the mid-plane 12° ply, the collocated resistance strain gage bonded to the surface of the pressure vessel, and the strain predicted using laminate plate theory. The two strain readings agree with each other, and with the theory to within 3%. However, the agreement is not so good (~13%) for the Bragg grating sensor embedded so that it is not parallel to the local reinforcing fibers. In this case, the difference between the Bragg grating and resistance strain gage data is a result of the transverse strain sensitivity in the optical fiber sensors caused by contact stresses between the optical fiber and the reinforcing fibers. Numerical analysis and strength of materials analysis were used to confirm this conclusion.

SUMMARY

This paper has, in a brief way, tried to convey the basic operating principles of Bragg grating sensors. Then several emerging application areas in composite structures that take advantage of the unique qualities offered by optical fiber sensors were touched upon, followed by a more detailed discussion an of application involving filament wound composite pressure vessels. While optical fiber sensors have come a long way over the last decade, they still require some effort before they will be as familiar to stress analysts as resistance strain gages. More

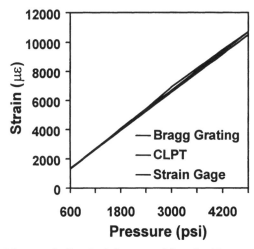

Figure 6. Strain Measured by the Bragg Grating Sensor Embedded Parallel to the Mid-Plane 12° Ply and the Resistance Strain Gage.

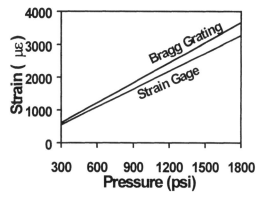

Figure 7. Strain Measured by the Bragg Grating Sensor Embedded Parallel at 0° Degrees Relative to the Mid-plane 12° Ply and the Resistance Strain Gage.

manufacturers of sensors and the associated instrumentation are required to diversify the product base, and to drive down prices. The manufacturers that are currently selling fiber optic sensors need to characterize their technology better, and learn to discuss their technology in jargon that is familiar to resistance strain gage users. These issues are almost obvious, and certain the manufacturers are addressing them. There are also several important technical issues that must be addressed, including dealing with the thermal apparent strain sensitivity and transverse apparent strain sensitivity of Bragg grating sensors, and taking better advantage of optical multiplexing techniques.

ACKNOWLEDGEMENTS

The financial support under SBIR Phase I Contract No. DAAH01-96-C-R060 is gratefully acknowledged

REFERENCES

Chang, C. C. and J.S. Sirkis, "Multiplexed Optical Fiber Sensors for Air-frame Repair Patch Monitoring," to appear in Experimental Mechanics.

J. R. Dunphy, G. Meltz, and W. W. Morey, "Multi-Function, Distributed Optical Fiber Sensor for Composite Cure and Response Monitoring," FIBER OPTIC SMART STRUCTURES AND SKINS III, San Jose, pp. 116-118, 1990

Foedinger, R. C., and Repper, C. J., "Embedded Fiber Optic Sensors for Filament Wound Composite Structures," Phase I SBIR Report (Contract No. DAAH01-96-C-R060), 1996.

Henriksson, A., and Brandt, B., "Design and Manufacturing of an EFPI Sensor for Embedded in Carbon/Epoxy Composites," Proc. OFS, 1994.

Melvin, L., personal communication, 1996.

Morey, W. W., Meltz, G., and Glenn, W. H., "Fiber Optic Bragg Grating Sensors," PROCEEDINGS OF FIBER OPTIC AND LASER SENSORS, SPIE Vol. 1169, 1989.

Morey, W. W., Dunphy, J. R., and Meltz, G., "Multiplexed Fiber Bragg Grating Sensors," PROC. OF DISTRIBUTED AND MULTIPLEXED FIBER OPTIC SENSORS, SPIE Vol. 1586, 1991.

Morey, W. W., Ball, G., and Singh, H., "Applications of Fiber Grating Sensors," Fiber Optic and Laser Sensors XIV, SPIE Vol. 2839, 1996.

Sirkis, J. S., and Chang, C. C., "Embedded Fiber Optic Strain Sensors," to appear in the SEM MANUAL ON EXPERIMENTAL METHODS FOR MECHANICAL TESTING OF COMPOSITES, (C. Jenkins, Ed.).

Udd, E., Corona, K., Slattery, K. T., and Dorr, D., "Fiber Grating System Used to Measure Strain in a 22 ft Composite Utility Pole," Proc. SPIE Vol. 2271, pp. 125137, 1996.

Improved Active Tagging Non-Destructive Evaluation Techniques for Full-Scale Structural Composite Elements

ZAO CHEN,[1] DR. VICTOR GIURGIUTIU,[2] DR. CRAIG A. ROGERS,[3]
DR. ROBERT QUATTRONE[4] AND JUSTIN BERMAN[5]

ABSTRACT

An on-site particle tagging non-destructive evaluation (NDE) technique has been developed for health monitoring of a full-scale structural element of glass-fiber reinforced plastics (GFRP) composites in which conventional NDE approaches are not very effective. Unlike conventional passive tagging NDE inspection, the technique uses an electromagnet exciter to interrogate tagged composites. A laser Doppler vibrometer is used for high-speed and non-contact surface vibration detection. An accept-reject criteria based on the signature pattern difference of a healthy and a damaged structure is then applied to extract an index of the health of the structural elements. The experimental results of the active particle tagging inspection shows a variation in the dynamic response of the specimens when defects and/or damage are presents. The variation could be used to diagnose and to monitor the integrity of materials.

INTRODUCTION

In high volume civil engineering constructions and industrial applications, the use of advanced reinforced composite materials can provide elegant solutions to difficult engineering problems. Unfortunately, the physical attributes of the composite materials present problems for the accurate detection and evaluation of internal flaws. This is especially true for the GFRP composites, which are electrical insulators and, hence, non-conductive. These difficulties have created a need for new NDE techniques optimized specifically for GFRP composites materials since conventional NDE methods are not very effective. A new NDE technique, dynamic characteristics evaluation using magnetic interrogation (DCEUMI), was shown to be non-intrusive, capable of fast data acquisition, and sensitive to a wide range of common composite flaws (Rogers, *et al.*, 1995; Zhou, *et al.*, 1995).

---

[1]Center for Intelligent Material Systems and Structures, Virginia Tech, Blacksburg, VA 24061-0261
[2-3]Formerly at Center for Intelligent Material Systems and Structures, Virginia Tech, Blacksburg, VA 24061-0261, presently at Mechanical Engineering Department, University of South Carolina, Columbia, SC.
[4-5]US Army Corps of Engineers, Civil Engineering Research Laboratories, Champaign, IL.

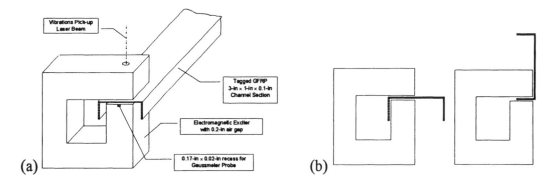

Figure 1. Layout of the active tagging interrogation of C-channel section using an 1.5-in × 1.5-in working section electromagnetic exciter and a laser beam for vibrations pick-up: (a) general layout; (b) placement options for the C-channel section.

The authors (Giurgiutiu, *et al.*, 1996) found that, with GFRP composites, a presence of damage could be detected by signature-pattern identification of the frequency response and natural frequency peaks. In this paper, the DCEUMI technique, which utilizes a powerful electromagnetic exciter and a laser Doppler vibrometer is developed for detection of internal defects and/or damage for full-scale GFRP composite-elements inspection.

PROPOSED CONCEPT AND METHOD FOR FULL-SCALE NDE INSPECTION EQUIPMENT

The proposed concept is presented schematically in Figure 1. A 3-in × 1-in × 0.1-in C-channel tagged-composite section is passed through the narrow air gap of an electromagnetic exciter. Under the influence of the electromagnetic field, the tagged composite material is locally excited at high-frequency. The resulting vibrations response is picked up by a laser beam. The processing of the vibration response in correlation with the high frequency magnetic excitation yields meaningful information regarding the integrity of the composite material, possible delamination, cracks, and other defects.

The method of performing NDE on the full-scale GFRP composite C-channels is straight forward and consists of passing the channels through the interrogation window of the equipment. Since the method is non-contact, several passes may be applied without affecting the part being inspected.

EXPERIMENTAL TECHNIQUES

The principal improvements of this technique over that previously reported by Giurgiutiu, *et al.* (1996) are the replacement of the piezoelectric force gage fixed to the tagged composite specimen with a laser Doppler vibrometer for the non-contact sensing and the development of a powerful custom-made electromagnetic exciter. The advantages of the new system are as follow:

1. The external impact on the tested specimens in successive evaluation is minimized. Since the laser detection is on non-contact, no extra mass is attached to the composite part being inspected.

2. Because the scanning laser head is remote from the composite elements being inspected the new inspection method can be applied to on-site and in-field quality control, e.g. while the composites parts are in the fabrication process, or while they are on the construction site.

3. Since exciting the structure and monitoring the response are done at the same position, the method can be used to simultaneously detect the presence of defects and to determine their location and size. In addition, the effect of boundary conditions on the evaluation process is minimized high-order frequencies and modal parameters are used.

The technique consists of the excitation of the vibration modes with an electromagnetic exciter. The response is sensed by the laser Doppler vibrometer. Input and response signals are fed into an FFT analyzer, and the frequency response function over the desired frequency span is displayed, in real time, on the oscilloscope and stored in the computer memory. Subsequent processing of the signal and of the frequency response curves yields the inherent characteristics of the specimen (natural frequencies, modal parameters, etc.)

SPECIMENS AND INSTRUMENTATION

Tagged and untagged GFRP composite specimens containing 4% by weight of resin lignosite powder were produced by Creative Pultrusions, Inc. as 3-in × 1-in × 0.1-in C-channels section ladder rails. In some specimens, delaminations were created using 1.5-in diameter, 4-mil Mylar film inserts.

The instruments consisted of three main components, exciting instruments, measuring instruments, and signal processing instruments:

The exciting instruments included a signal generator, filters, power amplifiers and an electromagnetic exciter with a small hole. Through this hole, a laser beam was pointed to sense directly the specimen region being magnetically excited. The excitation magnetic field was generated by an alternating current (AC) and a direct current (DC) in the different solenoid coils of the electromagnet. The DC coils created a static bias flux density to magnetize the particles and maximize the excitation to improve signal-to-noise ratio and suppress the non-linear frequency component, while the AC coils created an alternating magnetic flux density to produce vibrations of the particles.

The measuring instruments consisted of the laser Doppler vibrometer and a Gaussmeter. The measuring instruments accomplish the conversion of the physical quantity into an electrical quantity. After conversion and conditioning of the physical data, the test data could be either displayed, using oscilloscopes, or recorded and stored in the computer for further signal processing.

The signal processing instruments consisted of a Fast-Fourier-Transform (FFT) analyzer and a computer. In our experiments, the signal generator, the filter, and the Fast-Fourier-Transform (FFT) analyzer were united into one instrumentation system constructed around a Macintosh Quadra 950 computer. Both the input, measured with the Hall probe of a Gaussmeter, and the output, measured by the laser Doppler vibrometer, were passed to the frequency analyzer. Processing of these signals yielded frequency response curves. Search for the anomalies in the frequency response curves and comparison of natural frequencies of the evaluation specimens with an established accept-reject criteria would assess integrity of composite elements.

TEST PROCEDURES

Because NDE by DCEUMI is based on the fact that internal defects will generally result in changes in structural stiffness which leads to changes in the natural frequencies and in the frequency response signature, magnetic excitation is used to cause the tagged specimens to vibrate in order to measure the vibration characteristics. The responses induced by this vibration are measured with the laser Doppler vibrometer. The velocity and flux density signals from the laser Doppler vibrometer and from the magnetic field probe are input to the frequency response analyzer. The time record of the structural response to the magnetic excitation is converted to the corresponding frequency spectrum. The resulting frequency spectrum indicates the natural frequency peaks and frequency response signature pattern. The frequencies of the test specimen are readily identified from the peaks of the spectrum. The detection of defects can be done by comparison of the changes in the frequency responses of specimens. In this study, we concentrated our attention on the examination of the high-order mechanical resonance frequencies of the excited specimen as a possible way to identify local defects and thus develop an effective NDE technique.

CALIBRATION

The system needs to be calibrated using control specimens regarding of accept-reject criteria. The frequency response signature pattern of the control specimens will be stored in PC memory as reference. Meanwhile a meaningful definition of what is a "comforting" and a "non-conforming" material condition needs to be achieved.

SPECIMEN SUPPORT

Since the method utilizes high-frequency excitation, only the small-wavelength vibration modes are excited. Hence, the method is relatively insensitive to boundary conditions, as long as they satisfy the StVenant principle. The specimen can be supported either as a cantilever beam, or on end supports. To satisfy the StVenant principle for a typical defect length of 1-in, a distance of approximately 1-ft (i.e. > > 1-in) between the interrogation window of the equipment and the nearest specimen support is recommended.

RESULTS AND DISCUSSIONS

The experiments have been performed for proof-of-the-concept of full-scale composite elements inspection. The experiments have been carried out with sinusoidal signal sweep. Broad-band random excitation was also tried, but it was found that, due to some limitations of the equipment, the resulting structural response was not powerful enough to give a good signal processing result. Hence, harmonic force excitation with frequency sweep was used. Thus, the frequency response of the specimen was determined. The response signal obtained during excitation of the specimen over a frequency sweeping range of 500 - 1500 Hz was found to contain several natural frequency peaks. The frequency response of control specimen was compared with this of specimen having simulated defects in the form of delamination. Figures 2a presents the frequency responses of the control and delaminated specimens. It can be noted that the presence of the defects created significant changes in the shape of the frequency response and in the location of the resonant frequency peaks.

Note that frequency responses not only depend on the geometry of specimens but also on material stiffness. For specimens with the same geometry made from the same material, differences in frequency response signature may occur due to the defects. For the specimen without defects, the first few resonant peaks of the response (Figure 2a) were greater than that of delaminated specimen. Since each specimen was tested under the same experimental conditions, these changes could be directly correlated with the presence of the defects in the specimen. The resonant frequencies of the delaminated specimen were significantly lower than those of the control specimen.

When driven by the magnetic field, the embedded particle sensors interacted with their host matrix and generated measurable signatures of the structural response which could be interpreted as structural information about damage. The sensory signature of the frequency response from the tagged specimen could be extracted as a result of the interaction between embedded particles and their host matrix. Relationships between the responses of tags and important physical and structural parameters have been experimentally studied previously to understand the fundamental physics and mechanisms involved in using the active tagging method (Rogers, *et al.* 1995).

Figure 2b presents the frequency response of the control specimen when excited at different locations along its span. As expected, performing the measurement at different locations resulted in variations of the frequency response curve. The reason for this lies in the mode shapes of the structural modes. Therefore, the sensors may pick up a dominant mode at one location and another dominant mode at a different location. However, as seen in Figure 2b, there is consistent pattern of resonant frequency peaks that is found at all locations, though the peak amplitudes may differ somehow. This aspect needs more research, such that carefully conducted calibration and training experiments can be devised to develop an effect and reliable defect detection procedure for full-scale composite element inspection in a real-life environment.

The overall result of these preliminary tests indicate that the active tagging technique of dynamic characteristic evaluation using magnetic interrogation (DCEUMI) has been found to be effective and sensitive enough to detect the simulated defects.

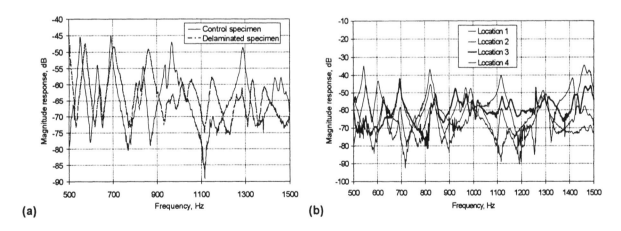

Figure 2   Comparison of frequency response curves: (a) Control *vs.* delaminated specimens; (b) variation with location on a control specimen.

CONCLUSION

The ferromagnetic active tagging NDE technique based on the use of an electromagnetic excitation in conjunction with laser Doppler vibrometer and a computer-based FFT analyzer has been developed for the inspection and evaluation of full-scale GFRP composites elements. The preliminary experimental results presented in this paper indicate that the proposed technique is effective in exciting the tagged composite specimens in the design frequency range 500-1500 Hz. The experiments have also shown that the proposed technique is sensitive enough to detect simulated delaminations defect purposely places in the specimen by the manufacturer. The graphs shown in Figure 2a indicate clear differences in the frequency response between the control and the delaminated specimens. In the presence of local defects (1.5-in diameter, 4-mil Mylar film inserts), the resonant frequency peaks presented a clearly identifiable shift towards lower frequency values. This fact can be directly related to the lowering of the local plate-bending stiffness associated with the presence of delaminations in a laminated composite material structure.

Preliminary investigation of the repetitiveness and consistency of the method has shown that the frequency response spectrum appears to be highly repetitive when measured at the same location, but it can show some variations from location to location. The latter phenomenon is to be expected due to the wavelength of the structural mode shapes, and its influence on the smallest size of detectable defect, together with calibration and training experiments, needs to be investigated in further research and correlated with the excitation frequency range.

ACKNOWLEDGMENTS

The authors gratefully acknowledge the financial support of the U. S. Army Corps of Engineers, Construction Engineering Research Laboratories, through contract No. DACA88-94-D-0021-002 and of the Composites Institute of the Society of Plastic Industry.

REFERENCES

1. Rogers, C. A., *et al.*, 1995, "Concept for Active Tagging of Reinforced Composites for In-Process and In-Field Non-Destructive Evaluation", *USACERL Technical Report*, March, 1995.
2. Zhou, S., Chaudhry, Z., Rogers, C. A., and Quattrone, R., 1995, "An Active Particle Tagging Method Using Magnetic Excitation for Material Diagnostic", *Proceedings of the 36th AIAA/ASME/ASCE/AHS/ASC Structures, Structural Dynamics, and Materials Conference*, New Orleans, April 9-12, 1995, Paper # AIAA-95-1105-CP.
3. Rogers, C. A., *et al.*, "Results of the Active Tagging Experiments with Ferromagnetic and Magnetostrictive Sensors", *CIMSS Technical Report to the US Army CERL sponsor*, Virginia Tech, February, 1996.
4. Rogers, C. A., *et al.*, 1995, "Results of the Application and Experimental Validation of the Active and Passive Tagging Concepts using Industrial Composite Samples", *CIMSS Technical Report to the US Army CERL sponsor*, Virginia Tech, February, 1996.
5. Giurgiutiu, V., Chen, Z. and Rogers, C. A., 1996, "Passive and Active Tagging of Reinforced Composites for In-Process and In-Field Non-Destructive Evaluation", *Proceedings of the 1996 SPIE Conference on Smart Structures and Materials*, 26-29 February, 1996, San Diego, CA, SPIE Vol. 2717, Paper # 2717-31, pp. 361-372.
6. Sun F. P., and Rogers, C. A., 1996, Disbond Defection of Composite Repair Patches on Concrete Structures Using Laser Doppler Vibrometer". *Proceedings of the 1996 SPIE Conference on Smart Structures and Materials*, 26-29 February, 1996, San Diego, CA, SPIE Vol. 2717.

Tagged Composites Mechanical Properties: Investigation and Correlation with Fractographic Examination

DR. VICTOR GIURGIUTIU,[1] JERONE GAGLIANO,[2] DR, ROBERT QUATTRONE,[3] JUSTIN BERMAN[4] AND JOHN VOYLES[5]

ABSTRACT

The mechanical properties of smart composites with embedded ferromagnetic tagging were examined. Flexure and tension tests were performed per ASTM standards. Weakening of the fiber/matrix interface resulting in a reduced interfacial shear strength, and reduced mechanical properties due to magnetite tagging powders was initially observed. Macroscopic and microscopic observations (fiber pull-out length, matrix adhesion remains on the fiber surface, etc.) were corroborated with the observed mechanical properties. An improved tagging system, using dried lignosite powder, gave much better mechanical properties, and indicated that adequate materials and process control can produce good tagged GFRP composites.

EXPERIMENTAL PROCEDURE

Smart composite samples "tagged" with magnetic substances (magnetite and lignosite solution) were tested. Pultruded C-channel profiles used as ladder rails (1"×3.5", color code yellow) were manufactured by Creative Pultrusions, Inc., and have a five layer internal architecture. The outside and center layers were thin random weaved mats. Thicker unidirectional pultruded E-glass fiber layers were contained between the mats in a vinyl ester resin. The tagging consisted of 2% ferrous oxide (magnetite) particles dispersed predominately in the outside two layers. Testing was done in a Instron 4505 machine with 1 kN and 20,000 lb capacity load cells run under displacement control through the GPIB interface using a Macintosh Power PC and LabVIEW software. The mechanical data was downloaded, stored and analyzed in the same computer. ASTM type specimens were

[1]Formerly at Center for Intelligent Material Systems and Structures, Virginia Tech, Blacksburg, VA 24061-0261, presently at Mechanical Engineering Department, University of South Carolina, Columbia, SC

[2]Summer undergraduate research student, National Science Foundation Science and Technology Center for High Performance Polymeric Adhesives and Composites, Virginia Polytechnic Institute and State University, Blacksburg, VA

[3-5]U.S. Army Corps of Engineers, Civil Engineering Research Laboratories, Champaign, IL

prepared from the C-channels and instrumented with CEA-06-125UW-350 strain gages.

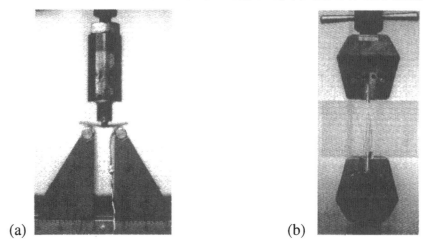

(a)   (b)

Figure 1   Test fixtures: (a) flexure test fixture; (b) tension test fixture

Three-point bending tests were done per ASTM D790 (Figure 1a). Specimen dimensions were $60{\times}25{\times}2.4$ mm^3. Ten specimens of each type were used. Each specimen was strain-gauged in the center, and then centered on the 3-point bend support, with the strain gage on the under side. The cross-head applied the load at 1 mm/min until crack initiation.

Tension test were done per ASTM D3039 (Figure 1b). Six tension specimens of each type were cut to $100{\times}13{\times}2.4$ mm^3 dimensions from the same structural C-channel members and strain gauged on both side at the center position. The specimens were placed into vise grips (using 60 grit sandpaper to increase friction) and loaded to failure at 1.5 mm/min load rate.

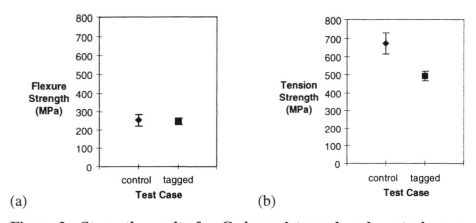

(a)   (b)

Figure 2   Strength results for C-channel tagged and control composites: (a) flexural strength; (b) tensile strength

MECHANICAL TESTING RESULTS

The load, displacement, and strain data was processed to yield the strength data (Figure 2). Figure 2a shows a 3% decrease in flexural strength of the tagged composite, which can be considered within the experimental error. This conclusion is similar to that of Linares (1996), and of Quattrone, Berman, and Voyles (1996). Figure 2b shows a significant decrease in tension strength (26.9%) and in the elongation (25%).

Table 1. Comparative results of the mechanical strength data.

		Linares (1996). Reichhold Chemicals	Quattrone, Berman, and Voyles (1996). CERL	Giurgiutiu and Jerone (1996) Virginia Tech
Flexure Strength[1]	Control	82.1	67.6	36.3
(ksi)	Tagged	78.2 (-4.6%)	65.7 (-2.7%)	35.3 (-2.8%)
Tension Strength	Control	112.0	105.7	96.0
(ksi)	Tagged	72.6 (-35.2%)	72.1 (-31.7%)	70.1 (-26.9%)

[1]Flexure strength variation from investigator to investigator is attributed to specimen length effects.

This decrease is consistent with the finding of Linares (1996) who reported a 35.2% decrease in strength and a 36% decrease in elongation, and of Quattrone, Berman, and Voyles (1996) (31.7% decrease in both quantities). A summary of all these results is presented in Table 1. The conclusion of Table 1 is that tagging with magnetite powders significantly reduces the tension strength but not the flexure strength. The explanation of these difference lies in the effect of interlaminar shear strength and of internal fiber architecture of the C-channel profile, where the high-strength glass fiber tows are placed in the middle of the thickness, while the outer layers are made of low-strength glass mat and veil material. Figure 3 shows the post-failure aspect of the untagged (control) and tagged tension specimens. It is apparent (Figure 3a) that the aspect of the untagged (control) specimen corresponds to wide spread catastrophic failure at high energy levels. Figure 3b shows the tagged specimen, which seems to show a "clean" cut in the outer veil and slippage between the thin outer mats (which has higher tagging concentration) and the inner unidirectional fiber layers. In fact, the load transfer path between the loading grips and the inner high-strength core of the specimen has been interrupted by the loss of adhesion in the interface.

(a)

(b)

Figure 3   Comparative macroscopic aspects of tension failure: (a) control specimen; (b) tagged specimen

MICROSCOPIC OBSERVATIONS

Fiber matrix adhesion can be correlated to the pull-out length of a broken fiber (Figure 4a). Following Giurgiutiu *et al.* (1996), the equations controlling the fiber pull-out process are:

$$P(x) = 2\pi \cdot r_f \cdot x \cdot \tau_0 \quad , \qquad P_{max} = \pi \cdot r_f^2 \cdot X_f \quad , \qquad \delta = \frac{r_f \cdot X_f}{2 \cdot \tau_0} \qquad (1)$$

The pull-out fiber breaks when the built-up load $P(x)$ reaches the ultimate fiber load, P_{max}. The pull-out fiber length varies inversely to the fiber-matrix bonding strength. Since the control and tagged composite were constructed from the same materials, the predominate variable was the interfacial shear stress τ_0. Equation (1) indicates that a longer pull-out length, i. e., a greater δ, indicates a lower τ_0, and a weaker fiber-matrix interface. Therefore, according to this fiber-pull-out argument, the fibers resulting from the composite with a lower strength (i. e. the tagged composite specimens), should be longer and "cleaner" than the fibers from the control composite specimens.

To verify the correlation with the fiber-pull-out-length, the specimens were analyzed at the microscopic level. The microscopic work was done with a ISI SX-40 Scanning Electron Microscope (SEM). Each specimen was gold coated using a sputtering machine. Figures 4b and 4c are representative micrographs for cracked surface edge in both specimens. The fibers are 25% longer in the tagged composite, which, according to Equation (1) infers a 25% decrease in interfacial shear strength.

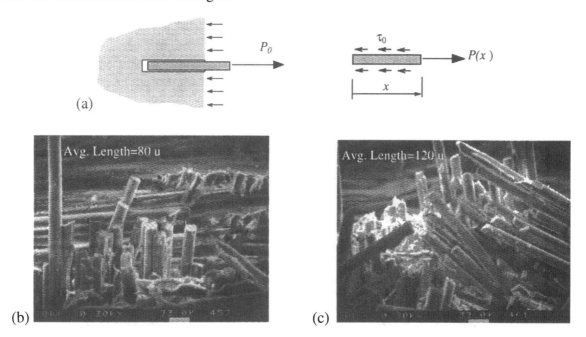

(a)

(b)   (c)

Figure 4   Microscopic examination of pull-out fibers at ×300 magnification: (a) Schematic drawing of pull-out fiber force and interfacial shear stress; (b) control composite specimens; (c) tagged composite specimens.

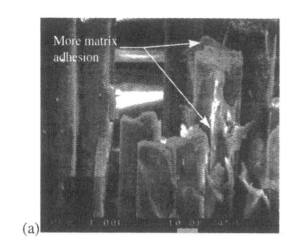 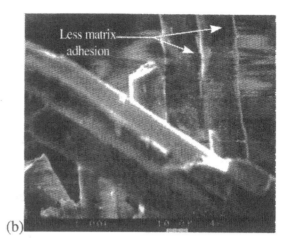

Figure 5   Microscopic examination of pull-out fibers at x1000 magnification: (a) control composite specimens; (b) tagged composite specimens.

Since the bonding of the resin to the fibers is weaker in the tagged composite than in the control, it is expected that the tagged fibers should have less matrix adhered to them. This difference is evident at higher microscopic magnification (×1000). Figure 5 compares the surface morphology of the untagged (control) composite (Figure 5a) with that of the tagged composite (Figure 5b) which presents significantly lower adhesion on the fibers surface.

IMPROVED MECHANICAL PROPERTIES THROUGH LIGNOSITE TAGGING

A tagged composite with greatly improved mechanical properties was recently presented and tested (Pauer, Gauchel, Klett and Troutman, 1996). This improved tagged composite material used lignosite powder (4% by weight of resin) during the pultrusion process and produced C-channel profile ladder rails dimensionally similar to those previously tagged with magnetite particles, but colored with blue pigment.

Mechanical testing of these improved tagged composites, were performed at the US Army CERL facilities. The results, shown in Table 2, indicate the advantages of the improved tagging formulation. As with the previous tagged material, the differences between the "tagged" and "untagged" flexure strengths is within the experimental error. As compared with the previous tagged material, the difference in the tensile strength is much smaller (10%), thus indicating the advantages of the lignosite tagging formulation.

Table 2.   Comparative results of the mechanical strength data for the improved tagged composite formulation.

	Untagged (control)	Tagged
Flexure strength	68.5±0.5 ksi	66.3±2.3 ksi
Tensile strength	61.0±1.0 ksi	54.7.0±1.0 ksi

CONCLUSIONS

The most significant finding of these tests lies in establishing the micro-mechanical mechanism that lead to the substantial (25-35%) decrease in the tension strength observed in the tagged composite samples using the magnetite tagging formulation. This difference was explained in view of the internal fiber architecture and fiber/matrix and interlaminar bonding strength. The load transfer path under tension loading, from machine grips, through the outside mat and veil layers to the inner high-strength unidirectional fiber layers was identified to be primarily adhesive. With a weakened adhesion strength, the load could be not properly transmitted to the inner high-strength layer, and the overall tension strength of the composite decreases. This difference in bonding strength could be correlated with the different aspect of the failed specimens, as illustrated in Figure 3. Further examination at microscopic level revealed further correlation. The tagged composite fibers had less resin still attached to the broken fibers, and the pull-out length of these broken fibers were, on average, 25% longer than the control fibers. This finding suggests a 25% decrease in the interfacial shear stress, which corresponds to a 25% decrease in strength, similar to the mechanical testing results.

The fact that the flexure strength did not present a similar significant decrease was also explained. In the flexure test, the loading was more directly distributed in the high-strength inner layers, since the thin and weak outside layers are not expected to carry much load. Hence, the further weakening of the outside layers due to tagging did not significantly affect the overall flexure strength of the composite. Additionally, as determined by Quattrone, Berman and Voyles, (1996) the tagging was not uniform across the thickness, and hence only affects the outer, already weak, layers. Further weakening of these layers by the application of tagging does not significantly modify the resulting flexure strength.

The latest results obtained with an improved tagging formulation utilizing lignosite material have shown that most of the initial problems have been alleviated. The decrease of the tensile strength has been reduced to 10%, thus indicating that adequate materials and process control can produce good tagged GFRP composites.

Acknowledgments

The authors gratefully acknowledge the support of the U. S. Army CERL, contract DACA88-94-D-0021-002, of the Composites Institute of the Society of Plastic Industry, and of the NSF-STC: High Performance Polymeric Adhesives and Composites, Contract DMR 9120004.

References

Linares, F., 1996, "Composites Institute Smart Composites Project." Reichhold, interoffice memorandum, March 22, pp. 1-2.

Quattrone, R., Berman, J, and Voyles, J., 1996, "Tagged Composite Distribution/Fracture Studies", *Smart Tagged Composites Meeting,* US Army CERL, Champaign, IL, July 2, 1996.

Giurgiutiu, V., Reifsnider, K. L., and Rogers, C.A., "Rate-Independent Energy Dissipation Mechanisms in Fiber-Matrix Material Systems", *Proceedings of the 37th AIAA/ASME/ASCE/AHS/ASC Structures, Structural Dynamics, and Materials Conference,* Salt-Lake City, UT, April 15-17, 1996, Paper # AIAA-96-1420-CP, pp. 897-907.

Pauer, R., Gauchel, J., Klett, M., and Troutman, D., 1996, Private communication, US Army CERL / Composites Institute Smart Tagged Composites Meeting, Blacksburg, VA, , October 29, 1996.

Addressing Fire Performance Standards: Creating a Doorway to Civil Engineering Markets— A Panel Discussion

NOTE: No formal papers were submitted for this session.

SESSION 21

Processing and Recycling: Two Sides of the SMC/BMC Coin

Supercritical Fluid (SCF) Application of SMC Primers: Balancing Transfer Efficiency and Appearance

JEFFREY D. GOAD AND JAMES HANSEN

ABSTRACT

The process of launching a new technology within a particular industry requires that the technical merits, quality improvements, and impact on economics be defined through laboratory experiments, modeling, and plant trials. Once the first customer is established within a given industry, the technology can provide its own testimonial. A technology being developed for application in the SMC industry is the use of carbon dioxide for spray applied primers. The use of carbon dioxide as an atomization tool has demonstrated a significant potential for quality improvements, VOC reductions, and economic benefits for Automotive SMC primers. The development of this process has included experiments to reformulate the paint, define the operating parameters, and optimize application methods to exceed industry standards for performance while making no change to general industry practices.

The use of carbon dioxide in spray painting is a Union Carbide patented process under the registered business trademark the UNICARB® System. In its simplest description, the technology involves using carbon dioxide as opposed to conventional diluents to reduce the viscosity of polymer or resin systems. This mixture in turn can be spray applied. The resulting characteristic of this spray has been coined "Decompressive Atomization". The act of compressing carbon dioxide stores energy which is released in the spray. This energy overcomes the binding forces of viscosity, cohesion, and surface tension to provide atomization that is characteristic of automotive finishing.

Transfer efficiency, the ratio of solids applied to solids consumed, is a large part of the economic analysis for any painting process. Being able to predict and manipulate transfer efficiency becomes an integral part of promoting the technology. With this in mind, experiments and plant trials were conducted to compare conventional air spray applications to the carbon dioxide process. This analysis has led to further experiments to define the effects of certain parameters on transfer efficiencies.

Union Carbide Corporation, 3200 Kanawha Turnpike, P.O. Box 8361, South Charleston, WV 25303

OVERVIEW OF TECHNOLOGY

The technology works under a very simple premise. A formulator changes only the solvent phase of a resin system, therefore leaving the dry film properties unaffected. This provides little risk to the applicator for properties such as adhesion, hardness, and etch resistance. With little or no fast solvent or diluent remaining in the formulation, the resin system is now at much higher viscosity and solids than the corresponding conventional product and is termed a concentrate. This concentrate is then mixed with carbon dioxide as depicted in Figure 1 via special equipment, heated, pressurized and sprayed. The conditions are typical of airless type processes, but the results are characteristic of improved air spray technology.

SIMPLIFIED CONCEPT   ATOMIZATION COMPARISON

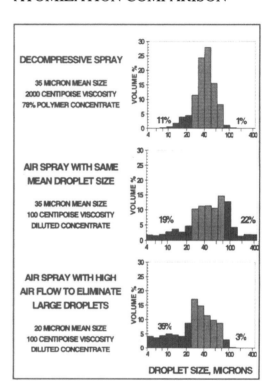

Figure 1   Figure 2

In this technology there are many parameters that can potentially affect transfer efficiency. Spray pressures, fluid temperatures, carbon dioxide concentration, spray gun distance, fan pattern width, flow rate, and spray gun path all contribute to transfer efficiency. These variables coupled with outside influences such as spray booth conditions and part configurations make for complicated predictions that involve many interactions. This is why typically transfer efficiencies can only be compared on a relative basis. The carbon dioxide process demonstrates conditions favorable to enhanced transfer efficiency. Compared to air spray, the carbon dioxide spray process has a narrower droplet size distribution as seen in Figure 2, which allows for optimizing transfer efficiencies with less sacrifice to appearance.

This data in Figure 2 was generated by Purdue University in an atomization study with an acrylic-melamine system. It demonstrated that the Decompressive Carbon Dioxide Spray will have fewer large particles which contribute to poor appearance given the same mean droplet size for the typical air spray. Conversely, if the atomization air for air spray is adjusted to reduce the number of large particles, the result is an increase in the very small droplets which causes transfer efficiency to suffer.

EXPERIMENTAL

To prove the technology for a potential customer, four variables were chosen to demonstrate how transfer efficiency can be manipulated within the painting process. Gun distance from the part, fan width, pre-orifice size, and carbon dioxide concentration were selected. Because part configuration plays a significant role in the effects of these variables, a challenging test piece was chosen to amplify any significance that may be seen. Rather than test on large flat pieces where transfer efficiency is easy to maximize, a part with intricate detail and several different plains of coverage was chosen. The SMC window frame construction of a rear door liftgate was used for the test. This part has historically demonstrated low transfer efficiencies in the range of 20% with a non-electrostatic application. To calculate Transfer Efficiency the following equation was used:

$$\% \text{ T.E.} = \frac{\text{(weight of coating on part)}(100)}{\text{(volume of fluid sprayed)(weight fraction solids)(specific gravity)}}$$

Table 1: Designed Experiment Results

Run	CO2 (wt. %)	Distance (inches)	Fan Width (inches)	PreOrifice Opening (inches)	Transfer Efficiency (%)	Flow Rate cc/min	Appearance Rating
1	27.0	9	6	0.0094	28.32	154	5
2	27.0	9	6	0.0094	29.40	196	4
3	27.0	9	8	0.0094	20.90	160	5
4	20.0	9	8	0.0094	27.59	178	2
5	20.0	6	4	0.0094	43.57	180	2
6	20.0	12	6	0.0094	18.38	165	1
7	20.0	12	8	0.0094	21.77	184	3
8	23.5	6	6	0.0094	38.99	175	2
9	23.5	12	8	0.0094	13.72	164	1
10	27.0	6	8	0.0094	24.47	160	5
11	27.0	12	4	0.0094	27.78	156	3
12	20.0	6	8	0.0108	28.00	218	2
13	20.0	12	6	0.0108	20.31	234	1
14	23.5	9	4	0.0108	30.90	169	3
15	27.0	6	4	0.0108	47.56	200	4
16	27.0	12	8	0.0108	19.01	194	3
17	27.0	6	4	0.0094	40.86	156	4
18	20.0	9	4	0.0108	41.35	226	2
19	23.5	6	8	0.0108	21.80	231	3
20	27.0	12	4	0.0108	34.18	192	2

Four experiments were conducted initially to establish ranges and confirm experimental procedures were appropriate, runs 1 to 4 in Table 1. An optimal experimental design was developed by adding 16 experiments. Three levels of distance, fan width, and carbon dioxide concentration and two different preorifice sizes were studied. One liftgate was painted for each run. Transfer efficiency and flow rate were measured for each set of experimental conditions. The panels were rated for general appearance using a scale of 1 to 5 with 1 and 2 indicating substandard to production, 3 meeting minimum requirements, and 4 and 5 exceeding minimum requirements.

RESULTS

The data was analyzed using multiple regression analysis to quantify the effects of the study variables. The regression models summarized the results and provided a way to evaluate different operating conditions, see Table 2. Contour plots of the predictions show the effects of the influential variables.

Table 2: Regression Equations

%TE = 30.268-.379*CO2-2.265*Dis-3.702*TpW+.405*CO2*Dis+.674*Dis*TpW-2287.7*TpW*PrO
(-2.7)*   (-13.1)   (-13.9)   (6.0)   (6.1)   (-5.9)
Flow Rate = 185.04 -3.44*CO2 + 32808*PrO
(-5.9)   (-12.1)
Appearance = 3.45 + 3.47*CO2 - 0.188*Dis - 0.108*Dis*Dis
(6.7)   (-3.3)   (-2.7)

Where: CO2 =(CO2 Concentration - 23.5), Dis = (Part-to-Gun Distance - 9), TpW = (Tip Width - 6),
PrO = (Preorifice Size - 0.009967)
* Student's t-ratio for the coefficient

All four study variables had a significant impact on the transfer efficiency. The two variables with the largest effect were the fan width and the distance the gun was from the part. The carbon dioxide concentration and the preorifice size influence the transfer efficiency primarily through interactions with distance and fan width respectively. Figures 3 to 6 show that transfer efficiencies greater than 40% can be achieved with gun-to-part distances less than 9 inches and fan width's less than 5 inches. Higher carbon dioxide concentrations lower the transfer efficiency and larger preorifice openings increase the transfer efficiency.

In the experiment parts were hung back-to-back just as they would be in a production plant. Only one part was used as the target to simulate the effects of only one spray booth. Two spray booths that are mirror images of one another are actually used to paint both parts. The amount of paint deposited on the opposing or non-target part was also measured to determine the amount of incidental coverage. This incidental coverage is necessary for conductivity and cosmetics. The data demonstrates that regardless of the transfer efficiency, about 5% of the paint was deposited on the non-target part during each experimental run.

The appearance rating increased with increasing carbon dioxide concentration. There was a strong interaction between the carbon dioxide concentration and gun-to-part distance. Figure 7 shows the predicted appearance for combinations of carbon dioxide concentration and distance. The plot shows that good appearance ratings (4's and 5's) are achieved at a distance of 8 inches and carbon dioxide concentrations greater than 26%.

The flow rate is a function of the preorifice size and the carbon dioxide concentration. Flow rates between 160 and 220 cc/min can be achieved at any acceptable concentration by adjusting the preorifice size. These rates were applicable to the test piece and target production line, but the same trends will apply for higher flow rate applications.

CONCLUSION

Transfer efficiencies greater than 40% can be achieved with a non-electrostatic primer application for robots utilizing the carbon dioxide spray. The results of the designed experiment confirm that indeed there is usually a trade-off between appearance and transfer efficiency, but that the impacts can be predicted. This project demonstrated that full utilization of fan shaping and flow control are critical for maximizing transfer efficiency while maintaining above standard appearance. It is the properties inherent in the carbon dioxide spray process such as narrow particle size distribution, particle size as a function of location in the fan, and particle velocity as a function of location in the fan that allow manipulation of variables to maximize transfer efficiency without sacrificing appearance. Of course these effects become less significant with an electrostatic application but the technology still provides benefits when compared to electrostatic air spray.

Plant trials were conducted to confirm the data generated and to test the predictions. In this trial transfer efficiencies met with the trends for predicted values while appearance characteristics exceeded normal production as follows:

Table 3: Spray Technology Appearance Comparison

Spray Technology	Profilometer	Gloss	Wavescan	Film Build	Conductivity
carbon dioxide	6.9	60	6.0	1.2 mils	162
conventional air	8.0	63	6.3	1.02	154

Transfer efficiency is, without question, influenced by many variables. It is recognizing those variables which are controllable and have the greatest impact that is critical for any new technology to be competitive in the marketplace. The carbon dioxide spray process has proven to be effective for transfer efficiencies thereby reducing VOC emissions by more than the face value of spraying higher solids materials. This is achievable without sacrificing, and in some cases improving, appearance and performance of the current resin system.

REFERENCES

Senser, D.W., et al. "A Comparison Between the Structure of Supercritical Fluid and Conventional Air Paint Sprays". Proceedings of the Seventh Annual Conference on Liquid Atomization and Spray Systems, Bellevue, WA, (May 1994)

BIOGRAPHIES

Jeffrey Goad is a Senior Chemist for Union Carbide. He has 8 years experience with emphasis on applications and has developed commercial products for the wood furniture and automotive plastics industries.

Dr. James Hansen is a Senior Statistician with Union Carbide. He has 22 years experience and his expertise in statistics has been applied to a variety of commercial areas including coatings applications and process chemistry.

Figure 3:   **Contour Plot of Transfer Efficiency**
CO2=21 PreOrf=0.0094

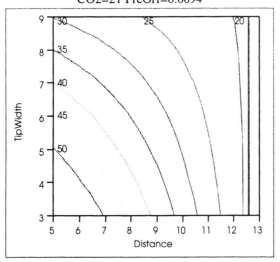

Figure 4: Contour Plot of Transfer Efficiency
CO2=21 PreOrf=0.0108

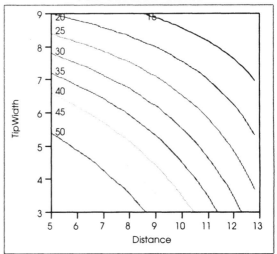

Figure 5: Contour Plot of Transfer Efficiency
CO2=27 PreOrf=0.0094

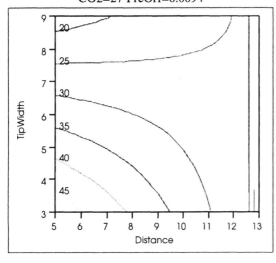

Figure 6: Contour Plot of Transfer Efficiency
CO2=27 PreOrf=0.0108

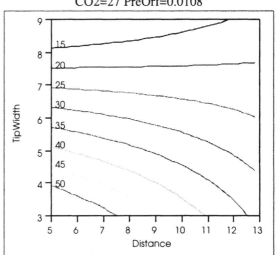

Figure 7: Contour Plot of Appearance

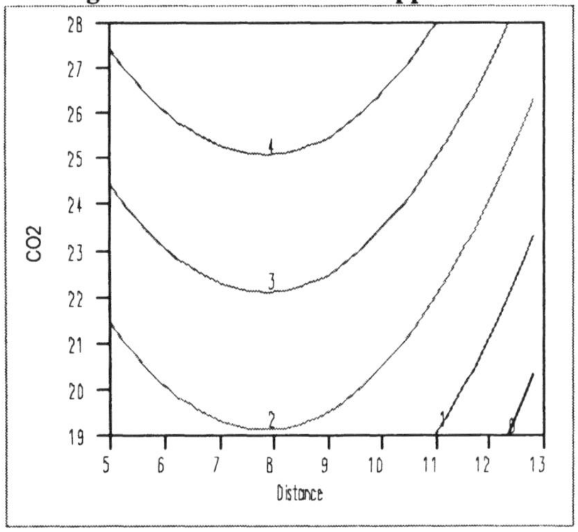

Figure 8: Contour Plot of Flow Rate

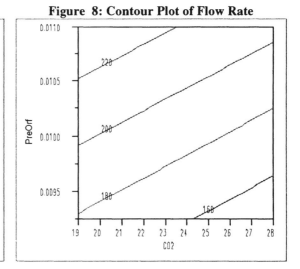

Acrylic BMC for Artificial Marble

HIROSHI KATO, KAZUNORI KOBAYASHI, TAKUJI KOYANAGI AND **RYOZOU AMANO**

ABSTRACT

This report describes a new molding compound called Acrylic BMC for artificial marble and a new method of hot press molding of this material. The Acrylic BMC features the excellent following; because of its MMA as main ingredient, Acrylic BMC has outstanding physical properties of durability, transparency and gloss; Acrylic BMC allows us to process artificial marble into complicated figures as unit structure, for instance, counter top with kitchen sinks; further, the hot press molding insures a higher productivity than the cast molding; technically, the high reaction conversion by hot press molding stabilizes artificial marble quality.

When compared with the artificial marble made from UPE resin, the acrylic resin outperforms UPE resin in terms of various resistance performances of thermal shock, hot water and weathering. However, some demerits inherent in the acrylic resin artificial marble as shown below has discouraged itself to expand its own market in Japan so far: only thirty percent of the market share. These demerits include the following; the casting process induces a lower productivity; the casting process equipment costs are much higher, because we must control molding of the acrylic resin with extraordinary curing reaction and high curing shrinkage rate at curing; further, the acrylic resin is twenty percent more expensive than UPE resin.

Thus, INAX's new material for the acrylic artificial marble and a new method of hot press molding will open a new horizon for measuring up to current consumers' needs and wants for much better function and appearance of sanitary products.

INTRODUCTION

Artificial marble is made of UPE resin or acrylic resin. In Japan, in 1996, the market of UPE artificial marble is thirty billion yen and that of acrylic artificial marble is thirteen billion yen. In general, acrylic artificial marble is better in resistance for discoloration against sunshine and UV-ray and for heat shock than UPE artificial marble. But acrylic artificial marble is only made by process of cast molding

Hiroshi Kato, Kazunori Kobayashi, Takuji Koyanagi and Ryozou Amano, INAX Corporation, 3-77, Minatom-achi, Tokoname, Aichi, 479, Japan.

TECHNICAL POINT FOR ACRYLIC BMC

Technical points for acrylic BMC are as follows.

1. Control of the compound viscosity.

It is necessary to rise the viscosity of compound because you can mold by hot press process. Although we added alkaline earth oxides and hydroxide to acrylic compound, the viscosity does not rise on account of hydroxyl group or carboxyl group is not included to PMMA in resin. So we investigated to add the carboxyl group to PMMA network. The quantity of addition is very important for molding and properties of artificial marble. FIG.1 and FIG.2 shows the increase in quantity of carboxyl group rises up the viscosity, but, it also lowers flexural strength. The degree of the viscosity rising depends on the shape of pressing products. The deep-shaped products require the high viscosity of compound. We decide the proper quantity for each products.

2. Control of the evaporation of MMA monomer.

Styrene Monomer evaporate at 145℃, on the other hand, MMA monomer evaporates at 100℃. For hot press molding like: SMC and BMC, molding temperature is usually from 130℃ to 150℃. Considering the boiling point of MMA monomer, molding temperature should be low down. But we

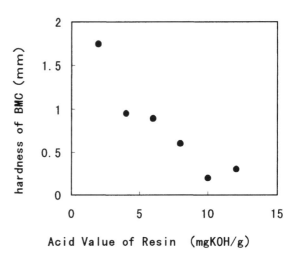

FIG. 1 acid value of resin vs. thickening

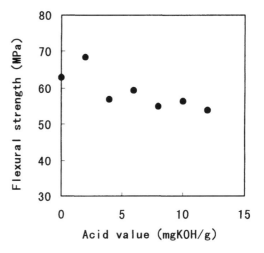

FIG. 2 Relationship of acid value of resin and flexural strength

use the steam to control the MMA molding temperature and the temperature range that we can control is limited more than 100℃. So we set the molding temperature above the boiling point of MMA monomer. We decide the molding temperature is 120℃ at the face side mold and 110℃ at the back side mold. And we limit the degree of vacuum at closing the mold. Further more we add wax, to prevent the evaporation of MMA monomer We have got confirmation the effect of wax against the evaporation.

3 .Control of the flow performance of BMC in mold.

The molecular weight of PMMA in acrylic resin is higher beyond as compared with the alkyd oligomer in UPE resin. So the flow performance of acrylic BMC at molding is worse than BMC of UPE resin. We tried to improve the flow performance of acrylic BMC by the control for syrup formation. FIG.3shows that the higher the molecular weight of PMMA is, the lower flow performance is. We estimate the flow performance by using the degree of gloss. FIG.4 shows the higher the polymer fraction is, the lower flow performance is .

4.Control of shrinkage.

FIG.5 shows the molecular weight of polymer dose not affect the shrinkage of BMC. FIG.6 shows the higher the polymer content of acrylic resin is, the lower the shrinkage rate at curing is. Generally, the shrinkage rate of acrylic BMC is more than that of BMC of UPE.

RESULT AND DISCUSSION

We draw FIG. 7 due to the fact that we confirm at examinations. And we tried to design the syrup formulation and compound formulation. The other hand, as for molding condition, we decided many

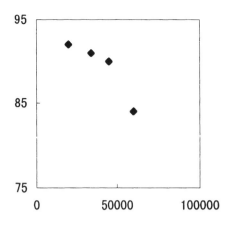

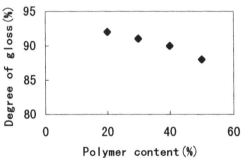

FIG. 4 Polymer content vs. degree of gloss

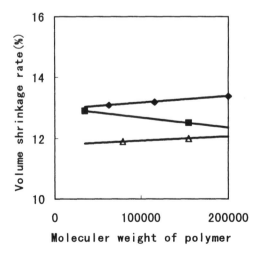

FIG. 5 Moleculer weight of polymer vs. volume shrinkage rate

items that should be controlled, for example, temperature of hot pressing, speed of closing mold, degree of vacuum at closing mold, charge pattern of BMC, etc. In conclusion, for each products, we should arrange the viscosity, the design of the syrup and compound and molding condition. For that purpose it is necessary to clear the requirement and shape of products.

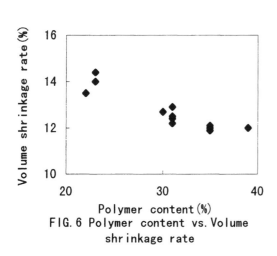

FIG. 6 Polymer content vs. Volume shrinkage rate

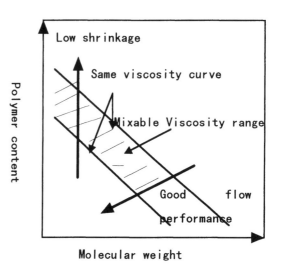

FIG. 7 Design of acrylic resin

Warpage in Injection Molded FRP: Establishing Causes and Cures Using Numerical Analysis

**STEFAN KUKULA, MAKOTO SAITO, NAOKI KIKUCHI,
TADAAKI SHIMENO AND AKIO MURANAKA**

ABSTRACT

Retail price wars in the computer and consumer electronics market have led to massive cost reductions in case manufacture. Injection molded FRP costs have been reduced by using thinner walls. However, this has increased the risk of warpage. Whilst this can be predicted for given conditions, little work has been done on the inverse problem; how to achieve minimum warpage. With product lifetimes as short as six months, reducing time spent in the analysis loop by improving the accuracy of the 'first guess' is critical. A major cause of warpage in short fiber FRP is believed to be anisotropy in thermal expansion coefficients due to local fiber orientation. We have carried out a combined analytical and experimental research program, examining the effects of a wide range of structural and manufacturing variables on warpage of injection molded FRP. These included wall thickness, gate position and mold temperature. The analysis used an in-house system to examine flow induced effects on material properties and mechanical warpage for given injection conditions. These results were compared with samples from a specially designed mold which allowed variation of overall dimensions, thickness, gate position and layout, temperature and molding speed. The aim is to establish a reference catalog for the causes of FRP warpage, identifying preventative design measures and reducing the time needed for design iteration.

INTRODUCTION

Over the past decade representative products of a multi billion dollar business have appeared in the high street shops, and in our pockets and briefcases. If the fifties and sixties were the age of the 'white good,' items for our home to make living easier, then the eighties and nineties are the age of the 'light good,' carryable items for entertainment and work. The personal stereo has been joined by the personal CD player, the electronic personal organizer, the personal phone and the notebook computer. The housings of these items need to be light, yet must withstand the severe wear and tear that personal use, or abuse, can subject them to.

Injection molded FRP has become a popular choice for these items, since it has a good strength to weight ratio, and can be used to form complex shapes. However, as the market has grown, competition has intensified. This has had two major effects.

Firstly, in an effort to reduce material costs, housing walls have become thinner. This increases the risk of warpage. Specialized analysis packages have become available to predict

Stefan Kukula, Makoto Saito, Mechanical Engineering Research Laboratory
Naoki Kikuchi, Tadaaki Shimeno, Polymer, Chemical and Bio Technology Laboratory
Akio Muranaka, Electronics and Information Division, Polymer and Composites Department
Kobe Steel Ltd., 1–5–5, Takatsukadai, Nishi–ku, Kobe, Japan 651–22

warpage in FRP (Zheng et al., 1996), but these require another step in the product design cycle.

Secondly, in order to increase product differentiation in an overcrowded market, constant change has become the order of the day. Product lifetimes of six months or less are not uncommon, and this can put enormous strain on engineering resources, especially in the later stages of the design cycle. The current trend towards concurrent design techniques increases this 'squeezing' effect (Figure 1). For some products the timescales can be compressed still further, from months to weeks, weeks to days, and days to hours.

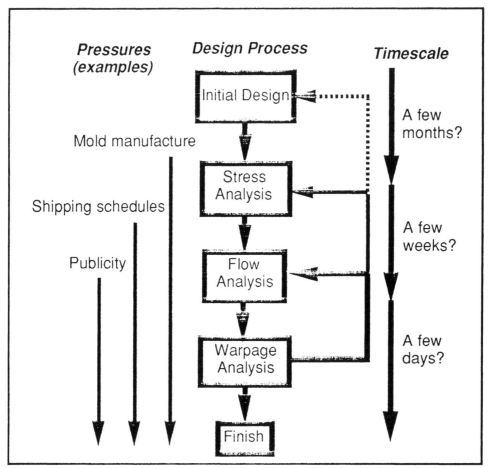

The design process and the analysis squeeze

Figure 1.

Since a closer 'first guess' of a low warpage design will reduce the number of iterations required during the design process, there is clearly still a place for engineering intuition even in the computer age. Surprisingly, however, there are few 'rules of thumb' available for trouble shooting warpage in designs, or guidelines available for producing low–warp designs.

The aim of our current work is to use a combination of experimental moldings produced using a novel adjustable mold, and an analysis method developed in-house to investigate the relationships between factors causing warpage, and produce guidelines for designers who wish to produce low–warp designs.

SIMULATION

PREDICTION OF WARPAGE

Warpage occurs in both reinforced and unreinforced injection molded plastics. In unfilled plastics the main causes are differential cooling, molecular changes such as molecular orientation through flow and crystallinity, and mold restraints. For fiber reinforced plastics the major cause of warpage is anisotropy in thermal expansivity due to flow induced orientation of the fibers (Zheng et al., 1996), although the factors important in unfilled polmers also play a part. Thus an understanding of the causes of warpage in FRP requires an understanding of the factors affecting fiber orientation. These are broadly summarized in figure 2.

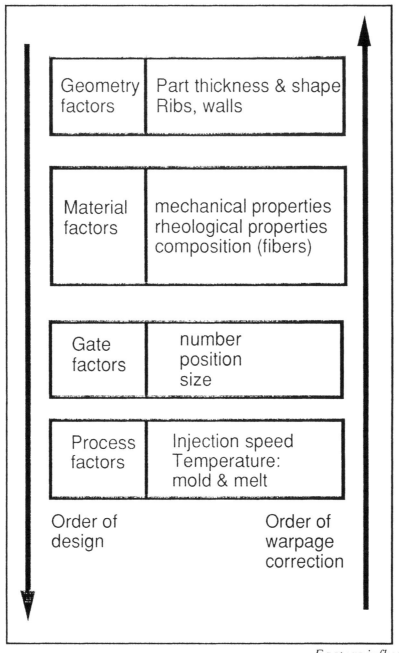

Factors influencing warpage

Figure 2.

It is noticeable that out of the factors which influence fiber orientation, and hence warpage, those which are the first to be fixed during the design stage are often the last to be altered when looking for solutions to warpage problems. There appears to be a requirement to get them as close as possible to optimal at an early stage of the design cycle.

Whilst computer modeling using finite element and finite difference techniques has transformed many areas of engineering, it should be remembered that the full name of both methods also includes the word 'approximation.' Warpage analysis especially requires a chain of assumptions and approximations which could lead to severe accuracy problems if one link in the chain is incorrect. For example,

✳ *Warpage analysis* requires	local property prediction and assumes	a dominant warpage mechanism, correct choice of linear elastic, plastic or buckling analysis
✳ *Local property prediction* requires	prediction of fiber orientation and assumes	a relationship between local fiber orientation and mechanical properties e.g. (Hull, 1987)
✳ *Prediction of fiber orientation* requires	prediction of material flow direction and assumes	a relationship between flow and fiber orientation e.g. (Folgar and Tucker, 1984)
✳ *Prediction of flow* requires	prediction of material properties under process conditions and assumes	a model for flow behavior, such as shear flow

ANALYSIS METHOD

The software used is described in detail in (Kukula et al., 1996), and is based on the statistically modified laminate model. Basically, the flow analysis results from a standard commercial injection molding simulation package are used to build up, element by element, a structural model suitable for input into a commercial structural analysis package. The rules and assumptions used for building the model are derived experimentally. The molding is modeled as a three layer skin–core–skin laminate, with the fibers aligned parallel to the flow in the skin layers, and perpendicular to the flow in the core layer. The distribution of fibers within these layers is a statistical constant value derived from experiments. The version of the code described in (Kukula et al., 1996) assumed a constant layer thickness; the current version allows the surface layer thickness to vary according to molding thickness and other factors, as determined experimentally. Although initially intended to determine variations in stiffness due to flow induced anisoropy, the anisotropy in thermal expansion coefficient can be modelled the same way (Datoo, 1991), and coupled with an analysis of the cooling of the part after removal from the mold to analyze warpage.

RESULTS

As described earlier, results obtained from warpage analyses are extremely dependent on assumptions made at each stage of the process. In order to gain understanding of the assumptions made, initial analyses considered purely theoretical variations in material properties and geometry.

For example, figure 3 below shows the change in warpage assuming different thicknesses of skin layer. This would correspond to the effects of different material viscosities and changes in injection speed varying the depth of the aligned region of the material.

Figure 4 shows the effect for one example material of moving a central gate towards one corner.

Figure 5 demonstrates the difference between results assuming a linear elastic warpage behavior and buckling behavior. Both deformed shapes have been observed in practice, for different materials. It also appears that moving the position of the injection gate can, in some cases, change the mode from linear elastic to buckling, as shown in figure 7.

We have now obtained material data from the first series of experiments using the test mold described in the next section, and future results will be compared with experimental observations.

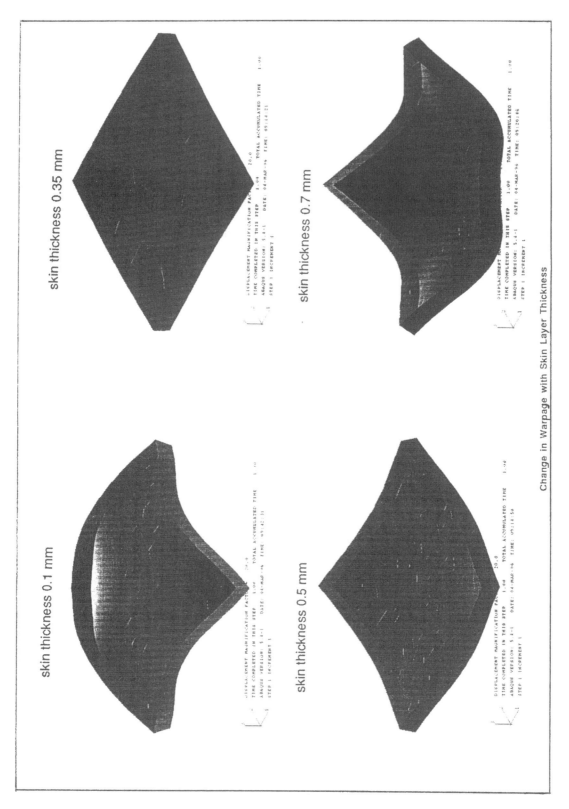

Warpage for centrally gated case, with different skin thicknesses assumed (total thickness 1.5 mm)

Figure 3.

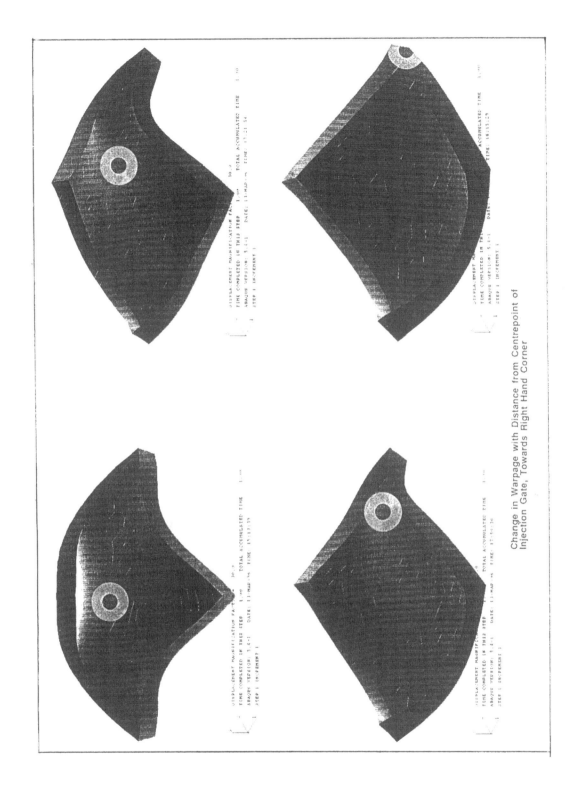

Effect of varying gate position (white ring shows gate position)
Figure 4.

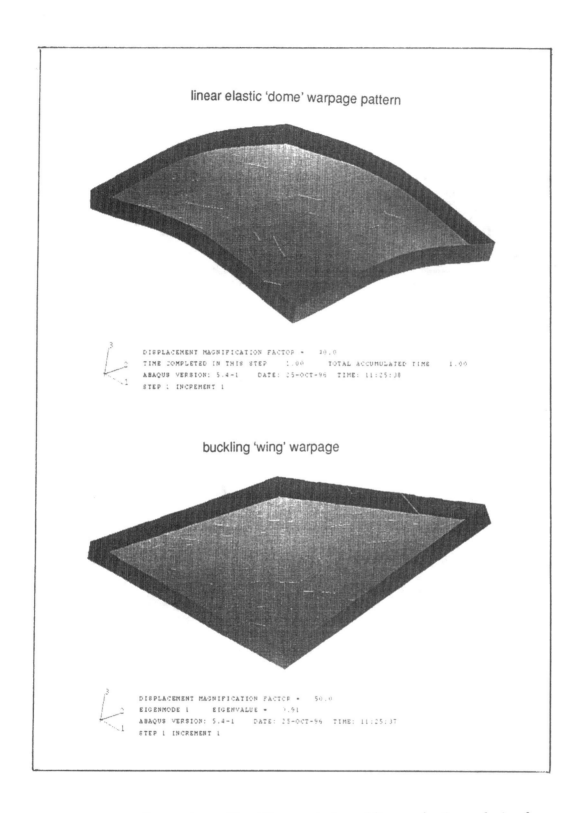

linear elastic 'dome' warpage pattern

DISPLACEMENT MAGNIFICATION FACTOR = 30.0
TIME COMPLETED IN THIS STEP   1.00   TOTAL ACCUMULATED TIME   1.00
ABAQUS VERSION: 5.4-1   DATE: 25-OCT-96 TIME: 11:25:38
STEP 1 INCREMENT 1

buckling 'wing' warpage

DISPLACEMENT MAGNIFICATION FACTOR = 50.0
EIGENMODE 1   EIGENVALUE = 1.91
ABAQUS VERSION: 5.4-1   DATE: 25-OCT-96 TIME: 11:25:37
STEP 1 INCREMENT 1

Comparison of buckling analysis and linear elastic analysis of **warpage**
Figure 5.

MOLDING

MOLD

The test moldings carried out used a specially designed mold of a simple shallow box, which is a common shape for computer cases, shown in figure 6. As well as varying the process conditions and material, this novel mold allowed us to vary the part thickness, part wall height, shape (square/rectangular) and the number, size and location of gates.

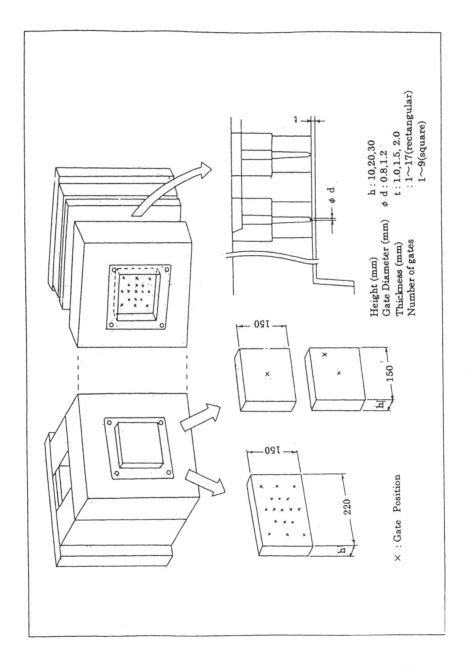

Diagram of test mold

Figure 6.

CASES

A wide range of molding conditions, materials and geometries have been used to test both warpage characteristics and moldability. Examples of test series are given in the table below.

TABLE I — TYPICAL TEST SERIES

Mold geometry 150 mm x 150 mm x 10 mm wall
Gate Position: 'a' is center gated, 'b' is off-center gate

Material	Thickness mm	Gate Position	Gate Diameter mm	Cylinder Temp. °C	Mold Temp. °C
PA66/GF30% wt	1.5	a	1.2	295	80
		b			
	2.0	a			
	2.0	b			
	3.0	a			
	3.0	b			
PC/GF30%wt	1.5	a+b	1.2	320	95
	2.0				
	3.0				
PP/GF20%wt	1.0	a	0.8	245	50
	1.5				40
	1.5				50
	1.5				60
	2.0				50
	3.0				50

RESULTS

Despite the relative simplicity of the mold compared to practical designs a wide variety of warpage types has been observed. Typical are the 'dome' shaped warpage and the 'wing' shaped warpage.' Moving the gate can change one form of warpage into the other, as shown in figure 7.

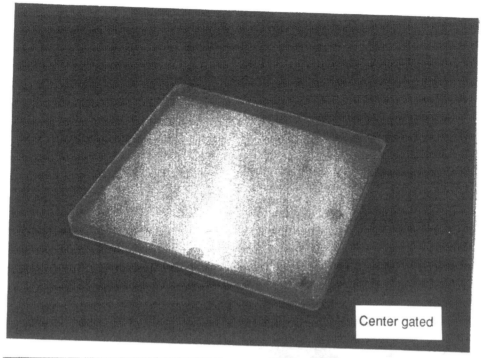

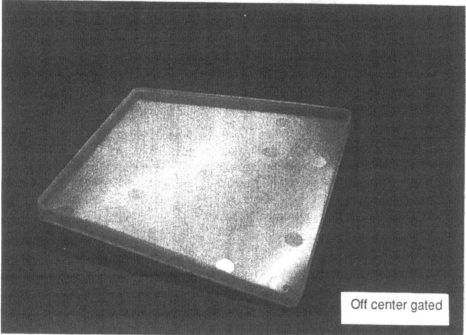

Warpage for central and off–center gating

Figure 7.

CONCLUSIONS

The warpage of injection molded FRP is clearly a complex phenomenon dependent on a wide range of factors. Prediction using computer analysis techniques is possible, but takes place at a late stage in the design cycle, when major alterations may not be practical.

The adjustable test mold has proven to be a valuable tool, both in demonstrating the variation of warpage with material, geometry and process conditions, and in generating material data for use in analysis programs. Using the results of the experiments as a guide, we are using the derived material properties in analyses tailored to each particular case. We have found that dropping the assumption that all materials follow the same rules in all cases allows more accurate prediction of warpage modes, although it does mean that a systematic method of building up experience must be followed.

At the moment the program is in an early stage, but we believe that our systematic testing, coupled with computer analysis predictions, should help to make future analyses more accurate. We hope to reduce the number of design iterations needed to reduce warpage to acceptable levels for thin walled injection molded products by providing a case by case basis for comparison, and work towards establishing a set of guidelines for warp resistant design.

REFERENCES

Folgar, F.P. and C.P.Tucker. 1984. "Orientation Behavior of Fibers in Concentrated Suspensions," *Journal of Reinforced Plastics*, 3:98

Datoo, M.H. 1991. "Mechanics of Fibrous Composites," Elsevier Science Publishers Ltd.

Hull, D. 1987. "An Introduction to Composite Materials," Cambridge University Press, London.

Kukula, S., M.Saito, Y.Kataoka and T.Nakagawa. 1995. "The Statistically Modified Laminate Model for Fiber Reinforced Injection Moldings — A Practical Way to get Flow and Structural Analysis Packages on Friendly Terms," 51st Annual Conference, Composites Institute, The Society of the Plastics Industry, Session 21–B: 1–22

Zheng, R., N.McCaffrey, K.Winch, H.Yu and P.Kennedy. 1996. "Predicting Warpage of Injection Molded Fiber–Reinforced Plastics," *Journal of Thermoplastic Composite Materials*, Vol. 9: 90–106

BIOGRAPHIES

STEFAN KUKULA

Stefan Kukula graduated from Cambridge University, U.K., with a B.A. in Engineering. After working in the U.K. nuclear power industry he entered the Imperial College of Science, Technology and Medicine, London, receiving an M.Sc. in Composites from the Centre for Composite Materials. and a Ph.D. from the Aeronautics Department for work on modeling crack growth in fiber reinforced plastic laminates. He joined Kobe Steel Europe Ltd. in Guildford, U.K., in 1992, transferring to the Mechanical Engineering Research Laboratory of Kobe Steel Ltd. in Kobe, Japan, in March 1993. His particular interest is the optimization of structural performance in fiber reinforced plastic products.

MAKOTO SAITO

Makoto Saito graduated from Osaka University in Solid State Physics with the M.S. degree. Upon graduating, he joined Kobe Steel. Before his current job, he was engaged in determining the reliability of metal materials using fracture mechanics. Also, he has experience in the development of a self-actuated shutdown system for a fast breeder reactor at the Power Reactor and Nuclear Fuel Development Corporation between 1985 and 1989. He is currently a senior researcher at the Mechanical Engineering Laboratory and partially engaged in structural analysis of composite products.

NAOKI KIKUCHI

Naoki Kikuchi joined Kobe Steel after graduating from Kyushu University with an M.S. degree in Physics. He has been engaged in fiber reinforced plastic (FRP and FRTP) processing research and development for about ten years and spent one year researching polymer electrolytes in secondary lithium batteries at the University of Pennsylvania in 1993. He is currently a researcher in the Chemical, Polymer and Bio Technology Laboratory, engaged in material development and application of flow/warp analyses to notebook type personal computer housings.

TADAAKI SHIMENO

Tadaaki Shimeno graduated from the University of Tokyo in Mechanical Engineering. He has been engaged in fiber reinforced plastic processing research for about six years at Kobe Steel. He is currently a researcher in the Chemical, Polymer and Bio Technology Laboratory, engaged in material development and application of flow/warp analyses to notebook type personal computer housings.

AKIO MURANAKA

Akio Muranaka graduated from Hiroshima University in Chemical Engineering. He has been engaged in processing research and design of mold tooling for fiber reinforced plastic for about seven years at Kobe Steel. He also had experience of designing products using fiber reinforced plastic before joining Kobe Steel. He is currently a technical manager in the Electronics and Information Division, Polymer and Composites Department of Kobe Steel.

Resins and Additives: Components for Achieving High Performance Composites

High Performance Modified-Phenolic Piping System

JOIE L. FOLKERS and RALPH S. FRIEDRICH

ABSTRACT

Many different resin systems are used to fabricate fiberglass pipe, including epoxies, vinylesters, polyesters, furan and others. A resin which has been of particular interest to both producers and users of fiberglass pipe has been phenolic. This interest has been generated primarily by the fire endurance and combustion properties of the phenolic resins. To date, the practical use of phenolic resin has been limited due to handling, curing, bonding and pressure performance of the finished product.

A breakthrough in technology has occurred from the combined efforts of engineers and chemists within Ameron which has the potential to significantly broaden the use of phenolic resin usage, particularly in the piping industry. This technological advance has been made possible by the modification of phenolic resin with polysiloxane.

Typical problems associated with straight phenolic systems include difficulty in controlling the cure, water content and its effect in the cured resin, brittleness and poor ability to bond to the laminate. The polysiloxane modified phenolic system has shown characteristics of lower water content, higher elongation, improved weathering and improved adhesive bond strength. Process optimization has also been conducted, along with some novel product development, to enhance the fire and pressure performance of the pipe system. The modification has not degraded, and in many cases improved, the chemical resistance, fire resistance and low smoke and toxicity characteristics inherent in the base phenolic system.

Phenolic pipe is now ready to perform the way designers and engineers have hoped it would be able to do. This paper describes the resin and pipe system which has been developed and the testing which has been performed to characterize this product.

RESIN DEVELOPMENT

Phenolic resins are made by reacting phenol and formaldehyde in the presence of a catalyst. An excess of one or the other of the materials is characteristic, along with the

Joie L. Folkers, Ameron International, 1004 Ameron Road, Burkburnett, TX 76354
Ralph S. Friedrich, Ameron International, 4671 Firestone Boulevard, South Gate, CA 90280

selection of an acidic or an alkaline catalyst to product either a phenolic novolac resin or a phenolic resole resin. For filament winding applications, where a low viscosity resin is required, the phenolic resole is the only option.

Unfortunately, phenolic resoles contain a significant percentage of water and are thus lower in performance than the phenolic novolacs. In addition, since the phenolic reaction is a condensate reaction, additional water is generated as the resin cures.

By proper selection of the polysiloxane and suitable processing, the phenolic resole can be modified to produce a higher performance polymer. This modification can be done by either a mix process or by an in-situ process. Due to the low viscosity of the polysiloxane, a lower initial water content of the resole resin can be used. An added benefit is that the siloxane reaction takes place more rapidly in the presence of water. A truly symbiotic relationship exists between the components of the modified resin.

PERFORMANCE CHARACTERISTICS OF PHENOLIC RESINS

The identified advantages of phenolic resins as they apply to composites are as follows:

- Excellent high temperature resistance
- Dimensional stability
- Low smoke generation and toxicity in fire
- Good thermal insulator
- Low viscosity (resoles)

The identified disadvantages of phenolic resins are as follows:

- Poor bondability
- Brittle, resulting in low impact resistance and pressure performance
- High water content, resulting in laminate implosion when heated
- Low bonding or wetting compatibility with fibers

As previously indicated, the polysiloxane modification has improved the characteristics of the phenolic resins. The characteristics of the polysiloxane add to the performance level of the phenolic. The two most significant of these are the compatibility of the material to fiber wetting and heat resistance in extreme conditions. The resulting improvements are realized in the modified resin:

- Adhesion with epoxy and modified epoxy adhesive
- Burst performance of pipe
- Impact resistance
- Heat resistance (particularly above 800°C)
- Processing and cure characteristics

The siloxane modification serves to flexibilize and toughen the phenolic matrix while not degrading or improving its physical performance.

PRODUCT CHARACTERIZATION

ADHESIVE PERFORMANCE

To quantify the improved bonding seen with the siloxane modified material, a series of lap shear tests were performed with laminate coupons with both modified and unmodified resins. Epoxy adhesives, both modified and unmodified with siloxane was also used. Table I shows the results of this testing and the dramatic improvement realized.

TABLE I

Laminate Material	Adhesive Material	Average Shear Strength (psi.)	Standard Deviation	Mode of Failure
Unmodified Phenolic	Unmodified Epoxy	580	160	Adhesive
Modified Phenolic	Unmodified Epoxy	924	57	Cohesive
Unmodified Phenolic	Modified Epoxy	731	90	Adhesive
Modified Phenolic	Modified Epoxy	1046	93	Cohesive

Not only did the shear strength of the bond improve, the standard deviation was lowered and the mode of failure was shifted from being associated with the bond to the adhesive itself.

PIPE PRESSURE PERFORMANCE

In short-term pressure tests conducted according to ASTM D1599, hoop stress values over 50,000 psi. have been realized with the modified phenolic matrix. This performance has been seen on unlined pipe specimens. Many of the epoxy pipe products on the market do not have this level of performance.

IMPACT RESISTANCE

The impact resistance of composite pipe is a somewhat difficult property to measure. Test methods vary with different profiles of the impact "tup" and what is defined to be the failure of the specimen. For the testing performed in this study, a 2.4 inch diameter steel ball weighing 2 pounds was dropped from various heights. After impact, the pipe specimen was pressurized to 25 psi. with air and immersed in water to detect leaks. This method is used as a "worst case" detector or damage. As is seen in Table II, a 40% increase in impact resistance is realized.

TABLE II

Matrix Material	Drop Height (in.)	# Leaks / Tests
Unmodified Phenolic	20	0 / 4
	25	0 / 4
	30	3 / 4
	35	4 / 4
	40	4 / 4
Modified Phenolic	20	0 / 4
	25	0 / 4
	30	0 / 4
	35	0 / 4
	40	2 / 4

OTHER PROPERTIES

Material properties typically required for qualification of the piping system have also been defined. These include tensile strength and modulus, Poisson's ration, coefficient of thermal expansion, thermal conductivity and others. Each of these properties is dependent on fiber orientation and laminate construction. Details of these characteristics would only be pertinent to the specific construction for each composite so will not be detailed in this paper. It is noted, however, that the interface between the modified phenolic resin and the fiber is not the governing factor in establishing the physical property performance. Those properties which are fiber dominated in typical epoxy products are fiber dominated with the modified phenolic. Likewise for matrix dominated properties.

Weathering resistance of phenolic composites is also enhanced by this modification. In accelerated exposure testing, gloss retention of the modified material is seen to be above 80% as compared to less than 60% for unmodified material under the same conditions.

FIRE PERFORMANCE

The most significant aspect of the modified phenolic piping system is its resistance to and performance in fire. This feature is important in applications such as offshore oil production platforms where corrosion resistance, light weight and fire safety are major considerations in design and economic decisions. Corrosion by-products within metallic piping systems have caused system failures when nozzles have been plugged. Exposure to sea water of most standard piping materials is so damaging that its use is precluded. Alloy steels or other metals, many of which have melting points below the temperature of the fire, have been the only options to this point for systems which have the potential to be exposed to fire insult. Fiberglass piping has been used in non-critical service on ships and on platforms, but has not been extensively been used for fire protection. U. S. Coast Guard regulations have disallowed its use, until now, for fire protection piping.

IMO LEVEL 3 TESTING

The International Maritime Organization has established test methods for fire evaluation for shipboard piping. The governing document for this work is Resolution A.753(18). This specification, adopted nearly identically by ASTM (F1173-95), gives the fire characteristics and endurance requirements for normally dry and normally filled piping systems. The modified phenolic piping system has been tested and certified by *Det Norske Veritas* to have met the requirements for Level 3 applications.

JET FIRE TESTING

In oil and gas production areas where flammable gas is stored under pressure in a vessel, the most severe fire scenario potential exists, namely a jet fire. Piping which would be exposed to this type of fire insult must be able to withstand a high level of heat flux as well as the flame temperatures of the fire itself. Agencies such as Lloyd's Register, SINTEF Energy in Norway and UKOOA (United Kingdom Offshore Operators Association) have contributed to a fire testing protocol to measure pipe performance in this situation.

An enhance version of the modified phenolic piping system has been developed for such cases. In order to provide the thermal insulation for this pipe, a series of alternating layers of composite material and thermoplastic tape are wound over the standard pipe. This construction provides a disbondment mechanism for the outer layers. When the fire is engulfing the jacketed pipe, a char is formed in the exterior layer of composite. This provides a first means of insulation. As heat is absorbed by the body of the pipe, the outer layer of thermoplastic tape melts and vaporizes, leaving an air gap between the outer and the next layer of composite. This also provides an insulation surface for the inner pipe, along with the heat absorbed through the vaporization process. This phenomenon is repeated for the consecutive layers of the pipe.

Typically, the pipe is required to endure a 5 minute period in the jet fire without any fluid inside the exposed pipe. After this period, flowing water is introduced. At this point, degradation is virtually halted due to the heat transfer made possible by the water flow.

The low thermal conductivity and capacitance of fiberglass composites is seen as an advantage in this situation as well. Consider the effect of a 1200°C propane fire on a metallic, dry piping system. The entire wall thickness of the pipe would readily (in less than 1 minute) reach the fire temperature. Should the pipe survive structurally, this red-hot pipe will now have water pumped into it for the purpose of extinguishing the fire. Until an adequate amount of water is introduced to drop the temperature of the pipe below 212 °F, the potential exists to create a flash-boiler of the pipe, causing an instantaneous pressure spike and potential explosion of the system. In the best case, high pressure steam and scalding hot water will be spewing from the deluge nozzles for the first few moments of the life of the system.

CONCLUSION

A practical, high performance, phenolic composite piping system has been made possible by a breakthrough in resin technology involving the modification of phenolic resole resin with a poly-siloxane. This pipe system will have particular application in areas where corrosion resistance, light weight and low smoke and toxicity characteristics are important. Certified testing has been conducted and proven the high level of performance for this system. Regulatory work is being conducted to expand the allowable scope of composite materials due to the enhanced characteristics of this product. Application of this technology to others areas where phenolic resins are attractive is anticipated.

REFERENCES

Friedrich, R. S. 1996. "Bondstrand® PSX® Piping for Offshore Fire Protection", Ameron International, South Gate, CA

Folkers, J. L. 1996. "Fire Endurance Testing of Bondstrand® Series 2000M-FP and Siloxane Modified Phenolic Piping According to International Maritime Organization (IMO) Resolution A.753(18), Adopted 4 November 1993 and A. S. T. M. F1173-95".

1996. "Investigation of the Toxic Potency of the Combustion Products of Phenolic-Siloxane Pipe in Accordance with Article 15, Part 1120, New York State Uniform Fire Prevention and Building Code, Final Report, Southwest Research Institute, San Antonio, TX.

BIOGRAPHIES

Joie L. Folkers is the Engineering Manager of the Fiberglass Pipe Division - USA for Ameron International. He received his B. S. M. E. from Northwestern University in 1975. Questions or requests to him can be sent to Ameron International, 1004 Ameron Road, Burkburnett, TX 76354. Telephone number: (817) 569-8646, Fax number: (817) 569-2764.

Ralph S. Friedrich is the Vice-President of Corporate Research and Engineering for Ameron International. He received his B. S. M. E. from Cal State - Fullerton in 1971. Questions or requests to him can be sent to: Ameron International, 4671 Firestone Boulevard, South Gate, CA 90280. Telephone number: (213) 357-6760, Fax number: (213) 357-6850.

Novel Thermally Triggered Catalyst/Promoter Additives for Unsaturated Polyester, Vinyl Ester and Epoxy Resin Cure

STEVEN P. BITLER, MARK A. WANTHAL, DAVID A. KAMP,
PAUL A. MEYERS AND DAVID D. TAFT

ABSTRACT

Novel side-chain crystalline (SCC) polymer additives have been developed incorporating catalyst or promoter functionality that can be thermally triggered at preset activation temperatures. These SCC polymer additives provide shelf-life and pot-life extension for a variety of thermoset systems including:

(a) cobalt/other metal accelerators for unsaturated polyesters or vinyl esters

(b) imidazole and tertiary amine catalysts for epoxies, and

(c) tin and tertiary amine catalysts for polyurethanes.

These SCC polymer additives provide a controlled catalyst/promoter thermal triggering mechanism tunable within 1°C of any temperature between 0-100°C. Conventional and new catalysts/promoters can be physically or chemically bound to these SCC polymers providing temperature controlled catalyst/promoter delivery. These polymer additives are available as small particle size (< 10 micron) easy to disperse powders.

Unsaturated polyester formulations with thermally triggered cobalt polymer additive and MEK peroxide exhibit dramatically extended room temperature pot-life while yielding comparable 80-120°C cure times, Tg and modulus values at both 30 and 120°C versus cobalt naphthenate control. Latent thermally triggered polymer additive catalysts with 15-30% imidazole moieties, melting between 50-100°C allow economic epoxy system formulation with 1-6-month 40°C shelf stability, 80-120°C reactivity and comparable mechanical properties versus imidazole control.

These latent catalyst/promoter polymer additives provide shelf-life and pot-life extension opportunities in shop and field applications for SMC, BMC, reinforced injection molding, filament winding/pultrusion, and other polyester, vinyl ester, and epoxy resin composite systems.

Landec Corporation, 3603 Haven Avenue, Menlo Park, CA 94025

INTRODUCTION

Side-chain crystalline (SCC) polymers marketed under the trade name *Intelimer®* Polymer Additives have been developed that exhibit sharp, well defined, reproducible melt transitions. These tailored melt transitions function as thermal switches that control catalyst or promoter availability in a variety of systems including thermosets. Thermal switch control has been designed into these materials through a process where aliphatic side-chains attached to the polymer main chain crystallize at preset temperatures that can be adjusted by varying the length of the aliphatic side-chains. Figure 1 shows the melt temperature variation for a series of acrylic polymers where the side-chain length has been varied from 12 to 22 carbons resulting in sharp melt temperature control from 0 to 68°C respectively. Figure 2 shows more precise temperature switch control as a function of varied comonomer content in preparing the SCC polymer. In this case, an acrylic copolymer was prepared from octadecyl acrylate (18 carbon side-chain) and docosanyl acrylate (22 carbon side-chain) monomers exhibiting tailored melt temperatures along the continuum from 48 to 68°C as a function of the monomer ratio. Figure 3 shows a DSC thermograph of an SCC polymer exhibiting a melt onset at 53°C and peak at 56°C. The materials are unique due to their sharp melt transitions and ease of melt temperature control allowing sharp thermal transitions custom designed for specific applications. Figure 4 shows a pictorial representation of an SCC polymer in a solid or crystalline state that upon heating to the switch temperature becomes a molten fluid. This thermal switch from crystalline solid to molten fluid can be used in product applications including (a) gating active ingredient (e.g. catalyst) availability or release during the transition from low permeability to high permeability barrier, (b) triggering adhesive to nonadhesive character and (c) influencing viscous flow in response to temperature resulting in process improvements.

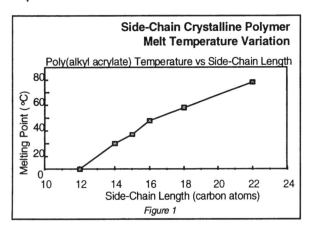

Side-Chain Crystalline Polymer Melt Temperature Variation

Poly(alkyl acrylate) Temperature vs Side-Chain Length

Figure 1

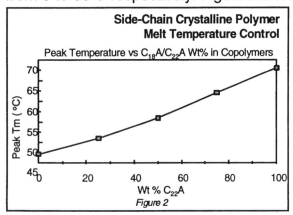

Side-Chain Crystalline Polymer Melt Temperature Control

Peak Temperature vs $C_{18}A/C_{22}A$ Wt% in Copolymers

Figure 2

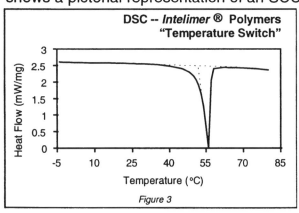

DSC -- *Intelimer* ® Polymers "Temperature Switch"

Figure 3

These *Intelimer®* materials can be applied to many thermoset chemistries through incorporation of a wide range of catalysts and promoters that can be made available to accelerate cure reactions in response to temperature.

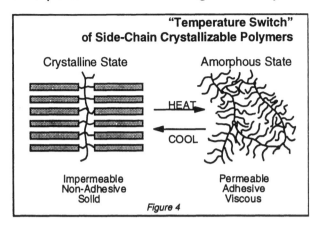

"Temperature Switch" of Side-Chain Crystallizable Polymers

Crystalline State   Amorphous State

HEAT →
← COOL

Impermeable Non-Adhesive Solid   Permeable Adhesive Viscous

Figure 4

Application of these SCC polymers to thermoset systems has been accomplished through development of latent catalyst materials that exhibit improved pot-life and/or long shelf-life at the mix or storage temperature while delivering fast reactivity at cure temperature. This has been applied across the broad thermoset industry including:

Thermoset Chemistry	Latent Catalyst/promoter
Unsaturated Polyester	cobalt, tertiary amine
Vinyl Ester	cobalt, tertiary amine
Epoxy	imidazole, tertiary amine
Polyurethane	tin, tertiary amine

These materials have been designed as (a) physical mixtures of catalyst/promoter with SCC polymers designed to extend pot-life or (b) polymer bound catalyst/promoter (PBC) materials designed to improve pot-life and/or shelf-life. The physical mixtures have catalysts/promoters locked in the crystalline array resulting in dramatic reduction in catalyst/promoter release below the switch temperature compared to release above the switch temperature. The polymer bound catalyst/promoter systems have the active catalyst/promoter chemically attached to the SCC polymer to prevent catalyst/promoter leakage into the reaction medium. Figure 5 shows a pictorial representation of the SCC-PBC system where (a) the SCC polymer portion controls switch temperature, viscous flow, polarity, hydrophobicity, etc. (b) the group linking the catalyst to the SCC polymer influences the system through its polarity and its potential to coaccelerate the thermoset reaction and (c) the choice of catalyst influences the reaction speed and selectivity.

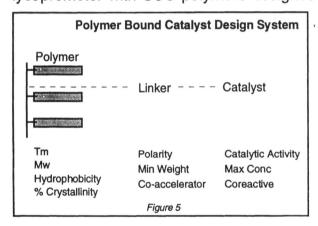

Polymer Bound Catalyst Design System

Polymer

– – – – – – Linker – – – Catalyst

Tm	Polarity	Catalytic Activity
Mw	Min Weight	Max Conc
Hydrophobicity	Co-accelerator	Coreactive
% Crystallinity		

Figure 5

We expect that these materials will find thermoset application by allowing formulators and users to develop systems that utilize:

(a) the same level of catalyst/promoter as used today resulting in extended pot-life and/or shelf-life while maintaining similar reactivity at elevated temperature,

(b) higher levels of catalyst/promoter than used today resulting in dialed-in improvements in pot-life or shelf-life but dramatic improvement in reactivity at the cure temperature, and

(c) catalysts/promoters that can not be used today due to instability at mix temperature but through the use of this new technology exhibit workable pot-life and dramatic improvement in cure rate at cure temperature.

RESULTS AND DISCUSSION
UNSATURATED POLYESTER (UPE) AND VINYL ESTER (VE) SYSTEMS

Development of UPE and VE resin systems that exhibit improved pot-life at 25°C while exhibiting equivalent reactivity at cure temperature can be accomplished through the use of latent cobalt, copper, vanadium, manganese, tertiary amine and other promoters depending on the choice of organic peroxide. These developments may result in:

- extension of pot-life from minutes to hours/days
- improved reaction control during processing
- faster mold cycles
- reduced styrene oxidation to benzaldehyde
- alternatives to β-diketones

Figure 6 shows a pictorial description of a polymer having aliphatic side-chains and polymer attached cobalt that exhibits improved pot-life while accelerating reactivity at cure temperature.

Table 1 provides UPE formulation data designed to test improvements resulting from new latent cobalt vs free cobalt naphthenate. In this test, Ashland Chemical's Aropol 2036 was utilized (31% styrene) with methylethylketone peroxide (MEKP). We observe a dramatic improvement in pot-life at 25°C from 16 minutes for

TABLE 1.   Cure of Unsaturated Polyester with MEKP and Cobalt Accelerators

	A	B	C	D	E
Aropol 2036 (31% Styrene)	100g	100g	100g	100g	100g
MEKP	1.25g	1.25g	1.25g	1.25g	1.25g
Co Naphthenate (6% Co)		0.6g			
Co-SCCP matrix (3.5%)			1.0g		
Co-PBC (4.5% Co)				0.8g	
Co-PBC (2.3% Co)					1.56g
Gel time @ 20 °C, min	8200	16	250	2700	2600
DSC					
Time to 50% cure @ 80 °C	37	2.5	2.5	3.1	3.0
Time to 90% cure @ 80 °C	75	17	15	11	9.0
RDA					
G' @ 120 °C, dynes/cm^2	3.5E+7	5.9E+7	5.5E+7	5.7E+7	5.6E+7
G' @ 30 °C, dynes/cm^2	7.2E+7	7.7E+7	5.7E+7	6.5E+7	6.2E+7
Tg (tan delta), C	107	124	122	123	123

free cobalt naphthenate to over 2500 minutes in the formulation with SCC polymer bound cobalt while delivering accelerated reactivity at 80°C as measured by DSC analysis and rheological cure analysis on a Rheometrics Dynamic Analyzer. A key observation is the greater than 150 times improvement in pot-life at 20°C where the latent cobalt system pot-life begins to approach the pot-life of the no cobalt control. The latent cobalt cured samples were observed to exhibit similar elastic modulus and tan delta (approximating Tg) compared to free cobalt naphthenate control. Figure 7 shows the relative cure rate for an UPE system with resin, MEKP and one of the following (a) no cobalt promoter, (b) free cobalt naphthenate, or (c) SCC polymer based latent cobalt. The cobalt containing systems exhibit accelerated reaction rate compared to the no cobalt control. All systems achieved similar final elastic moduli in this test. The SCC based latent cobalt system combines the benefits of fast reactivity at cure temperate with pot-life stability. In general, cobalt naphthenate is supplied as a solution in mineral spirits containing 6% cobalt. For this reason, we have designed most of these latent cobalt materials with 2-6 wt% cobalt though we have prepared systems containing 0.5 to 14 % cobalt.

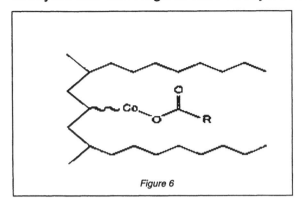

Figure 6

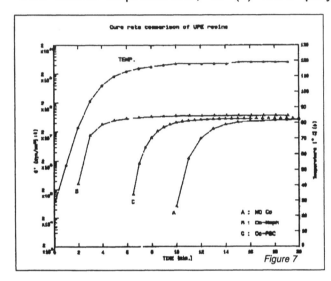

Figure 7

Table 2 provides VE formulation data. In this test, Dow Chemical's Derakane 411-350 (45% styrene) and D470-300 (35% styrene) were used with MEKP. Pot-life improvements compared to cobalt napthenate control at 20°C from 27 minutes to over 2600 minutes (411-350 system) and 70 minutes to over 8000 minutes (D470-300 system) were observed resulting in about 100 times improvement in pot-life while maintaining comparable reactivity at

TABLE 2.	Cure of Vinylester Resins with MEKP and Cobalt Accelerators							
	A	B	C	D	E	F	G	H
Derakane 411-350 (45% Styrene)	100g	100g	100g	100g				
Derakane 470-300 (35% Styrene)					100g	100g	100g	100g
MEKP	1.25g	1.25g	1.25g	1.25g	1.25g	1.25g	1.25g	1.25g
Co Naphthenate (6% Co)		0.6				0.6		
Co-PBC (4.5% Co)			0.8g				0.8g	
Co-PBC (2.3% Co)				1.56g				1.56g
Gel time @ 20 °C, min	>7000	27	3000	2600	>7000	70	>8000	>8000
DSC								
Time to 50% cure @ 120 °C	7.4	4.9	2.6	3.9	7.6	2.3	2.8	1.4
Time to 90% cure @ 120 °C	17	20	9	15	20	18	19	14
RDA								
G' @ 120 °C, dynes/cm²	1.1E+6	6.4E+6	2.0E+7	3.0E+7	6.5E+7	6.5E+7	6.5E+7	8.6E+7
G' @ 30 °C, dynes/cm²	7.1E+6	7.4E+7	7.6E+7	7.1E+7	6.3E+7	8.3E+7	6.3E+7	8.1E+7
Tg (tan delta), C	107	107	117	-	-	-	-	-

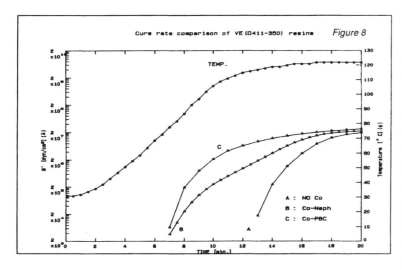

Figure 8

120°C as measured by DSC cure and rheological cure analysis. The latent cobalt cured samples were observed to exhibit similar elastic modulus and tan delta (approximating Tg) compared to free cobalt naphthenate control. Figures 8 and 9 show the relative cure rates for VE systems with 411-350 or D470-300 resin respectively mixed with MEKP and one of the following (a) no cobalt promoter, (b) free cobalt naphthenate or (c) SCC polymer based latent cobalt. While both cobalt containing systems exhibit similar reaction rates, both much faster than the no cobalt control, the new latent cobalt-PBC material also delivers pot-life extension similar to the no cobalt control.

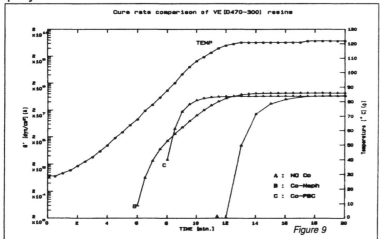

Figure 9

DSC cure analysis was performed in isothermal fashion by heating the cell to either 80 or 120°C followed by quickly positioning the sample and starting the evaluation. Due to the method used, some variation in reproducibility is observed leading us to surmise that the reaction rates are similar for the latent cobalt and the free cobalt naphthenate. The rheological cure was measured on 2.5 cm diameter parallel plates by heating from 30 to 120°C at 10 and 20°C/min observing G' at 120°C, cooling to 30°C, observing G' at 30°C and then heating from 30 to 180°C at 10°C/min and observing tan delta that is proportional to Tg. The latent cobalt samples yield comparable modulus and tan delta (Tg) values versus cobalt naphthenate control in both the unsaturated polyester and vinyl ester formulations.

EPOXY SYSTEMS

Latent imidazole catalysts have been developed for epoxy resin systems that result in pot-life extension, shelf-life extension and reactivity in the 80-150°C range. These latent imidazoles are provided in two forms: (a) polymer bound imidazole for storage stability and (b) encapsulated imidazole for pot-life extension. Figure 10 shows a pictorial description of an SCC polymer with polymer bound imidazole functionality. The pot-life and shelf-life, rate of reaction and temperature of reaction can

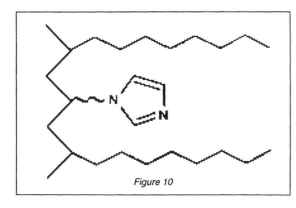

Figure 10

be optimized through appropriate choice of catalyst level for a given formulation. For example, epoxy systems formulated by mixing 100 parts of Shell's Epon 828, 80 parts MTHPA and 6.3 parts of latent imidazole *Intelimer®* XE-7004 (16% active imidazole) resulted in formulations exhibiting the following storage viscosity rise (a) from 6 to 60 poise in 60 days at 25°C and (b) from 6 to 330 poise in 60 days at 40°C. Sole cure laboratory test formulations with Epon 828 and 3 phr effective amine have exhibited six months shelf-life.

APPLICATIONS

We expect these pot-life improvements to allow the use of higher concentrations of catalyst/promoter than used today in FRP applications to reduce total reaction time resulting in higher productivity and higher throughput for a given piece of capital equipment. In addition we expect the improved latency to result in reduced waste and reduced labor costs associated with mixing multiple batches of formulation throughout the day with existing materials.

Applications like sheet molding compound generally require some storage period while viscosity builds through the crosslinking reaction of the MgO with functional groups on the resin. This viscosity build yields an uncured reinforced mixture that resists undesired flow during the hot molding process thereby preventing the fiber from showing through or thinning at the edges. If the viscosity build step requires 24 hours then, the latent catalyst must meet that time delay also. We observe systems with these latent cobalt catalysts that have the potential of greatly reducing cure time while meeting the pot-life requirement during the uncured viscosity build stage. Further, unwanted side reactions like cobalt assisted oxidation of styrene to benzaldehyde may be reduced when the cobalt is locked in the crystalline polymer array.

Filament winding applications employing vinylester resins where reduced cure time at reaction temperature is desired can be designed using these latent cobalt materials (a) without adversely influencing the pot-life of the coating bath or (b) even increasing the pot-life of the coating bath while increasing overall throughput. Similarly, epoxy/anhydride filament wound applications should benefit from the latent imidazoles. These latent catalysts should help ensure consistent bath viscosity throughout the shift also minimizing the need for multiple batch mixing during a work shift while potentially providing more consistent resin applied to the final product.

CONCLUSIONS

We have demonstrated the following:

(1) development of a wide variety of latent catalysts addressing unsaturated polyester, vinyl ester and epoxy systems demonstrating the technology's breadth and applicability,

(2) pot-life and shelf-life extension at 25 and 40°C,

(3) comparable cure rate at reaction temperature versus control systems, and

(4) comparable final properties versus control systems.

ACKNOWLEDGMENT

The authors would like to thank Mr. Terry Edwards for rheological test and evaluation.

REFERENCES

Gum, W. F., Riese, W. and Ulrich, H. 1992. "Reaction Polymers," Hanser Publishers, New York, Oxford University Press

Kia, H.G. 1993. "Sheet Molding Compounds, Science and Technology," Hanser Publishers, New York

AUTHOR BIBLIOGRAPHIES

Steven P. Bitler is currently Director of Polymer Development. He received his B.S. in chemistry from Boston College followed by a Ph. D. in chemistry from the University of Southern California. He developed polymeric materials at the SRI International prior to joining Landec in 1988.

Mark A. Wanthal is currently Project Leader of Formulated Products. He received a B.S. in chemical engineering from Tulane University. He has been active in the epoxy industry for many years holding development positions at Magnolia Plastics, Dexter Corp. Adhesive and Structural Materials Division, and Arcor Inc. prior to joining Landec in 1995.

David A. Kamp is currently Project Leader of New Polymer Synthesis. He received his B.S. in chemistry from the University of California at Los Angeles followed by a Ph. D. in chemistry at the University of Oregon. He developed polymeric materials at General Electric Company and Raychem prior to joining Landec in 1990.

Paul A. Meyers is currently Senior Project Engineer. He received his B.S. in chemistry from California State University at Hayward. He has developed polymeric and fine chemical materials at Dow Chemical prior to joining Landec in 1988.

David D. Taft is currently Chief Operating Officer. He received his A.B. in chemistry from Kenyon College followed by a Ph. D. in chemistry from Michigan State University. He has held a number of executive and development positions at Ashland Chemical, General Mills Chemical, Henkel Inc. and Raychem Inc. prior to joining Landec in 1993.

New Development in Low Profile Additives

N. UJIKAWA, M. TAKAMURA, F. B. LAURENT AND C. B. BUCKNALL

ABSTRACT

Novel PS/PVAc and PS/PMMA block copolymers were prepared using special organic peroxides, which have several peroxy bonds per molecule. The two contrasting segments of these block copolymers have differing solubility parameters, and in solution form micelles. Consequently during all stages of blending with solutions of other polymers, one chain of the block copolymer is always compatible with the surrounding solution. The novel copolymers act not only as primary LPAs but also as compatibilizers for other LPAs. They provide excellent shrinkage control, mechanical strength and surface quality. Electron microscopy of specimens stained with ruthenium tetroxide is used to study the morphology of cured blends, and the effects of micelle formation on low profile behavior.

INTRODUCTION

Composite materials based on glass fiber-reinforced polyester resins have shown remarkable growth over the past few decades, partly as a result of improvements in resin formulation. A volume shrinkage of 6-10% in the UP matrix during curing, for example in sheet molding compounds (SMC), causes several problems, including poor surface quality, warpage, sink marks, internal cracks and lack of dimensional stability. In order to reduce this shrinkage and obtain final products having excellent properties, several thermoplastics such as polystyrene (PS), poly (vinyl acetate) (PVAc), poly (methyl methacrylate) (PMMA), styrene/butadiene/ styrene copolymer (SBS) and polyurethane have been used as low profile additives (LPAs). These additives are classified into two groups according to their blend morphology: initially single-phase systems and two-phase systems. In the former, there is no phase separation in the liquid state before curing: an example is PVAc. In the latter, the LPA is insoluble in the liquid resin, and forms a distinct second phase: typical examples include PS and SBS. All of these materials have advantages and disadvantages as LPAs. PVAc has good compatibility with UP, and gives molding products having good dimensional stability, but lacks pigmentability, mechanical strength and hot water resistance. On the other hand, molding compounds containing PS and SBS have good pigmentability, but the PS and SBS additives have poor compatibility with UP, and therefore separate rapidly from the blend with UP during curing.

N. Ujikawa, M. Takamura, NOF Corporation, Aza-Nishimon, Taketoyo-cho, Chita-gun, Aichi-ken, 470-23, Japan
F.B.Laurent, C. B. Bucknall, Cranfield University, Cranfield, Bedford MK43 0AL, UK

This paper introduces novel PS/PVAc and PS/PMMA block copolymers (Modiper®) prepared by using special organic peroxides which have several peroxy bonds per molecule. These block copolymers consist of two segments having different solubility parameters. Therefore, in blends with UP resin, the less miscible blocks of the additive molecules will aggregate to form a spherical structure (a micelle) before and during curing. The thermodynamic stability of the blend, shrinkage control, surface roughness and mechanical properties of the molded products will be discussed.

Many researchers have studied the shrinkage control mechanism in UP blends (Pattison et al., 1974, 1975; Atkins et al., 1976; Atkins, 1988; Mitani et al., 1989; Bucknall et al., 1991; Kinkelaar and Lee, 1992). The effects of micelle formation on the low profile behavior of the novel block copolymers will be discussed in the light of this earlier work.

EXPERIMENTAL

MATERIALS

The Modiper® grades used in this study are listed in Table 1. They were compared with three conventional LPAs: PS (Diarex HF-77;Mitsubishi Chemical), PVAc (LPS-40AC;Union Carbide) and SBS (Kraton D; Shell Chemical). All LPAs were dissolved in 70wt% styrene before mixing with UP. The UP used in this study is a highly reactive resin designed for SMC applications. It consists of maleic anhydride, fumaric acid and propylene glycol, and has a MW of 156 per C=C bond. The resin is a solution of 65wt% UP in styrene. The initiator was TBPB (t-Butyl peroxybenzoate; NOF Corp.).

CURING PROCEDURE

Stability of blends

Blends were made by mixing the UP and LPA solutions together, with mechanical stirring, to give a UP/LPA ratio of 55/45 parts by weight (pbw). Each mixture was poured into a 18mm diameter glass test tube to a height of 100mm. The test tubes were kept at 25°C, and on the assumption that there is no migration of the solutes between the two phases, the degree of separation was calculated by dividing the height (mm) of the interface above the bottom of the tube by 55 (mm).

Curing conditions

A typical SMC formulation is shown in Table 2. The UP, LPA, TBPB and other ingredients were mixed together to form the SMC and then allowed to thicken for one day at 40°C before molding. They were molded at 152°(lower mold) / 146°C (upper mold) for 180 seconds using a 100 ton compression molding press. A Micromet Cure Analyzer was used to evaluate the cure parameters of gel time (GT; CP3) and cure time (CT; CP4). A mold having 10 mm diameter and 5 mm depth was used to measure dimensional changes on curing, and another 150x100x4 mm mold, with a class A surface finish, was used to assess surface quality and prepare specimens for mechanical testing.

MICROSCOPY

Hot stage microscopy was used to determine the volume fraction of the LPA phase in the cured blend with UP, and to observe phase separation and void formation during curing. A tiny droplet of the mixture of UP, LPA, styrene and TBPB was spread between two glass cover slips and placed on the hot stage. Optical microphotographs (magnification x200) were taken quickly, and the mixtures were then cured at 140°C for 30 min. The volume fraction was calculated from the observed area of the LPA phase in the micrographs.

Scanning Electron Microscopy (SEM) and Scanning Transmission Electron Microscopy (STEM) were carried out on fracture surfaces and ultra-thin sections respectively. Etching and staining techniques were used in order to distinguish the different phases in the cured blends.

RESULTS AND DISCUSSION

STABILITY OF MIXTURES OF LPA WITH UP

Modiper is a grade of shrinkage modifiers, based on the A-B type block copolymer structure, in which the ratio of PS/PVAc can be varied (90/10, 70/30 and 50/50 wt phr). Because of their differing solubility parameters, the chains comprising the Modiper molecules aggregate to form a micelle structure which minimises the interfacial area between the PS and PVAc phases. The PS segment should be compatible with styrene because its solubility parameter is close to that of styrene, whereas mixing with UP should cause phase inversion (Figure 1). From Table 3, it can be seen that mixtures of Modiper with UP are stable over long periods, while both PS and SBS separate from UP mixtures immediately. These separations can be prevented by adding Modiper as a compatibilizer.

Table 4 shows the sizes and volume fractions of the LPA phase in the mixtures with UP after mechanical stirring. Compared with PS and SBS, the sizes of the SV10B and 10A Modiper phases are very small. When either SV10B or S501 is used as a compatibilizer for SBS, the size of the LPA phase decreases as the amount of Modiper increases. A further remarkable observation is that the volume fraction of the Modiper phase (SV10B, 10A) to the continuous phase is greater than the concentration at which it was added, and is also higher than the volume fractions of SBS and PS at the same additive concentration. This suggests that the micelles formed by Modiper absorb both styrene and some polyester into their structure. Thus, during all stages of blend preparation, one chain of the block copolymer is always compatible with the surrounding solution. This confers good stability to the mixtures and prevents bleeding during cure.

SHRINKAGE CONTROL, SURFACE ROUGHNESS AND MECHANICAL PROPERTIES

Typical properties of SMC molded products are listed in Table 5. Blends containing plain PVAc show superior shrinkage control and surface properties, but inferior mechanical properties and color. On the other hand, PS provides good pigmentability, while SBS, which has elastomeric segments in its molecule, gives good shrinkage control. However, in both PS and SBS blends, molded products have inferior surface properties. Modiper shows an excellent balance of properties, and specific characteristics can be obtained by selecting from the various grades. These block copolymers can be used not only directly as LPAs, but also as compatibilizers for other LPAs. Molded products modified with S501 (20)/SBS (80) exhibit excellent shrinkage control with improved mechanical properties and surface quality. From Figure 2, it can be seen that molded products show zero shrinkage at Modiper concentrations between 10 and 13.5 wt%. The S501 grade shows superior shrinkage control to SV10B. This difference may be due to the higher PVAc content of S501, which influences the morphology of the cured UP.

MORPHOLOGY

Figure 3 is a micrograph of a cured blend of UP with 8 wt% SV10B, obtained using hot stage microscopy. Particles of the LPA phases were dispersed on a scale of <10μm before curing. Within 5 or 6 minutes from the onset of curing, many voids were created inside the LPA phase. By contrast, in case of S501 (added at >8 wt%), the LPA-rich phase became continuous, and the polyester precipitated as discrete particles. About 2 or 3 minutes from the onset of curing, the mixture became densely white in appearance. In both blends, no further change in morphology was observed after 30 minutes.

Figure 4 shows an SEM micrograph at 16 wt% S501. The morphology is phase inverted. Most of polyester nodules are fractured, and form cohesive spheres with an average diameter of 1 μm. Some nodules remain intact as whole spheres, and some contain subinclusions. In order to obtain more detail of the morphology, and to identify the Modiper micelles, STEM techniques were used. Figure 5 shows STEM micrographs of cured blends, at 16 wt% S501, both unstained (Fig 5a) and stained with ruthenium tetroxide (RuO4)(Fig 5b). This compound stains PS (Trent et al.,1983; Vitali and Montani, 1980) but also potentially has the ability to stain the polyester resin, which contains a proportion of styrene units. Although there are no directly relevant data in the literature, RuO4 is expected to have little effect upon PVAc, because it does not stain PMMA, which is similar in chemical composition (Trent et al., 1983). The large white areas in Figure 5(b) are obviously polyester resin, a point that is confirmed by separate etching experiments, which are not described here. These large white areas are more continuous than the corresponding light grey areas in Figure 5(a), where the section is unstained, and where contrast is due to differences either in thickness or in electron density. This comparison indicates that the thin dividing lines between the near-circular resin domains in Figure 5(a) consist of PVAc, which apparently is not stained by RuO4. The grey regions with lighter spots seen in Figure 5(b) are similar to those observed in many block copolymers, which suggests that they are regions of block copolymer, consisting of a continuous, darker, stained PS phase, in which are embedded small spheres of PVAc (the lighter-colored spots), with dimensions in the order of 30 nm. It appears most likely that the white regions in Figure 5(b) are areas in which the PVAc phase of S501 has cavitated. Theory suggests that the result is not a simple hole, but a network of 30 nm PS spheres connected by fibrillated PVAc.

THE LOW PROFILE BEHAVIOR OF BLENDS EXHIBITING A MICELLE STRUCTURE

Chan-Park and McGarry (1995) showed the morphologies produced in crosslinked polyesters by three contrasting types of LPAs: compatible, incompatible, and slightly compatible. The initial mixtures of Modiper with UP show inhomogeneous structures. In the case of SV10B, which has 10 wt% PVAc in its molecule, two phases are formed. The volume fraction of the Modiper exceeds the added concentration, and is larger than those of PS and SBS. This indicates that the Modiper is preferentially swollen with styrene, and possibly also contains some polyester. On curing, microvoids are created inside the Modiper, probably through cavitation of the PVAc phase. However, it is possible that a core region composed of styrene-swollen polystyrene chains from the block copolymer remains soft enough to cavitate while the surrounding resin is curing, under the influence of dilatational stresses set up as a result of resin shrinkage.

By contrast, Modiper S501, which is a 50/50 PS/PVAc block copolymer, shows microphase separation, and above 8 wt% S501 gives a co-continuous phase morphology after curing. Microvoid formation in the PVAc phase of S501 is suggested by the STEM study. At low concentrations of S501 (2%, 4%), the LPA is dispersed in the continuous UP phase. It follows from these observations that the morphology of cured UP blends can be controlled by the weight ratio of PVAc to PS chains in the PS/PVAc block copolymer molecules. It is well known that a combination of PVAc and $CaCO_3$ gives a synergistic effect in the shrinkage control observed during SMC molding (Bucknall et al., 1991). Higher performances can be obtained in SMC molding by the use of these new block copolymers, having PVAc chain in their molecules.

CONCLUSIONS

1. PS / PVAc andPS/PMMA block copolymer molecules aggregate to form a micellar structure in mixtures with UP, and the PS core is preferentially swollen with styrene. This can lead to cavitation in a PS-rich phase. Alternatively, at higher PVAc concentrations, the PVAc phase is the site of cavitation.

2. Mixtures of the block copolymers with UP are very stable for long periods. The copolymers act not only as LPAs but also as compatibilizers for other LPAs.

3. SMC molding products modified with these block copolymers have an excellent balance of properties such as shrinkage control, mechanical strength and surface quality.

4. Staining with ruthenium tetroxide is an effective method for studying the morphology of blends containing UP, PS and PVAc phases, and investigating the mechanism of low profile behavior.

REFERENCES

Atkins, K. E. Koleske, J. V. Smith, P. L. Walker, E. R. and Matthews, V. E. 1976 "Mechanism of Low Profile Behavior," 31st Annual Tech. Conf., Reinforced Plastics/Composites Institute. Paper 2-E

Atkins, K. E. 1988. "Low-profile Behavior," Polymer Blends, London, D.R. Paul and S. Newman, Academic Press. 2:391-413

Bucknall, C. B. Partridge, I. K. and Phillips, M. J. 1991 "Mechanism of Shrinkage Control in Polyester Resins Containing Low-profile additives," Polymer, 32:636-640

Bucknall, C. B. Partridge, I. K. and Phillips, M. J. 1991 "Morphology and Properties of Thermoset Blends made from Unsaturated Polyester Resin, Poly (Vinyl Acetate) and Styrene," Polymer, 32:786-790

Chan-Park, M. B. and McGarry, F. J. 1995 "The Low Profile Mechanism in Sheet Molding Compound," J. Adv. Mater., 27(1):47-58

Kinkelaar, M. and Lee, L. J. 1992 "Development of a Dilatometer and its Application to Low-shrink Unsaturated Polyester Resin," J. Appl. Polym. Sci., 45:37-50

Mitani, T. Shiraishi, H. Honda, K. and Owen, G. E. 1989 "Mechanism of Low-profile Behavior by Quantitative Morphology Observation and Dilatometer," 44th Annual Conf., Composites Institute, The Society of the Plastic

Pattison, V. A. Hindersinn, R. R. and Schwartz, W. T. 1974. "Mechanism of Low-profile Behavior in Unsaturated Polyester Systems," J. Appl. Polym. Sci., 18:2763-2771

Pattison, V. A. Hindersinn, R.R. and Schwartz, W. T. 1975. "Mechanism of Low-profile Behavior in Single-phase Unsaturated Polyester Systems," J. Appl. Polym. Sci., 19:3045-3050

Trent, J. S. Scheinbeim, J. I. and Couchman, P. R. 1983 "Ruthenium Tetraoxide Staining of Polymers for Electron Microscopy," Macromolecules, 16(4):589-598

Vitali, R. and Montani, E. 1980 "Ruthenium Tetroxides as a Staining Agent for Unsaturated and Saturated Polymers," Polymer, 21:1220-1222

BIOGRAPHIES

Norihisa Ujikawa

Norihisa Ujikawa received his B.S.degree in organic chemistry from Nagoya University. He joined NOF Corporation in 1975 and has been working in the Fine Chemicals & Polymers Research Laboratory. He is Research manager in charge of curing agents and LPAs for UP.

Masumi Takamura

Masumi Takamura received his B.S.degree in applied chemistry from Nagoya Institute of Technology. He joined NOF Corporation in 1989 and has been working in the Fine Chemicals & Polymers Research Laboratory. He is responsible for the development of LPAs.

Clive B. Bucknall

Clive Bucknall received his BA in chemistry, and his PhD and ScD degrees in Materials Science, from Cambridge University. After eight years in the plastics industry, he joined Cranfield in 1967. He is now Professor of Polymer Science and Head of the Advanced Materials Department at Cranfield University.

Francois B. Laurent

Francois Laurent received his Diplome d'Ingenieur in Mechanical Engineering from ESIM, Marseilles, and his MPhil in Materials from Cranfield University in 1994. His thesis at Cranfield was entitled 'Shrinkage Control in Polyester Resins using Modiper Additives'. Since graduating, he has been working for a consultancy firm in Sao Paulo, Brazil.

Table 1 Grade of new block copolymers and the typical properties

Modiper	Composition (PS / PVAc)	Viscosity [1] (P at 25 $^{\circ}$C)	Notes
SV10B	90 / 10	18	
SV10A	90 / 10	35	Containing carboxyl groups
S101	90 / 10	3	Lower viscosity
SV30B	70 / 30	20	
S501	50 / 50	40	
MS10B	90 / 10	18	PS / PMMA

1) Dispersion in 70wt% styrene (typical data)

Table 2 The formulation of SMC

Material	Formula (wt% in total SMC formulation)
UP	18
LPA	15
TBPB	0.5
Calcium carbonate	65
Zinc stearate	1.3
Magnesium oxide	0.4
PBQ	0.1
Glass fiber (1-inch, chopped)	10

Table 3 Stability of the mixtures of LPA and UP

(Degree of separation,%)

LPA / Time (day)	SV10B	SV30B	S501	SV10B / PS (20 / 80)	S501 / SBS (20 / 80)	PS	SBS	PVAc
(5 hrs)	0	0	0	0	0	85	100	0
(8 hrs)	0	0	0	0	0	100	100	0
1	0	0	0	0	0	100	100	0
14	0	0	0	5	0	100	100	0
28	4	0	0	6	0	100	100	0

Table 4 Size and propotion of the dispersed LPA phase in the mixture of LPA and UP (initial additive amount of LPA (in 70wt% styrene) is 27%)

LPA	Dispersed LPA phase	
	Size (μm)	Volume fraction (%)
SV10B	3	51
SV10A	3	37
S501	microphase separation	
MS10B	42	31
SBS	36	27
SBS (96) / SV10B (4)	33	39
SBS (84) / SV10B (16)	9	39
SBS (96) / S501 (4)	14	46
SBS (84) / S501 (16)	5	47
PS	80	29

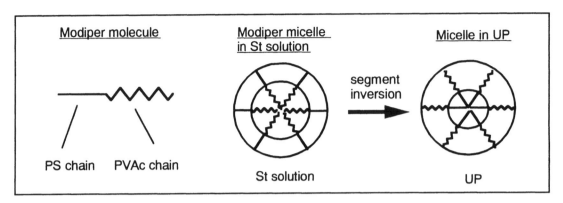

Figure 1 Changes in micelle structure during blending

Table 5 Typical properties of SMC molding products

LPA	SV10B	SV10A	SV30B	S501	MS10B	S501 / SBS (20 / 80)	PS	SBS	PVAc

Cure analysis
  GT(CP3, sec)
  CT(CP4, sec)
	41.1
	107.4
		43.8
		103.3
			42.2
			104.9
				43.3
				107.5
					47.5
					105.9
						43.8
						101.2
							41.6
							101.7
								46.4
								101.7
									46.9
									116.8

Shrinkage control (%)	0.00	+0.03	+0.01	+0.03	0.00	+0.05	-0.05	+0.05	+0.09

Mechanical properties
  Flexural strength (MPa)
  Flexural modulus (MPa)
	7.4
	1010
		6.8
		1020
			8.6
			990
				7.1
				950
					7.9
					1030
						8.4
						940
							8.4
							1070
								8.1
								900
									6.4
									790

Surface properties
  Average roughness (μm)
  Gloss (%)
	0.11
	86
		0.12
		85
			0.08
			90
				0.14
				81
					0.08
					88
						0.24
						64
							0.48
							69
								0.35
								47
									0.06
									95

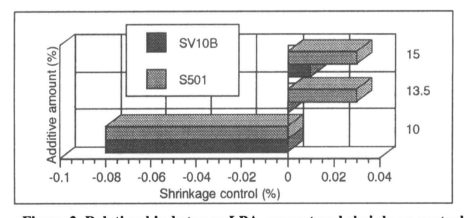

Figure 2 Relationship between LPA amount and shrinkage control

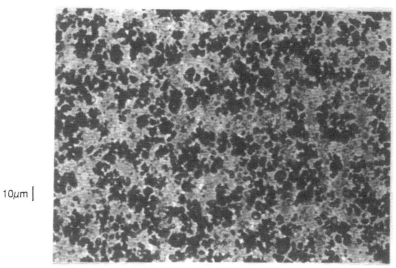

10μm

**Figure 3 Optical micrograph of UP / SV10B cured blend at 8wt% Modiper
Isothermal curing at 140°C. Magnification X200**

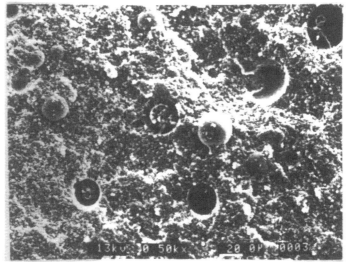

**Figure 4 SEM micrograph of fracture surface obtained from UP /S501
cured blend at 16wt% Modiper. Magnification X500**

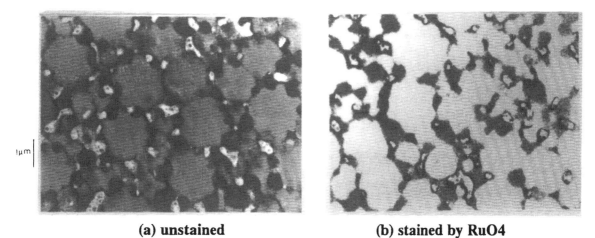

1μm

(a) unstained   (b) stained by RuO4

**Figure 5 STEM micrograph of cured UP modified with 16wt% Modiper
S501. Magnification X15500**

Development of High Molecular Weight Unsaturated Polyester Resin

EIICHIRO TAKIYAMA, YOSHITAKA HATANO AND FUMIO MATSUI

I. Summary

A new method has been developed by which number average molecular weight(Mn) of an unsaturated polyester resin can be increased up to two to five times of the original(i.e.5,000-15000).

Note: For convenience,an unsaturated polytester and its resin are represented by UP in this paper.

It has been confirmed that all of the physical and chemical properties of cured resins improve with increase in Mn of the resins.

The following items are key points to obtain desirable UP.

(i)Molecular weight of UP whose end groups are hydroxyl groups.

   The higher the molecular weight, the better the result.

(ii)Deglycol reaction with appropriate catalyst

(iii)Degree of depressurization during deglycol reaction

   It has been confirmed that the higher the degree of depressurization, the higher the molecular weight of UP.

(iv)Selection of optimum reaction temperature

(v)Selection of optimum stabilizer

In this paper discussion is made centered around an outline of the physical and chemical properties of cured high molecular weight UP(HUP).

II. Introduction

Generally speaking, the molecular weight of currently available UP ranges roughly between 1,500-2,500,or in the range of an oligomer. Also, it is widely known that its properties(chemical, electrical, thermal, water and chemical resistance)improve as the molecular weight increase.

However, it is almost impossible to obtain HUP whose molecular weight is higher than the avove-mentioned range due to gelation of the reaction solution as long as UP is synthesized through simple esterification of an acid and a glycol.

Also, it is thought that deglycol reaction at high temperature (250°C) with a

Eiichiro Takiyama, Yoshitaka Hatano, Fumio Matsui

SHOWA HIGHPOLYMER Co.,Ltd., Kanda Chuo Bldg. Kanda Nishiki-Cho 3-Chome,Chiyoda-ku,Tokyo,Japan

catalyst employed in the synthesis of a thermoplastic resin such as poly(ethylene terepthalate) is not applicable to the synthesis of HUP due to problems associated with gelation.

In fact, we do not recall any resin of this type having been introduced to the marketplace by anyone other than the HUP developed and introduced in 1994 by SHOWA HIGHPOLYMER Co.,Ltd.

The company has developed the method to increase UP's molecular weight in excess of 5,000 and has looked into advantages that such HUP offers. Such advantages are taken up in the ensuing pages.

III. Synthesis of HUP and Measurement of Molecular Weight

III-1. Raw Materials

The following commercially available chemicals were used as raw materials without any treatment:

Propylene glycol

Phthalic anhydride

Isophthalic acid

Maleic anhydride

Fumaric acid

Tetraisopropyltitanate

Phosphorous acid

Styrene

Hydroquinone

III-2. Synthesis of HUP

The following is a typical example of the synthesis of HUP, however,some differences show up depending on the type of glycol, the unsaturation degree of acid, the kind of stabilizer,etc.

Synthesis Procedure

7.0kg of propylene glycol and 6.6kg of isophthalic acid were charged into a 30-liter stainless steel flask equipped with an agitator, a partial condenser, a gas introducing tube, and a thermometer for esterification at 180-190°C under nitrogen gas flow, then the reactor temperature was decreased to 150°C when the reactant acid value reached 27.1. Then, 4.7kg of fumaric acid and 2g of hydroquinone were added. Further esterification was carried out at 190-195°C. When the reactant acid value reached 9.1, 50g of tetraisopropyltitanate and 8g of phosphorous acid were charged into the flask, then deglycol reaction was carried out at 190-195°C under 7-10torr at the beginning and 0.9torr at the final stage.

During this reaction samples necessary for measurement of molecular weight and other HUP properties were taken and test samples for measurement of physical and chemical properties were prepared as 50wt.% solution in styrene.

Their molecular weight was measured by GPC and the properties, by JIS standard.

III-3. Molecular Weight Measurement by GPC

Equipment: Shodex GPC(Showa Denko K.K.)

Eluent: C₄H₈O(THF;tetrahydrofuran)
Sample column: KF-805Lx2
Column temperature: 40°C
Flow rate: 1.0ml/min.
Detector: Shodex RI-71
Control sample: Polystyrene(Showa Denko Shodex Standard)
Calibration curve: See Fig.1

IV. Investigation

IV-1. Changes of HUP molecular weight and distribution

Changes in HUP molecular weight during the deglycol reaction stage of the synthesis of the samples are shown in Fig.2-5. Fig.2 and Fig.3 are bar charts showing respectively changes in Mn and Mw with the passage of time during the deglycol reaction.

A noticeable differences in the changes of Mn and Mw was observed. Mn maintained a constant value after reaching a peak, while Mw kept on increasing throughout the reaction.

Changes in Mw and Mn of high molecular weight saturated polyester(HSP) are shown in Fig.4 and Fig.5 respectively. When the changes of HUP are compared with those of high molecular weight saturated polyesters, the differences between the two were decisive.

Changes in HUP molecular weight distribution with the passage of time during the deglycol reaction are shown in Fig.6-13. It was observed that as the time passed the distribution spread into a higher molecular weight range. Beginning with reaction at 6 hours, a "shoulder" began appearing on the left side of the curve.

Relationship between Mn/Mw and deglycol reaction time is shown in Fig.14. The ratio increases exponentially as time passes and finally the reaction solution takes on the form of a gel.

IV-2. Molecular Weight-Properties

In general, the following properties show improvement as molecular weight increases.
  (i) Faster cure rate (Fig.15)
  (ii) Increased viscosity of solution in styrene (Fig.16)
  (iii) Improved flexural strength and modulus (Fig.17-18)
  (iv) Higher heat distortion temperature (Fig.19)
  (v) Improved toughness (Fig.20)
  (vi) Improved water and chemical resistance (Fig.21-22)

IV-3. High Toughness Resin

Conventional UP is generally hard and not elastic, which makes it unquestionably inferior to epoxy and vinylester resins in physical and chemical properties.

Althogh UP has been commercially used in some applications, it is limited only to semihard and soft grades but their presence has never drawn any attention of any significance because of their properties.

The newly-developed HUP is quite different from conventional grades of UP and offers properties far superior to those of any known resin.

The following information is on such high toughness resin producible at relatively low cost.

Propylene glycol   105mol.
Phthalic anhydride   25
Adipic acid   26
Maleic anhydride   49
Mn: 5,000
Mw: 6,400
Acid value(KOHmg/g): 9.2
Cast product properties (casting resin:40%styrene solution)
 Tensile strength (kg/mm^2): 7.0
 Elongation (%): 4.2
 Heat distortion Temperature (°C): 92
Recipe for laminating
 Resin solution   100 parts
 Aerosil 200   2
 TR (thixotropic agent) 3
 Cobalt naphthenic   0.4
 Dimethyl aniline   0.03
Conditions for test piece preparation
 #450 glass mat 7-ply
 5mm thick after curing (2hr.at 120°C)
 Glass content: 35-37%

Fig.20 is an S-N diagram obtained by repeated tensile fatigue tests of FRP test pieces molded from different resins under the same conditions. The resins are vinylester resin, high toughness HUP (ortho-type) and conventional UP (iso-type).

HUP based on phthalic anhydride and adipic acid exhibits an anomalously high toughness exceeding that of the vinylester resin.

IV-4. Water and Chemical Resistance

HUP exhibits an outstanding water and chemical resistance thanks to the increased molecular weight. To confirm such improvement several types of HUP given in Table 1 was synthesized and their cast products were evaluated. Synthesized for the evaluation were four types, namely, ortho- and iso-types synthesized with and without deglycol reaction.

Hot water resistance was evaluated by measuring the time taken for an internal crack to develop as a result of hydrolysis, and the results are given in Fig.21. Chemical (alkali) resistance was checked by using a 10% NaOH solution and results are given in Fig.22.

In both cases distinct improvements in the properties brought about by an increase in molecular weight were observed.

At the same time, it was found that the iso-type product exhibits high hot water resistance nearly equal to that of the ortho-type one. This is quite helpful in producing a high performance resin at low cost as iso-type chemicals are generally priced higher than ortho-type ones.

IV-5. Effect of Varying Degrees of Unsaturation

Effects of varying degrees of unsaturation on resin properties were investigated by comparing HUP with conventional UP. Both resins were of the propylene glycol-isophthalic acid-fumaric acid system. The results are given in Fig.23 and Fig.24.

It is clear that property improvements are remarkable, especially with less reactive resins.

IV-6. Styrene Content-Molecular Weight

In casting, the molecular weight of HUP and styrene content were varied while the solution viscosity was kept within the range of 7-8 poise, and the effects of the variations on the properties of the resin and cast products were investigated. The results are given as follows:

Cure rate   Fig.15
Heat distortion temperature   Fig.25
Flexural strength   Fig.26
Flexural modulus   Fig.26
Charpy impact strength   Fig.27
Hot water resistance   Fig.28

The test pieces for hot water resistance were prepared for bulk molding compounds. One of the resins contained 25wt.% of styrene and UP whose Mn was 2,140. The other resin contained 35wt.% of styrene and HUP whose Mn was 7,460, while the viscosity of both resins was 430 poise.

With the exception of the test pieces for hot water resistance, the styrene content was increased from 39wt.% to 53wt.% in order to keep the viscosity constant as the molecular weight increased.

Although no noticeable change was observed in mechanical properties, the cure rate increased. Also, heat distortion temperature and water resistance improved.

V. Conclusion

V-1. Synthesis

A new method to synthesize high molecular weight unsaturated polyester resins has been developed and a new type of high molecular weight (Mn:5,000-15,000) unsaturated polyester resin has been obtained.

V-2. Properties

By increasing molecular weight of an unsaturated polyester, practically all properties of its cast product improved. Noted improvements are as follows:

(i) Heat resistance (heat distortion temperature): (up by 20-30%)

(ii) Impact strength (up by 20-30%)

(iii) Toughness

(iv) Water and chemical resistance

V-3. Casting resin

No deterioration in product properties was observed even at high styrene content.

V-4. Improved resin

Synthesis of a high toughness and high performance resin that was heretofore unavailable has become possible.

V-5. Cost reduction

The new method enables the supply of high performance resins to be made at low cost. Through this method a new type of high performance resin with properties being practically equal to those of an iso-system resin can be obtained, using ortho-system raw materials. In general, ortho-type chemicals are less expensive than iso-type ones.

Tab.1 Polymer for Evaluation

Sample Type	Ortho-Type		Iso-Type	
	Ortho-1	Ortho-2	Iso-1	Iso-2
Reaction - Procedure	(1 Stage) Esterification	(2 Stage) Esterification Deglycolization	(1 Stage) Esterification	(2 Stage) Esterification Deglycolization
Mn	1600	4600	2900	5700
Mw	4200	25300	9900	33000
Mw/Mn	2.2	6.5	3.4	5.8
Styrene Content (wt %)	50	50	50	50
Acid Value	17.9	1.0	13.5	7.9

Composition Ortho-Type: Propylene glycol/105mole
Phthalic anhydride/50mole Maleic anhydride/50mole
Iso-Type: Propylene glycol/105mole
Isophthalic acid/50mole Fumaric acid/50mole

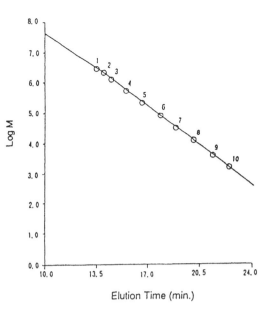

Fig.1 GPC Calibration Curve

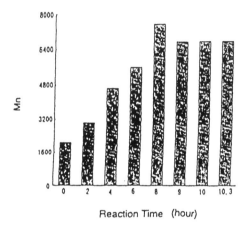

Fig.2 HUP Molecular Weight (Mn) Changes
with Passage of Reaction Time

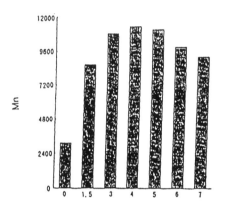

Fig.4 HPE Molecular Weight (Mn) Changes
with Passage of Reaction Time

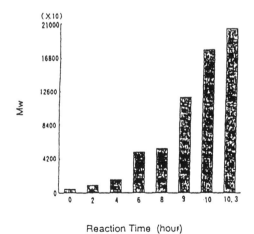

Fig.3 HUP Molecular Weight (Mw) Changes
with Passage of Reaction Time

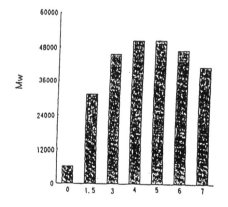

Fig.5 HPE Molecular Weight (Mw) Changes
with Passage of Reaction Time

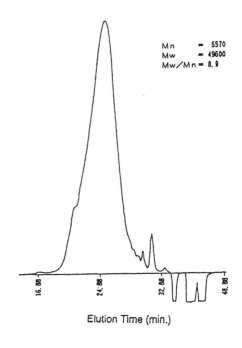

Mn = 5570
Mw = 49600
Mw/Mn = 8. 9

Elution Time (min.)

Fig.9 HUP Molecular Weight Distribution
at 6 Hours of 2'nd Stage Reaction

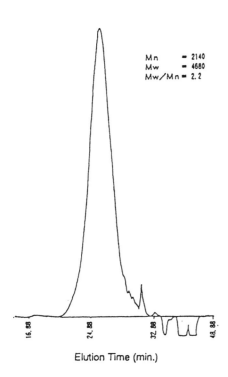

Mn = 2140
Mw = 4680
Mw/Mn = 2. 2

Elution Time (min.)

Fig.6 HUP Molecular Weight Distribution
at Start of 2'nd Stage Reaction

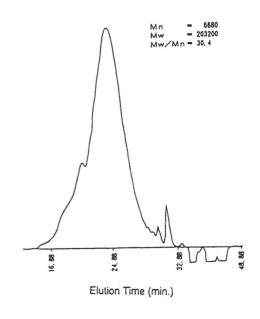

Mn = 6680
Mw = 203200
Mw/Mn = 30. 4

Elution Time (min.)

Fig.13 HUP Molecular Weight Distribution
at 10.3 Hours of 2'nd Stage Reaction

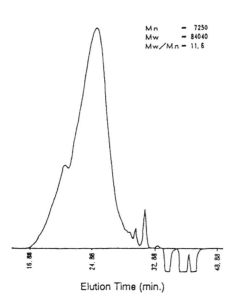

Mn = 7250
Mw = 84040
Mw/Mn = 11. 6

Elution Time (min.)

Fig.10 HUP Molecular Weight Distribution
at 8 Hours of 2'nd Stage Reaction

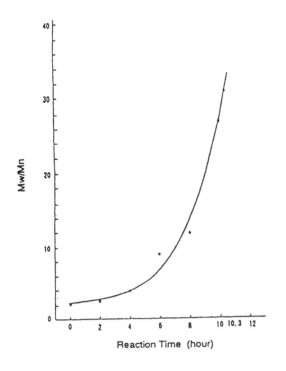

Fig.14 Mw/Mn Changes
with Passage of 2'nd Stage Reaction Time

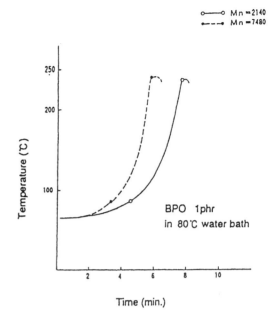

Fig.15 Cure Rate

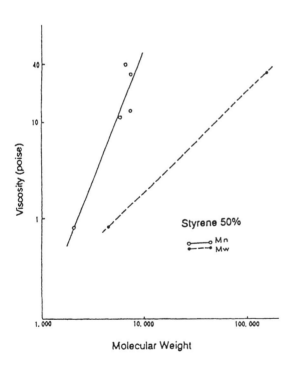

Fig.16 Relationship between Molecular Weight
and Viscosity

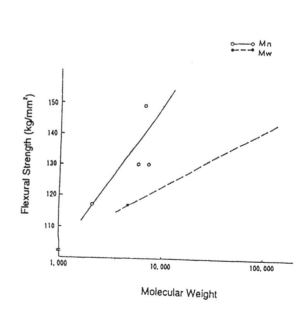

Fig.17 Relationship between Molecular Weight
and Flexural Strength

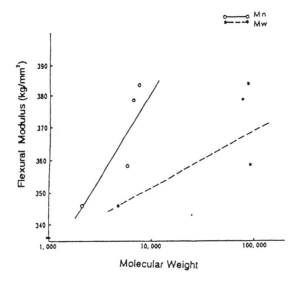

Fig.18 Relationship between Molecular Weight
and Flexural Modulus

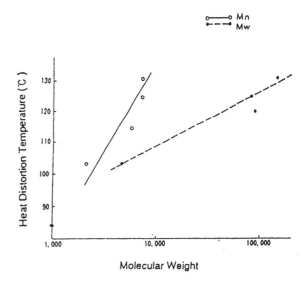

Fig.19 Relationship between Molecular Weight and
Heat Distortion Temperature

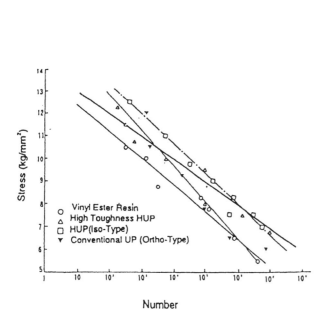

Fig.20 S-N Curve (Tensile-Fatigue)

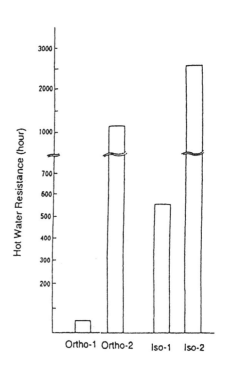

Fig.21 Hot Water Resistance

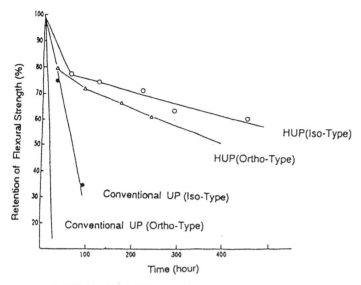

Fig.22 Alkali (10%NaOH,99°C) Resistance

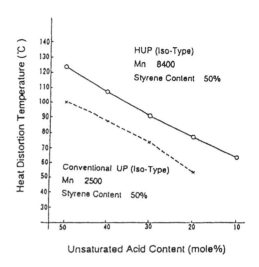

Fig.23 Influence of Unsaturation
on Heat Distortion Temperature

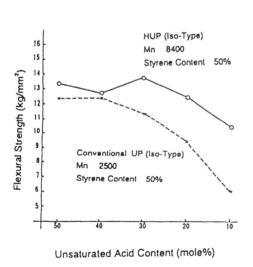

Fig.24 Influence of Unsaturation
on Flexural Strength

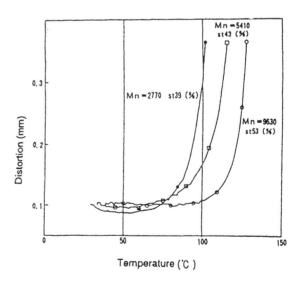

Fig.25 Relationship between Molecular Weight (Mn)
and Heat Distortion Temperature

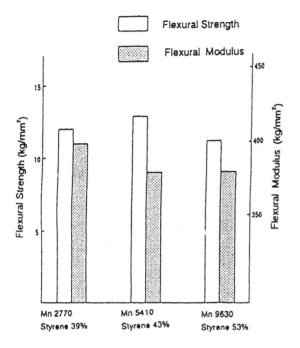

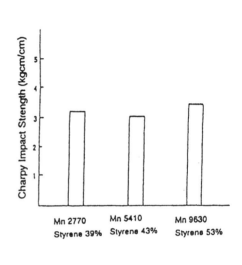

Fig.26 Influence of Molecular Weight (Mn)
and Styrene Content on Flexural Strength

Fig.27 Influence of Molecular Weight (Mn)
and Styrene Content on Charpy Impact Strength

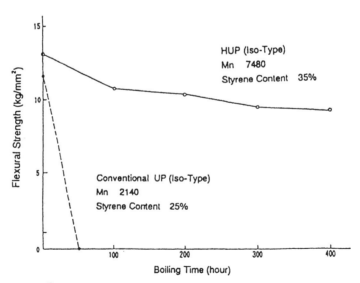

Fig.28 Influence of Molecular Weight (Mn)
and Styrene Content on Hot Water Resistance

Introduction to Phenolic Resin—Application, Processability and Behaviour

PETER FALANDYSZ

Introduction

Phenolic resins are used extensively in the manufacture of laminates Orientated Strand Board and other similar products.

Phenolic resins are also compounded into phenolic molding compounds used for compression and injection molding these materials are used extensively in electrical, automobile and appliance applications.

Liquid phenolic resins have been used in the FRP composite industry for many years and their role is expanding.

Why would phenolics be used?

What kind of applications are best suited to the use of phenolics?

What are the advantages (and disadvantages of phenolic resins)?

What are the perceptions and reality of phenolic resins?

How easy or difficult are they to process?

What kind of processes are suited to phenolic resins?

What advances have taken place?

What does the future hold for phenolic resins?

Why would a phenolic resin be chosen in preference to a polyester or vinyl ester?

Phenolic resins are inherently fire retardant <u>without</u> the use of fillers such as ATH or additives (many of which produce toxic fumes). They have excellent fire, smoke and smoke toxicity properties.

The composite part produced using phenolic resin therefore has all of the properties of low flame, smoke and toxicity, and this fact is critical in many applications that have demanding flame/smoke toxicity specifications.

Peter Falandysz, Composites by Design, 38 Francis Street S., Kitchener, Ontario N2G 2A2.

TABLE I

| | MASS TRANSIT | | |
	REQUIREMENT	PAINTED CELLOBOND FRP.	
SPECIFIC GRAVITY	-	1.4	
LC50	-	61	
E 162			
Is	<35	0.85	
E 662		NON FLAMING	FLAMING
Ds (1.5)	<100	0.1	0.6
Ds (4.0)	<200	9	15
Max Ds	-	30	51
Time To Max Ds (mins)	-	16	14

Applications

Some applications that currently use phenolic resins successfully and some potential applications that are being seriously considered with phenolic resins are:

Aircraft interiors
Ventilation Shafts
Conduit
Railcar passenger seat backs and bottom panels
Railcar window masks
Railcar food galleys and washrooms
Aircraft washrooms
Grating for offshore rigs and Industrial application

Growth

Some of the growth areas are:
- Mass Transit interiors
- Grating and piping in offshore high fire risk applications
- Industrial Ducting
- Tunnelling
- Marine (Navy, Ferryboats, Cruise ships)
  Interior panels, etc.

- Automotive - heat shields
- Construction - paneling, cladding

Advantages

The advantage of using phenolic resins for composites are very definitely the much superior performance regarding flammability, smoke generation, toxicity, high temperature resistance (392°F continuous) and reduced weight. A typical polyester composite has a specific gravity of 1.8 to 1.9. A typical phenolic composite has a specific gravity of 1.40.

How does a phenolic composite compare to a polyester composite in physical strength.

There are some differences. Design and the key properties that are required are of course integral to choosing the correct matrix.

TABLE II

*Mechanical Properties with alternate Liquid Composite Molding (LCM) resins:		
	Phenolic	Polyester
Tensile	20,000	23,000
Flexural	40,000	42,000
Flex Modulus	1.50	1.40
Notched Izod	18.0	15.0
Specific Gravity	1.50	1.50
Water Absorption	1.0	0.40
Glass%	40	40
*information courtesy of MFG. Ashtabula.		

Perceptions/Realities of Phenolic Resins

Acid Cure corrodes tooling

Aluminum or carbon steel tooling cannot be used however, composite tooling, 316 stainless steel tooling, chrome plated tools, nickel plated tools can and are being used successfully.

Modification kits are available to use on standard equipment to allow for spray up of phenolic resins (cost $3000 approx).

Phenolic resins cannot be colored

Only black or brown, phenolic resins are not pigmentable.

Shelf Life

At room temperature, 1 month is OK. Increase in viscosity is seen at 5 to 6 weeks. Phenolic resins can be stored 3 to 6 months @ 40° to 50°F.

Post Cure

Post curing is needed and recommended typically 1 to 3 hours @ 140° to 180°F. Post curing is performed off the mold.

Processes suitable for phenolic resins

Filament Winding
Hand Lay Up
Spray Lay Up
Resin Transfer Molding
SCRIMP
Pultrusion

These processes are being used today to produce phenolic composites for a variety of applications.
SCRIMP is becoming an important process for phenolic composites.

Advances in phenolic resins

- Phenolic compatible gel-coats are available
- Toughened phenolic resins are available for use in pre-preg manufacture
- Low viscosity resins (300 cps.)
- Delayed action catalysts (low corrosivity)
  5 mins. to 5 hours (Esters)
- Thixotropic phenolic resins are available
- Surface Paste
  Thixotropic phenolic based surface paste can be brushed or sprayed in place. This produces a smooth void free surface ready to be painted without further finishing.

Summary

The technology of phenolic resins is clearly not static.
New resins are being developed to meet the challenges of the marketplace.
Increased awareness of these user friendly systems will allow for wider acceptance.

Phenolic resins do require some modifications to the process as compared to other resin systems, however phenolic resins do offer some key advantages.

Phenolic resins produce a safe and cost effective composite, a distinct advantage in many applications.

Formulation Optimization of an Acrylic Modified Polyester Resin for Use in Resin Transfer Molding

LOUIS ROSS AND MATTHEW KASTL

ABSTRACT

A low profile acrylic modified polyester resin has recently been developed by Interplastic Corporation. This resin was used in a study with the objective of producing a formulation that could be molded into excellent surface quality parts without the use of a gel coat. Flat plaques were molded using the resin transfer molding (RTM) process in a tool that has one side made with plate glass. This allowed surface quality based on a Diffracto TPS-2 to be determined accurately. Surface quality as well as physical properties were determined as a function of filler, low profile additive, and fiberglass content. The low profile additive is a saturated polyester dissolved in styrene. Higher levels of thermoplastic improved surface quality, but reduced Barcol hardness. As filler content increased, the surface quality improved, but as expected, the density also increased. Based on this investigation, a formulation that provided an optimum balance of surface quality and physical properties was determined.

INTRODUCTION

Low profile RTM resins have become widely available in the past five years. The major reason for their growth in the industry is that there are more RTM applications where surface quality is important. The goal of our effort was to develop a system that produced primer-ready parts without the use of gel coat.

The two methods investigated to reduce surface waviness were the use of low shrink resins with high filler content and the use of low profile systems similar to compression molding formulations. The former system failed to provide a surface quality smooth enough so that the gel coat process could be eliminated. Systems using thermoplastic shrink control additives were prone to exhibit porosity and had insufficient cure. Recently, a patented acrylic modified resin has been developed and found to provide improved surface quality when used in conjunction with thermoplastic additives. This system also exhibited improved cure and physical properties, as well as negligible porosity. In this investigation, the objective was to optimize the resin/filler/thermoplastic mixture in order to provide a commercially viable formulation.

Louis Ross and Matthew Kastl, Interplastic Corporation, 1225 Walters Blvd., St. Paul, MN 55110-5145.

EXPERIMENTAL

Materials

The resin used in this experiment, CoREZYN[R] 44-BZ-9606, is a patented low shrink modified acrylic polyester thermoset designed for methyl ethyl ketone peroxide cure. The filler used is J. M. Huber W 4 calcium carbonate selected because of its particle geometry which allowed greater filler loading with reduced viscosity. The thermoplastic shrink control additive, CoREZYN[R] 20-LP-7510, is a styrenated saturated polyester resin product manufactured by Interplastic Corporation. It was selected for previous success in providing improved surface appearance and shrink control. Multiple plies of Owens-Corning M8610 1.50 ounce/ft^2 continuous strand mat were used as the fiber reinforcement.

Tooling

A 12 inch by 20 inch RTM tool built with a plate glass face and heated surfaces was used to simulate ideal mold conditions. The tool heat was supplied by conductive fabric electrically controlled by a low voltage rheostat.

Equipment

A pressure pot style RTM injector was used to pump the catalyzed mixture into the tool. A Diffracto TPS-2 surface analyzer was used to evaluate the waviness of the molded panels.

Designed Experiment

Design-Expert software available from Stat-Ease, Incorporated, was used to define a mixtures experiment based on the component levels found in Table 1. Panels were molded at two temperature ranges: ambient (approximately 25°C) and 60°C. These tool temperatures were selected because they are the typical ranges for RTM, and also to study the thermoplastic's behavior.

Initially, 20 experiments were performed per the guidelines set up by the design software. Each panel was allowed to cure in the mold for 30 minutes beyond its peak exotherm as determined by a thermocouple. The molded panels were allowed to age a minimum of 48 hours before the various testing began. Additionally, certain test measurements were taken before and after a 45 minute 93°C postcure in order to study the effects that the intense heat of paint baking would have on molded parts.

TABLE 1 - Mixtures Experiment Levels

Component	Minimum Level (%)	Maximum Level (%)
Resin	30	70
Calcium Carbonate	10	60
Thermoplastic Additive	10	25

In each experiment, three plies of the continuous strand glass mat were used which yielded a glass content of approximately 12% by volume. The glass content was reported in percent by volume because the content ranged from approximately 10% to 28% by weight because of the change in specific gravity of each mixture. No surfacing veil was used in these experiments in order to get a uniform composite. Additional inhibitor was required in some of the tests to extend the gel time of the resin systems when molded at 60°C.

Once all the data was accumulated and entered into the design software, the outcome was checked to insure there was a minimum of 95% reliability. The data then underwent an optimization process to predict the best mixture for lowest surface waviness value, lowest specific gravity, highest Barcol hardness, and best flexural strength.

RESULTS/DISCUSSION

Surface

A Diffracto TPS-2 surface analyzer was used to quantify surface quality of the molded RTM panels. The design response surface found in Figure 1 displays the image of the mixtures experiment for ambient cure, non-postcured test panels, as well as the pattern calculated from data input. For ease of interpretation, axis X1 will always be resin, X2 filler, and X3 thermoplastic additive.

Figure 2 is a three dimensional image of Figure 1. As can be seen, the lowest areas on the wire diagram are on either end of the design surface.

Figure 2 shows that under ambient cure, non-postcured conditions a low surface waviness value can be achieved using 30% resin, 53% filler, and 17% thermoplastic additive. On the opposing end of the response surface, a low waviness value can also be obtained using 70% resin, 15% filler, and 15% thermoplastic additive. This type of response surface may be of interest in some cases; however, for this study, the other responses must also be considered in order to locate the optimum performing mixture.

Similarly, data collected from panels molded at 60°C before and after postcuring were entered into the design software and plotted in Figures 3 and 4. These figures both show that a mixture of 62% resin, 13% filler, and 25% thermoplastic additive should result in low surface waviness values. Also, little change was seen after postcuring, which may indicate acceptable cure.

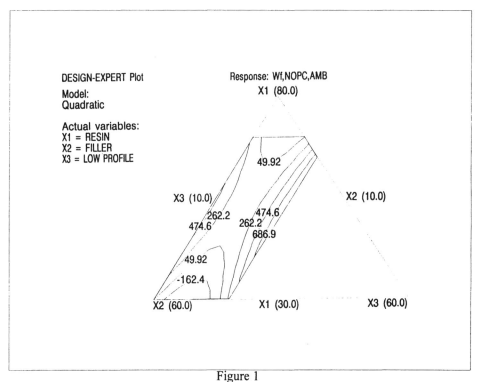

Figure 1

Two Dimensional Plot of Surface Smoothness (Wf) for Ambient Cured Non-Postcured Mixtures of Resin, Filler, and Thermoplastic Additive

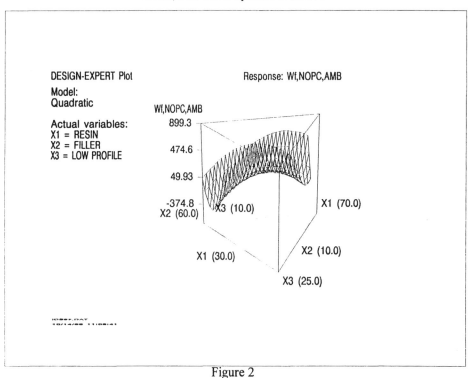

Figure 2

Three Dimensional Plot of Surface Smoothness (Wf) for Ambient Cured Non-Postcured Mixtures of Resin, Filler, and Thermoplastic Additive

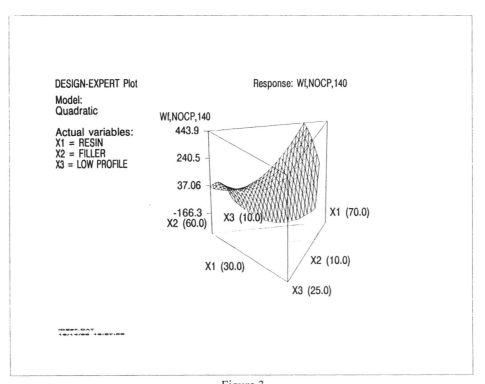

Figure 3

Three Dimensional Plot of Surface Smoothness (Wf) for 60°C Cured, Non-Postcured Mixtures of Resin, Filler, and Thermoplastic Additive

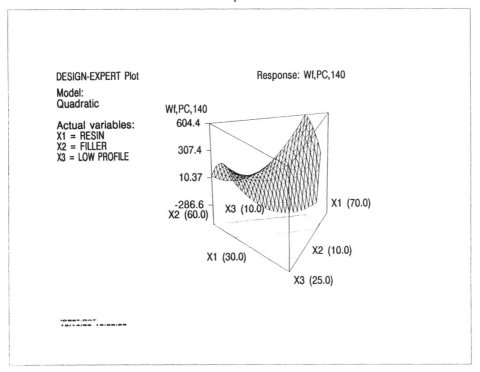

Figure 4

Three Dimensional Plot of Surface Smoothness (Wf) for 60°C Cured, Postcured Mixtures of Resin, Filler, and Thermoplastic Additive

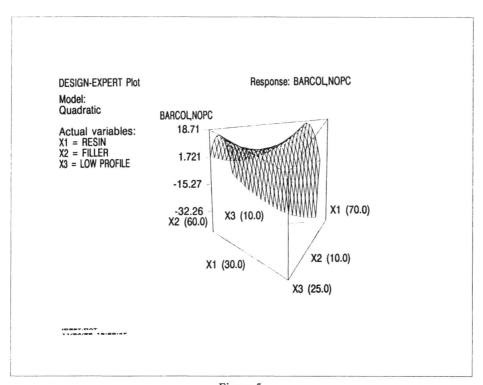

Figure 5
Three Dimensional Plot of Barcol Hardness of Non-Postcured Mixtures of Resin, Filler, and Thermoplastic Additive

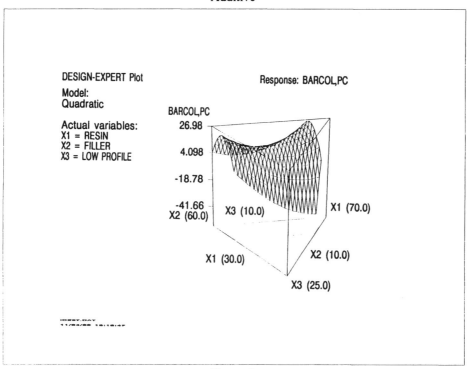

Figure 6
Three Dimensional Plot of Barcol Hardness of Postcured Mixtures of Resin, Filler, and Thermoplastic Additive

Barcol Hardness

Using the guidelines of ASTM D 2583, response surfaces were developed from the design software for Barcol hardness using a Barber Colman 934-1 gauge. Figures 5 and 6 display the plots of non-postcured and postcured panels. In these diagrams, it can be seen that the greatest Barcol hardness is found along the edge where there is the least thermoplastic additive. Also, the shape of the response surface was not significantly affected by postcuring possibly indicating that the parts cure well without additional baking. The predicted mixtures for highest Barcol hardness are 30% resin, 54% filler, 16% thermoplastic additive, and also, 70% resin, 19% filler, and 11% thermoplastic.

Physical Properties

Flexural strength (ASTM D 790) response surfaces were also developed. Figure 7 shows the response surface and, as can be seen, the areas of greatest strength are on opposing ends. The mixtures for peak flexural strength were found to be 70% resin, 15% filler, and
15% thermoplastic additive, or 30% resin, 52% filler, and 18% thermoplastic additive. But as stated earlier, this experiment involves several responses, all of which must be optimized.

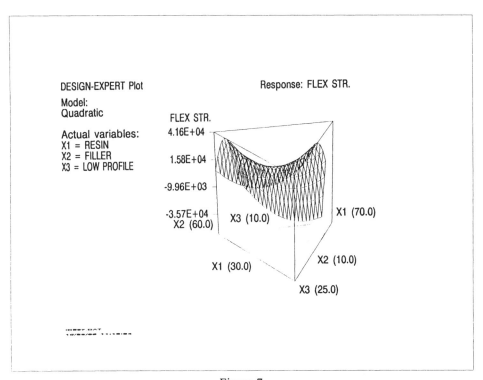

Figure 7
Three Dimensional Plot of Flexural Strength (ASTM D 790) of Mixtures of Resin, Filler, and Thermoplastic Additive

Specific Gravity

The fourth response studied was specific gravity. This characteristic was examined because molded part weight is an important aspect for many end users. Following the procedure of ASTM D 792, the specific gravity of each of the molded panels was determined and entered into the design software. Figure 8 shows the response surface for specific gravity. Expectedly, when filler content increases, specific gravity increases.

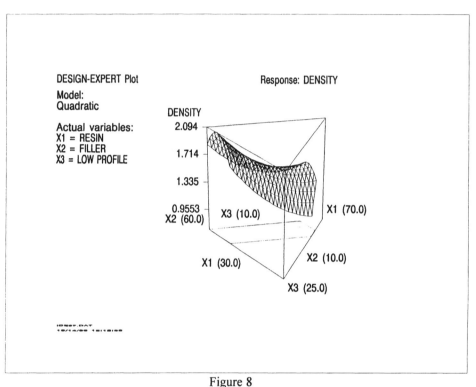

Figure 8
Three Dimensional Plot of Specific Gravity (ASTM D 792) of Mixtures of Resin, Filler, and Thermoplastic Additive

Optimization

The optimization process built into the design software involves manually choosing the response limits and/or ranges for response data to be evaluated. In this case, during the optimization process some adjustments in setting target limits were necessary. Because three or more response limits were involved, primary and secondary targets were selected.

In this optimization, a maximum surface waviness value of 100 was selected as the primary target.

Specific gravity, Barcol hardness, and flexural strength were selected as secondary targets, and their limits were adjusted in order to find areas where a mixture met the criteria.

In Figures 9 through 12, the unshaded areas displaying flags indicate the zone on the response surface which meets the designated criteria. Note that the unshaded areas and mixtures change as the molding conditions change from ambient to 60°C cure and from non-postcured to postcured.

The optimization limits for the non-postcured, ambient cure system seen in Figure 9 are: surface waviness of 100 maximum, Barcol hardness of 20 minimum, specific gravity of 1.55 maximum, and flexural strength of minimum 111 MPa. The flag in Figure 9 pinpoints the unshaded area. The value on the flag represents the predicted surface waviness value for the area, and the legend displays the mixture where the flag is pinpointed. In other words, 68% resin, 19% filler, and 13% thermoplastic should meet the designated criteria and result in a 32.31 waviness value.

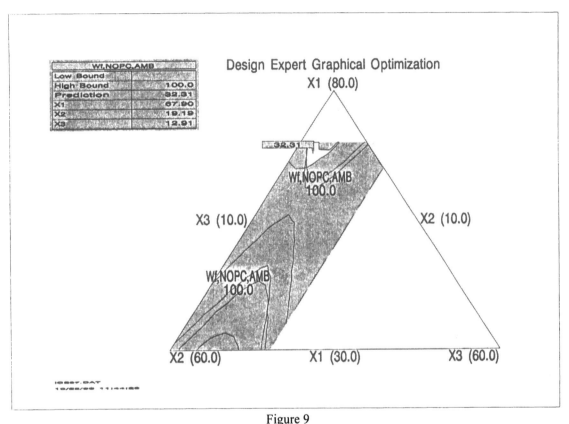

Figure 9
Optimization Plot of Ambient Cured, Non-Postcured of Mixtures of Resin, Filler, and Thermoplastic Additive

Figure 10 displays how the optimum mixture area changes after the panels plotted in Figure 9 have been postcured. Here the flagged area has moved to an area showing a mixture of 53% resin, 37% filler, and 10% thermoplastic additive. The flag now reads 82.69 for surface waviness, and in order to get any unshaded area on the plot, the Barcol limit had to be reduced to 1 and the flexural strength limit dropped to 28 MPa. These results indicate that it may be difficult to get acceptable properties if the parts are exposed to the heat of a paint oven.

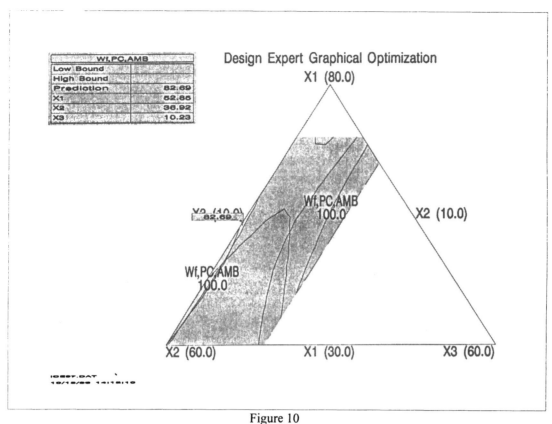

Figure 10
Optimization Plot of Ambient Cured, Postcured of Mixtures of Resin, Filler, and Thermoplastic Additive

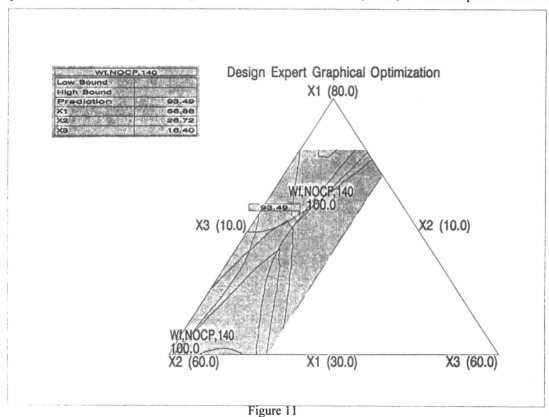

Figure 11
Optimization Plot of 60°C Cured, Non-Postcured of Mixtures of Resin, Filler, and Thermoplastic Additive

The optimization process for the panels that were molded at elevated temperature appeared to be more straightforward. The flagged areas in Figures 11 and 12 both are adjacent. The predicted optimum mixture for parts to be molded at 60°C is approximately 55.50% resin, 27.50% filler, and 17% thermoplastic additive. The criteria of 100 maximum surface waviness is met with the guidelines for Barcol of 10 minimum, specific gravity of 1.55 maximum, and a flexural strength of 69 MPa minimum.

Once the predicted results were established, additional panels were molded to validate the results of the design optimization.

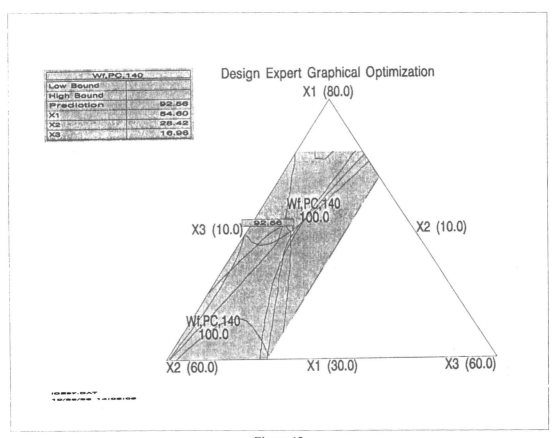

Figure 12
Optimization Plot of 60°C Cured, Postcured of Mixtures of Resin, Filler, and Thermoplastic Additive

Conclusion

In this mixtures experiment, surface quality, Barcol hardness, flexural strength, and specific gravity were the response variables. The components of the mixture were modified acrylic polyester resin, calcium carbonate filler, and saturated polyester thermoplastic resin.

We found that the optimized mixtures were related to the molding temperature. The mixtures experiments which were performed at ambient temperature and then postcured failed to meet the minimum requirements of 100 maximum surface waviness, a maximum specific gravity of 1.55, Barcol hardness of 5 minimum, and flexural strength of 69 MPa

Table 2 - Optimum Formulations

	% RESIN	%FILLER	%THERMOPLASTIC
AMBIENT CURE, NON-POSTCURED	68	19	13
AMBIENT CURE, POSTCURED	53	37	10
60°C CURE, NON-POSTCURED	57	27	16
60°C CURE, POSTCURED	55	28	17

minimum. In order to get any optimal zone on the response surface for the ambient cured, postcured experiments, the Barcol hardness limit had to be reduced to 1 and the flexural strength limit lowered to 28 MPa. In general, the responses that had the greatest effected on the optimization process were specific gravity and Barcol hardness. The optimized formulations for the experiments are found in Table 2.

BIOGRAPHIES

Louis R. Ross

Louis R. Ross received a Bachelor of Science degree in Chemistry in 1975 from the University of Akron and a Ph.D. in Polymer Science from the University of Akron in 1984. He has worked in the fiberglass reinforced plastics industry for the past 14 years and is currently Technical Director of Silmar Resins Division of Interplastic Corporation.

Matthew C. Kastl

Matthew C. Kastl is a Senior Resin Chemist with Commercial Resins Division of Interplastic Corporation. He joined Interplastic Corporation in 1979 after receiving a degree in Plastics Technology from North Hennepin Technical College, Brooklyn Park, Mn.

Innovations in Pultrusion Design, Processing and Control

Evaluation of the Quality and Consistency of Open-Bath Wet-Out Techniques for Pultruded Composites

ELLEN LACKEY AND JAMES G. VAUGHAN

ABSTRACT

The quality of fiber/resin wet-out is known to affect the properties of composite materials, but wet-out is often overlooked when pultrusion process parameters are considered. Possible reasons for this are that the evaluation and control of the quality of wet-out are difficult. The objectives of this study are to observe and evaluate the quality and consistency of wet-out achieved using various open-bath wet-out techniques with the pultrusion process. Unidirectional fiberglass/epoxy and graphite/epoxy samples were examined in both the transverse and longitudinal directions using a scanning electron microscope. These samples were examined to evaluate the bond consistency at the fiber/resin interface and the uniformity of fiber distribution, both over the cross-section and along the length of the part. These observations were compared with measured mechanical properties from the samples to evaluate the connections between wet-out, appearance at the fiber/resin level, and mechanical property data. Example micrographs and mechanical property data are presented for samples produced using various open-bath wet-out techniques. Knowledge gained from this study will be used to improve wet-out techniques for pultruded composites.

INTRODUCTION

The pultrusion process is an automated, cost-effective method for the production of composite materials. The basic open-bath pultrusion process for a thermoset resin involves pulling continuous reinforcement, in the form of mat or roving, through a resin bath to coat the reinforcement with resin, through preform plates to shape the material and remove excess resin, and through a heated die to cure the resin and form the finished product. Even though the pultrusion process appears to be simple, process variables and variable interactions are known to influence the mechanical properties and overall quality of pultruded composites (Lackey, 1993).

Wet-out, the exposure of dry, reinforcing fiber to resin, is an essential element of the pultrusion process. Wet-out techniques can have a significant influence on the microstructure of a composite, and it has been demonstrated that microstructural features such as voids

---

Ellen Lackey and James Vaughan, University of Mississippi, 201 Carrier Hall, University, MS 38677

(Bowles, 1992), fiber distribution (Karam, 1991), and fiber alignment (Wisnom, 1990) can affect the mechanical properties of composites. However, the process variable of wet-out has received very little attention. In the pultrusion process, wet-out methods can be divided into two general categories — closed injection methods and open-bath wet-out methods. Some reports concerning the use of proprietary resin injection methods (Blosser, 1994; Dawson, 1995; Beckman, 1995; Dawson, 1996) have been published, but these reports do not provide details concerning these wet-out techniques. Open bath wet-out techniques are more commonly used, but few studies (Bijsterbosch, 1990; Chandler, 1993; Lackey, 1994) have examined them.

Possible reasons that wet-out methods are often overlooked when pultrusion process parameters are considered include the fact that the evaluation and control of the quality of the wet-out are difficult. The objectives of this study are to demonstrate methods to examine wet-out and to use the methods to observe and evaluate the quality and consistency of wet-out achieved using various open-bath wet-out techniques with the pultrusion process.

EXPERIMENTAL PROCEDURE

Pultruded composites were produced using various open-bath wet-out methods in the University of Mississippi Composite Materials Laboratory using a PTI Pulstar 804 pultruder. A 2.54 cm x 0.318 cm (1 inch x 0.125 inch) flat profile was produced for all experiments. Epoxy resin was used as the matrix in all experiments. Both AS4-W 12K unidirectional graphite fiber and PPG 2001 112 yield E-glass reinforcements were examined. An individual tow of the AS4-W 12K graphite fiber is composed of approximately 12,000 fibers that are about 7 μm in diameter, and an individual end of the 112 yield E-glass is made of approximately 3,920 fibers that are about 24 μm in diameter. The examination of different fiber types was necessary because differences in permeability and surface reactivity are seen for different types of fiber.

A number of open-bath wet-out methods were used in these experiments. These methods were selected to range from poor wet-out to sufficient wet-out so the effect of wet-out quality could be evaluated. A summary of the wet-out methods that were examined is shown in Table I. Resin baths designated Bath A, Bath B, and Bath C, which are described in Table I, are shown in Figures 1, 2, and 3, respectively. Resin Bath A has combs that are designed to

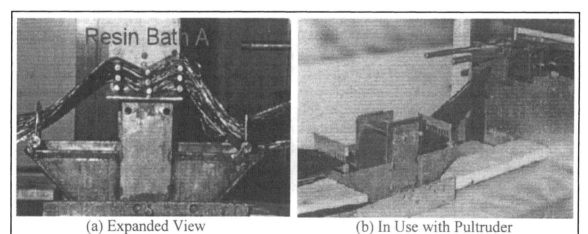

| (a) Expanded View | (b) In Use with Pultruder |

Figure 1. Views of Resin Bath A showing (a) the separation of the fiber through the pressure bars and (b) the bath in use on the pultrusion machine.

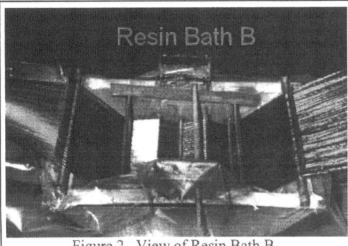
Figure 2. View of Resin Bath B.

maintain fiber alignment and to separate the fibers so resin can enter the interior of the fiber bundle. Both Resin Bath A and Resin Bath B have pressure bars that the resin-coated fibers travel over and under to work resin into the fiber. Resin Bath C, a flow-through resin bath, also has pressure bars to work resin into the fibers; however, fibers follow a straight path through this bath and are not separated by combs or diverted over and under these bars.

TABLE I. OPEN-BATH WET-OUT METHODS EXAMINED	
No Bath	Resin was poured on the fiber bundle as it entered the first preform plate. No additional mechanical working of the resin-coated fiber was performed. Designed to give poor wet-out.
Resin Bath A	Dip-type resin bath that used combs and pressure bars to separate the fiber bundle and work resin into the fiber bundle. Designed to give good wet-out.
Resin Bath B	Dip-type resin bath that used pressure bars to work resin into the fiber bundle. Bath is shallower and wider than Bath A. Designed to give good wet-out.
Resin Bath C	Flow-through resin bath in which fibers followed a straight path through the resin bath. Pressure bars were used to work resin into the fiber bundle. Designed to give good wet-out with limited fiber movement.

After the composites had been produced using these wet-out methods, mechanical property tests and microscopic examinations were used to evaluate the quality of the wet-out obtained. Mechanical property test conducted included four-point flexural strength tests (ASTM D790), three-point flexural strength tests (ASTM D790), and short-beam shear tests (ASTM D2344). Transverse tensile tests, which were designed to cause the samples to fail parallel to the fiber/resin interfaces of the samples so the extent of resin penetration into the fiber tow could be examined, were also conducted. In order to have sufficient area to grip the 2.54 cm

Figure 3. Resin Bath C, flow-through bath, in use on the pultrusion machine.

wide test coupons so they could be loaded perpendicular to the fiber direction using an MTS universal test machine, composite tabs were glued to the test coupons. A JEOL 6100 SEM was used to examine cross-sections and fracture surfaces of the composite samples. Standard metallographic polishing techniques were used to prepare the sample cross-sections for examination. Oxford Link ISIS software and Noesis Visilog software were used for digital image acquisition and analysis.

RESULTS AND DISCUSSION

Examination of the composite samples revealed that microstructural features and mechanical property results were related to the wet-out methods used to produce the pultruded composites.

CROSS-SECTION EXAMINATION

Cross-sections of graphite/epoxy and fiberglass/epoxy composites were examined. A portion of the 2.54 cm x 0.318 cm cross-section of a fiberglass/epoxy sample produced using Resin Bath A is shown in Figure 4. The entire 0.318 cm (0.125 inch) thickness of the sample is visible in the figure. Areas of variation in fiber distribution are readily apparent. Resin Bath A relies on fiber path variations to mechanically work resin into the fiber bundle as the resin-coated fiber travels over and under pressure bars. It is believed that the amount of fiber movement associated with this wet-out method contributed to the variation seen in the micrograph in Figure 4. A portion of the cross-section of a fiberglass/epoxy composite produced using Resin Bath C, the flow-through bath that employs a straight fiber path, is shown in Figure 5. Some variations are seen in the fiber distribution in this sample, but

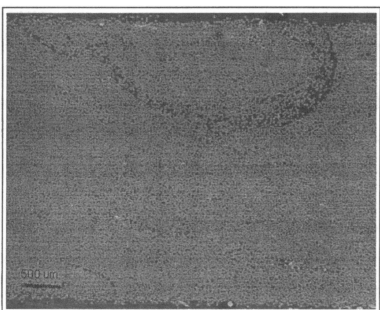

Figure 4. Cross-section of a fiberglass/epoxy sample produced using Resin Bath A showing pronounced variation in fiber distribution.

these variations are less pronounced than the variations seen in Figure 4. When examined at higher magnification, variations in fiber diameter of the fiberglass can also be seen, as shown in Figure 6. Though these variations in fiber diameter are not controlled by the wet-out method, these variations are significant because the fiber diameter can affect things such as the permeability of the fiber bundle and the pressure rise that occurs as the resin-coated fiber bundle enters the pultrusion die.

To obtain a better understanding of the variation seen over the entire 2.54 cm x 0.318 cm (1 inch x 0.125 inch) cross-section, higher magnifications were used and micrographs were pieced together to show the entire 0.318 cm thickness of the cross-section. A portion of such a composite micrograph of a fiberglass/epoxy sample produced using Resin Bath C is shown in Figure 7. An examination of the change in fiber distribution that occurred along the length of the pultruded composite was also examined by sectioning the samples at 0.636 cm (0.25 inch) intervals along the length of the part. These samples were then mounted, polished, and examined using the SEM. The composite micrograph in Figure 8 is a view of the cross-section of the sample taken 0.636 cm

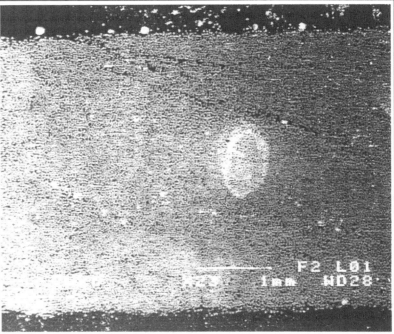

Figure 5. Cross-section of a fiberglass/epoxy sample produced using Resin Bath C showing less pronounced variation in fiber distribution.

down the length of the part from the sample shown in Figure 7. A comparison of Figure 7 and Figure 8 shows that the fiber distribution changed along the length of the part even though the flow-through resin bath that employed a straight fiber path was used. This reveals the complex nature of the fiber distribution variations seen in pultruded composites.

Cross-sections of graphite/epoxy composites were also examined; however, because the diameter of the AS4-W 12K graphite fiber is much smaller than the diameter of the 112 yield E-glass fibers, relatively high magnifications

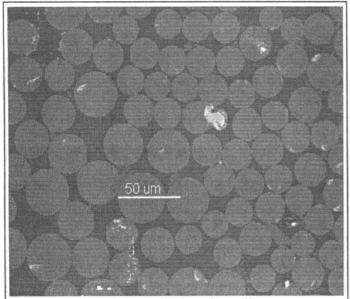

Figure 6. Variations in E-glass fiber diameter.

must be used in order to see the individual fibers. Similar variations in fiber distribution are seen in the graphite/epoxy samples, but the higher magnifications that are required make it more difficult to view the distribution over large areas of the cross-sections of these samples.

Figure 7. Portion of the cross-section of a 2.54 cm x 0.318 cm E-glass/epoxy composite showing fiber distribution variation.

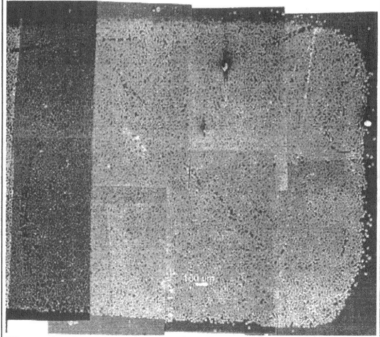

Figure 8. Cross-section portion that corresponds to the section shown in Figure 7 that was taken from a sample 0.636 cm down the length of the part. A comparison to Figure 7 shows fiber distribution changes that occur along the length of the pultruded composite.

Mechanical properties and failure appearances of the composites were compared so the effect of a wet-out method could be evaluated.

E-glass/Epoxy Composites

Microscopic examination of transverse tensile fracture surfaces of the E-glass/epoxy samples revealed that samples which exhibited good mechanical properties generally had well-coated fibers and little evidence of clean fiber pull-out. Samples with poor mechanical properties had areas of clean fiber pull-out and fibers with little resin coating them. Examples of these observations are shown in Figures 9a and 9b, and the mechanical properties of these

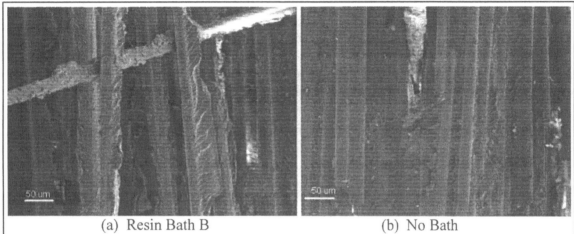

(a) Resin Bath B   (b) No Bath

Figure 9. Views of E-glass/epoxy fracture surfaces showing the effect of wet-out method.

samples are shown in Table II. These samples were produced under similar pultrusion conditions and contained 68.2% E-glass by volume. The mechanical properties of the sample shown in Figure 9a, which was considered to have good wet-out, are in the range normally expected for this composite system. The "no bath" wet-out method used to produce the sample shown in Figure 9b was expected to result in poor wet-out because nothing was done to work resin into the interior of the fiber bundle; the poor mechanical properties, regions of uncoated fiber, and type of break seen for the flexural strength samples confirm this. The type of break of the three-point flexural strength samples, illustrated in Figure 10, was also shown to correlate to wet-out quality for the glass/epoxy composites. Center bend failures were seen for samples that had sufficient wet-out, and center split failures were seen for poor wet-out samples. Interlaminar shear associated with the

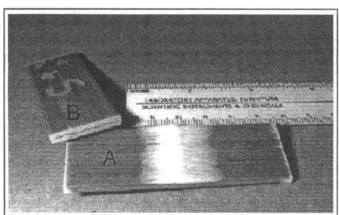

Figure 10. Failure types from three-point flex tests of E-glass/epoxy samples. A — Center bend associated with good wet-out. B — Center split associated with poor wet-out.

center split fracture would be expected if the fiber/resin interface was weak due to poor wet-out. This correlation provides a good quality-control evaluation tool which can be used for preliminary evaluation of wet-out quality.

TABLE II. MECHANICAL PROPERTIES OF E-GLASS/EPOXY SAMPLES		
Wet-out Method	Three-Point Flexural Strength (MPa)	Short-Beam Shear Strength (MPa)
Resin Bath B	1332 MPa	74.9 MPa
No Bath	1073 MPa	45.6 MPa

Graphite/Epoxy Composites

Similar examinations and comparisons were made for AS4-W 12K graphite/epoxy composites, but the results for the samples with graphite reinforcement were not as straightforward. Mechanical properties of two graphite/epoxy samples are shown in Table III. Low magnification views of these samples, which are shown in Figures 11a and 11b, show that more loose fibers are present for the sample with poor mechanical properties; however, differences are not readily apparent at higher magnification. Higher magnification views of transverse fracture surfaces of the graphite/epoxy composites referenced in Table III are shown in Figures 12a and 12b, and it is apparent that both samples have numerous regions of fibers that are not well-coated with resin. These AS4-W 12K/epoxy graphite samples also did not show the correlation between wet-out and the flexural test break types seen for the fiberglass/epoxy samples. However, preliminary tests on samples produced using Zoltek PANEX 33-0048 and 0160 graphite as reinforcement indicate that this may not be the case for all graphite/epxoy composites. Additional experiments are currently being conducted for further study of this.

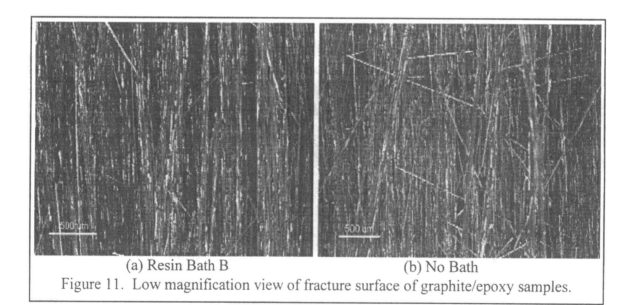

(a) Resin Bath B   (b) No Bath

Figure 11. Low magnification view of fracture surface of graphite/epoxy samples.

TABLE III. MECHANICAL PROPERTIES OF AS4-W 12K GRAPHITE/EPOXY SAMPLES			
Wet-out Method	Fiber Volume	Four-Point Flexural Strength (MPa)	Short-Beam Shear Strength (MPa)
Resin Bath B	63.8%	1502 MPa	78.7 MPa
No Bath	62.2%	375 MPa	49.7 MPa

SUMMARY AND CONCLUSIONS

The results of this study demonstrate that wet-out techniques can affect the micro-structure and mechanical properties of pultruded composites.

Micrographs of the cross-sections of E-glass/epoxy composites showed that significant variations can occur in the fiber distribution over the cross-section and down the length of the part. Wet-out techniques have the potential to affect the degree of fiber distribution variation seen for the composites; however, variations seen even short distances down the length of the part demonstrate that this will be difficult to control.

Wet-out was also shown to affect the mechanical properties and fracture appearance of pultruded composites. Wide variations in wet-out quality could be identified through examination of the extent to which resin coated the fiber and the type of failure seen for flex test samples for E-glass/epoxy samples. These variations in wet-out quality were also reflected in the mechanical properties of the pultruded E-glass/epoxy composites. A similar connection was seen for the mechanical properties of the AS4-W 12K/graphite samples; however, the connection between microstructure and wet-out was not as evident for these samples.

Data from this study can be used as a basis to improve pultrusion wet-out techniques. The development of additional techniques to quantify the variations described here will allow more subtle variations of wet-out to be evaluated.

ACKNOWLEDGMENTS

The authors would like to acknowledge the support of NSF/EPSCoR grant number EPS-9452857 and the State of Mississippi.

REFERENCES

Beckman, J. J., et al. 1995. "A Method for Producing Ladder Rail Using High Pressure Injection Die Impregnation," 50th Annual Conference, Composites Institute, The Society of the Plastics Industry, Inc., pp. 6A1-5.

Bijsterbosch, H. and R. J. Gaymans. 1990. "Pultrusion Process: Nylon 6.Glass Fibres," *IMechE*, pp. 101-104.

Blosser, Randy, John Florio, Jr., and Prasad Donti. 1994. "Continuous Resin Transfer Molding™ of High Quality, Low Cost, Constant Cross Section Composite Structural Elements," *39th International SAMPE Symposium*, pp. 1961-72.

Bowles, Kenneth J. and Stephen Frimpong. 1992. "Void Effects on the Interlaminar Shear Strength of Unidirectional Graphite-Fiber-Reinforced Composites," *Journal of Composite Materials*, Vol. 26, No. 10.

Chandler, H. W., B. J. Devlin, and A. G. Gibson. 1993. "Impregnation Mechanics for the Melt Pultrusion of Thermoplastic Matrix Composites," *Composites Modelling and Processing Science Volume III — Proceedings of the 9th International Conference on Composite Materials (ICCM/9)*, Madrid, pp. 737-744.

Dawson, Donna K. 1995. "Wind and Water Power," *Composites Technology*, July/August, pp. 28-31.

Dawson, Donna K., ed. 1996. "Injection Die Pultrusion," *Composites Technology*, January/February, pp. 23-30.

Karam, G. N. 1991. "Effect of Fibre Volume on Tensile Properties of Real Unidirectional Fibre-Reinforced Composites," *Composites*, Vol. 22, No. 2, March, pp. 84-88.

Lackey, E., D. Theobald, and J. G. Vaughan. 1994. "Effect of Fiber/Resin Wet-Out on Pultruded Composites," *Proceedings of the 39th International SAMPE Symposium*, pp. 183-193.

Lackey, E. and J. G. Vaughan. 1993. "Effect of Pultrusion Process Parameters on Flexural Strength," *Proceedings of the 38th International SAMPE Symposium*, pp. 461-470.

Wisnom, M. R. 1990. "The Effect of Fibre Misalignment on the Compressive Strength of Unidirectional Carbon Fibre/Epoxy," *Composites*, Vol. 21, No. 5, September, pp. 403-407.

BIOGRAPHIES

Ellen Lackey — Dr. Lackey is an Assistant Professor of Mechanical Engineering at the University of Mississippi. She has been involved with composite materials research for the past six years in areas such as mechanical and physical property characterization, optimization of manufacturing through the use of statistical experimental design techniques, microscopy, and development of composite manufacturing techniques.

James G. Vaughan — Dr. Vaughan received his B.S. degree in Electrical Engineering in 1971, and his M.S. and Ph.D. in Materials Science and Metallurgical Engineering (1974, 1976), from Vanderbilt University. He worked with the Bendix Corporation before joining the University of Mississippi in 1980. Dr. Vaughan serves as Professor of Mechanical Engineering and Associate Dean of the School of Engineering at the University of Mississippi. He has worked extensively with pultrusion over the past ten years and has developed a pultrusion laboratory at the University where he conducts various experimental and analytical studies of the pultrusion process.

Low Cost Methods for VOC Abatement in the Pultrusion Plant and for Compliance with the Clean Air Act

JEFF MARTIN AND DOUGLAS S. BARNO

ABSTRACT

The SPI and CFA Trade Associations have reached an agreement with the federal Occupational Safety and Health Administration (OSHA) to voluntarily limit worker's exposure to styrene in the workplace to 50ppm starting in July 1997. In order to implement this action, SPI and CFA have agreed to educate and train their membership on the steps they can take to be in compliance with OSHA. There are numerous procedures that can be implemented by processors to effect their compliance. In this case the processors simply need to learn about these steps.

The same can be said for compliance with the EPA on the steps that processors need to take to be in compliance with the Clean Air Act guidelines relative to the exhaust air that is expelled to the atmosphere. Since President Bush signed this law into effect in November 1990, processors have delayed taking many steps until a later time. Well, later is NOW! EPA is drafting its recommendations and guidelines (expected by the Fall of 1997) for a Maximum Achievable Control Technology (MACT) Standard that will have to be implemented or reflected in your permits by May 15, 1999 nearly two years from now. Pultruders, emitting over 10 tons of styrene or other VOC's, who make up only 6% of the total pounds of composites processed will, nevertheless, have to be prepared to be in compliance with the MACT Standards.

This paper discusses technologies that can collect VOC/styrene laden process air within the plant and treat it prior to exhausting to it to the atmosphere. There are many technology choices. This paper discusses several of the newer technologies that show cost-effective promise.

Jeff Martin, President, Martin Pultrusion Group, 207 N. Hayden Pkwy Hudson, OH 44236 USA
Douglas S. Barno, Vice President, VVK Weege of North America Inc.
905 River Road, Granville, OH 43023 USA

INTRODUCTION

If you emit more than ten tons of styrene or other VOC's into the atmosphere per year you are a major emitter of HAP's, Hazardous Air Pollutants, and you must presently report the amounts by VOC category. Effective on May 15th 1999 you will be further required to install maximum achievable control technologies or reflected in your permitting as determined by EPA. This will happen in 1999 even if EPA has not published a standard!

In the Fall of 1997, EPA expects to have their first draft of these standards written. They are obligated by Congress to meet this deadline. They expect to have Standards published by late 1998. EPA has already said they expect to be late in publishing these standards. Whenever these standards are published they will cover all composite processes, including pultrusion.

Pultruders probably put a smaller amount of styrene into the air than any of the other composite processes. It has been estimated that this can be as low as 2% of the total pounds of resin mix processed. Processing 1,000,000 pounds of mixed resin with a 2% loss will exceed the 10 ton per year limit and make you a major source for HAPs.

As a pultruder you may have participated with the Pultrusion Industry Council in an effort to quantify the effect of air flow over a resin bath. The conclusion was that there was a direct correlation between air flow and the percentage of styrene loss. The more the air flow the larger the loss of styrene from the resin. These tests were conducted in enclosures built by MMFG in Bristol, Va. As a result of this containment test, EPA is considering enclosures for use by pultruders. Along with this they are considering resin injection for dies to minimize the amount of styrene that is exposed to the facility and subsequently to the environment. However, to date, few have been able to achieve acceptable, competitive processing speeds with resin injection. While a pultruder may find himself in compliance with EPA with injection, he may find himself out of business shortly after that. Fortunately, EPA must consider the economic impact of any of the technologies they require as part of MACT.

EXISTING ABATEMENT TECHNOLOGIES

In Europe generally, and in Germany in particular, the FRP and environmental industries have had nearly a decade to deal with the complex issue of styrene emissions and have developed superior, cost effective technologies to address the problem. For example, in Germany, the "TA-Luft" laws specify a maximum of 23ppm styrene in emitted air. Local regulators can, and do, establish lower levels of styrene with no additional technical or scientific justifications. This possibility also exists in the U.S. with States free to establish standards lower than those mandated by the Federal EPA.

In 1992, in Germany a market investigation identified the existing styrene emission technologies that had some usefulness to reduce styrene. These technologies were:

1) Thermal Processing:
- Incineration
- Catalytic
- Regenerative Thermal Oxidizer

2) Adsorptive Procedure:
- Activated Carbon
- Polyad

3) Oxidizing Processing:
- UV-Oxidation

4) Biological Procedure:
- Bio Filter System

While most of the above systems had not been used for cleaning styrene emissions, the capital cost alone has made this equipment of these types very expensive to purchase. Most had very high energy costs for operation. Investigators then turned their attention to bio filter systems, especially after they learned that this technology was being used in the pharmaceutical industry for cleaning dissolvents, which required a water-soluble and dust free method.

It was determined that all bio-filter systems worked on the same basic principle:

1. Bacteria live in tanks or containers which are filled with adsorbing material such as shredded wood or bark products, compost, heather or sinter stone. It was obvious that due to styrene's low solubility in water that this system would not be ideally suited for biological emission cleaning.

2. Styrene and other dissolvents must be adsorbed, because bacteria can only absorb the pollutant in association with some form of filter material or under water.

3. Styrene is susceptible to polymerization. Their filter material is made of wood which likewise represents a biopolymer. Therefore, there is a threat of there being a resultant styrene-wood-polymer, which can not be decomposed by the bacteria. Bacteria prefer monomers.

4. After approximately 3-to-5 years of operation, you end up with no compost, but rather styrene-wood-waste polymer, which requires disposal at high costs.

The data available for styrene, indicates that with emissions containing styrene, more than double the filter volume is required, in this case, with a 2,000 m^3 volume flow, approx. 500 m^2 of filter area is required. Few companies have the necessary space for such filter areas. In addition, with aerodynamic drag through the filter media, high energy costs would be required making this process less attractive from an economic standpoint. With alternative media this picture could change dramatically.

The team which conducted the market survey concluded that none of the existing styrene removal technologies listed above would be satisfactory for general use in the FRP composites industry. The engineers then turned their attention to identifying new styrene removal technologies that would be technically-superior and commercially viable in comparison to the available offerings. The development team established the following criteria for a "successful" styrene removal system:

Criteria for a Successful Styrene/VOC Abatement Technology

1. The maximum amount of styrene should be removed up to 99%.
2. Styrene emissions should be accomplished via the activity of micro-organisms, high-intensity energy, etc.
3. The decomposition of styrene or other VOC's in the reactor should be 100% controllable.
4. No wastewater would be generated by the process.
5. No solid waste or toxic sludge would be generated by the process.
6. No subsequent secondary contamination of the environment would be permitted.
7. The degree of absorption and system performance must be guaranteed.
8. The capital cost of the system must be low.
9. The operational costs of the system must be low.
10. The entire system must be "environmentally friendly."

Using the above criteria, three different styrene removal technologies were selected for subsequent development and demonstration. These include:

1. Bio-Scrubbers
2. Bio-Filters with new high efficiency media
3. Photo-Catalytic conversing

A decision was taken to concentrate initially on development and demonstration of the bio-scrubber approach for styrene removal. The results of a four year development program on bio-scrubbers is summarized below.

THE WEGA BIO-SCRUBBER

The first product which was developed for the German FRP composites industry is a revolutionary styrene/VOC abatement system called the Wega Bio-Scrubber. This technology capitalized on two complimentary and proven techniques: absorption of soluble gases and bio-digestion by micro-organisms. Because this is the most fully developed of the new German technologies, its features and benefits are most fully understood. Following is a brief summary of the development of the Wega Bio-Scrubber technology.

It was determined that research work would be conducted on a specialized biological emissions cleaning system as it appeared that only that technology had the ability to meet the criteria set forth above. Bio-scrubbers employ microorganisms (bacteria) that are only suitable for industrial use if certain process conditions exist. These conditions include; PH, temperature, concentrations, thermodynamics, process, haldane relationship, processing, reactor design, product engineering, and kinetics (e.g., oxygen, nutrients, environment, temperature).

When the most suitable cultures were found for this purpose, thanks to the most modern measurement and sensing technology, a first small styrene emissions cleaning system was built. This model, which operated for more than two-years years, cleaned air flows of 200 m³/h of styrene in a laboratory setting. During this time, many tests were carried out and data collected. When this program of tests was concluded in the summer of 1994, it was determined that a production system should be built to the following specifications:

1. a volume flow of 20,000 m^3/h emissions
2. and a mass flow of 4 kg/h styrene (= 47 ppm/h styrene concentration)
3. residual content of 5ppm/h styrene in the emitted air.

This system was constructed and was put into operation in August 1995. It consists of a scrubber and a reactor with integrated data space. In the meantime, the third generation Wega styrene/VOC emissions cleaning system will be ready for commercial introduction in Europe and the U.S. in the first quarter of 1997. The production of sewage sludge is no longer any problem in this fully automatic bio-scrubber system which is fully controlled by means of computers and data communication. In the third-generation Wega-system, the microbes succeed in completely transforming pollutants into carbon dioxide (CO_2) and water (H_2O) via a "cold biological burning".

THE OPERATION OF THE WEGA BIOSCRUBBER

PHASE I
 The emissions from an industrial operation containing styrene and other VOC's enters below the absorption stage. The scrubbing fluid traveling downward against the upward air flow removes the styrene and other VOC's from the air. The emissions exit the column in the form of cleanly scrubbed gas. A sensor continually monitors the condition of the emitted air and, acting on this information, the computer controls adjust the process as required.

PHASE II
 The scrubbing fluid is transferred from the absorption chamber to the fermenting tank.

PHASE III
 The styrene and other VOC's are biologically degraded in the fermenting tank. The desorbed scrubbing fluid is pumped to the fluid distributor and is then again ready to be used for absorption.

PHASE IV
 After the degrading phase is completed, additives are applied to regenerate the bacteria and the scrubbing medium. The medium is returned to the absorption chamber.
 The entire bio scrubbing system is equipped with nearly thirty sensors and measuring devices.

The heart of the data acquisition is based on a PC-station. All measurements are gathered here. The computer also controls secondary functions such as refilling evaporation loss, water quality, adding the regenerative medium and the system security. The data is communicated to a central technical support laboratory via modem, in order to continually monitor the function of systems, or to make adjustments to a system when necessary. Therefore, maintenance is restricted to a minimum, namely to the refilling of regeneration medium, checking the scrubbing fluid and mechanical functions.

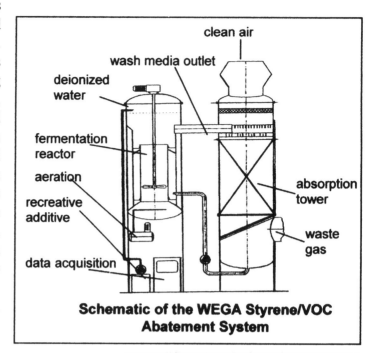

Schematic of the WEGA Styrene/VOC Abatement System

WEGA BIO-SCRUBBER SYSTEM COST RANGE

The cost for the Wega Bio-Scrubber System does not include the ducting for the exhaust air within the plant to get to the bio-scrubber. This equipment, its cost and installation is provided by a number of national firms, but it can be provided by local heating and air conditioning contractors as well. While each pultrusion plant offers its own conditions, existing ducting and run distances, these systems need to be engineered to handle the quantity of air necessary to remove enough styrene without drawing styrene from the resin mix.

Styrene needs to be drawn from the pultrusion process at the resin bath location and at the exit end of the die where hot styrene fumes are frequently visible coming off the part. This can be done with ducting from above the machine or a plenum system below the resin bath, near the floor (see picture below). With styrene heavier than air, a plenum system or ducting at the edge of the resin bath generally proves more effective.

Ducting by a Scandinavian pultruder draws styrene fumes from the resin bath and exit end of the die

Estimates for the cost of the ducting to the bio-scrubber generally run about $500.00 per foot of manufacturing floor width. If you have 10 pultruders in a 100 foot wide plant lay-out, the cost to duct for styrene pick-up will add about $50,000.00 to the cost of the Wega System.

One pultruder uses a plenum in the floor to capture the heavy styrene laden air.
Photo Courtesy: Stoughton Composites, Inc.

We should applaud the efforts of the SPI/CI staff who are working hard to make certain that EPA considers and adds the cost of plant air ducting to their calculations when mandating a MACT standard. EPA is required, by law to do this, but this cost can get lost as incidental in the total cost calculations. EPA, with the oversight of CI staff, will include this in their calculations to make certain that they are not superimposing a MACT standard that exceeds the amount their MACT can cost per pound of VOC/Styrene treated.

All Wega System pricing is based on the volume flow of air required to handle your process. This is the reason the completion of a Class A questionnaire is so important. This questionnaire can be obtained from any company representing VVK Weege NA.

The price range can be from $250,000 for a system to handle 37,000 cubic feet per hour (615 cubic feet per minute), to approximately $950,000 for a system to handle 1,500,000 cubic feet per hour (25,000 cubic feet per minute).

SUMMARY OF THE WEGA BIO SCRUBBER SYSTEM PERFORMANCE

The new Wega Styrene/VOC Abatement System meets all of the needs of the FRP composites industry on the following basis:

1. The power of absorption has a minimum of 80% and in most cases can be raised to 99%.
2. With the appropriate stirring capacity and ventilation, degradation of the styrene is strictly aerobic.
3. Sludges nor solids are not developed. In an organic catalytic process, the microorganisms transform the styrene into carbon dioxide and water.
4. As soon as the correct biomass concentration (200-300 kg/m^3) has been reached, the microbiological degrading stems are forced, in an active phase, to biomineralise the styrene and, during the rest phase, to regenerate themselves and the scrubbing medium.
5. This bio-scrubbing method's closed system is completely friendly to the environment.
6. The operating costs are extremely low. For a system with a volume flow of e.g. 20,000 m^3/h, all that is needed is an 11kW installed line for power, a water loss of 200 l/d and additives of approx. 20,--DM/day (approx. $15 US/day).

7. Maintenance costs are extremely low and are restricted to the refilling of nutrients and the periodic checking of scrubbing fluid and all functions.
8. The systems are engineered for the specific emissions and the pollutant content of the facility in question. Wega styrene/VOC abatement systems can be created for a volume flow of 1,000 m³/h up to 100,000 m³/h or more.

SECOND AND THIRD GENERATION WEGA TECHNOLOGIES

Readers will recall that two additional styrene removal technologies were identified by the investigating team, i.e; Bio-Filters and Photo-Catalytic conversion. Following is a brief summary of the status of development and demonstration of these technologies.

NEW WEGA BIO-FILTER TECHNOLOGY

Bio-filters were the second technology to be investigated. Bio-filters have a well-established record of performance in a number of industries. As previously stated, the primary problems with bio-filters include the large area required for adsorbing media, variability in process performance due to temperature, humidity, etc. and the tendency of organic bio-media to polymerize over time.

The new Wega Bio-Filter uses a proprietary high-density fiber media which developed in association with the Danish Technical Institute of Copenhagen. Compared to traditional bio-mass media (bark, peat, etc.) this high-density fiber provides nearly two orders of magnitude greater surface area to volume and permits compact bio-filter footprints. This, in turn, allows for the bio-filter media to be contained in small spaces with optimum temperature and humidity conditioning of the media to maximize process efficiency. The new media also presents a much lower surface area to volume ratio to the airflow, enabling energy costs to be reduced dramatically.

The resulting system makes bio-filtration a viable candidate for styrene remediation for the first time.

WEGA PHOTOCATALYTIC TECHNOLOGY

The third technology involves the use of light energy to destabilize hydrocarbon compounds. When styrene or other complex hydrocarbons are subjected to intense light energy of selected wave length or frequency (Hz), the light energy strips electrons from the styrene molecule, destabilizing it for a short period of time. During this period of instability, and in the presence of certain inexpensive mineral catalysts, styrene can be reduced to water (H_2O) and carbon dioxide (CO_2).

A patented new photocatalytic system for styrene has been developed and will be demonstrated for the first time at the 1997 Composites Institute's International Composites Exposition (ICE), 27-29 January 1997 in Nashville, Tennessee. The Wega Photocatalytic system has no moving parts, is extremely low cost to operate and achieves 100% reuse of the mineral catalyst.

CONCLUSION

The technology development curve for styrene removal in Europe is extremely aggressive. Perhaps this is because the FRP composites industry in Europe has been highly regulated for nearly a decade and the European approach to process engineering tends to be holistic in nature. It is reasonable to expect that new styrene removal techniques will continue to be developed. It is the role of technology specialists such as VVK Weege to identify emerging styrene removal technologies for development and demonstration in North America.

Styrene occurs naturally in nature. Within 12 hours styrene is naturally broken down into CO_2 and H_2O within the environment. With the signing of the Clean Air Act, we will not be permitted the luxury of waiting the 12 hours for nature to do its job. Bacteria, like that used by the Wega Bio-Scrubber System occur in nature as well. With some help in the form of training from man this bacteria will breakdown the styrene/VOC into CO_2 and H_2O within 10 seconds of exposure. When performing this natural action on VOC's the bacteria produces no residual waste. In fact, it is 100% free of any waste. The system is relatively low in cost to purchase and inexpensive to operate on a daily basis.

When compared to the cost for incineration systems that initially cost $3,000,000 to $5,000,000, and more, and which have high daily operational costs, any microbiological system offers many advantages to composite processors.

Pultruders come in all sizes, but many are small in size. While some of those will fall below the size required to install equipment to MACT standards, numerous others will be burdened with this effort and expense. While some pills are hard to swallow, this microbiological pill may offer many of us some alternatives that are at least more tolerable financially, while fully meeting our obligations under the Clean Air Act.

BIOGRAPHIES

Jeff Martin, is president of Martin Pultrusion Group, a firm representing companies world-wide who produce hardware and supply technology for use in the pultrusion process. He has 30 years experience in the sales and marketing of reinforcements, pultruded products, equipment and tooling all involving the pultrusion process.

Jeff is a member of the Society of the Plastics Industry's (SPI) Composite Institute (CI) division, where he has served as a member of the Board of Directors for four years and Conference Chairman for their Annual Conference. He has managed the Pultrusion Sessions of that conference for the past 20 years and was instrumental in the establishment of the Pultrusion Industry Council (PIC). He is also a member of the European Pultrusion Technology Association (EPTA). Jeff has published extensively and lectured on the pultrusion process and its markets, world-wide.

Douglas S. Barno, specializes in market, application and business development of fiber-reinforced polymer (FRP) materials. He has over 30 years experience in the reinforced-plastics and composites industry, including international assignments in Europe, the Middle East and Africa.

Doug is the principal in his own consulting firm, the DSB Group, N.A. He is vice president of VVK Weege of North America, Inc., a start-up firm with advanced air pollution control technology technology for the FRP composites industry. Doug is also a member of the American Concrete Institute (ACI), American Society of Civil Engineers (ASCE), and is the FRP composites industry's representative to the Construction Materials Council (CONMAT) of the Civil Engineering Research Foundation (CERF).

A Material/Processing Approach to Density Reduction and Cost Savings for Pultruded Rods and Profiles

STEVEN G. KATZ AND DARRYL A. PAYNE

ABSTRACT

Until the early 1970's pultrusion was only possible using continuous rovings. Pultruded rods and bars were manufactured with fiberglass percentages between 70 and 80% by weight. These high glass loadings were necessary to completely fill out the die cavity and minimize the void content of parts. The early 1970's saw the development of profile shapes using continuous strand mat and rovings. While the main reason for the introduction of mat was to improve transverse tensile properties and reduce axial shear, it also had a corollary benefit of reducing glass loadings to between 40 and 50%, thereby reducing material cost.

A certain amount of roving is needed in solid rod and bar and in profiles to handle the machine load of the process, but frequently that amount is less than the amount the pultruder must insert just to fill the cavity. With fiberglass having the highest density of those components used in a pultrusion and generally the highest cost per pound, there is always pressure on the pultruder to reduce glass loading. Pultruders who tried spun or texturized rovings reduced glass contents but the higher cost of these materials frequently offset the change in part density. Other rigid core materials such as wood and closed cell foam are being used commercially for volume displacement, but they are accompanied by additional manufacturing difficulties such as pre-shaping and interfacial delamination, and they are limited to large parts with simple geometries. With continuing market pressure for reducing weight and improving economics, the need still exists for an efficient method to reduce reinforcement content of pultruded products.

A method developed by ISORCA has been demonstrated to increase the space a continuous roving occupies in the pultrusion die and hence in the finished part. The method involves expanding the roving and keeping the filaments apart by means of a combination of mechanical expansion of thermoplastic microspheres and chemically binding them. The result is a part that is controllably lower in density and cost than conventional pultrusion, allowing parts to be made to strength performance specifications without excessive reinforcement. The treatment process works with ordinary rovings and can be added to an existing line with a modest investment in equipment.

OBJECTIVE

The objective of our development program is to lower the weight and reduce materials cost of pultruded composites by substituting expandable plastic microspheres for glass fiber and resin where the design and performance specifications will allow.

The authors are at ISORCA, Inc., Box 414, Granville, OH 43023

BACKGROUND

Earlier work (Bastone and Payne, 1995) in this technology focused on the RTM process and loading glass microspheres onto a thermoformable mat using a binder resin. Technical success was achieved with ISORCA's Lo-D™ Mat, resulting in nearly a 20% lower part density, but for economic reasons the program has not met widespread commercial acceptance.

The present program focuses on the use of thermoplastic microspheres in the pultrusion process. Early indications are that both weight reduction and cost savings are realistically achievable.

FEASIBILITY TRIALS

The intent of our initial trials was to see if we could successfully replace the rovings in the neutral axis of a solid pultrusion with specially treated rovings containing a high loading of expandable plastic microspheres. In our laboratory we treated approximately 1,200 meters of conventional 113 yield pultrusion roving by dipping it into a bath with Expancel brand thermoplastic microspheres suspended in a binder emulsion. The roving passed under a spreader bar which spread out the filaments so that they could be thoroughly wet out by the emulsion. The roving then passed through an oven where the water was removed and the microspheres were expanded, spreading the filaments apart. The expanded roving, relatively stiff compared to untreated roving, was then wound onto spools for subsequent introduction into the pultrusion process. We also processed some roving in the foregoing way but without expanding the microspheres. The intent was to demonstrate that they could be expanded on-line before going into the die, or perhaps expanded in the die.

Working with Martin Pultrusion Group, Inc. we conducted trials at Pultrusion Dynamics, Inc. in Northeastern Ohio. A pultrusion die having a rectangular cavity 1.00 x 1.15 cm was selected for the initial work. We used a conventional pultrusion grade medium reactivity isophthalic polyester resin with 20 phr alumina silicate (clay) filler.

EXPERIMENTAL PROGRESS

In our first trials four conventional rovings were removed from the center of a rectangular pultruded rod and replaced with treated rovings. Centering and parallelism of the treated rovings were major problems, but were solved by the use of forming guides that allowed insertion of the treated strands directly into the center of the die. Bleed-back of resin at the die entrance still resulted in some random movement of fibers, but the treated rovings remained essentially in the center of the profile.

The second major problem was microsphere stripping. Keeping the microspheres from coming off during handling and subsequent processing ("dusting") was a key to making this technology successful. An unacceptably large fraction of the microspheres was further lost to the resin and could be observed to be dripping off the guide plates and the die entrance. This problem was addressed by using other binder emulsions that provide better adhesion of the microspheres to the glass fibers. Emulsion candidates tried were commercially available: (1) vinyl acrylic copolymer, (2) vinyl acetate ethylene copolymer, (3) acrylic emulsion, and (4) a special emulsion. Based on overall performance, the vinyl acrylic copolymer was chosen as the best candidate because: (1) it allows good expansion of the microspheres; (2) it displays no dusting; (3) it provides good resin wet-out; and (4) it is not completely soluble in styrene. Subsequent trials using the vinyl acrylic copolymer demonstrated success in eliminating the dusting and microsphere stripping observed earlier. Microscopic examination of the resin dripping off the die entrance showed virtually no trace of microspheres.

In addition to being able to use roving treated with microspheres and wound onto a spool for later use, on-line expansion capability was also explored. The treated but unexpanded roving was run through a tunnel oven to expand the microspheres and then into the pultrusion die. Although oven control was problematic, the roving did expand before entering the die. Backward integrating the process to take ordinary roving, impregnate it, and expand it on-line appears feasible and is a subject for further development effort.

OPTICAL EXAMINATION

In order to examine the distribution of microspheres, we used a technique borrowed from optical mineralogy. We had wafer thin cross sections made of the pultruded part and mounted on glass microscope slides. Before mounting, the part was subjected to a vacuum and a blue dyed epoxy was allowed to permeate the part, filling in air voids with the easily seen epoxy.

The thin sections were examined under transmitted light at 100x magnification. Individual glass filaments could be seen as small circles with polyester resin between them. Samples of standard pultruded parts with ordinary rovings showed a distribution of filaments that was dense and fairly uniform throughout except for a few areas that appeared to be resin rich. Samples of pultruded parts with microsphere treated rovings appeared quite different. The microspheres appear as dark brown, somewhat irregular blobs, and considerable air is evident where the microsphere-treated rovings did not wet out. Using this technique we could track the distribution of the treated rovings and determine the effect of different experiments as the development effort progressed.

FURTHER TRIALS

Subsequent trials were conducted to demonstrate the feasibility of running continuously for an extended period of time with no operational difficulties and to generate data that could be used to indicate density and cost savings potential. The first step was to reduce the amount of conventional rovings in the profile from 36 to 34 to 32 to 30 to 28 ends. As the amount of glass decreased, surface properties and corner radii began to deteriorate. At 28 rovings the profile was of very poor quality. A single treated roving was then introduced, and the improvement in quality was noted. Subsequent additions of treated rovings yielded further improvement, but it was apparent that additional optimization will be required to pultrude profiles that meet performance specifications and surface requirements. Further experimentation in our laboratory on plastic microspheres in resin clearly demonstrated that the microspheres may be softened excessively by the styrene at pultrusion temperatures. The use of more styrene-resistant microspheres may be desirable.

The treated roving is tightly bonded with the binder emulsion, so very little wet out by resin can take place. It was therefore pointless to run the treated roving through the resin bath before going into the pultrusion die. Additional work is planned to address wet-out resistance.

Additional work is also planned to determine whether the microspheres could be expanded by the heat of the pultrusion die, but given the expansion temperatures of the microspheres, it is unlikely that existing microspheres can be expanded before the resin gels. Nevertheless, it may be possible to formulate a microsphere that would allow such a phenomenon to happen.

DENSITY AND COST REDUCTION

Some of the data we generated were used to build a spreadsheet model of the density and cost reduction and could be used to predict target values. Experimentally derived data on the roving bare glass and sizing are in Table I. Untreated conventional 113 yield roving was put into a muffle furnace to determine loss on ignition. Measurements of the bare glass weight and sizing weight were converted into grams per meter.

TABLE I. EXPERIMENTALLY DERIVED ROVING DATA

Burn off Untreated Roving	g/m	Wt. pct
Weight with sizing	4.37	100.00
Weight of bare glass	4.34	99.35
Weight of sizing	0.03	0.65
Burn off Treated Roving "A"	**g/m**	**Wt. pct**
Weight of Treated Roving	4.89	100.00
Weight of bare glass	4.34	88.68
Weight of sizing	0.03	0.58
Weight of binder + microspheres	0.53	10.74
Weight of binder (75%)	0.39	8.06
Weight of microspheres (25%)	0.13	2.69

Next, a sample of treated roving "A" was burned off. The bare glass and sizing were subtracted from the initial weight, leaving the weight of binder plus plastic microspheres. The proportion of microspheres to binder solids in the application emulsion was 25/75 by weight. The resulting data are shown in Table I.

Next, three samples of the pultruded profile about 5 cm long were burned off, leaving the bare glass and the clay filler used in the resin. The bare glass was subtracted, leaving the weight of the clay. The sizing, microsphere binder, and microspheres were subtracted, leaving the weight of the pultrusion resin. The volume of the part was measured and extrapolated to a part 1 meter long. Then the specific gravity was calculated. The results are shown in Table II for a typical pultrusion with 36 ends, with 32 ends, and with 28 normal ends plus 4 microsphere-treated ends. The specific gravity decreases from 1.95 to 1.87 to 1.70 g/cc, a reduction of about 13 percent. Using these data to build a predictive model, we have established a reasonable target weight reduction of about 30%. We hypothesize that we could do so with 20 conventional rovings and 6 treated rovings, assuming our optimization efforts are successful. We would require more volumetric displacement than we have so far achieved, using higher microsphere loading, getting more expansion per microsphere, or using different microspheres with greater expansion. Our model assumes more expansion per microsphere.

TABLE II. EXPERIMENTALLY DERIVED WEIGHT AND VOLUME DATA ON PULTRUDED PROFILES.

	Units	Sample 1	Sample 2	Sample 3	Target
Standard ends	ends	36	32	28	20
Treated ends	ends	0	0	4	6
Weight of 1 m length	g/m	219.98	211.28	191.69	155
Weight of bare glass	g/m	156.16	138.81	138.80	113
Weight of clay	g/m	7.32	8.81	4.10	3
Weight of sizing	g/m	1.01	0.90	0.90	0.7
Weight of binder	g/m	0.00	0.00	1.58	2.4
Weight of microspheres	g/m	0.00	0.00	0.53	0.8
Weight of resin	g/m	55.49	62.76	45.78	35
Volume of 1 m length	cc	112.70	112.70	112.70	112.70
Specific gravity of part	g/cc	1.95	1.87	1.70	1.37
Reduction in part sp.gr.	pct			13	30

TABLE III. COST INFORMATION ON PULTRUDED PROFILES.

Component	Cost ($/kg)			
Glass roving	1.76			
Resin mix	1.76			
Binder resin	2.15			
Microspheres	11.57			
		Sample		
Cost for 1 m length	1	2	3	Target
of composite				
Glass fraction	0.2772	0.2464	0.2464	0.2002
Resin fraction	0.1108	0.1262	0.0880	0.0670
Binder fraction	0.0000	0.0000	0.0034	0.0051
Microspheres	0.0000	0.0000	0.0061	0.0091
Total material cost	0.3880	0.3726	0.3438	0.2814
Cost reduction from Sample 1		4	11	27

Finally, our model includes the cost assumptions on the components on a weight basis (Table III), and estimates the cost of each component fraction in a 1 meter long pultruded piece. The materials cost decreases from 38.8¢ to 37.3¢ to 34.4¢, a reduction of about 11 percent. At the targeted specific gravity the material cost is reduced over 27%, a significant amount for a process which is highly sensitive to material cost. We believe this cost savings ought to be of considerable interest to the pultrusion industry and would justify further optimization efforts.

PATENT STATUS

A provisional patent application has been submitted to the U.S. patent office, giving ISORCA, Inc., protection to disclose this technology in detail to others who may have an interest.

CONCLUSIONS

Feasibility trials have clearly demonstrated the potential for density and cost reduction using microspheres to displace resin and glass fibers in the center of a solid pultrusion profile. The method involves expanding the roving mechanically on-line with microspheres and setting a chemical binder, keeping the filaments apart. The result is a part that is controllably lower in density and cost than conventional pultrusion because parts can be made to strength performance specifications without excessive reinforcement. The treatment process works with ordinary rovings and is suitable for a pultruder to add to an existing line with a modest investment in equipment.

More work is planned to improve resin wet-out of the treated rovings, to optimize the type and amount of microspheres, to achieve satisfactory mechanical and surface properties, and to optimize cost and density reduction. However, the technology appears to be feasible and the initial work reported here opens the path for further development efforts.

REFERENCES

Bastone, A.L. and D.A. Payne. 1995. "Density Reducing Reinforcements for Liquid Molding Processes," The Society of the Plastics Industry, Inc., Composites Institute, 50th Annual Conference Proceedings, 7B:1-3.

ACKNOWLEDGMENT

We are grateful to Jørgen Petersen of Expancel, Casco Products AB of Sweden for his support of this program, and to Andrew Bastone, Thomas Philipps, Reid Billig, Jeff Martin, and Joseph Sumerak for their helpful suggestions and encouragement.

BIOGRAPHIES

Steven G. Katz has been ISORCA's VP/Operations since 1991. He has a technical background and strong business management experience. Prior to his tenure with ISORCA, he spent 10 years with Owens Corning, mostly in process control, and 3 years with a scientific instrument company. He holds a Ph.D. in Geosciences and an M.B.A. degree.

Darryl A. Payne has been ISORCA's Product Development Manager since 1992. He has broad chemical applications experience in sizings, coatings, and other areas relating to glass fiber manufacturing. Before ISORCA he spent 24 years with Owens Corning and four years with Manville (now Schuller). Darryl has a B.S. in Chemistry.

A Mass Produced Trans-Tibial Prosthesis Made of Pultruded FRP

MAKOTO SAITO, HIROKI NAKAYAMA, SHUUJI HAGIWARA, SEISHI SAWAMURA,
MIKIO YUKI, ICHIRO KITAYAMA, RICHARD REED AND BILL CARROLL

Abstract

Using a funds of Japanese non-governmental organization, a mass produced prosthetic component was developed in order to reduce the cost of use in the areas of conflict. On the basis of a new theory for smooth walking, equivalent performance to that of a conventional western based expensive component was obtained using a rigid component. This made it possible to build the artificial leg using pultruded material. A drastic reduction in cost was obtained without sacrificing performance or safety.

Introduction

Several million people in areas of conflict, such as East Asia, Bosnia or parts of Africa require prostheses. However, current prosthetic components, developed mainly in western countries, are extremely expensive in comparison with the funds available in these areas. There are various humanitarian aid programs in the world, but the number of the prostheses they can provide is limited due to their price. The major reason for the high price is the small scale production methods where major resources are invested in making a clever imitation of the human leg or in meeting the precise requirements of western countries. On the other hand, very simple components, like a piece of wood or hand made bamboo components are used in the rural areas, where no safety standards are applied. There is a hopeless gap between western counties and the areas of need.

At the request of a Japanese non-governmental organization for aid for physically handicapped people in the Pacific Asian area, the authors have developed a mass produced prosthetic component for this region by using pultruded FRP for a drastic reduction in cost, where a priority was placed on the performance and durability (safety). The component is composed of a skeleton structure and a cosmetic cover made of plastic, where a special design was developed for the skeleton structure in order to reduce the cost without sacrificing the walking performance.

The procedure to determine the optimum design of the prosthetic component based on FEM structural analysis and analysis of the walking pattern will be focused on in the present paper.

Makoto Saito, Hiroki Nakayama, Shuuji Hagiwara:Mechanical Eng. Res. Lab., Kobe Steel Ltd., 1-5-5 Takatsukadai, Nishiku, Kobe, Japan 651-22   Seishi Sawamura, Mikio Yuki, Ichiro Kitayama: Hyogo Rehabilitation Center   Richard Reed, Bill Carroll: Glastic Corporation

Conventional prosthetic component

A general below-knee prostheses is composed of three components; a foot section and trans tibial bone (main below knee bone) and a socket which fixes the bone to the remaining human leg.(See Fig. 1) The socket must be custom made in order to fit each patient. However, the other components can be commonly used for wide range of patients and they can potentially be replaced by mass produced products. Generally a simple aluminum pipe is used for the trans tibal bone covered with cosmetic material which imitates the human leg. The major functions for smooth walking are performed by the foot section.

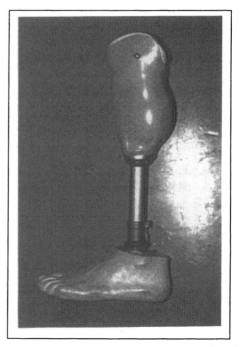

Conventional below-knee prosthesis

Figure 1.

Fig. 2 shows the typical motion or deformation of the conventional artificial foot section during the standard walking mode. Two major functions can be observed; one is the shock absorption during the heel contact process(Fig.2-1) and the other is a sufficient flexibility around the toe finger joint during the toe-off process. (Fig. 2-2) For athletic use, a special design is required for the heel structure to absorb severe dynamic load 1), however, the most important function is the toe-off process for general patients (especially in the areas of conflict), since the top priority is ordinary moderate walking. Various inventions have been made up to this point in the history of the artificial foot. Fig. 3 indicates a cross section of a typical artificial foot component. It can be seen that a wooden block is encased in a solid foot shaped rubber block, where stress can be transferred from the bone to the foot section through the wooden block. Correct flexibility can be obtained around the end of the wooden block as in the human foot. This flexibility is believed to be the most important factor in the toe-off process. However, such a hybrid structure is thought to be expensive and have durability problems around the interface between the block and rubber.

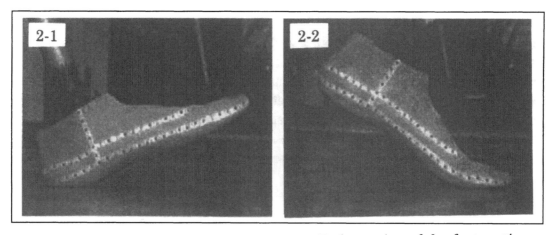

Deformation of the foot section

Figure 2.

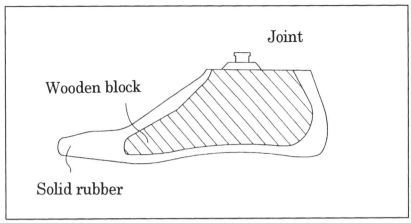

Figure 3. *A cross section of the foot section*

Basic design of the component

A rigid single structure was used as a strategy in order to resolve both durability and price problems, for which a new approach for smooth walking, to replace the flexibility around the toe joint, was required. Careful observation of the walking motion of the conventional foot component suggested that "the trace of the ankle" during the walking phase could be another key factor for smooth walking.(See Fig. 4) The trace can be obtained with a rotational motion of the foot component and the optimum trace can be changed by modifying the bottom configuration of the foot component.

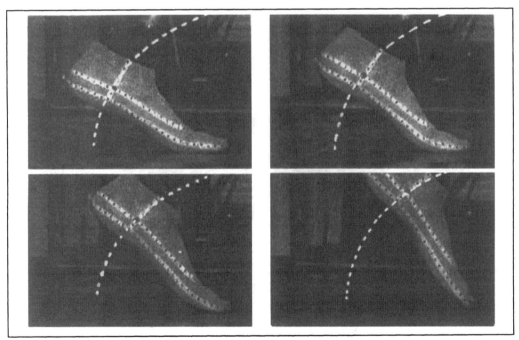

The trace of the ankle during the toe-off process of the conventional foot component

Figure 4.

Figure 5 shows a basic design of the rigid foot component where major features are;

  (1)Simple rigid shell structure with constant thickness
  (2)Constant cross section in the width direction
  (3)The trace of ankle motion can be changed by changing R and L

These features makes it possible to produce a foot component using a slice with constant length of pultruded material.

  An examination of walking was carried out to verify the "trace of ankle" theory using test samples made of aluminum sheet, where the optimum R and L were determined in order to obtain the same trace as that of the conventional foot component. Figure 6 shows the change in the trace of the ankle with R and L. The results of the examination revealed that an equivalent walking performance was obtained with the ankle trace basis approach.

  In addition, an arch shell was introduced in the upper shell so that the inclination of the pipe can be adjusted. The adjustment of the inclination is indispensable for western based foot components. A simple sheet of FRP was attached to the heel as a shock absorber during the heel contact process. The sheet was processed by compression.

  The foot shape looks similar to that of a standard   human foot; however, it should be stressed that the dimensions are substantially different from that of the standard human foot so as to obtain a particular trace of ankle motion, as shown in the figure.

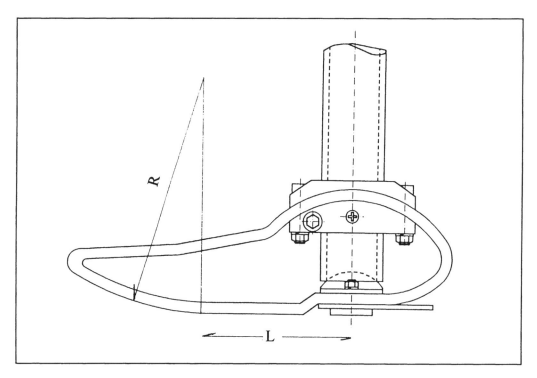

Basic design of the rigid foot component

Figure 5.

Pultrusion

In order to minimize the cost, standard process conditions were employed. Since preliminary stress

analysis indicated that the major stress applied to the shell of the component was bending, the thickness of the shell was limited to 5 mm in order to avoid interlaminate failure. Four 0°/90° fiber glass fabric layers and random mats were impregnated with slight 0° roving for pulling. Vinylester was used for easy pulling and strength. The volume fraction of the 0°/90° fiber glass fabric was about 30%. The estimated strength of the pultruded material was 20kg/mm^2 in the normal direction to the pultruding direction.

Modification for the strength

The load bearing capacity of the component was examined using FEM analysis and mechanical tests, where ISO 10328 test standard (2) was employed. Note that as of now, we have no authorized test standard for prosthetic components. The ISO 10328 is still in draft form but will be published soon as the first authorized test standard for prosthetic components. Figure 6 indicates the result of the FEM stress analysis showing the stress distribution of the component under static failure test conditions, the most severe loading condition defined in the standard draft. Peak stress was observed around the tip of the toe. (Fig.6-1). The maximum stress was about twice that of the strength of the material.

Reinforcements or stiffeners were suggested to reduce the peak stress. Figure 6-2 shows the stress distribution of the component reinforced with two vertical stiffeners whose positions were optimized by the analysis. It was found that the maximum stress can be reduced to the standard strength of the pultruded material.

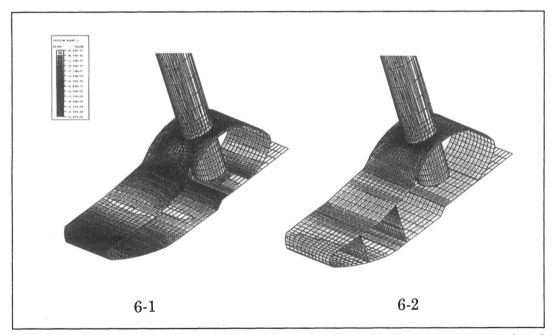

6-1   6-2

Results of analysis showing the stress distribution of the component under the toe-off process. Left : Prototype Right : Reinforce with stiffeners

Figure 6.

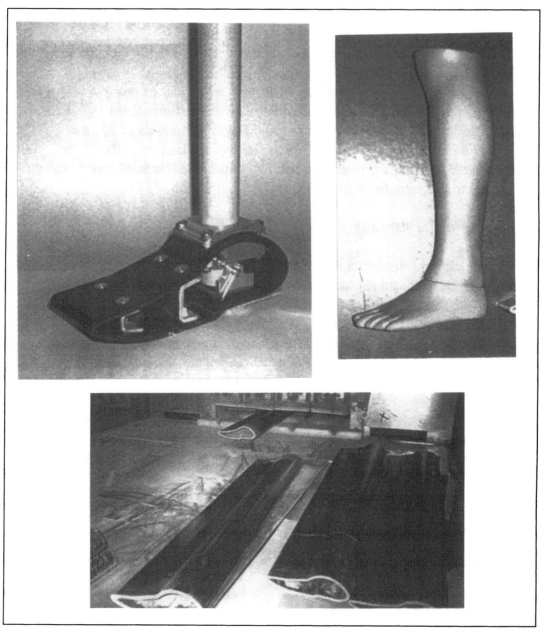

Fabricated rigid component, cosmetic cover and pultruded material.

Figure 7.

Since the load for the fatigue test defined in the standard draft is about 1/3 of that for the static failure test, the component is expected to pass the fatigue test automatically.

Since the funds available for the die were limited, the stiffeners can not be united into a part of the shell. Instead, aluminum stiffeners were used in the development program, as shown in Fig. 7.

Results of mechanical test

Static and dynamic(fatigue) tests were carried out on the skeleton of the prosthesis (foot section and pipe), on the basis of the ISO 10328 standard. The results are summarized in Table 1. It was found that the static strength of the component meets the ISO standard for the patients whose weight is less than 100 kg. As for the fatigue strength, for each loading condition, no damage was observed for 3 million loading cycles which is the maximum number of cycles expected for a three year period in the field.

Table Ⅰ －Results of mechanical tests

Test mode	ISO/Condition Ⅰ (Heel contact)	ISO/Condition Ⅱ (Toe-off)
Static failure	5500 N/ 3360 N*	3400 N/3019 N*
Fatigue**	>1500 N/1330 N*	>1250 N/1200 N*

* The minimum eligibility requirements shown in the ISO 1-328
    test standard draft for patients with the weight less than 100 kg.
** Number of loading cycle was 3,000,000 cycles

Performance

Figure 7 shows the fabricated skeleton structure and the cosmetic cover made of polyethylene. This prosthesis was applied to a patient and walking performance was examined as shown in Fig.8. The number of the patients in the test was limited but it was found that equivalent performance to that of

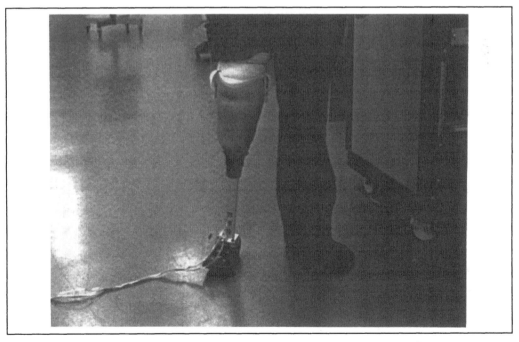

A view of the walking test using the component.

Figure 8.

a current expensive foot component was obtained. After a three month laboratory scale walking test, 100 sets of the component were delivered to Thailand for the field test, where a major aim of the test is the durability in the region. Since people in the rural area in East Asian countries often walk without shoes, serious abrasion or cracking is widely observed in conventional flexible foot components. Significant improvement in durability is expected for the skeleton. However, some improvement may be required for the cosmetic cover based on results of the field test which will be concluded in a few years.

Additional advantages were found for the component. Due to the shell structure, the weight of the foot component was significantly reduced in comparison with that of conventional foot components composed of solid rubber with wooden blocks. It was found that the reduction in weight of the foot section was a "Holy Grail" for patients, especially elderly people or children, because a slight change in the weight of the foot section causes a substantial decrease in the resistance of the swing motion of the below-knee segment, due to the distance from the knee joint. This indicates that the foot component is suitable even for patients in western countries. If carbon fiber is used instead of glass fiber, it is clear that further reduction in weight can be expected. In addition, aluminum stiffeners can also be replaced by part of the shell processed during the pultrusion process, if further funds can be used for the die.

Cost

As a result of using mass produced pultrusion materials, the cost for material for the skeleton is expected to be less than $10 a unit. Even if the machining and fabrication is carried out in western countries, the total cost will be $50 a unit, including aluminum parts for 10,000 units order. If necessary, the cosmetic covers may cost $10 -20 a unit. Exact costs of the conventional components are unknown but we believe that the component is significantly cheaper since the price of the current components are in the range of several hundreds to a thousand US dollars.

Even though the humanitarian aid programs for physically handicapped people in areas of conflict, are suffering from shortages of skilled people to fit the components to the patients, it is believed that such an inexpensive component will be a great help to the aid program. It is hoped that composites technology will narrow the big gap between western countries and the areas of need.

Summary

Using funds provided by a non-governmental organization in Japan, a mass produced inexpensive prosthetic component has been developed for humanitarian aid to areas of conflict. The features are;

1. A rigid structure was employed on the basis of a new theory for smooth walking in contrast to conventional flexible components.
2. The component is composed of a piece of inexpensive pultruded material.
3. A significant reduction in cost was made possible without sacrificing the walking performance of the western based conventional components.

4. Safety of the component was examined using FEM structural analysis and mechanical tests on the basis of an ISO test standard draft.

5 . An increase in the durability and reduction in weight are additional advantages in comparison with conventional components.

References

1) Yasuhiro Hatsuya,MD, 1987. "Current State of Prosthesis In Asia", So-go Rihab., Vol.15, No.2, 105-109

2)Thamrongrat Keokan,MD, 1981 "Prosthetic and Orthotic Work in Thailand", Asian Conference on Prosthetics and Orthotics '81   Proceeding 46-49

3) Ernst M.,1983, "A Design for Active Sports" Orthtics and Prosthetics, vol.37, No.1, 25-31

4) Draft International Standard   ISO/DIS 10328-1 ~ 8

Biographies

Makoto Saito
He graduated from Osaka University in Solid State Physics with M.S. degree. Upon graduating he joined Kobe Steel. He was engaged in determining reliability of metal materials using fracture mechanics. He is currently a senior researcher at the Mechanical Engineering Research Laboratory and engaged in design and reliability studies of composites products using structural analysis techniques.

Shuuji Hagiwara
He graduated from Toyohashi College for Engineering in 1992 and joined Mechanical Engineering Research Laboratory, Kobe Steel. He is engaged in the development of composites products.

Hiroki Nakayama
He graduated from Tohoku University in 1994 and joined Kobe Steel in the same year. He is engaged in the FEM structural analysis.

Seishi Sawamura M.D.
He graduated from the Department of medicine, Kobe University in 1955. After much experience as a medical doctor in the rehabilitation of disabled people, he is currently Director of Hyogo Rehabilitation Center, and President of ISPO(International Society for Prosthetics and Orothotics) He raised funds for the present development program.

Mikio Yuki

He graduated from the College of the National rehabilitation Center for the Disabled Prosthetics and Orthotics in 1985. He is currently a researcher in the Prosthetics and Orthotics Department , Hyogo Rehabilitation Center.

Ichiro Kitayama

He graduated from the Department of Engineering, Kobe University in 1978. He is currently a senior researcher at the Hyogo Rehabilitation Center.

Bill Carroll

He is presently R&D manager for the Glastic Corporation. He has a B.S. in Chemistry and M.S. in Polymer Science. He has 20 year experience in fiber reinforced polymer matrix composites, including product and process development, testing and analysis. Currently involved in the development of advanced composite, pultruded composite and laminates, as well as laboratory supervisor.

Cure Systems for Pultrusion: Multiple Peroxide vs. Promoted

TED M. PETTIJOHN

ABSTRACT

Pultrusion and other heat assisted processes have historically relied on multiple peroxides to cure unsaturated polyester and vinyl ester resins. Those systems usually contain peroxides which require special refrigerated storage and the cure is often difficult to control since the peroxide concentrations must be balanced with respect to each other.

Promoted systems like those used in room temperature cure processes have been ignored by most processors due to short pot life. This paper compares multiple peroxide cures with promoted cures and announces the discovery of a new promoted system which can be used for processes like pultrusion that require long catalyzed pot life.

INTRODUCTION

Pultrusion is a process used for the production of composite profiles such as rod, tubing, ladder rails, and rebar reinforcements. In the process, glass fiber and mat are continuously pulled through a pre-initiated (pre-catalyzed) resin bath then through a heated die. The resin cures in the die to produce a profile which is cut to desired length. See Figure 1.

Organic peroxides are used to initiate (or catalyze) resin cure. They are organic compounds that decompose in the presence of heat and/or promoters to form free-radicals. The free-radicals start a cross linking process within the resin syrup. Styrene or other monomers contained in the syrup polymerize and they react with the resin to produce a complex network that gels, then cures to form the final product.

Dr. Ted M. Pettijohn, Witco Corporation, PO Box 1439, Marshall, Texas 75671-1439

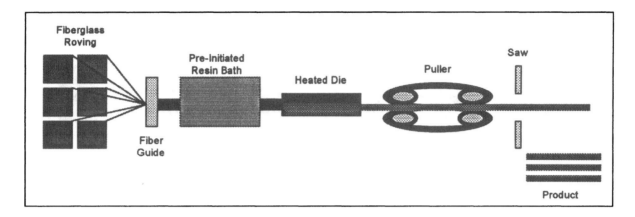

Figure 1. Schematic of the pultrusion process.

EXPERIMENTAL

MATERIALS

The heat activated promoter (HAP) is Witcat™ Pro-15 (Witco). The peroxide initiators like Esperox® 5100 (t-amyl peroxybenzoate or TAPB) are commercially available from Witco. A general purpose isophthalic pultrusion resin (Reichhold 31020-03) was used in this study. All other types of unsaturated polyester and vinyl ester resins can be used.

HOT BLOCK EXOTHERM EXPERIMENTS

A hot block tester at 250° F was used to evaluate resin cure characteristics via a modified SPI procedure (SPI, 1981). When used, promoter solutions were first blended into 50 g resin samples. Peroxide was then added and the sample was allowed to equilibrate in temperature for 15 to 30 min. (Safety Note: Never mix a promoter directly with the organic peroxide. A violent reaction could occur. Always mix one into the resin first and then add the other.)

A 5 mL aliquot of the sample was injected into the heated block. The resin temperature profile was recorded through gel and peak exotherm. Figure 2 shows a typical hot block exotherm curve. Gel time was taken at 10° F above the block temperature. Peak exotherm is the maximum temperature achieved during the cure process. Exotherm time is the time at which peak exotherm is reached.

POT LIFE

The remaining hot block sample prepared above was left at ambient temperature (ie. 72-77° F) and checked periodically for signs of gel. Initiated resin pot life was taken as the first sign of gel formation in the sample.

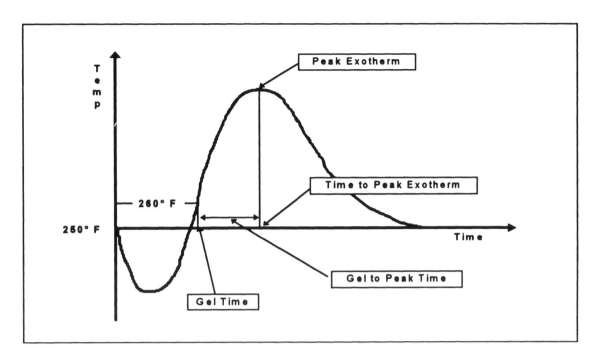

Figure 2. Typical hot block exotherm curve.

RESULTS AND DISCUSSION

MULTIPLE PEROXIDE SYSTEMS

The state-of-the-art initiator system used in pultrusion consists of a multiple peroxide mixture. Most use a combination of two or three peroxides. Pultruders generally mix a low temperature peroxide with medium and high temperature peroxides to achieve desired performance, ie. cure characteristics, line speed, pot life, etc. Table I lists examples of each peroxide type.

Multiple peroxide systems work in a stepwise fashion. When the uncured resin/glass matrix first contacts the heated die, the low temperature peroxide starts to decompose at a high rate. The medium and higher temperature peroxides are slower to react. As the low temperature peroxide decomposes and its free radicals react with monomer and resin, heat is given off and exotherm builds quickly. That exotherm aids in the decomposition of the higher temperature peroxides. They complete the cure after the low temperature peroxide is decomposed.

All peroxides must be treated with care. They are very reactive chemicals. Each peroxide has recommended storage and handling procedures. For safety and product integrity reasons, the low and medium temperature peroxides used in pultrusion must be kept cold. This often creates problems for pultruders without refrigerated storage facilities or whose supplier cannot ship refrigerated materials. In either case, those pultruders are limited to the use of the higher temperature initiators that are slower to react. They are forced to use higher initiator concentrations and/or slower line speeds to obtain the desired cure.

TABLE I. Multiple peroxide pultrusion initiator system.

Peroxide Type (10 hr - $t_{1/2}$ temp)*	Multiple Peroxide System - Example Peroxides	Recommended Storage Temperature (maximum); Physical Form
Low Temperature (47-50° C)	di-(4-t-butylcyclohexyl)-peroxydicarbonate (DTBHPD) or	60° F (16° C); Solid
	t-butyl peroxyneodecanoate (TBPN)	32° F (0° C); Liquid
Medium Temperature (68-72° C)	t-amyl peroxy-2-ethylhexanoate (TAPO) or	60° F (16° C); Liquid
	2,5-dimethyl-2,5-di (2-ethylhexanoylperoxy)-hexane (DMDOPH)	60° F (16° C); Liquid
High Temperature (100-105° C)	t-butyl peroxybenzoate (TBPB) or	100° F (38° C); Liquid
	t-amyl peroxybenzoate (TAPB)	100° F (38° C); Liquid

*Note: Ten hour half-life temperature (10 hr - $t_{1/2}$ temp) is a benchmark commonly used in the peroxide industry to compare the reactivity of one peroxide to another. In general, lower half-life temperatures correspond to more reactive peroxides and to lower shipping and storage temperatures. The above half-life measurements were determined using 0.2M benzene solutions.

PROMOTED SYSTEMS

Recently, Witco discovered a new promoted cure system that eliminates the need for low temperature, hazardous peroxides. Promoters are used extensively for room temperature cure in other composite fabrication techniques like spray-up, lay-up, and casting in industries such as tub and spa, marine, and polymer concrete. Commonly used promoters consist of metal salts of cobalt, iron, nickel, and manganese that chemically react with the peroxide to produce free radicals (Sosnovsky and Rawlinson, 1970, 1971). Promoted systems have not been used extensively in pultrusion due to poor resin pot life. With most promoters, the pot life of catalyzed resin is very short since the peroxide decomposes in the absence of heat. For example, pot life can be as short as a few hours with promoters such as cobalt naphthenate. Witco's new system involves a proprietary "heat activated" promoter (HAP) which is specifically designed to produce good cure performance while still maintaining resin pot life. See Table II.

The new promoted cure system (patent pending) consists of a HAP and only one, high temperature peroxide. The HAP is designed to assist high temperature cures commonly used in pultrusion. It is used in very low levels. The recommended range is 0.10 to 0.40 parts per hundred parts resin (phr). A loading of 0.15 to 0.20 phr is best for most applications.

As stated above, the best initiator used with the HAP is a high temperature peroxide such as t-amyl peroxybenzoate (TAPB). TAPB can safely be shipped, stored and handled at temperatures up to 100° F. This makes it much safer to use than low and even medium temperature peroxides.

TABLE II. Pot life comparison for catalyzed resin (1.5 phr peroxide, 0.15 phr promoter) at room temperature.

Initiator System	Pot Life
Multiple Peroxide [DTBHPD/TAPO/TBPB (1:1:1)]	20-24 hrs
Heat Activated Promoter (HAP) with TAPB	20-24 hrs (or longer)
Cobalt Naphthenate Promoter with TAPB or TBPB	8 hrs (or less)

Resin cure performance of the new heat activated promoter system depends primarily on the peroxide concentration in the resin. The gel time of the resin decreases with increasing TAPB concentration. Normal TAPB addition levels range from 1.0 to 3.0 phr. For a general purpose isophthalic pultrusion resin, 1.25 to 2.00 phr produces cure rates/line speeds similar to a typical three peroxide cure system. Table 3 contains a comparison of resin cure performance for typical multiple peroxide cure systems and the new HAP system. In this comparison, a loading of 1.25 to 1.75 phr TAPB and 0.15 phr HAP solution produce results similar to multiple peroxide cures (at 1.50 phr total peroxide level). As the peroxide (TAPB) levels are increased in the promoted system, faster cures (and line speeds) are possible. Peak exotherm temperatures are also slightly lower with the promoted cure system.

TABLE III. Resin cure performance[a] comparison between a multiple peroxide initiator formulation and the Heat Activated Promoter (HAP) system.

Initiator Type	Initiator Conc.	Gel Time (min)	Exotherm Time (min)	Peak Exotherm (° F)
Multiple Peroxide System	1.50[b] phr	0.75	1.28	340 - 360
Heat Activated Promoter (HAP) System[c]	1.00 phr TAPB	0.92	1.400	330 - 350
Heat Activated Promoter (HAP) System[c]	1.50 phr TAPB	0.77	1.24	330 - 350
Heat Activated Promoter (HAP) System[c]	1.75 phr TAPB	0.71	1.19	330 - 350
Heat Activated Promoter (HAP) System[c]	2.00 phr TAPB	0.67	1.14	330 - 350
Heat Activated Promoter (HAP) System[c]	2.50 phr TAPB	0.61	1.09	330 - 350
a Hot block gel data; 250° F hot block; general purpose isophthalic pultrusion resin.				
b Combination of DTBHPD/TAPO/TBPB peroxides in a ratio of 1:1:1.				
c 0.15 phr HAP solution; The peroxide initiator is t-amyl peroxybenzoate (TAPB).				

The catalyzed resin pot life of the HAP system is similar to the three peroxide system. In both cases, the pot life is about 1 day (20 - 24 hrs) under ambient conditions. (Refer to Table II.) At higher levels of the promoter and peroxide or at higher resin temperatures, pot life may be shorter.

The long pot life observed with the new system is a result of the unique HAP. The reactivity of the peroxide with the HAP in the resin bath is very slow. However, when the catalyzed resin enters the heated die, the promoter is freed to react with the peroxide. Free radicals are produced. The free radicals initiate cross linking and exotherm builds. As the exotherm increases, non-promoted peroxide decomposes along with promoted to help complete the cure of the resin.

CONCLUSIONS

A new heat activated promoter system has been developed that is an alternative to the multiple peroxide based initiator systems. When compared with multiple peroxide cures, the new cure system using a heat activated promoter is generally
- safer since it eliminates the need for the more hazardous, lower temperature peroxides,
- easier to use because the cure is controlled by varying just one peroxide rather than several, and
- has the potential for increased line speeds with long catalyzed pot life.

ACKNOWLEDGMENTS

Ron Pastorino's advice and suggestions were invaluable to the discovery and commercialization of this technology. I would also like to acknowledge Mike Wells, my co-inventor on the heat activated promoter system.

REFERENCES

Sosnovsky, G. and D. J. Rawlinson. 1970. "Metal-Ion Catalyzed Reactions of Peroxyesters" in *Organic Peroxides*, Vol. I, D. Swern (Editor), Wiley, New York, 585-608.

Sosnovsky, G. and D. J. Rawlinson. 1971. "The Chemistry of Hydroperoxides in the Presence of Metal Ions" in *Organic Peroxides*, Vol. II, D. Swern (Editor), Wiley, New York, p. 153-268.

SPI Resins Committee Test Procedures. 1981. "Procedure for Running Exotherm Curves--Using the Block Test Method" Reinforced Plastics/Composites Institute, New York, p. 11-12.

BIOGRAPHY

Ted M. Pettijohn, PhD is Manager of the Organic Peroxides' Plastics Technology Group for Witco Corporation. He has more than 25 US patents and is an inventor of the heat activated promoter technology.

Methods, Materials, Testing and Emissions Control for Open Molding

Reaction Spray Molding Using Polyurethane and Polyurethane Technology

RODNEY D. JARBOR

INTRODUCTION

As we all know, urethane technology has been around since the late 1950's. However, it's really only been in the past 10 - 15 years that urethane-related polymer systems have achieved true acceptance. Today, urethanes are a force to compete against in a wide variety of industries. As with any high-tech product, when it's progressed past the incubation stage and has been whole-heartedly embraced by key players in the business community, growth is inevitable. In the case of urethane technology, that growth has accelerated exponentially.

Across dozens of markets, urethanes are in a period of rapid growth. They are versatile polymers that can be formulated to be anything from soft elastomers to rigid structural plastics to high performance coatings. This makes urethane the most versatile polymer in the coating and plastics industries.

This booklet will illustrate how Reaction Spray Molding (RSM) systems based on urethane technology will assist you to dramatically increase productivity...reduce labor costs...meet strict environmental regulations...improve product quality and expand your composite versatility. While urethane coatings, elastomers and structural resins presently represent only a small segment of the open mold fabrication industry, these

Rodney D. Jarbor, Futura Coatings, Inc.

products are posed to grow significantly before the close of the decade. I say this for many reasons.

For starters, there are environmental regulations and concerns which are limiting the use and growth of gelcoats and unsaturated polyester fiberglass layup. These systems contain styrene and volatile organic content levels far in excess of coming regulations.

Second, polyurethane and polyurea systems have been able to quickly respond to VOC regulations, due to their solventless nature. After all, many of these products contain zero VOC's. What's more, polyester/polyurethane systems have cut styrene levels by over half, putting them in a more attractive position for open mold fabricators.

CHEMISTRY

The versatility of the urethane chemistry adds to the difficulty of understanding the various categories and related terminology. Therefore, a brief overview of the chemistry will clarify and explain some of the complex and various categories of polyurethane (see Figure 1.)

THE POLYURETHANE SYSTEM

The term "polyurethane" has become a very broad term to basically describe compounds based on the reaction of various isocyanates, compounds based on various reactants of resins and composites with specific physical properties. For the most part, prior to the 1980's, the reactant was primarily a hydroxyl terminated polymer known as a polyol resin. There are many types of polyol resins but commonly the polyol is based on typically a polyether or a polyester (saturated) polymer.

THE POLYUREA SYSTEM

However, since the late 1970's to early 1980's, the proliferation of other reactants has caused the formulation of terminology to more specifically describe the reactant resins that is being combined with the isocyanate. For instance, the reaction of an isocyanate with an amine terminated polymer is described as a polyurea system. It should be noted that the term polyurea is beginning to be used too broadly. The proper use of the term polyurea should be to describe a system where the reactant resins are formulated using all amine terminated resins on the Component B side. Today there are systems which have blends of both amine terminated polymers and hydroxyl terminated polymers and, the resulting polymer should be described as a "polyurethane/polyurea system" and not grouped in the category of simply "polyurea".

Polyurea coatings and elastomers have some unique properties in comparison to polyurethanes. First, even though polyurethane reactions can be catalyzed to be as fast as 5-20 seconds, polyurea reactions are inherently fast without catlyst. Furthermore, most polyurethanes are sensitive to excessive humidity or moist surfaces and can produce an unwanted microcellular end product. The amine terminated polymers react so quickly and preferentially with the isocyanate component, that the isocyanate does not have a chance to react with humidity or moisture and, therefore, foaming or a microcellular end product does not typically occur. This can be advantageous in humid conditions.

THE POLYESTER/POLYURETHANE HYBRID SYSTEM

Another category or polyurethane is known as a polyester/polyurethane hybrid. This term refers to a system which is a reaction between an isocyanate and an unsaturated polyester/styrene system. The uniqueness of this technology is that there are actually two reactions going on simultaneously. During the cure cycle, the isocyanate reacts with the hydroxyl function of the polyol resin to develop a very high molecular weight linear polymer, and with the assistance of a peroxide, the system reacts with the unsaturated portion of an unsaturated polyester polyol to add strength and stiffness by creating a crosslinking network (see Figure 4). The physical properties and end uses of polyureas, polyurethanes and hybrid polyester/polyurethanes systems vary from low flexural modulus to high flexural modulus. These details will be discussed later in this paper.

ALIPHATIC VS. AROMATIC

Just as there is specific terminology used to describe urethanes based on using different reactant resins with isocyanates, there is terminology used to describe the reactant resins with different types of isocyanates. There are two common classifications for isocyanates: aromatic and aliphatic. These terms describe the main chemical backbone of the isocyanate component. An aromatic isocyanate is a polymer based on benzene or aromatic rings, and the significance of this is that the molecule has unsaturation in the polymer. A polymer containing an aromatic ring or an unsaturation ring in its backbone is not as color stable as saturated molecules; therefore, various degrees of yellowing can occur. Examples of common aromatic isocyanates are MDI (diphenylmethane diisocyante) and TDI (toluene diisocyanate). Urethane systems based on MDI tend to significantly yellow in exterior applications. However, while there may be color change and some surface chalking, a well-formulated MDI urethane will not degrade or embrittle. Pigmenting the urethane system with color such as black, tan, yellow, red, orange, or green will mask the color change and provide an acceptable appearance and performance in exterior applications.

An aliphatic isocyanate is a polymer based on a linear saturated molecular structure or a saturated cyclic ring structure. Due to the fact that the aliphatic polymer has no unsaturation, it is not easily attacked by UV and, therefore, has outstanding color stability and gloss stability. Typically, aliphatic urethane systems are used to formulate high quality coating which provide superior weathering over all existing coatings systems. Aliphatic urethanes are used for high end automobiles, aircraft, boats and industrial applications (see Figure 5).

COATINGS, ELASTOMERS AND STRUCTURAL RESINS TERMINOLOGY

And, finally, other terminology used in this paper will be coatings, elastomers and structural resins. A coating, a term common to the industry, is a layer of material which is used to provide a protective or decorative finish to a substrate. Typically, coatings contain solvents so they can be sprayed out in thin film thicknesses of 2-6 mils. They are designed to be very UV resistant and provide a weatherable finish.

An elastomer is a material or polymer that can be elongated when stressed to many times its original length and which then returns to the original dimension when the stress is relieved. Urethane technology allows a wide variety of elastomers to be formulated with elongation ranging from 50-700%. In the urethane family, this wide range of elongation sometimes makes it difficult to decide when to describe a system as a coating and when to describe it as an elastomer. Typically, materials are considered an elastomer when they have elongation greater than 100%. It is not unusual that a highly flexible coating be designated an elastomeric coating.

Structural plastics are systems that have sufficient flexural modulus or stiffness to maintain a particular molded form which is self-supporting. As in the case of coatings and elastomers, there is a gray area of when to describe a system as an elastomer and when to describe it as a structural resin. While there is no official industry definition, for the purpose of this paper, I will use a crossover point in flexural modulus of approximately 20-50,000 psi (130-340 MPA). A material will have sufficient structural properties to hold a form at this stiffness.

LATEST TECHNOLOGY IN SPRAY APPLIED URETHANES

There are seven areas of open mold fabrication that polyurethane technology can benefit.

STRUCTURAL RESINS/REACTION SPRAY MOLDING

With the increased regulations concerning volatile emissions and the health-related issues of styrene, many fabricators are looking for alternatives to conventional spray-up FRP/polyester systems. A new group of urethane and urethane-based structural products are being developed for the open-mold fabrication process. This new technology is being identified as reaction spray molding (RSM). The characteristics of the RSM process is that the system is based on two-component, instant curing structural isocyanate based polymer systems that can be sprayed into a wide variety of molds and then can be demolded in a matter of minutes compared to the 3-4 hour cycle time of FRP polyester. Most of the systems typically have no volatile organic compounds (VOC) and respond to the need to replace or supplement styrene-containing resins.

At the present time, there are three types of isocyanate-based polymers that can be used as REM systems. First there is the structural "polyurea" RSM system. The structural polyurea systems are normally low modulus systems which can be formulated with a stiffness of structural modulus in the range of 10,000 to 110,000 psi. This is the ideal range for structural polyurea. The maximum flexural modulus that can be achieved is 110,000 psi because most polyurea systems become very brittle above this range. The polyurea systems are recognized for their outstanding impact resistance and toughness, heat sag/heat distortion resistance and low temperature flexibility. The polyurea systems can be used on applications which require overall toughness and impact resistance such as automobile/truck bumpers, ground effects and snowmobile hoods.

While polyurethane RSM systems can be formulated to have a flexural modulus to partially overlap the polyurea range, typically under 100,000 psi. Structural "polyurethane" systems are better suited in a modulus range between 100,000 to 250,000 psi. (unreinforced). The structural polyurethane provides properties similar to plastics like ABS, polystyrene, acrylic and other thermoplastics and can be used for numerous general purpose applications.

The polyester/polyurethane hybrid RSM system has characteristics of both unsaturated polyester resin systems and polyurethane systems. The typical flexural modulus range for the hybrid system is 200,000 - 450,000. The unsaturated polyester moiety provides the high flexural modulus or stiffness properties. The polyurethane moiety provides toughness and fast reaction properties. Together, the system has the best features of both polymers.

The polyester/polyurethane hybrid systems are finding many applications in the replacement of polyester FRP in many small to medium size open mold fabrications. Lower density versions of the hybrid can economically be used to reinforce products like thermoplastic spas or to produce strong, lightweight composites without the need for chopped fiberglass.

Another use for the hybrid system is as a gelcoat enhancer. The hybrid can be sprayed between a polyester gelcoat and the polyester and chopped glass to eliminate the print through and provide a "Class A" surface (see Figure 6).

The open mold fabrication industry is in trouble with environmental factors. Companies in the United States and other industrial countries are faced with air quality regulations on the styrene emissions which are seriously restricting their growth with styrene/polyester resins. Reaction spray molding is gradually finding excellent acceptance in the open mold fabrication market because it brings a new set of tools and capabilities into a mature market of FRP. Furthermore, RSM allows fast mold turnover, allows the material to be engineered for the specific application, and, many times, provides the same end result that a more expensive process provides. In the next 5 years polyurethanes, polyureas and polyester/polyurethane hybrid systems will make significant headway in the fabrication industry.

RSM Systems can accomplish the following:

-Cure in seconds and can be demolded in minutes, dramatically improving productivity.

-Have a wide variety of physical properties from flexible elastomers to rigid structural systems. They can be easily tailored to specific application engineering and performance requirements.

-Significantly reduce or eliminate high styrene VOC and other environmental concerns.

-Unlike polyester/fiberglass systems, produce little airborne dust when cut or sanded.
-Produce up to 35% lighter weight composites.

STRUCTURAL HYBRID FOAM SYSTEM

A new, spray-applied hybrid polyester/polyurethane foam system based on a reaction between an unsaturated polyester and an isocyanate (polyurethane) will allow new composites versatility and open up countless opportunities. The unsaturated polyester portion of this polymer provides the stiffness and hardness and the polyurethane portion provides the fast reaction and toughness properties.

This polyester/polyurethane hybrid structural foam can be supplied in a variety of different densities. Compared to other types of spray foam or foam board, stock systems, the unique hybrid system has higher stiffness, flexural modulus, flexural strength and compression strength. In addition, compared to wood and polyester FRP, the hybrid foam system also has better screw-holding strength and significantly lower water absorption. The hybrid foam core technology can be used in conjunction with FRP/polyester or structural urethane laminates to produce high strength-to-weight ratio composites. When the hybrid foam system is used in conjunction with solid structural skins, a composite can be produced which is up to 25% lighter than a FRP/polyester composite.

BARRIER COAT/GELCOAT ENHANCER

In the open mold fabrication industry there are three problem conditions which call for a barrier coat. First, applications with complex detailed molds have a high tendency to have air voids and defects since consistent roll out can be difficult. Secondly, problems can occur in which polyester/fiberglass composites blister in immersion applications due to the wicking characteristics of fiberglass laminated behind the gelcoat. A final problem is glass pattern transfer defect can occur in the gelcoat from a polyester laminate which uses a heavily textured reinforcing mat.

A newly developed polyester/urethane hybrid system can be sprayed at 30-40 mils between a gelcoat and polyester/fiberglass laminate to eliminate the three problems outlined above. This hybrid system cures in 15-30 seconds which produces an instant resin-rich barrier layer. This fast cure layer produces a "Class A" type of surface profile, reduces or eliminates roll out defects, reduces cycle time and improves blister resistance.

URETHANE GELCOATS

Aliphatic urethane in-mold coatings/gelcoats are also finding opportunities in the open mold FRP industry. There are polyester/aliphatic urethane hybrids that have created a new dimension in the gelcoat market. Polyester gelcoat is a term for the unsaturated polyester or barrier coating used at 20 mils (0.5mm) in an open mold prior to applying a chopped or mat fiberglass reinforced unsaturated polyester system. Typically, the high quality neopentyl polyester gelcoats commonly used in the boating industry begin chalking and losing gloss within 12-18 months. As a result, these gelcoats require regular waxing and continuous maintenance to maintain a glossy surface appearance.

Hybrid urethane gelcoats are now available that combine the high weather resistance and flexibility properties of the urethane chemistry and hard surface and composite capability of the unsaturated polyester technology. The advantage of this hybrid gelcoat is that it forms a very hard, yet impact resistant surface and has the weatherability equal to a high quality automotive finish (see Figure 7).

WEATHERING

Another key advantage of this technology is that while a typical polyester resin/fiberglass system will not adhere to a conventional urethane in-mold coating, the polyester/polyurethane hybrid gelcoat has excellent adhesion with an unsaturated polyester resin and vinyl ester resin, structural polyurethane resin and epoxy resin systems.

The driving force today is quality. The open mold industry has had to rely on polyester gelcoat as the weather resistant barrier. While polyester has served a useful purpose, it cannot stand up to the new quality standards. For the first time, the open mold fabrication industry has the ability to meet these new weathering standards with a new generation of urethane gelcoats.

Urethane gelcoats are designed to be sprayed up to 20 mils and to provide a hard, durable surface similar to conventional gelcoats. The cure times can be adjusted between 5-45 minutes. There are a number of characteristics that are gained from the urethane portion of the system. The urethane portion contributes a more flexible, crack resistant gelcoat that has weatherability comparable to an automotive finish with excellent gloss and color retention. The polyester portion of the system contributes the properties of hardness and the compatibility with polyester/fiberglass composites.

The urethane gelcoats are targeted typically for exterior applications that require high weatherability, scratch and abrasion resistance and flexibility, such as boating, architectural building trim and components or for various transportation-related uses.

OPEN MOLD SOFT PART URETHANE COMPOSITE

Presently, elastomeric urethane in-mold coatings are beginning to be used on high-end Japanese and European automobile as dashboards and door panel composites as an alternative to the conventional vinyl dash system which is the prevalent system used today. The new solventless aliphatic urethane systems are designed to produce a "10 year quality performance", provide a leather-like feel, have no loss of flexibility and does not contain plasticizer that can fog the windows.

It is now possible to produce cost-effective automotive quality soft padded dash and door panel-type parts with existing polyester or epoxy open molds with a three step process. The first step of the process uses a fast-cure flexible urethane in-mold coating that was originally developed for the automotive industry. The aliphatic urethane is inherently flexible and typically cures in 1-2 minutes.

The second step of the process uses a spray-applied flexible foam which is applied at 1/4 to 1/2 inch thickness. The spray-applied flexible foam cures in 1-2 minutes and is available in a variety of different densities and firmness.

The final step is to back up the flexible foam with a structural support. A fast cure structural urethane or hybrid polyester/polyurethane system is then sprayed over the flexible foam. The structural systems have a flexural modulus comparable to ABS.

The complete system can be sprayed in 10-15 minutes from start to finish and can be demolded within 2-10 minutes after the last step.

Presently the system is being used in applications such as instrument panels and door panels for various types of boats and vehicles, for producing soft bath tubs, soft hair salon sinks, soft water slides and a variety of safety padding applications.

EQUIPMENT-PROCESSING OF SOLVENTLESS URETHANES

Due to the very fast reaction rates of the solventless polyurethane and polyurea systems, it is necessary to use plural component equipment for application. In order to insure consistent and successful spray applications, there are certain minimum specifications that the equipment needs to meet. The sprayability of a plural component urethane is directly related to its viscosity. Two equipment related factors that can control the sprayability as it relates to viscosity of the components are spray pressure and material temperature. As the viscosity of a system increases, the pressure and/or material temperature needs to be modified to provide satisfactory sprayability.

EQUIPMENT SPRAY PRESSURE

First, the equipment should have a spray pressure capability of 1600-3500 psi (112-245 bars). The viscosity of the system determines the minimum pressure required to properly mix and atomize the material. Table 2 shows typical pressure required to spray various viscosities.

*The above pressure recommendation was obtained using a Gusmer H-3500 and a GX-7/400 gun. The preheaters and hose heaters were set at 140°F (60°C).

**In systems with material viscosities above 1500 cps, the material will need to be heated by recirculation of drum heater to 100-110°F (37.8 - 43.3°C). The equipment preheaters need to be increased to 160 - 170°F (71 - 76°C).

SYSTEM HEATING

From practical field experience, it appears that a plural spray unit should have 6-8,000 watt in-line heaters for each component. This amount of wattage provides adequate capacity to handle materials that are as cold as 50°F (10°C). If the heater wattage is less than 6,000 watts, the material should always be conditioned to a minimum of 75-80°F (23.9 - 26.7°C).

In addition to the primary heaters, the hose needs to be heated and insulated right up to the gun. Heated hoses are required to maintain the temperature of the material. Unheated whip hoses can cause cold spells and unpredictable/inconsistency problems.

Gusmer has a number of plural component spray units available which fit the parameters mentioned above. Gusmer's H-3500 and FF-3500 are two quality spray units which provide precision heat and pressure control and require low maintenance.

THE PLURAL SPRAY GUN

The gun is the heart of the success for spraying fast polyurethanes, polyureas and polyester/polyurethane hybrids. As a result of the instant set times and static mix, solvent purge guns are not acceptable. For systems having gel times under 15-30 seconds, an impingement gun such as the Gusmer GX-7-400 is recommended.

Static mix solvent flush guns are satisfactory for urethane related products which have a potlife of over 60 seconds. However, most of the fast cure urethanes technically have a cure time in the range of 10-20 seconds. So this factor rules out those guns with most of the MDI based urethane systems.

CONCLUSION

As a final thought, I'd like to reiterate polyurethanes are simply one of the most versatile polymers available. No other polymer can offer properties that range from a low Shore A elastomer to a high Shore D plastic. The systems can easily be tailored to meet the criteria of a limitless number of end applications.

Besides the versatility and wide range of physical properties, polyurethanes are capable of being adjusted to comply with environmental regulations on VOC's. Now fabricators can rest assured their production is totally compliant with the world's toughest regulations.

Polyurethane systems provide solutions for the evolving quality-driven directions of the industry. Aliphatic polyurethanes provide the best weather stability of any coating presently available in the market-place. Flexible urethanes provide improved alternatives to vinyl dashes or conventional rigid coating systems. Fast cure systems improve cycle time and productivity dramatically. This lowers labor cost and reduces energy costs related to systems requiring high temperature ovens.

And, finally, the abilities of the polyurethane system allow the fabricator or processor to expand their company's ability and capabilities. It is common for the companies who buy plural component equipment to look for additional markets and products to produce from these unique materials.

It's clear that environmental and health regulations, quality, productivity and performance are the driving forces behind the fabrication industry. And spray-applied polyurethanes are positioned as the key replacement material for many existing coating, elastomer, and plastic applications. Add to that fact that urethanes are so versatile in their chemistry and their properties can be tailored to production needs, and it's obvious that urethanes possess all the strength necessary to insure success today and well into the next century.

Figure 1. Categories of Spray-Applied Urethane Systems

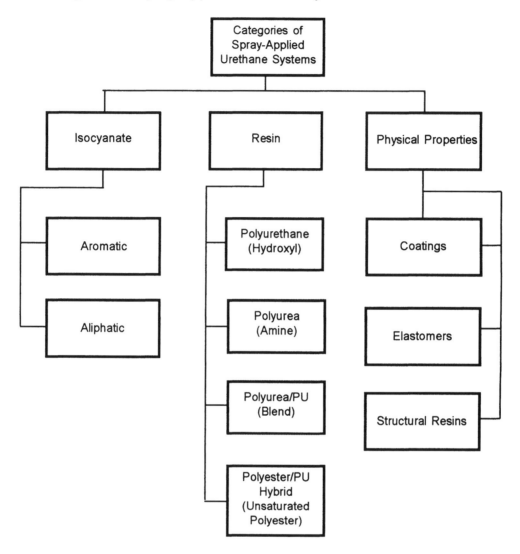

Figure 2. Polyurethane Reaction

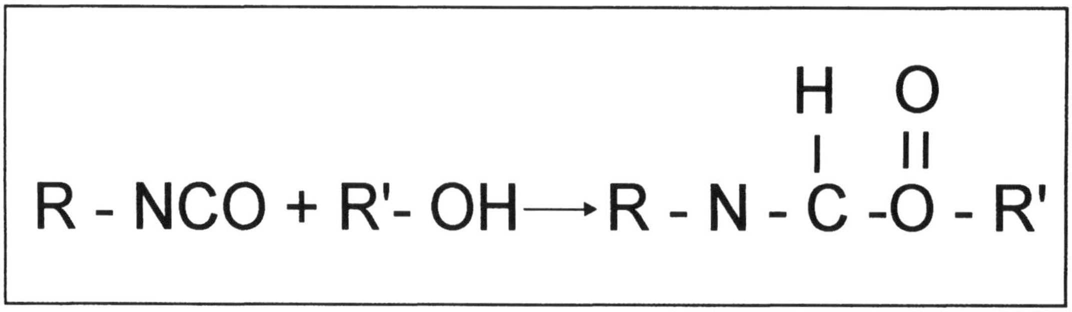

Figure 3. Polyurea Reaction

$$R - NCO + R'- NH_2 = R - NH - C - NH - R'$$

Figure 4. Hybrid Polyester/Polyurethane Reaction

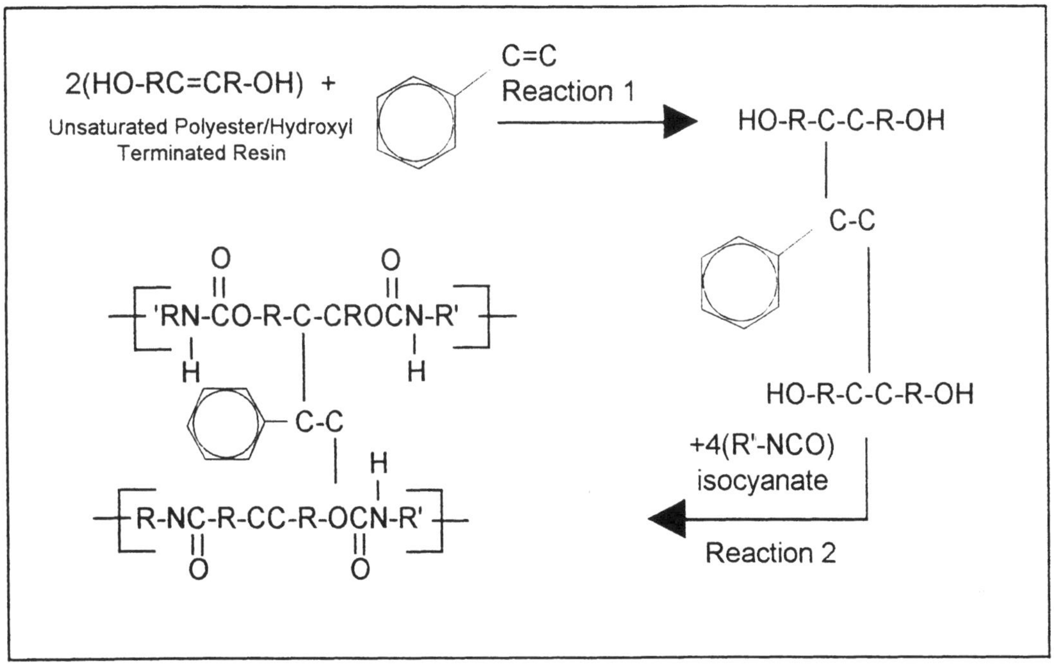

Figure 5. Aromatic and Aliphatic

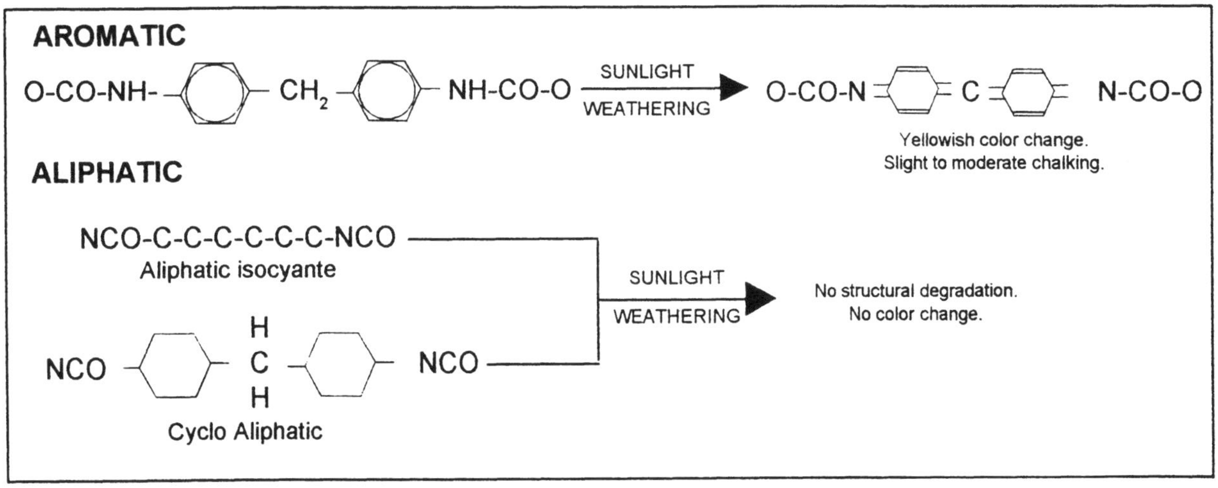

Figure 6. Structural Resins

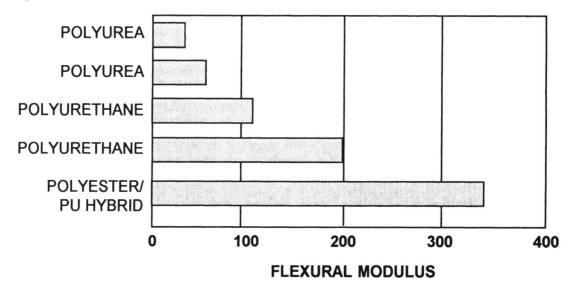

FLEXURAL MODULUS

Figure 7. ULTRACHROME 4005 WEATHERING

**GLOSS RETENTION
FLORIDA WEATHERING**

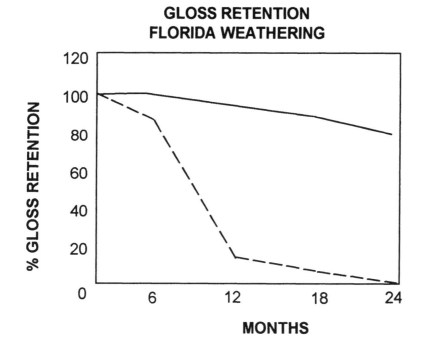

Figure 8. Spray Equipment

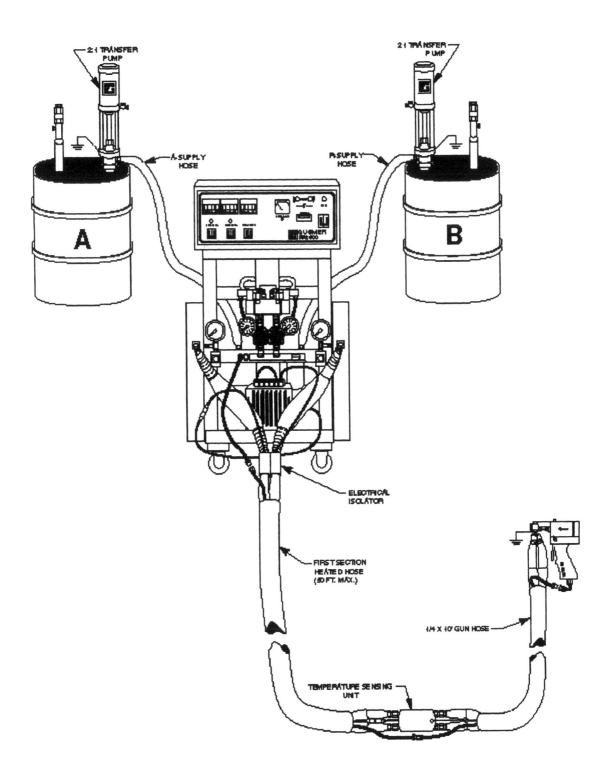

Exploration of Innovative Composite Joining Methods Using Abrasive Waterjet Machining

DAVID G. TAGGART, JOEL A. KAHN, THOMAS J. KIM AND MADHUSARATHI NANDURI

ABSTRACT

In fiber reinforced composite applications, the load bearing capacity of the composite structure is often limited by the strength of the joints, rather than the intrinsic material strength. As a result, numerous studies have examined a variety of methods for joining composite materials. Design criteria for bonded and bolted joining of composites have become well established. These methods are limited, however, by machining processes which provide straight cuts and circular holes. The relatively new machining technique, abrasive waterjet (AWJ) machining has been shown to produce high quality, complex shape machined surfaces. With this capability, a wide range of innovative joining techniques become feasible. This paper explores two of these techniques. In particular, AWJ is used to produce threaded joints and bolted joints with non-circular holes. Both finite element stress analysis and experimental measurement of composite joint strengths are presented. For the threaded joint investigation, a machining study was performed to identify parameters which provide a high quality thread geometry. For the bolted joint investigation, parametric studies of critical dimensions were performed. Correlation of analytical and experimental results reveals that AWJ provides the potential for achieving improved joint strength.

INTRODUCTION

In the design of components fabricated from advanced laminated composite materials, effective means of joining are required. Joint design for conventional metals often relies on local plastic deformation to reduce stress concentrations due to cutouts (Juvinall and Marshek, 1991). Since most composites do not exhibit significant plastic deformation, joint design becomes critical to the overall performance of the component. To take full advantage of the strength and stiffness of composites, effective joining methods are required. As a result, numerous joining methods have been investigated and associated design methodologies have been developed (Hart-Smith, 1978; Hart-Smith, 1994; Ramkumar, 1981). Existing joint designs are limited by difficulties in machining complex shapes without inducing significant damage to the composite laminate. As a result, current joining methods are limited to those formed by straight cuts and circular holes.

Authors' address: URI Waterjet Research Laboratory, Department of Mechanical Engineering and Applied Mechanics, University of Rhode Island, Kingston, RI 02881.

In recent years, a new machining technique, abrasive waterjet (AWJ) machining has been shown to be an effective tool for cutting laminated composites (Hashish, 1988; Arola and Ramulu, 1993; Ramulu and Arola, 1993). In abrasive waterjet machining, a high velocity stream of water is produced by high pressure (200 - 350 MPa (30,000 - 50,000 psi)) water accelerated through a fine orifice (.25-.40 mm (0.01 -.015 in) diameter). A vacuum created by a venturi effect acts to pulls abrasive into a mixing chamber and mixing tube (or nozzle) where the abrasive is mixed and accelerated in the high velocity water stream. The high velocity mixture of abrasive, water and air is capable of cutting hard materials such as metal, stone, and ceramics. The AWJ machining method is well suited for composite machining because cutting forces are very low and minimal localized heat is generated. With careful control of cutting parameters such as abrasive type, abrasive flow rate, water pressure, orifice size, nozzle size and traverse speed, excellent surface quality with no interlaminar damage can be achieved. In addition, through robotic manipulation of the mixing head, precision machining of complex geometries is possible.

The University of Rhode Island Waterjet Research Laboratory was established in 1984 through equipment grants from the waterjet and robotics industry. The lab is equipped with a Flow International model 9XD Waternife™ dual intensifier pump, a Paser II abrasive delivery system, a Flow model 414 shape cutting table and controller. In this study, the use of abrasive waterjet machining to produce innovative composite joints is explored. Two joining methods, threading and bolted joints using non-circular holes are considered.

MACHINING OF COMPOSITE JOINTS

Both joining methods considered require that appropriate cutting conditions be established. The optimal conditions provide high surface quality machined surfaces with no induced interlaminar damage. Also, appropriate cutting paths and speeds need to be determined and programmed into the robotic controller. These conditions were determined through a systematic trial and error investigation.

THREADED JOINTS

To machine composite threads, a lathe was used to rotate the composite shaft while the waterjet traversed parallel to the rod axis. By controlling the angular speed of the rod, the axial traverse speed of the waterjet and the distance from the jet to the center of the rod (the crossfeed), a spiral groove (or screw thread) is machined in the shaft. A schematic of this experimental configuration is shown in Fig. 1. In the initial study to identify effective AWJ machining parameters, a 1.27 cm (0.5 in) OD unidirectional carbon fiber reinforced epoxy rod was examined.

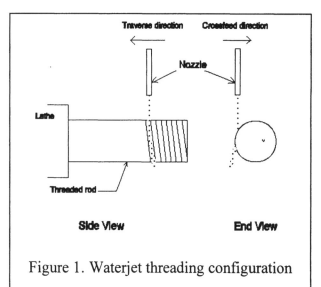

Figure 1. Waterjet threading configuration

™ Waternife is a trademark of Flow International, Kent, WA

It was found that the quality and dimensions of the thread can be controlled by adjusting the traverse speed, the crossfeed, the mixing tube nozzle bore and the orifice bore. In the mixing head, a fine stream of high velocity water is accelerated in the orifice and is then mixed with abrasive in the mixing tube or nozzle. It was found that a nozzle bore of 0.0762 cm (0.03 in) and an orifice bore of 0.0254 cm (0.01 in) gives a good quality thread profile. Traverse speeds of 5.08 - 6.35 mm/sec (12-15 in/min) were found to give good results. Details of this study are discussed by Sheridan et al., 1994. In Sheridan's work, a thread asymmetry was observed which contributed to higher stress concentrations and, hence, reduced the strength of the joint (see discussion below). Recent refinements including adjustments to the jet inclination, turning direction (upturning vs. downturning), standoff distance and crossfeed have been successful in producing thread geometries with no significant asymmetry.

BOLTED JOINTS

Characterization of bolted joint strength and failure was achieved by open hole tension and tensile pin loading on elliptical holes with aspect ratios (a/b=width/height) of 2.0, 1.0 (circular) and 0.5. A specimen width of 25.4 mm (1 in) and a hole area of 31.7 mm^2 ($\pi/64$ in^2) were evaluated. The corresponding hole dimensions (width/height) were 8.98/4.49 mm , (.354/.177 in), 6.35/6.35 mm (.25/.25 in), and 4.49/8.98 mm (.177/.354 in). The composite material evaluated was a 1.27 mm (0.05 in) thick graphite woven fabric / F655 bismaleimide resin laminate provided by Hexcel Corporation. Barton Garnet mesh 80 was used as the abrasive material. AWJ cutting conditions of 315 MPa (45,000 psi) water pressure, 0.76 mm (.030 in) nozzle diameter, 0.25 mm (.010 in) orifice diameter, 0.45 kg/min (1 lb/min) abrasive flow rate and a traverse speed of 12.7 cm/min (5 in/min) were found to provide high quality machined surfaces.

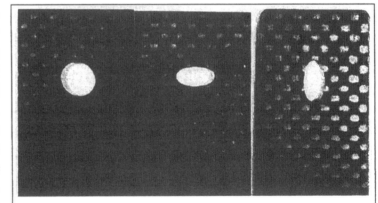

The lateral edges of the coupons were machined by straight cuts starting from a free edge of the laminate. All specimens were machined parallel to the weave direction. Machining of the elliptical holes was complicated by the need to first pierce through the laminate prior to cutting the elliptical hole. When the jet is first turned on, a short delay of 1 - 2 seconds is required until the abrasive becomes entrained in the water

Figure 2. Photograph of bolted joint coupons showing elliptical holes.

stream. During this interval, a high velocity pure water stream impinges on the surface of the laminate. Since little cutting occurs without abrasive, very high impingement forces are induced on the laminate surface. These forces induce interlaminar damage similar to that observed for out of plane low energy impact. To avoid this damage, a small diameter hole was pre-drilled using conventional tooling and the abrasive waterjet stream was initiated from this hole. An elliptical cutting path was then used to obtain the desired hole geometry. Figure 2 is a photograph of the bolted joint coupons. Future studies are planned to investigate laminate piercing techniques which will avoid the need for pre-drilling.

PERFORMANCE OF AWJ MACHINED JOINTS

Characterization of the AWJ machined threaded and bolted joints has been performed using both experimental determination of joint strengths as well as finite element analysis of various joint configurations.

THREADED JOINTS

To assess the load bearing capacity of threaded joints, a series of experiments was performed on 2.22 cm (0.875 in) OD, 0.953 cm (0.375 in) ID carbon fiber reinforced epoxy tubes. In these tubes, screw threads were machined using the AWJ machining parameters discussed above. A steel nut was the screwed on to the composite shaft. A test fixture was developed to load the threaded joint in tension in an Instron test machine. To evaluate the effect of composite layup on threaded joint strength, two configurations were evaluated. In the first configuration, the fibers were aligned with the axis of the shaft. The other configuration consisted a woven fabric which was wrapped around a cylindrical mandrel. The primary difference between these configurations is the circumferential orientation of approx. 50% of the fibers in the woven fabric reinforced shaft as compared to 100% axial fibers in the unidirectional fiber reinforced shaft. Also, the crimp in the weave provides a small amount of fiber reinforcement in the radial direction. This radial reinforcement is believed to enhance the strength of the composite threaded joint. The average maximum axial stress applied to the unidirectional and woven fabric reinforced shafts was 28.2 MPa (4,070 psi) and 47.4 MPa (6,850 psi), respectively.

To better understand the effect of thread profile and composite configuration on thread strength, a finite element analysis was performed. The analysis was performed using the software package I-deas™ to create an axisymmetric model which was analyzed using the finite element analysis program Abaqus™. The results of the finite element analysis (see Sheridan et al, 1995) reveals that the threads nearest the base of the nut carry the largest load. Due to the higher axial modulus of the unidirectional thread a slightly higher concentration of stress on the first (lowest) thread is induced. This effect leads one to predict a slightly higher strength for the woven fabric threads. Also, since the woven fabric has some radial fiber reinforcement, the woven fabric reinforced shafts can withstand higher thread shear stresses prior to failure. This effect is believed to be the predominant explanation for the higher strength observed for the woven fabric reinforced shafts.

It is of interest to consider the relation between the stress analysis predictions and the AWJ machining conditions. The asymmetry of the thread angle results in a single contact point between the steel nut and the composite thread. The stress analysis shows that this results in a stress concentration at the thread root where shear failure of the thread is initiated.

™Abaqus is a trademark of Hibbitt, Karlsson & Sorensen, Inc of Providence, RI.
™I-deas is a trademark of Structural Dynamics Research Corporation of Milford, OH.

Clearly, improved control of the thread angle will give a more uniform load transfer from the nut to the composite thread. This will reduce the stress concentration at the thread root and will therefore improve joint strength. Another observation is that failure initiates at the rounded corner between the thread root and the thread side wall. In this region, scanning electron microscopy revealed matrix damage and fiber pullout. This localized damage may act as a flaw and may contribute to premature failure of the thread. As mentioned above, recent refinements have led to improved thread geometries. Experiments are planned to determine if these improvements will lead to higher joint strengths.

BOLTED JOINTS

To characterize the strength of bolted joints with elliptical holes, two types of experiments were performed. To assess the strength reduction due to the elliptical stress riser, open hole tension tests were performed. The measured strengths were compared to the tensile strength of the unnotched tensile specimens. The elastic modulus and Poisson's ratio were determined using electrical resistance strain gages on the unnotched tensile specimens. These results, along with other estimated elastic properties are summarized in Tables I and II. The effect of elliptical holes on tensile strength was also investigated using finite element analysis.

The next experiment determined the bolted joint strength by inserting an elliptical steel pin machined to match the hole dimensions. The distance from the edge of the hole to the edge of the specimen, c, was varied. The results of these tests are shown in Figure 3. For all of the configurations evaluated, failure was observed to be a combination of bearing failure followed by a gradual shear out of the pin. It can be seen that increasing the pin aspect ratio, a/b, results in an increase in the failure stress. It is believed that this increase is associated with an increase in bearing area as the pin aspect ratio increases.

To better understand the effect of elliptical holes and bolted joint failure mechanisms, a plane stress finite element stress analysis was performed to determine the stress field in the composite laminate in the vicinity of the pin. For this analysis, the CAD/FEA package Algor™ was used to create the finite element meshes and Abaqus was used to perform the stress analysis and post-processing. The pin was modeled as a rigid frictionless contact surface. A far field stress equal to the experimentally observed failure stress was applied. The resulting stress fields show a large compressive stress at the top of the pin and relatively high shear and tensile stresses near the edge of the pin. These complex stress fields were evaluated by applying the Tsai-Wu quadratic interactive failure parameter (Tsai, 1992) given by

$$I_F = (\frac{1}{X_t} + \frac{1}{X_c})\sigma_{11} + (\frac{1}{Y_t} + \frac{1}{Y_c})\sigma_{22} - \frac{1}{X_t X_c}\sigma_{11}^{~2} - \frac{1}{Y_t Y_c}\sigma_{22}^{~2} + \frac{1}{S^2}\sigma_{12}^{~2} + 2F_{12}\sigma_{11}\sigma_{22}$$

where X_t is the lamina axial tensile strength, X_c is the axial compressive strength, Y_t is the

™Algor is a trademark of Algor, Inc., Pittsburgh, PA.

transverse tensile strength, Y_c is the transverse compressive strength, S is the shear strength, σ_{11} is the axial stress, σ_{22} is the transverse stress, σ_{12} is the shear stress and F_{12} is an interaction parameter taken to be 0. When I_F reaches the critical value of one, the failure criterion is satisfied and failure is predicted.

TABLE I. LAMINATE PROPERTIES

Axial Modulus, GPa (Msi)	59.5 (8.50)
Poisson's Ratio	0.052
Transverse Modulus (est.), GPa (Msi)	59.5 (8.50)
Shear Modulus (est.), GPa (Msi)	4.2 (0.6)
Axial Tensile Strength, MPa (ksi)	532 (76)
Transverse Tensile Strength (est.), MPa (ksi)	532 (76)
Axial Compressive Strength (est.), MPa (ksi))	280 (40)
Transverse Compressive Strength (est.), MPa (ksi)	280 (40)
Shear Strength (est.), MPa (ksi)	49 (7)

TABLE II. EFFECT OF ELLIPTICAL HOLES ON TENSILE STRENGTH

Aspect Ratio (a/b)	Tensile Strength	% of Laminate Tensile Strength
0.5	312 (44.6)	59
1.0	288 (41.1)	54
2.0	244 (34.9)	46

In analyzing the finite element results for the open hole tension specimens, it is observed that in all cases, the maximum stresses corresponding to measured failure loads exceed the Tsai-Wu failure criterion. The failure criterion is reached at stress levels of 34%, 19% and 13% for the cases of a/b=0.5, 1.0 and 2.0, respectively. Similarly, the results of the bolted

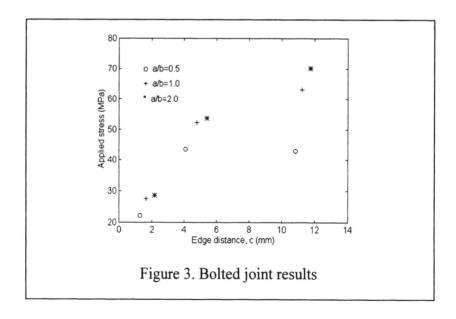

Figure 3. Bolted joint results

joint finite element analyses (Figures 4-6) show a finite volume of material where the failure criterion is exceeded. These discrepancies can be explained by the statistical nature of failure where small volumes of material can withstand stresses in excess of the laminate strength. Also, many strength properties for the laminate were estimated. Finally, interlaminar effects were not considered in the finite element analysis. Nevertheless, both the experimental and numerical results indicate a significant variation in strength with hole geometry.

CONCLUSIONS

It has been demonstrated that AWJ machining is an effective tool for threading composite rods and cutting non-circular holes in composite laminates. These geometries provide the potential for innovative joining methods for composite materials. The strength of these joints has been evaluated by both experimental and numerical studies. Further work is required to determine optimal joint geometries for particular laminates and applications.

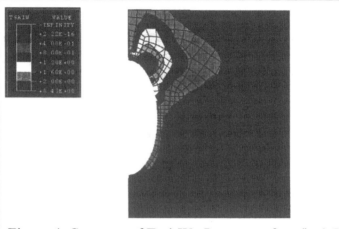

Figure 4. Contours of Tsai-Wu Parameter for a/b=0.5.

ACKNOWLEDGMENTS

The authors would like to thank Mr. Lawrence A. Girouard for providing materials for the threaded joint investigation and Hexcel Corporation for providing laminates for the bolted joint investigation.

REFERENCES

Arola, D. and Ramulu, M., 1993, "Mechanisms of Material Removal in Abrasive Waterjet Machining of Common Aerospace Materials," *7th American Water Jet Conference*, Seattle, Washington, pp. 43-64.

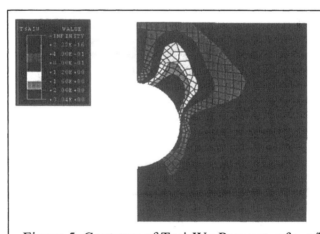

Figure 5. Contours of Tsai-Wu Parameter for a/b=1.

Hart-Smith, L. J., 1978, "Mechanically fastened joints for advanced composites - phenomenological considerations and simple analyses," *Fibrous Composites in Structural Design*, E. M. Lenoe, D. W. Oplinger, and J. J. Burke, Eds.., Plenum Press, NY, pp. 543-574.

Hart-Smith, L. J., 1994, "The key to designing efficient bolted composite joints," Composites, Vol. 25, No. 8, pp. 835-837.

Hashish, M., 1988, "Machining of Advanced Composites with Abrasive Waterjets," *Machining Composites*, Taya, M. and Ramulu, M., eds., ASME PED-Vol. 35, MD-Vol. 12.

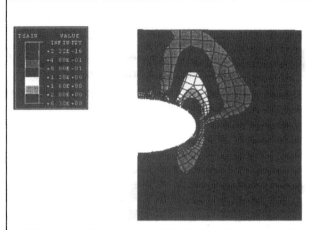

Figure 6. Contours of Tsai-Wu Parameter for a/b=2.

Juvinall, R. C. and Marshek, K, M., 1991, *Fundamentals of Machine Component Design*, John Wiley & Sons, New York, pp. 339-422.

Ramkumar, R. L., 1981, "Bolted Joint Design," *Test Methods and Design Allowables for Fibrous Composites*, ASTM STP 734, C. C. Chamis, Ed. ASTM, pp. 376-395.

Ramulu, M., and Arola, D., 1993, "Water Jet and Abrasive Water Jet Cutting of Unidirectional Graphite/Epoxy Composite," *Composites*, Vol. 24, No. 4., pp. 299-308.

Sheridan, M. D., Taggart, D. G., and Kim, T. J., 1994, "Screw Thread Machining of Composite Materials Using Abrasive Waterjet Cutting," *Symposium on Nontraditional Manufacturing Processes for the 1990's*, ASME, pp. 421-432.

Sheridan, M. D., Taggart, D. G., Kim, T. J. and Wen, Y., 1995, "Microstructural and Mechanical Characterization of Threaded Composite Tubes Machined using AWJ Cutting, 8th American Water Jet Conference, pp. 245-258.

Tsai, S., 1992, *Theory of Composites Design*, Think Composites, Dayton, OH, pp. 8-1 - 8-23.

BIOGRAPHIES

David Taggart is an Associate Professor in the Department of Mechanical Engineering and Applied Mechanics at the University of Rhode Island and serves as Associate Director of the URI Waterjet Research Laboratory. Joel Kahn is an undergraduate student in the Department of Mechanical Engineering. Thomas Kim is Dean of Engineering and a Professor of Mechanical Engineering and serves as Director of the URI Waterjet Research Laboratory. Madhusarathi Nanduri is a graduate student in the Department of Mechanical Engineering and Applied Mechanics and serves as Laboratory Manager of the URI Waterjet Research Laboratory.

Residual Compressive Strength of Hard Lay-Up Glass Fiber-Reinforced Plastic Laminates Due to Low Velocity Impact

IVAN VOLPOET, CHIHDAR YANG, AND SU-SENG PANG

ABSTRACT

With the increasing use of composite materials and especially composite laminates with organic matrixes, the problem of the impact becomes predominant. Low velocity impacts (less than 100 m/s) are considered as particularly dangerous because the damage is not always visible. In many cases, the level of impact which significantly affects the mechanical properties of a composite is much less than the impact which causes visible damage. There is a need to better understand the damaging phenomena to predict the loss of mechanical properties. The residual compressive strength of hand lay-up laminates, which are commonly used in the composites industry, due to low velocity impact was experimentally investigated. Boeing Specifications Standard (BSS 7260) was adopted during this study. The effects of impact velocity and impact energy on residual strength were experimentally investigated in this study. However, the velocity does not seem to have a predominant influence on the residual compressive properties.

INTRODUCTION

This project deals with the impact problem on composite laminate plates. The problem of the impact on isotropic materials has been very well researched. However, those results cannot be applied to anisotropic materials. Therefore, with the increasing use of composite materials, and especially composite laminates with organic matrixes, the problem of impact becomes predominant. Low velocity impacts (velocity less than 100 m/s) are considered as particularly dangerous because the damage is not always visible. The resulting damage is internal and localized, and thus may be left undetected and unrepaired. In many cases, the level of impact which significantly affects the mechanical properties of a composite is much less than the impact which causes visible damage. Low velocity impact occurs when tools are dropped on a structure during maintenance. A barely visible point of impact can contribute up to 60% loss in an aircraft compressive strength.

Mallick (1988) and Dorey (1984) surveyed the effect of impact on the residual properties. In their studies it was suggested that compressive strength, stiffness, and tensile strength decreased with the increase in the impact energy. This is the theoretical basis that both Boeing and SACMA standard tests specify the required impact energy as 35 joules (25.81 ft-lbs). However, after Olsson (1992) investigated the impact response of orthotropic composite plates, he concluded that the common method of relating impact response and damage to impact energy

---

I. Volpoet, C. Yang, and S.-S. Pang, Louisiana State University.

was inadequate since impactor mass and impact velocity were independent parameters. Cairns *et al.* (1988) found that the target mass properties became more important as the impactor velocity increases.

Because hand lay-up process is a very common process in the composites industry, the objective of this study is to experimentally investigate the residual compressive strength of hand lay-up laminates due to low velocity impact. Boeing Specifications Standard (BSS 7260) was adopted during this study. While the standard impact energy (35 joules) was used for all the impact specimens, several combinations of impactor mass and velocity were selected to determine the effects of each individual parameter.

Three types of hand lay-up laminates were used during this study: (1) mRmmRm, (2) mRmRmRm, and (3) MMMMM, where the symbols m and M stand for 0.75 and 1.5 oz./sq. ft. chopped strand mat, respectively, and R represents 24 oz./sq. yd. woven roving mat. The fiber was E-glass and the resin was Derakane 470. These laminates represent the three commonly used hand lay-up laminates. Their semi-transparency made detecting the damage mode and level of damage easier. In these types of hand lay-up laminates, the mechanical properties are mainly from the woven roving fabrics. The chopped strand mat layers, however, are used to bond layers of woven fabrics since chopped strand mats are more resin-rich and soft. If no chopped strand mats are used for a hand laid-up laminate the laminate would delaminate while it's cured. Therefore, it is important to study the difference in the residual compressive strength that exists between two hand lay-up laminates: (1) one layer of chopped strand mat between two layers of woven roving (mRmRmRm) and (2) two layers of chopped strand mat to bond two layers of woven roving (mRmmRm).

The first objective of this study was to obtain the residual compressive strength of the three types of hand laid-up laminate with three energy levels -- 17.5, 35, and 70 joules (12.91, 25.81, and 51.63 ft-lbs). Laminates without impact were also tested for their compressive strength in order to determine the strength reduction due to impact. The second objective was to highlight the influence of the impact velocity on the residual compressive strength. A constant impact energy of 35 joules (25.81 ft-lbs) which is specified in Boeing "Compression After Impact Standard Test (BSS 7260)" was adopted with three combinations of impactor mass and velocity. Depending on the results of the contact force and the residual compressive strength for these three impactor mass and velocity combinations, it was attempted to quantify the phenomenon and to investigate the mechanisms of the failure mode. The delamination was also quantified and its influence on the compressive strength was studied.

EXPERIMENTAL SETUP AND PROCEDURE

The BSS (Boeing Specification Standard) 7260 was followed during this study. This standard describes the procedures for compression testing of continuous fiber reinforced polymer matrix composite laminates. This standard applies primarily to unidirectional tape and fabric laminates. It will be extended for the MMMMM laminates. This standard includes tests on laminates with and without holes or impact damage. It is used to establish uniform practices for compression testing. The SACMA (Suppliers of Advanced Composite Materials Association) Recommended Test Method for Compression After Impact Properties of Oriented Fiber-Resin Composites was also followed because it specifies the testing procedures and the characteristics of the plate.

IMPACTOR

The impactor was designed in accordance with the BSS 7260 specifications. Its variable mass was made possible with the eight attachable weights of 0.454 kgw (one pound) for each and the main impactor, which had a weight of 1.62 kgw (57.22 ounces). The diameter of the hemispherical tip was 1.59 cm (0.625-inch). It is steel made and the tip was hardened. The specified energy for a 0.5-cm (0.2-inch) plate is 35 joules (25.81 ft-lbs). In order to measure the impact force, the contact duration, and the energy, two semi-conductor strain gages (BLH SR4, with a gage factor of 130) were cemented close to the impactor tip at the same cross-sectional area, 180 degrees apart. The response was recorded by a digital oscilloscope (Nicolet 4094C) through a Wheastone bridge and amplifier.

DROP WEIGHT TOWER

The drop weight tower consisted of two steel guide rods attached to the frame, which support the impactor holder. A timer unit consisting of two photodiode detectors was used for measuring the time taken by the impactor to travel through a fixed distance. The impact velocity can then be determined by the recorded time and the distance. A device has been designed on the impactor holder to prevent a secondary impact of the target. The base was made to follow the specifications. Brass bushings were used on the impactor holder to reduce the noise caused by the vibration.

SPECIMEN

The BSS 7260 preconises to use a 6" × 4" × 0.2" plate. At least one inch wide strip was removed from the edge of the original plate in order to have a uniform thickness. All machined surfaces had to be smooth and free of nicks, scratches, or other defects. The thickness of the specimen had to be recorded as the average value of four thickness measurements taken around the impact area prior to testing. Fiber volume was determined in accordance with SRM10. The following table represents the three types of plates tested.

TABLE 1 FIBER CONTENT OF TEST SPECIMENS

Laminate	Fiber Content (% in Weight)	Fiber Content (% in Volume)
mRmmRm	32.95	16.05
mRmRmRm	40.08	20.64
MMMMM	33.29	16.25

m: 0.75 oz/ft^2 chopped strand mat ;   M: 1.5 oz/ft^2 chopped strand mat
R: 24 oz/yd^2 woven roving mat

COMPRESSION AFTER IMPACT FIXTURE

A compressive loading fixture was used to prevent general buckling of the specimens during the loading. The fixture was made by Wyoming Test Fixtures Inc. The compression machine was a 25.5 ton Instron machine. The testing speed was 0.05 inch/minute.

ENERGY TRANSMITTED

The reduction in impactor energy can be represented by the change in kinetic energy as

$$\Delta E = E_o - E_f = \frac{1}{2} m (V_o^2 - V_f^2) \tag{1}$$

where E and V are the kinetic energy and velocity of the impactor and the subscripts o and f denote the beginning and end of the impact, respectively. The linear momentum and impulse relation can be written as

$$\int_0^{t_f} F(t) \, dt = m (V_o - V_f) \tag{2}$$

where $F(t)$ is the force history during the impact. By defining E_a as

$$E_a = V_o \int_0^{t_f} F(t) \, dt \tag{3}$$

the energy transmitted to the target can be determined as

$$\Delta E = E_a \left(1 - \frac{E_a}{4E_o} \right) \tag{4}$$

RESULTS AND DISCUSSION

A total of 45 specimens were tested with three different energy levels. Using Eq. (4), the energy transmitted to the target was found to be 77%. Other results are described in the following sections.

MATRIX CRACKING AND DELAMINATION

The peanut shape of the delamination area was observed on the mRmRmRm and the mRmmRm specimens while no particular shape of delamination occurred on the MMMMM. The matrix cracking followed the map of damages presented in the theoretical section also, except for the MMMMM which did not show any matrix cracks. For the two other materials, the matrix cracks were oriented in the fiber direction or toward the corner of the plate. They were located at the impact center, where delamination occurred, and in two triangular zones bounded by the diagonals of the plate. It can be seen from Figs. 1 and 2 that the larger the matrix cracking the larger is the delamination. It is harder to see the correlation on Fig. 2 because it was a lot more difficult to measure with a good accuracy the delamination area located at the centered layer of the mRmRmRm plate, problem that did not exist with the mRmmRm.

COMPRESSION AFTER IMPACT

During the compression, local buckling appeared after a critical load which makes the

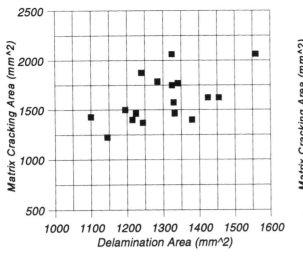

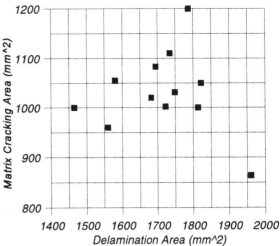

FIGURE 1 mRmmRm LAMINATE   FIGURE 2 mRmRmRm LAMINATE

compression curve deviate from linearity. After that, the delamination area grew quickly until breakage. The growth of the delamination has its major direction toward the lateral edges.

During this study, 35-joule impact was used for mRmmRm laminate while 17.5, 35, and 70-joule impact were used for mRmRmRm and MMMMM laminates. Figure 3 shows the compressive strength of these laminates before and after impact. It can be seen from these results that the loss of compressive strength is larger for the MMMMM laminate than for the mRmmRm and the mRmRmRm laminates. It can also be seen from Fig. 3 that the residual compressive strengths for mRmmRm and mRmRmRm were almost the same. Based on this result, one 0.75 oz/ft^2 chopped strand mat is sufficient to bond two woven roving mats and mRmRmRm is preferable than mRmmRm because the former provides a higher glass content and strength.

Figure 4 shows the residual compressive strength of the three laminates due to the impact energy of 35 joules with different impact velocities. It can be seen that the impact velocity did not seem to have a great influence on the residual compressive strength when the energy is restrained to be constant even some of the existing theories give a lower compressive strength with a larger

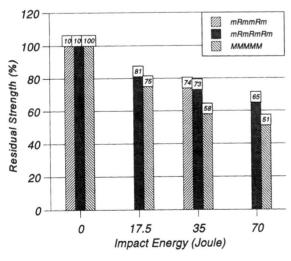

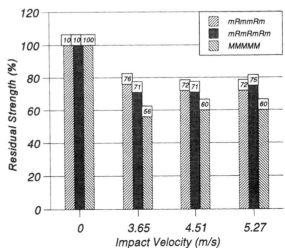

FIGURE 3 RESIDUAL STRENGTH VS.   FIGURE 4 RESIDUAL STRENGTH VS.
IMPACT ENERGY   IMPACT VELOCITY (35-JOULE ENERGY)

impact velocity. One of the explanations for this conflict is that a small range of velocity, which was due to the restriction of the weight drop tower in this study, might not be sufficient to highlight the influence of the impact velocity. If a larger velocity range can be adopted, the trend might be justified.

Therefore, one conclusion of those tests can be made as: "For this range of low velocities (3.5 to 5.5 m/s) it is not necessary to study the residual compressive strength of hand lay up composite laminates as a function of two independent parameters -- mass and velocity." It has been shown that the influence of the energy on impact was accurate enough to reach a good conclusion of the residual compressive strength of such materials, even if mass and velocity are mathematically independent in the theoretical studies.

CONCLUSION

This study dealt with the residual compressive strength of three hand lay-up composite laminates subjected to low velocity impact. It gave some results that an engineer can use to design composite structures. The first conclusion is that it is preferable to use laminates with woven roving fabrics than only chopped strand mat because woven roving fabrics's loss of compressive behavior is smaller than for the whole chopped strand mat's one. The second conclusion concerns the fabrication of the laminates with woven roving fabrics. It is no use to put two layers of mat to bound two woven roving fabrics layers. One is enough. The loss of compressive strength rates of both solutions is the same, in the range of our tests, but the laminates with one chopped strand mat as a bounding joint is more stronger than the other because it contains a higher rate of woven roving fabrics.

The third conclusion of this study concerns the influence of the velocity on the residual compressive strength. Even if both mass and velocity are mathematically independent in the impact problem, no influence have been shown in our experiments. When the energy is restrained to be constant and equal to 35 joules, the variations of impact velocity between 3.65 m/s and 5.27 m/s did not induce any remarkable changes of residual compressive strength.

ACKNOWLEDGMENTS

This research has been possible with the exchange program between Louisiana State University and the Ecole Nationale Superieure d'Ingenieurs de Constructions Aeronautiques (ENSICA), which is a French engineering school specialized in aerospace technics and which is located in Toulouse, France. This study was partially sponsored by the Louisiana Board of Regents under contracts LEQSF (1995-98)-RD-B-05 and LEQSF (1996-99)-RD-A-26.

REFERENCES

Olsson, R. "Impact Response of Orthotropic Composite Plates Predicted from a One-Parameter Differential Equation," *AIAA Journal*, Vol. 30, No. 6, 1992.
Abrate, S., "Impact on Laminated Composite Materials," *Appl Mec Rev*, Vol. 44, No. 4, 1991.
Boeing Specification Support Standard BSS 7260, "Advanced Composite Compression Tests," 1988.
SACMA Recommended Test Method for Compression After Properties of Oriented Fiber-Resin Composites.

90% Reduction in Styrene Evaporation from White Gelcoat

KARL RØDLAND AND ROAR MØRK

ABSTRACT

Use of a low styrene emission (LSE) gelcoat has been a goal for the GRP-industry over a long period of time.

Several test products have been tried over the years, and low emission has been obtained, but without maintaining acceptable adhesion properties to the back-up laminate.

In 1990/1991, a study of the use of alternative monomers to replace styrene began. It was found that partial styrene replacement gave some improvement to emission values, but it was only when combining both wax additives and monomers that really good results were obtained. Unfortunately, adhesion was still a problem. Subsequently the wax was replaced with another type of additive. In distinction from a wax this additive is able to react with the back-up resin, if the resin is applied within a reasonable time span.

This presentation summarizes the main results obtained using a system, based upon a low vapour pressure monomer and the new additive. Thus developing a LSE gelcoat with good adhesion properties and excellent performance profile.

INTRODUCTION

Styrene, as the well-known monomer for unsaturated polyester resins and gelcoats, has for years been subject to rather harsh regulations due to the possible health risk by exposure to the human being.

Jotun Polymer AS entered the market with LSE laminating resin in the early 70's. The development of a low styrene emission (LSE) gelcoat has been an aim for Jotun Polymer over a long period of time. However, the development progressed more rapidly in 1990, when a project investigating the possibility of using alternative monomers in polyester resin was initiated.

In this project more than 50 different monomers were investigated regarding properties like health/environment, water/weather resistance, physical/mechanical data, curing properties, solubility etc. This investigation gave the conclusion that a new monomer could be used to replace some of the styrene in a gelcoat.

Karl Rødland, Product Manager, Jotun Polymer AS, P.O. Box 2061, N-3235 Sandefjord, Norway
Roar Mørk, Senior Engineer, Jotun Polymer AS, P.O. Box 2061, N-3235 Sandefjord, Norway

This presentation will summarize the main results obtained using such a system to develop a LSE gelcoat with good adhesion properties and otherwise a good performance profile.

EXPERIMENTAL

Physical data were measured according to standard Jotun Polymer procedures.

1. Laminate construction for QUV, QCT, water bath and boiling test

Gelcoat applied in 500 microns wet film thickness on a glass plate with a draw down applicator. Catalyst: 1.5% NORPOL Cat.No.1.

As a back-up laminate, 4 layers of MK 510 powder bound mat 450 g/m² + iso polyester resin in a ratio of 3 to 7 were used. Post cure: 16 hours at 40°C.

2. Accelerated weathering: QUV (Cyclic Ultraviolet Weathering tester)

   1) QUV-B313
      UV-source   : UVB-313
      Cycle   : 8 hours UV at 70°C + 4 hours condensation at 50°C
      Exposure time : 120 hours

   2) QUV-A340
      UV-source   : UVA-340
      Cycle   : 8 hours at 60°C + 4 hours condensation at 40°C
      Exposure time : 1500 hours

Yellowing was measured with the DE (colour change) and B* (blue-yellow axis in LAB colour coordinates) on a Coloureye spectrophotometer. Gloss retention was measured using a Byk Labotron Pocket gloss (60°).

3. Water bath

Samples were immersed in water at 50°C. The occurrence of fibre pattern and blisters were monitored with time.

4. QCT

Samples were exposed to water condensation at 60°C and 100 % Related Humidity. The occurence of fibre pattern and blisters were monitored with time.

5. Boiling test

Samples were exposed to boiling water (100°C) for 100 hours. Visual evaluation of cracks and blisters, ranging from 0 = poor, to 6 = excellent.

6. Styrene/monomer emission

Prewaxed glass plates, 30 x 50 cm, were sprayed with gelcoat at 1500 microns wet film thickness. The plates were then placed on a balance measuring weight loss with time.

Weight loss in g/m² were then calculated by dividing the measured weight loss with the surface area of applied gelcoat.

Catalyst level   :   1.5 % NORPOL Cat.No.1
Temperature   :   ca. 23°C
Spray equipment: Applicator IPG 6000

7. Adhesion/DCB test

Gelcoat was sprayed with 1500 microns wet film thickness onto a glass plate. The gelcoat was allowed to cure for 12 hours, then a very thin aluminum foil was placed over the gelcoat's midsection, followed by spray application of roving and polyester resin. A 4 mm thick standard combi mat and iso polyester resin was applied as back-up. The purpose of the aluminum foil is to work as a crack propagator, ensuring a gelcoat/laminate delamination.

The laminate was demoulded after 24 hours, the gelcoat surface ground with 120 grade sading-paper to a matt appearance, and a similar laminate as described above was applied onto the gelcoat surface. This way the gelcoat is located between two equal laminates with the same stiffness. This laminate construction was used for all samples, if not marked otherwise.

The laminate was then divided into pieces of 25 x 200 mm, and special hinges mounted on the specimens.

The samples were then placed in an Instron tensile tester and force applied. Crack length were monitored until maximum/critical force was reached. Critical force (N), crack extension (mm), displacement (mm) and specimen width were then used to calculate the fracture energy (delamination strength) G_{Ic} after the following equation (1):

$$G_{Ic} = (3 \cdot Pc \cdot \blacktriangle)/(2 \cdot b \cdot a) \qquad (1)$$

where

G_{Ic} = Fracture energy/delamination strength in J/m²
P_C = Critical force at crack extension (N)
$\blacktriangle$ = Displacement between laminates where force is applied (m)
b   = Specimen width (m)
a   = Crack extension (m)

In order to avoid too large (non-linear) bending of the specimen arms and hence erroneous G_{Ic} values, the following inequality (2) needs to be satisfied.

$$\blacktriangle/2a < 0.2 \qquad (2)$$

A G_{Ic} value of 175 J/m² is accepted as a minimum value for acceptable adhesion.

A visual evaluation of fibre pullout was also performed on DCB specimens after delamination in the Instron machine. The results were evaluated from 0 to 100 % fibre pullout.

Test pieces for DCB- test method

a.

b.

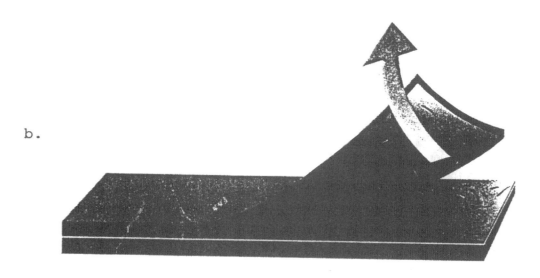

Fig. 1

Fig. 1 shows the test pieces for adhesion between gelcoat and GRP-laminate. Piece A. were then placed in an Instron tensile tester and force applied. Piece B. - a visual evaluation of fibre pullout.

RESULTS/DISCUSSION

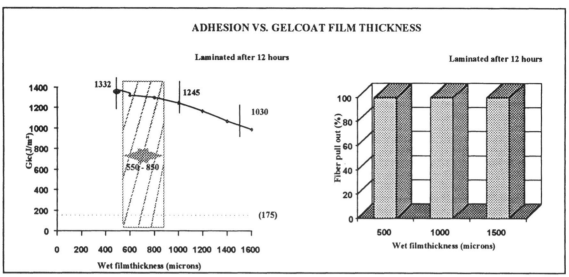

Fig. 2

Fig. 2 shows an adhesion comparison of LSE gelcoat sprayed with different gelcoat/film thickness. The gelcoat was laminated after 12 hours. It can be seen that gelcoat film thickness influences the adhesion, however the results are still above the critical value for adhesion, which is also indicated with a fibre pullout of 100 %.

It is also seen that different types of polyester resin and roving mat influences the adhesion, but the results are well above the critical value for adhesion. The powder is somewhat better than the emulsion mat. The gelcoats were laminated after 12 hours open time.

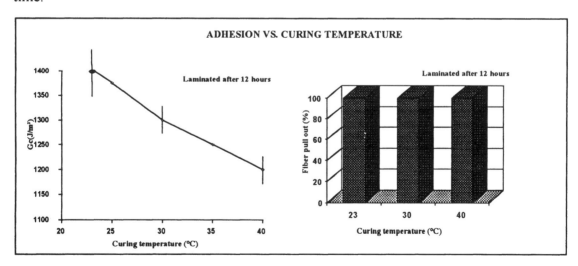

Fig. 3

Fig. 3 shows that the adhesion is not adversely affected with increased curing temperature, when the gelcoat is laminated within 12 hours. The fibre pullout is 100 %.

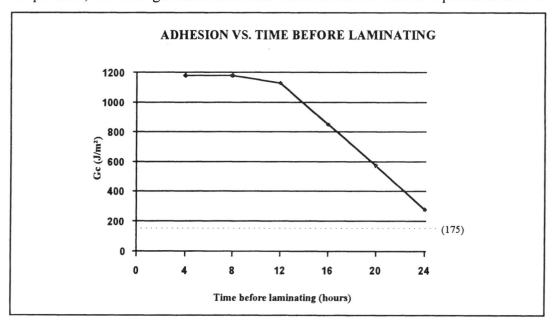

Fig. 4

Fig. 4 shows the need for laminating after 12 hours to be on the safe side when regarding gelcoat/laminate adhesion.

STYRENE/MONOMER EMISSION

Prewaxed glass plates, 30 x 50 cm, were sprayed with gelcoat at 1500 microns wet film thickness. The plates were then placed on a balance measuring weight loss over time. Weight loss in g/m² were then calculated by dividing the measured weight loss with the suface area of the applied gelcoat.

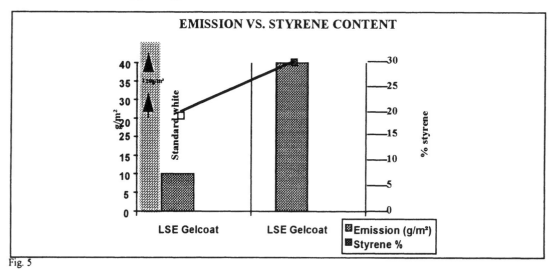

Fig. 5

Fig. 5 shows the styrene emission of the new white LSE gelcoat in comparison with white standard gelcoat, 10 g/m² (emission reduced by 90 % on white gelcoat). It is also seen that the emission varies, depending on different colour, or in other words on the level of colour paste present in the formulation. The reason for this is that a higher level of colour paste reduces the need for base resin in the formulation. This means that less styrene is introduced into the gelcoat. This effect is also illustrated in Fig. 5, where it is clearly seen that emission increases along with styrene content, when going from white to a clear formulation.

The formulation of LSE gelcoat white spray contains 20 % styrene, standard white spray contains 36 %. Brush formulation of LSE gelcoat white contains 22 % styrene, standard white brush contains 28 % styrene.

ACCELERATED WEATHERING

Accelerated weathering tests expose the product to an "artificial sun" and offers humidity according to specified standards (test method as described in JP QC Document No. 10.230.37, R110). The LSE gelcoat has been tested in Cyclic Ultraviolet Weathering Tester (QUV) with a UV-source: QUV-B313 bulb and with QUV-A340 bulb. QUV-B313 bulb has very high UV-intensity and the result will be much discolouration and little change of surface appearance (loss of gloss, chalking). QUV-A340 bulb has lower UV-intensity which gives less discolouration and more chalking and loss of gloss.

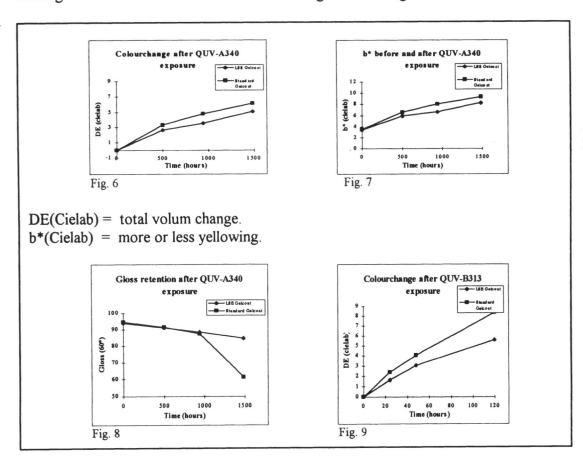

Fig. 6

Fig. 7

DE(Cielab) = total volum change.
b*(Cielab) = more or less yellowing.

Fig. 8

Fig. 9

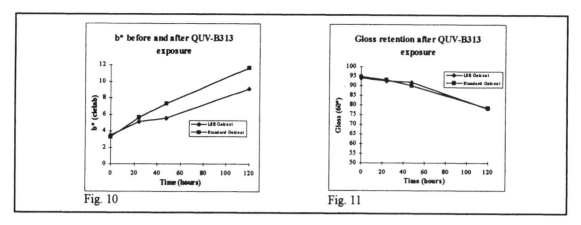

Fig. 10

Fig. 11

Figs. 6 through 11 show the results of the accelerated weathering tests.

Both QUV-A340 and QUV-B313 show that the LSE gelcoat is somewhat better than standard gelcoat. Both gelcoats tested were white formulations.

It is also seen that the QUV-A is more able to separate test samples when regarding gloss retention. In the QUV-A the LSE gelcoat is distinctly better than the standard gelcoat, while the QUV-B shows no significant difference in gloss retention after exposure.

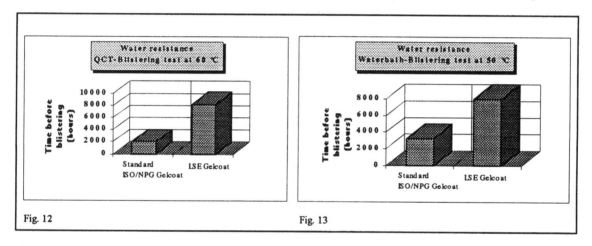

Fig. 12

Fig. 13

Figs. 12 and 13 the water resistance of LSE gelcoat is compared with an ISO/NPG gelcoat. Both were white formulations. Based on these tests, it is apparent that the LSE gelcoat is superior to ISO/NPG gelcoat, when it comes to blister and crack resistance. We can therefore say, that the LSE gelcoat has a water resistance better than high quality gelcoat ISO/NPG.

A Method for Fabrication of Structural Adhesive Joints

RONNAL P. REICHARD

ABSTRACT

Adhesive structural joints are typically formed by applying a bead of adhesive to one surface of a structural part or panel and placing a second structural part or panel onto the adhesive bead, compressing it to cover most or all of the joint surface. The joint is then secured (usually with mechanical fasteners or by clamping) until the adhesive has cured. The patent pending method presented here is a combination of adhesive/adhesive tape technology, involving the use of adhesive tape to temporarily secure and to create a channel between the structural parts to be joined. Liquid adhesive is then pumped into the channel to form a permanent bond. This allows multi-part structures to be "tacked" together quickly, checked for accuracy, and then permanently bonded later by injecting the adhesive. The technology eliminates adhesive drips and spills, provides a more consistent bond-line, and uses less adhesive. It also significantly reduces the amount of labor involved in bonding the parts.

INTRODUCTION

FRP dominates the small boat industry because hulls, decks and small parts can be molded inexpensively. These components have to be joined together to form the finished boat. Traditionally, components have been taped or tabbed together by hand lay-up of FRP (Pfund, 1996; Pattee and Reichard, 1988). Tabbing labor costs often exceed the labor costs to laminate the hull of small boats. Larger boats have many more structural components which must be joined together. This is a labor intensive process, quality varies considerably, and the resin used in the tabbing usually does not have good adhesive properties. Several builders have switched from hand lay-up FRP tabbing to adhesive bonding of structural joints in an attempt to reduce costs and improve quality (Galpin, 1996).

Ronnal P. Reichard, Ph.D., Structural Composites, Inc., 7705 Technology Drive, W. Melbourne, FL 32904

A joint government/industry project, funded under the MARITECH program, is developing a composite superstructure system for ships. One of the major problems being addressed in this study is fabrication cost. The key to minimizing the cost of an FRP superstructure is to minimize the labor and tooling costs. The approach utilized in this project is to mass produce simple elements, primarily flat panels, flat bonding plates and pultruded bonding angles, which can be quickly and easily joined to produce a variety of structural configurations.

Adhesive structural joints are typically formed by applying a bead of adhesive to one surface of a structural part or panel and placing a second structural part or panel onto the adhesive bead, compressing it to cover most or all of the joint surface. The joint is then secured (usually by clamping) until the adhesive has cured. Shortcomings associated with these techniques include difficulties in:

1) clamping the parts or panels in place while the adhesive cures;
2) controlling bond line thickness and width (critical to optimize strength of the joint);
3) limiting the adhesive to the joint area (adhesive "escaping" through the edges of the seams is wasted and often must be removed manually);
4) waiting for the adhesive to cure before preceeding with other fabrication steps.

Expensive equipment, labor intensive procedures, extra adhesive (an expensive product) and increased fabrication time are typically required to address these difficulties. Installation of mechanical fasteners and/or use of custom clamping devices are often required to secure the parts or panels while the adhesive cures. Bond line thickness and width are critical to obtain the optimum structure performance of the adhesive. This requires accurate positioning of the parts, including the spacing between the parts, and controlled application of the adhesive. The time, labor, and equipment necessary to obtain the required accuracy is a significant portion of the cost to produce an adhesive joint. Adhesive flows in its uncured state, thus it tends to flow in any direction in which it is not confined. As a result, adhesive tends to escape through the seams of the joint. This adhesive which escapes is wasted (20 to 50 percent wastage is typical in the marine industry), and often must be removed, usually manually with rags and solvents, creating hazardous waste. Removal after the adhesive is cured eliminates the hazardous waste, but requires considerably more labor and may require refinishing of the panel or part surface. Fast curing adhesives are available, but severely limit the time to apply the adhesive and position the parts, which typically increases rework and/or reject rates. An external curing agent, such as heat, can be used to accelerate the cure, but the heating equipment and energy can be expensive, and the heat may adversely affect the parts being bonded, especially plastic, fiberglass or painted parts. Room temperature cure adhesives are commonly used for many applications, but the cure time is typically on the order of hours.

The patent pending method presented here is a combination of liquid adhesive/adhesive tape technology, involving the use of adhesive tape to temporarily secure structural parts or panels and to create a channel between the structural parts or panels to be joined. Typically, two (or more) strips of adhesive tape would be applied to one part or panel, and the second part or panel would then be positioned and pressed into place. The thickness of the bond line is determined and controlled by the thickness of the adhesive tape, and the width of the bond is defined by the distance between the strips of adhesive tape. The adhesive is then pumped into the channel at some later time to form a permanent bond. The method has the following advantages:

1) The use of the adhesive tape provides a simple, low cost method of securing the parts or panels together prior to the application of the adhesive. This allows multi-part, multi-joint structures to be "tacked" together quickly, checked for accuracy, and then permanently bonded at some later time by injecting the adhesive;

2) The adhesive tape provides a consistent bond line thickness and width, thus providing optimum structural performance of the joint;

3) The adhesive tape effectively seals the edges of the joint, eliminating adhesive waste through drips and spills, and eliminates expensive clean up or rework;

4) The adhesive tape temporarily secures the joint, allowing other fabrication steps to proceed before the adhesive is injected or while the adhesive cures.

TEST METHODS

A series of joints were fabricated using this method and then tested for strength. The materials used to construct the base laminates and the joints are presented in Table 1. "T" joints, representative of bulkhead intersections and deck/bulkhead intersections, were designed, fabricated and tested. The "T" joints were fabricated from sandwich panels with 2 layers of NEWFC 2308 on each side of 1.0 inch thick balsa core. The top of the "T" was 24 inches long by 4 inches wide, while the stem of the "T" was 10 inches high by 4 inches wide. Three geometries were investigated: the "A" geometry, in which the stem of the "T" is spaced a small distance from the top panel, allowing optimum bonding of the end of the stem panel to the top panel; the "B" geometry, in which the stem of the "T" was fitted tightly to the top panel, preventing adhesive from fully penetrating between the end of the stem and the top panel; the "C" geometry, in which there is a gap of at least 0.5 in. between the end of the stem. Three thicknesses were investigated for the bonding angles: 2 layers of NEWFC 2308 (equal to the skin of the sandwich, approximately 0.125 in., designated as "2"); 4 layers of NEWFC 2308 (approximately 0.25 in., designated as "4"); 6 layers of NEWFC 2308 (approximately 0.375 in. Thick, designated as "6"). The test laminates are presented in Table 2. Two radii were investigated for the bonding angles, 3/8 inch and 1.0 inch, designated as "S" and "L" respectively. The eight combinations of the above which were tested are presented in Table 3 and graphically in Figure 1. Schematics of the test configurations are presented in Figure 2 and shown photographically in Figure 3. The stem was loaded 8 in. from the bonding surface of the top plate in the "T" bending test. The support span for the top plate in the "T" tensile and compression tests was 20 in.

A single set of sandwich panels were butt joined using bonding plates on each side to obtain a double lap joint. The support span for the double lap sandwich flexural test was 33 inches with the load applied at 1/3 points and is shown photographically in Figure 4. The samples were tested using ASTM D-393, method B.

Single lap shear joints were fabricated using a 2 inch overlap with either epoxy or urethane adhesive. Double lap shear joints were fabricated with both 2 and 3 inch overlaps using epoxy adhesive. The shear samples were tested using ASTM D-3165.

RESULTS

The results are presented graphically in Figure 5. The "T" bend test was used to select the best joint design. Fully bonded joints ("A") were stronger than hollow joints ("C") and the partially bonded joint ("B"). The larger radius ("L") was similar in strength to the smaller radius ("S") for hollow joints ("C"), but was stronger for fully bonded joints ("A"). Thicker bonding angles ("6") were stronger than thinner bonding angles ("4" and "2"). Based on these results, only thick bonding angles with small radius ("S6") were tested in tension and compression. The fully bonded geometry ("AS6") was significantly stronger than the hollow geometry ("CS6") for both tensile and compression. The epoxy single lap shear ("SLE2") carried a higher load than the urethane adhesive single lap shear ("SLU2"). The double lap shear with a 2 inch long joint ("DLE2") had a higher shear strength than the 3 inch long double lap shear ("DLE3"), although the failure load was higher for the 3 inch long double lap shear samples.

The results of the double lap sandwich flex test are not presented graphically, since only one geometry was tested. Failure (core shear) occurred at a skin stress of 25,000 psi.

DISCUSSION

Almost all of the failures in the "T" test series were core shear failures of the sandwich panels. An example of such failure is shown photographically in Figure 6. The difference in strength between the various geometries was due to the ability of the joint to distribute the load and the resulting stress concentrations. Thicker bonding angles are better able to distribute the load, thus providing higher joint strengths. None of the failures occurred at the corner of the bonding angle, thus the larger radius was effective only in helping to distribute the load over the sandwich panel. This is important, since the most efficient method of producing the bonding angles is pultrusion and there are many companies offering a wide range of standard angle products with small radius corners. Also, angle variation due to resin shrinkage increases as the corner radius increases, and the angle of the bonding angles must be accurate within 1-2 degrees for consistent joints and to minimize difficulties during assembly.

The lap shear test failures were predominantly delamination or tearing between the layers of the laminate, as opposed to adhesion failures at the surface of the laminate. Both the epoxy and urethane adhesives appear to have excellent surface adhesion. The epoxy adhesive has a higher modulus than the urethane adhesive, which may have effected the delamination of the laminate, leading to higher strengths. The longer double lap shear joint (3 in. overlap, "DLE3") had a slightly higher load to failure than the shorter double lap shear (2 in. overlap, "DLE2"), however since the failure was predominantly delamination of the laminate, and shear strength is a function of the area of the joint, the "DLE3" samples had a lower shear strength than the "DLE2" samples.

The double lap sandwich flex test demonstrated the ability of the joint to develop the full strength of the sandwich panel. The failures were all core shear related, rather than adhesive failures. The skin stress was in excess of 25,000 psi in every sample when failure occurred, thus the ability of the joint to transfer high stresses was also demonstrated.

The method of manufacturing adhesive joints presented here produces joints with excellent strength. It also has the potential to significantly reduce the overall fabrication cost of large composite structures.

ACKNOWLEDGMENTS

The author was supported during this study by funds from the MARITECH Program. ATI, Baltek, Interplastic and 3M supplied the materials for the study. Finally, a special thanks to the lab staff at Structural Composites for their ingenuity and perseverance in setting up and running these tests.

REFERENCES

Pfund, B., "Taping and Tabbing," <u>Professional Boatbuilder</u>, February 1996.

Pattee, W.D. and R.P. Reichard, "Hull-Bulkhead Construction in Cored RP Small Craft", *43rd Annual Conference, the Society of the Plastics Industry, 1988*.

Galpin, D., "A Guide for Matching the Product to the Job," <u>Professional Boatbuilder</u>, June 1996.

BIOGRAPHY

Dr. Reichard is the CEO of Structural Composites, Inc., an engineering R&D company focusing on FRP materials and structures. He is also the Technical Director of the Navy Center of Excellence for Composites Manufacturing Technology (CECMT) *Marine Composite Technology Center*, is a member of numerous professional societies, and is the author of more than 50 technical papers on FRP materials and structures.

TABLE I - MATERIALS

<u>Interplastic Corp.</u>
VE8121 vinyl ester resin

<u>Advanced Textiles, Inc.</u>
NEWFC 2308, 23 oz/sq.yd. (0,90) E-glass fabric stitched to 3/4 oz/sq.ft. chopped strand mat
NEWFC 3310, 33 oz/sq.yd. (0,90) E-glass fabric stitched to 1.0 oz/sq.ft. chopped strand mat

<u>Baltek, Inc.</u>
1.0 in. thick end-grain balsa core
2.0 in. thick end-grain balsa core

<u>3M Company</u>
Scotch-Weld 3549 B/A urethane adhesive
Scotch-Weld 2216 B/A epoxy adhesive
4466W polyethylene foam tape with rubber adhesive

TABLE II - TEST SAMPLE LAMINATES

"T" Joint Sandwich Panel Laminate
NEWFC 2308 (mat side out)
NEWFC 2308 (mat side toward core)
1.0 in. Balsa core
NEWFC 2308 (mat side toward core)
NEWFC 2308 (mat side out)
note: Vacuum compression molded
         with VE8121 vinylester resin

"T" Joint Bonding Angle Laminates

2 Layer Laminate
NEWFC 2308 (mat side down)
NEWFC 2308 (mat side up)

4 Layer Laminate
NEWFC 2308 (mat side down)
NEWFC 2308 (mat side up)
NEWFC 2308 (mat side down)
NEWFC 2308 (mat side up)

6 Layer Laminate
NEWFC 2308 (mat side down)
NEWFC 2308 (mat side up)
NEWFC 2308 (mat side down)
NEWFC 2308 (mat side up)
NEWFC 2308 (mat side down)
NEWFC 2308 (mat side up)
note: Hand lay-up with VE8121
         vinyl ester resin

Flex Test Sandwich Panel Laminate
NEWFC 2308 (mat side out)
NEWFC 2308 (mat side toward core)
2.0 in. Balsa core
NEWFC 2308 (mat side toward core)
NEWFC 2308 (mat side out)
note: Vacuum compression molded
         with VE8121 vinylester resin

Flex Test Bonding Plates
NEWFC 2308 (mat side down)
NEWFC 2308 (mat side up)
note: Vacuum compression molded
         with VE8121 vinyl ester resin

Single Lap Shear Laminate
NEWFC 3310 (mat side down)
note: Hand lay-up with VE8121
         vinyl ester resin

Double Lap Shear Laminate
NEWFC 2308 (mat side down)
NEWFC 2308 (mat side up)
NEWFC 2308 (mat side down)
NEWFC 2308 (mat side up)
note: Vacuum compression molded
         with VE8121 vinyl ester resin

Double Lap Shear Bonding Plate
NEWFC 2308 (mat side down)
NEWFC 2308 (mat side up)
note: Vacuum compression molded
         with VE8121 vinyl ester resin

TABLE III - "T" DESIGNS SUBJECTED TO TESTING

Geometries
"A" bondline between end of stem and top panel
"B" no bondline between end of stem and top panel
"C" hollow between end of stem and top panel

Bonding Angle Radius
"S" 3/8 inch radius
"L" 1 inch radius

Bonding Angle Thickness
"2" 2 plies of NEWFC 2308 (0.125 in.)
"4" 4 plies of NEWFC 2308 (0.25 in.)
"6" 6 plies of NEWFC 2308 (0.375 in.)

T-Bend Test
Set #1 - BS4, AS4, AL4
Set #2 - AS2, AS4, AS6
Set #3 - CS4, CL4, CS6

T-Compression Test
Set #1 - AS6, CS6

T- Tensile Test
Set #1 - AS6, CS6

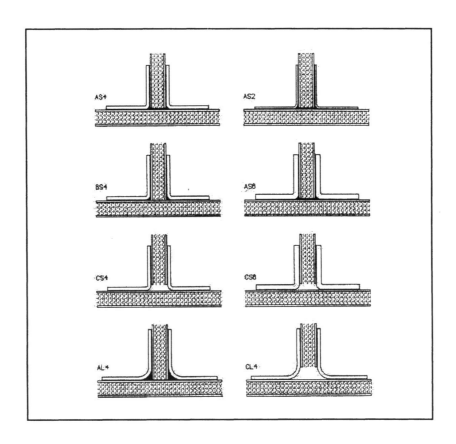

FIGURE 1 - "T" JOINT DESIGNS

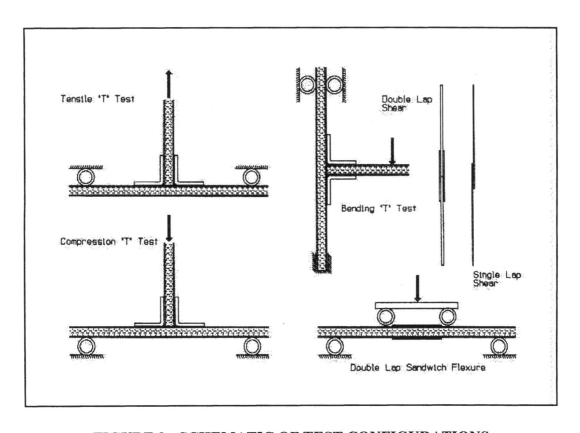

FIGURE 2 - SCHEMATIC OF TEST CONFIGURATIONS

FIGURE 3 - PHOTOGRAPH OF TEST CONFIGURATION

FIGURE 4 - PHOTOGRAPH SHOWING SUPPORT SPAN

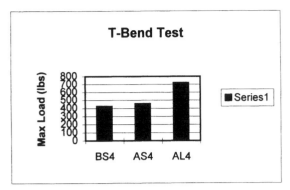

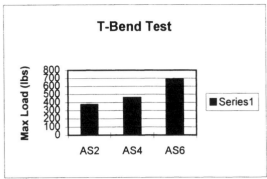

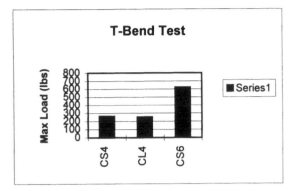

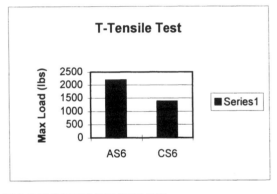

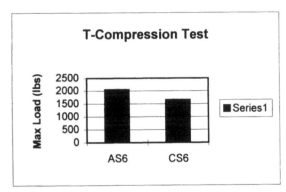

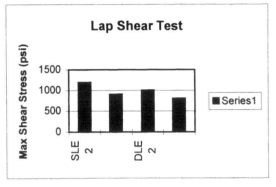

FIGURE 5 - COMPARISON OF TEST RESULTS

FIGURE 6 - "T" TEST SERIES CORE SHEAR FAILURE

New Technologies for Expanding Infrastructure Applications: Dams, Highway Systems and Railroads

Using the Highway Innovative Technology Evaluation Center to Bring New Proprietary Products into the Highway Market

DAVID A. REYNAUD

ABSTRACT

HITEC is a national one-stop, fee-based service center for the evaluation of innovative technologies, including new products, materials and services. Through HITEC, manufacturers can receive an impartial and fair evaluation of their technologies. This paper will trace the efforts to establish HITEC and discuss its present status, operation, and overall benefit to the highway and bridge market. Specifically, the paper will highlight the composite technologies that are currently participating in the HITEC program. Ultimately, the final, published evaluation reports increase recognition and acceptance by the public sector officials who are responsible for procuring, authorizing and specifying these products.

INTRODUCTION

The Highway Innovative Technology Evaluation Center (HITEC) opened for business in January of 1994 as a new service center to evaluate market-ready products for the highway and bridge markets. Applications are accepted from companies and entrepreneurs to evaluate innovative technologies and new products. HITEC hopes to assist the U.S. become competitive into the next century by evaluating a wide range of highway products. For these new technologies to be accepted by the highway community, it will be essential that credible, national evaluations be planned and conducted and that their results be made available to practitioners.

BACKGROUND

HITEC is a service center of the Civil Engineering Research Foundation (CERF), the research affiliate of the American Society of Civil Engineers (ASCE), which was created to introduce and promote the use of research findings. By serving as a coordinator, facilitator

David A. Reynaud, Highway Innovative Technology Evaluation Center (HITEC), 1015 15th Street, N.W., Ste. 600, Washington, D.C. 20005

and integrator, CERF forms public-private sector alliances to attack real problems and implement real solutions. One example of this mission is HITEC.

Very early in the HITEC planning effort, CERF concluded it was essential to learn as much as possible about barriers to innovation and perhaps more importantly, identify the key factors and elements which HITEC would need to incorporate into a successful program.

ROAD BLOCKS TO INNOVATION

In early 1993, CERF planned and conducted a nationwide needs survey of the highway community. Respondents reported that the most common obstacles preventing them from using new highway products included: procurement systems based on prescriptive standards and specifications; reliance on lowest initial cost selection criteria; regulatory constraints on the use of proprietary products; and evaluation processes which take too long, are too costly, or do not adequately address key concerns.

Traditionally, the existing highway user evaluation system has forced companies to demonstrate the acceptability of innovative products by paying for their testing and evaluation multiple times. In fact, according to the HITEC needs survey, over 60% of the entrepreneurs reported that they were required to have a different evaluation of their product for each user organization. Half of the respondents reported that, on average, they were required to complete more than twelve different evaluations of their product. Therefore, it has been extremely difficult for companies to justify making the effort to research, introduce and market innovative products to the highway community.

HITEC'S ROOTS

HITEC's roots can be traced to several national studies and reports over the past decade which highlighted the problem and recommended change. Most significantly, in 1992 a Transportation Research Board (TRB) task force specifically called for the creation of national clearinghouse for highway innovation. Coincidentally, this concept was consistent with plans by CERF to establish national test and evaluation centers to assist practitioners in moving research more rapidly into practice.

Both groups coordinated a workshop to explore the idea further. That workshop, held in 1992, brought together about 100 officials who represented a wide range of interests. Two days of discussions led to an affirmation of the need for HITEC, a preliminary action plan for it, and eventually, a cooperative agreement between CERF and the Federal Highway Administration (FHWA). The four-year cooperative agreement allows CERF to receive funds to staff the center and to cover the administrative overhead for its first years of existence.

SO WHAT IS HITEC?

HITEC is a national first-stop service center for the evaluation of innovative technologies. It accepts a wide range of new products, materials, equipment, processes and services for which no industry standards exist. Evaluations are not limited to U.S. technologies; the program is designed to improve the U.S. highway system utilizing the best technologies the private sector can provide, regardless of their origin.

HITEC recognizes that many other research and evaluation programs exist in the highway arena and duplicate testing is not necessary to achieve HITEC's goals. Instead, the HITEC process facilitates the use of existing product knowledge through expert panels and existing evaluation information.

HITEC does not necessarily mean "high technology," and we will not restrict our attention to just high-tech solutions. Our goal is to select products for evaluation which will solve real problems, meet real needs and make a real difference in the highway community. The process is general enough to be applied to anything from low-tech pothole fillers to the most sophisticated technologies.

THE HITEC PROCESS

HITEC is set up to accept applications from entrepreneurs and companies who are attempting to sell new and innovative products. Once an application is submitted, HITEC staff review it and determine whether the product meets the following eligibility criteria:
- the applicant owns or controls the use of the technology,
- the product is beyond the development stage and is truly "market-ready,"
- the technology has innovative features which cannot be readily evaluated against existing, widely-accepted standards or specifications, and
- the applicant is prepared to pay the fees and other associated costs for conducting the evaluation.

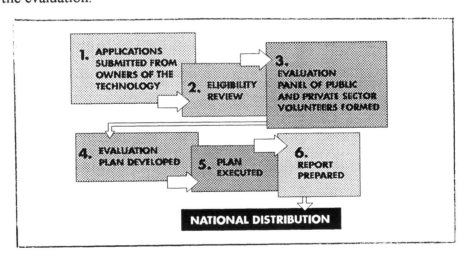

Figure 1.

Any applications not accepted will be returned to the applicants with explanations and/or referrals to other appropriate organizations, such as AASHTO's National Transportation Product Evaluation Program (NTPEP) or TRB's Ideas Deserving Exploratory Analysis (IDEA) program.

In all cases, HITEC accepts new products involving technologies which are not adequately covered by existing standards, and ideally which could satisfy high priority needs within the user community.

For those applications that are accepted, HITEC enters into a contract for services with the applicant to guide the product through the evaluation process.

Using a process modeled after the National Cooperative Highway Research Program, volunteer technical experts from the user community, academia, and the private sector, are brought together to establish an impartial consensus group for the evaluation of the specific product. HITEC Panels typically consist of about ten individuals to ensure geographic, technical and organizational diversity.

Regional representatives along with other county and municipal "users" are typically included on the panel to formulate the many questions that need to be answered regarding product performance as well as research experts to help formulate plans to determine how best to answer the "users'" questions. The result is a HITEC Evaluation Plan for the technology.

The scope of these plans vary considerably from product to product and may include tests and/or in-service demonstrations. In all cases, the evaluation plan is structured against what the product claims to be or do, and as appropriate, address issues of constructability, practicality, maintenance, safety, and environmental impact.

Once the plan is developed, the applicant is given time to review and decide whether he or she wishes to proceed with the evaluation since the applicant is asked to pay for all the costs to implement the evaluation plan. Because some applicants are not able to afford the total cost of the technical evaluation, especially the true small business or entrepreneur that traditionally has introduced many of the innovative products, HITEC provides applicants with information on known sources of funding as part of its clearinghouse role.

Once the evaluation plan has been finalized and funded, the plan is implemented. Testing or evaluation tasks will be subcontracted to the most appropriate facility. This might be a university research center, private laboratory or a federal or state transportation center. Eventually, an evaluation report is prepared by the Panel which documents the results of the evaluation of the product against the specific criteria specified in the HITEC plan. At that stage, HITEC works closely with the applicant to disseminate the results of the evaluation to the user community.

TECHNOLOGIES IN THE HITEC PROGRAM

As of October 31, 1996, over 50 different products are participating in the HITEC program. These products range from group evaluations of sophisticated technologies such as seismic isolators, to single client evaluations of valve box covers. The following types of products are in the program:

- Liquid bonding agent for pothole repair
- Nuclear gauge for water cement measurements
- Innovative sign display system to reduce glare
- Recycled rubber utility valve box cover
- Heated pavement system
- Corrosion-inhibiting concrete admixture
- Concrete pavers for utility cut repairs
- Asphalt pavement test device
- Composite dowel bars for concrete pavement joints
- Cementitious material for concrete patching and bridge repair
- Seismic isolation and energy dissipation devices for bridges
- Fiber reinforced polymer retroreinforcing and bridge strengthening systems

- Polycarbonate plastic stop sign
- Pre-cast segmental overpass system
- Sight and sound screening system
- Composite bridge fendering system
- Composite column wraps
- Bridge lock-up device
- Weigh-in-motion sensors
- Earth retaining systems

HITEC's ROLE IN INTRODUCING FIBER REINFORCED POLYMERS TO THE HIGHWAY MARKET

The HITEC process is uniquely suited to the introduction of new materials to this traditionally conservative marketplace. As shown above, currently there are four different applications of FRP technology under evaluation at HITEC: a bridge fendering system, column wrap systems for seismic retrofit, bridge retroreinforcing/strengthening systems and a rigid pavement joint dowel bar system. In each case, the HITEC evaluation panel consists not only of members of the user community, who were not previously knowledgeable about the benefits and performance attributes of FRP, but also technical experts from the FRP community itself. This collaborative approach to technical evaluation has resulted in the development of acceptance criteria, as well as planned demonstrations of composite technologies in several states, which will hopefully open the highway market for future applications of new FRP technology.

SUMMARY

The future of the U.S. highway system depends greatly on the introduction of innovative highway products to improve the quality, safety and reliability of America's infrastructure. As more and more companies invest in R&D and develop innovative solutions to our highway problems, they need an impartial mechanism to evaluate them and expedite their acceptance in the highway community. The HITEC process for evaluating technologies is broad-based enough to facilitate even the most sophisticated technologies transition into the market.

HITEC hopes to continue to work with the public and private sector to expedite the entry of new technology into the highway marketplace.

BIOGRAPHY

David A. Reynaud attended the University of Maryland and stayed in the Washington area to work for a construction contractor on the Metro station at the Pentagon. After serving as project manager on a sewage treatment plant and two subsequent Metro projects, David joined the Reinforced Earth Co., where he set prices, sent quotes, and closed contracts. In 1987, David joined two other former Reinforced Earth employees to form The Neel Co., which brought two retaining wall systems, T-WALL and Isogrid, and several other precast products to the marketplace. At HITEC since April of 1995, David has handled the development of evaluation plans for a diverse range of new products for the highway market.

Conceptual Design, Optimization, Development, and Prototype Evaluation of Composite-Reinforced Wood Railroad Crossties

JULIO F. DAVALOS, MICHAEL G. ZIPFEL, PIZHONG QIAO AND KIRA L. KALEPS

ABSTRACT

Due to in-service damage resulting in splitting and excessive wearing, over 12 million wood railroad crossties are replaced annually on Class 1 railroads in the U.S. at an approximate cost of $500 million. Therefore, a need exists to develop innovative means for improving the performance and service-life of wood ties. In this paper, the development and prototype evaluation of a wood tie wrapped by or encased in fiber-reinforced polymer (FRP) composite is discussed. Using glass fibers, epoxy resin, and a resorcinol formaldehyde primer, wood cores are reinforced with a relatively thin layer of composite by the filament winding process. The paper includes the conceptual design, modeling, optimization, and testing of prototype samples. A 3-D finite element (FE) model of a beam on elastic foundation is used to predict the response of the wood-composite crosstie, and the same model is used to conduct parametric studies of important design variables of the composite reinforcement. The FE models are combined with innovative optimization techniques to minimize the volume of composite while simultaneously minimizing the stresses in the wood core. Based on the optimization results, a final recommended design is proposed and prototype samples are manufactured and evaluated. To verify the predictions of the model, both wood and FRP-wood samples are tested under combined static and moisture loadings. The predicted linear response of the samples correlates well with experimental results. The combined modeling, optimization and evaluation study presented in this paper provides design guidelines for the development of prototype composite-reinforced wood crossties. The future commercial manufacturing and potential implementation of this new product are also discussed.

INTRODUCTION

To serve the transportation needs of the United States, both new and existing technologies are being explored to meet the increasing transportation demands of providers of freight

Julio F. Davalos, Department of Civil and Environmental Engineering (CEE) and Constructed Facilities Center (CFC), West Virginia University, Morgantown, West Virginia 26506-6103
Michael G. Zipfel, Lockheed Martin Astronautics, Denver, CO 80123
Pizhong Qiao, CEE and CFC, West Virginia University, Morgantown, WV 26506-6103
Kira L. Kaleps, Industrial Fiberglass Specialties, Inc., Dayton, OH 45414

services. Of particular interest is the renovation and improvement of the railroad system by upgrading the railroad track and supporting structure. Studies indicate that wood has been and will continue to be the material of choice for railroad crossties in the U.S. (Sonti et al., 1995). The primary reasons for using wood crossties are: (1) adequate availability of timber resources; (2) experience that track engineers and maintenance crews have with wood ties; (3) compatibility of existing track equipment with wood ties; and (4) ease of manufacturing and handling. Wood ties are cost-effective in relation to concrete and steel ties, and their field-performance has been reasonably acceptable. Wood crossties are typically made of chemically-treated hardwoods, and their main functions are to: (1) transfer and distribute the rail loads to the ballast, (2) secure the rails and maintain them to the correct gage-width, and (3) resist the cutting and abrading actions of bearing plates and ballast material. The damaging effects caused by bearing plates and ballast, namely plate-cutting and ballast abrasion, in combination with splitting of wood ties caused by stresses during air-drying have resulted in premature failure and high replacement rates of crossties. Over 12 million wood ties are replaced annually at a cost of $500 million. Therefore, there is a need to develop innovative means for improving the performance and service-life of wood ties.

Several efforts have been undertaken to improve the performance of wood ties. For example, the use of wood preservatives to prevent decay and dowels to minimize splitting; however, each action is targeted to improve only one or two specific problems. A feasible and innovative solution for improving the overall performance of a wood crosstie is to wrap or encase the tie with fiber-reinforced polymer (FRP) composites. The favorable characteristics of FRP composites can significantly improve the stiffness and strength of reinforced wood members (Davalos et al., 1992; Sonti and GangaRao, 1995). However, the primary concern with composite-reinforced wood ties is the interface bond strength and integrity under mechanical and moisture/temperature loads (Gardner et al., 1994; Barbero et al., 1994). The performance of the interface bond is influenced by the surface textures of the wood and wrapped composites and the characteristics of the resin used. Other important considerations with composite-wood ties include: (1) abrasion resistance of the composite at the tie/steel-plate and tie/ballast interfaces; (2) feasibility of an economical manufacturing process; and (3) compatibility of the reinforcing material with the wood preservative treatment. For the successful implementation of FRP-wood ties, these factors as well as the long-term performance should be investigated. In this study, wood cores are reinforced with a relatively thin layer of composite by the filament winding process. In the filament winding process, a tension force is applied to prestress the reinforcement. The wrapped crosstie is always under some degree of compression, which further prevents wood tie from splitting. The reinforcing material consists of glass fibers, epoxy resin, and a resorcinol formaldehyde (RF) primer, which is applied over the wood core to provide an integral bonding with the epoxy resin.

The objective of this paper is to discuss a combined modeling, optimization, and evaluation approach to characterize the performance of a prototype composite-reinforced (FRP) wood crosstie and to provide design recommendations for future commercial implementation of this product.

MODELING OF FRP-WOOD CROSSTIE ON ELASTIC FOUNDATION

A 3-D Finite Element (FE) model, which is based on a Winkler elastic foundation (Hetenyi, 1946), is used to analyze the performance of glass fiber-reinforced polymer (FRP) wood crosstie. The FRP-wood crosstie is modeled as a beam on elastic foundation, and the basic assumptions used in this model are: (1) linear foundation response, (2) constant foundation stiffness along the length of the beam, and (3) small deflections. The solution for a discrete beam on a Winkler foundation depends on the material and geometric properties of the beam and the foundation. An average foundation modulus, $K_n = 120$ lb/in^3 (Hetenyi, 1946), that represents the linear stiffness per unit width and length of the crosstie is used. The elastic modulus for wood crosstie is taken as 1.0×10^6 psi, based on No. 2 or No. 3 grade northern red oak material commonly used for crossties.

MATERIAL AND GEOMETRIC PROPERTIES

Assuming transverse isotropy for the wood material, the properties used to model the northern red oak crosstie are given in Table 1. The composite wrap consists of E-glass fiber and vinylester matrix. To simulate the filament winding process of wrapping, a fiber volume fraction of $V_f = 50\%$ and a fiber wrap angle of $\alpha = +/- 45°$ are used to model the FRP laminate. The laminate material properties in Table 1 are predicted from micro/macromechanics models (Davalos et al., 1996) and applied in the FE model by considering material orientations.

TABLE 1. MATERIAL PROPERTIES FOR FRP-WOOD CROSSTIE

	E_1 (10^3 psi)	E_2 (10^3 psi)	E_3 (10^3 psi)	G_{12} (10^3 psi)	G_{13} (10^3 psi)	G_{23} (10^3 psi)	υ_{12}	υ_{13}	υ_{23}
Wood	1000.0	70.225	70.225	62.500	62.500	17.860	0.186	0.186	0.432
FRP*	2166.0	2166.0	733.00	1663.0	1663.0	281.90	0.506	0.506	0.300

*In-plane properties refer to 1-2 plane in material coordinates.

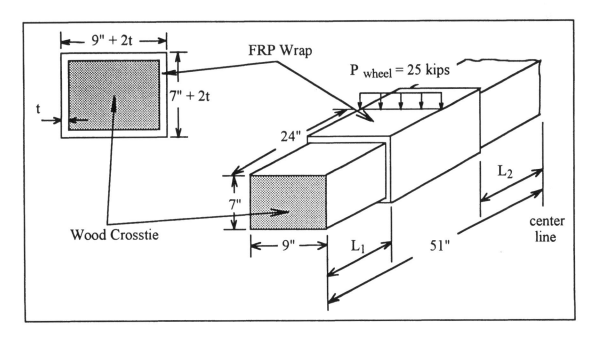

Figure 1. Geometric constants and design variables for the FRP-wood crosstie model

TABLE 2. COMPARISON BETWEEN THE 3-D FE MODEL AND THE WINKLER SOLUTIONS

Position on the Wood Tie	3-D FE Model	Beam on Winkler Foundation
Deflection at the end	0.380 in	0.383 in
Deflection at the railseat	0.503 in	0.480 in
Deflection at the center	0.437 in	0.452 in
Bending stress at the railseat	1450 psi	1786 psi
Bending stress at the center	684 psi	682 psi
In-plane shear stress to the left of load	- 163 psi	- 270 psi

Typical dimensions of a wood crosstie are 9 in wide by 7 in deep by 102 in long (Fig. 1), and the crosstie is considered to be supported by a 24 in deep foundation. For a crosstie beam, the longitudinal foundation stiffness per unit width is $k = bK_n = 1080$ lb/in^2. Based on the cooper E-80 load and accounting for impact factor, as specified by the American Railway Engineering Association (AREA), the loading used is a 25-kip wheel-load applied at each of the two railseats, which are located at 24 in from each end of the crosstie (Fig. 1).

FINITE ELEMENT MODEL

In this study, the Integrated Design Engineering Analysis Software (I-DEAS) is used to generate the finite element model of the FRP-wood crosstie on an elastic foundation. The wood crosstie is model as a beam using 8-node linear and orthotropic brick elements. To reduce the stresses in the wood beam, a composite wrap is used around the wood crosstie. The FRP reinforcement acts as an integrated laminated plate around the wood crosstie surface and is also analyzed using 8-node orthotropic brick elements. The bond between the FRP and wood is assumed to be perfect, and no interface delamination effects are considered in the FE model. The 24-inch deep elastic foundation is modeled with linear spring elements, and each of the two 25-kip wheel loads is discretized along the 9-inch width of the crosstie.

To verify the prediction accuracy of the FE model, the crossties are also analyzed by an explicit solution of an finite beam on a Winkler elastic foundation (Zipfel, 1996). The results in Table 2 indicate that the FE predictions compare favorably to those by the explicit solution. There are some differences with the explicit solution, because the 3-D FE model more realistically accounts for out-of-plane material properties. Also, the actual loading condition is better represented by the FE model than by the concentrated point loads in the explicit solution. However, both models predict stress levels in the wood crosstie that exceed the allowable limits specified for the timber grade of crosstie. The FE model facilitates the analysis and design of the FRP-wood crosstie and reduces the need for elaborate laboratory tests. The FE model is used in a preliminary study of the effect of the FRP reinforcement.

EFFECT OF FRP REINFORCEMENT

To demonstrate the effect of the composite wrap on the performance of the crosstie, the results for an FRP-wood crosstie on elastic foundation are compared to those for an unreinforced wood crosstie (Table 3). The FRP wrap, which is expected to be applied by the filament winding process, encases the wood core over its entire length with a thickness of 0.25 in, a fiber wrap angle of +/- 45°, and a fiber volume fraction of 50%. The results show a significantly improved performance for the composite-reinforced wood tie. The relative

TABLE 3. 3-D FE ELASTIC FOUNDATION MODEL OF WOOD vs. FRP-WOOD CROSSTIES

Application Position on the Tie	Wood	FRP-Wood	Performance Improved
Deflection at the end ($\delta_{x = 0"}$)	0.380 in	0.405 in	-
Deflection at the railseat ($\delta_{x = 24"}$)	0.503 in	0.493 in	-
Deflection at the center ($\delta_{x = 51"}$)	0.437 in	0.450 in	-
Relative Deflection ($\delta_{x = 24"} - \delta_{x = 0"}$)	0.123	0.045	63.4 %
Bending stress* at the railseat (x = 24")	1450 psi	856 psi	41.0 %
Bending stress* at the center (x = 51")	684 psi	403 psi	41.1 %
In-plane shear stress at the left of load	- 163 psi	- 129 psi	20.9 %
In-plane shear stress at the right of load	206 psi	158 psi	23.3 %

*Bending stresses for both analyses represent the stresses along the bottom surface of the tie.

deflection of the crosstie decreases by 63%, which indicates a stiffer crosstie after being wrapped by the FRP reinforcement. As shown in Table 3, there are also significant decreases in bending stress levels for the FRP-wood model: around 41% at the railseat and midspan. For shear stresses, the critical points occur around the loading points, and there is a 23% decrease in shear stress for the FRP-wood crosstie. Following this preliminary study, the FRP wrap is optimized.

DESIGN OPTIMIZATION OF FRP-WOOD CROSSTIE

As described above, the wood crosstie wrapped by a constant-thickness composite along its entire length may not be a cost-effective design. In this section, we use the 3-D FE model of a beam on elastic foundation to carry out the design optimization of the FRP-wood crosstie. The objective of the optimization is to simultaneously minimize the stresses in the wood core and the volume of composite wrap, which is equivalent to minimizing cost.

The FRP-wood beam considered is the same as the FE model on elastic foundation described above, except that the FRP wrap does not cover the entire length of the wood crosstie (Figure 1). The relative deflections, critical bending and in-plane shear stresses, and volume limit of FRP wrap are considered in the optimization process. The design variables include the FRP thickness (t) and FRP-wrap locations along the crosstie length (L_1 and L_2), which allow for either total crosstie encapsulation or a minimum FRP wrap around the railseat. Based on a parametric study, the orientation of the FRP wrap was taken as $\alpha = +/-$ 45°, which can be conveniently applied by the filament winding process. The functions for relative deflections and critical wood bending and shear stresses of the FRP-wood crosstie depend on the composite wrap thickness and locations along the length. The 3-D FE model is used to carry out parametric studies of design variables of the composite reinforcement, and the performance functions are defined through a global approximation technique at a number of design points. The approximate equations are then obtained by using a multiple linear regression technique to generate a power law as a function of the design variables (Qiao, Davalos and Barbero, 1996). Finally, the optimization problem is solved with the commercial program IDESIGN (Arora, 1989).

The optimum results for the design variables are: t = 0.20 in, L_1 = 1.50 in, and L_2 = 0.0 in. Considering practical applications and the flexibility of the filament winding process, a multi-thickness FRP wrap is proposed in the final design (Figure 2), which consists of t = 0.10 in

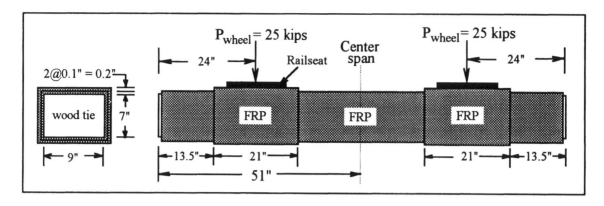

Figure 2. Final recommended design of FRP-wood crosstie

along the entire length and t = 0.20 in along the railseat. The critical bending stress (σ_x) in the final design is 1075 psi, which is about 26% lower than for unreinforced wood crosstie. The critical in-plane shear stress (τ_{xy}) is 175 psi, which is 15% lower than for the wood alone. The volume of the multi-thickness FRP wrap reduces by about 22% in comparison to the volume of the model with an FRP wrap t = 0.25 in along the entire length.

PROTOTYPE EVALUATION OF FRP-WOOD CROSSTIE

The manufacturing and experimental evaluation of a full-scale FRP-wood crosstie can be costly and time consuming. To verify the results of the FE model, prototype samples (3 ft long) are manufactured and tested under combined static and moisture loadings. The FE model, which is generated based on the full-scale crosstie FE model, is used to predict some testing results of the prototype samples.

MANUFACTURING OF FRP-WOOD SAMPLES

Northern red oak is chosen as the wood-core material for the crosstie samples, since it is commonly used in the railroad industry. The prototype samples were scaled down and manufactured with the dimensions of 1.75 in x 1.75 in x 36 in. Four lots of ten beams each were chosen from 60 samples and were conditioned to 12% moisture content (MC) in an environmental chamber. Each lot was representative of each other on a global scale, but not necessarily on a sample-by-sample basis. Two lots (20 samples) were encapsulated with glass FRP composite by the filament winding process at Industrial Fiberglass Specialties, Inc., Dayton, Ohio. The following FRP variables were defined: thickness t = 0.07 in, wrap angle α = +/- 45°, and fiber volume fraction V_f = 50%. The constituent materials consisted of fiberglass rovings and epoxy matrix. Before applying the FRP, the wood cores were coated with a resorcinol formaldehyde (RF) primer, which once cured provided an adequate interface for bonding of vinylester resin to wood (Dailey et al., 1996). The strain gages and lead wires were installed before the samples were covered by the primer and FRP wrap. The four lots of wood samples were used to conduct four sets of tests. Lot 1: wood alone at 12% MC; Lot 2: wood alone at fiber saturation point (FSP) of about 31% MC; Lot 3: wood-FRP at 12% MC, placed in a chamber for seven days after wrapping to allow for proper curing of the matrix; Lot 4: wood-FRP submerged in water for 60 days to allow moisture ingress into the laminate.

TESTING OF WOOD AND FRP-WOOD SAMPLES

The wood and FRP-wood samples were tested as simply-supported beams under 4-point bending. The deflections and tensile and compressive strains were recorded and compared with the FE model predictions within the linear range. Five FRP coupon samples were cut from the FRP-wood samples and tested in tension to obtain the longitudinal elastic modulus, E_x. The average value of $(E_x)_{coupon} = 0.909 \times 10^6$ psi obtained experimentally is in good agreement with the predicted value of Micro/macromechanics $(E_x)_{theory} = 1.064 \times 10^6$ psi. The material properties for the FRP were obtained analytically from micro/macromechanics models (Davalos et al., 1996). Both wood and FRP-wood samples were tested to failure and results for the corresponding sample lots are compared and discussed.

COMPARISON AND DISCUSSION OF RESULTS

The linear and failure responses for the four lots of beams is briefly discussed. As shown in Table 4, the FE maximum deflections and strains in the linear range agree closely with average experimental results for the wood and FRP-wood samples at 12% MC. Compared to wood alone, the FRP-wood beams show significant improvements in performance (23.8% decrease in maximum deflection and 15.0% and 17.5% decreases in compressive strains and tensile strains, respectively). There were noticeable effects on beam performance due to wood moisture contents (Table 5), but the reinforced samples are less affected by moisture than the wood samples (about 20%). The improved performance of FRP-wood samples at ultimate (failure) loads is shown in Table 6. The ultimate loads are 28% and 70% higher than for wood alone at 12% MC and saturation MC, respectively. Also, the wrapped samples are not as affected by moisture as the wood samples.

TABLE 4. DEFLECTIONS AND STRAINS OF WOOD AND FRP-WOOD SAMPLES AT 12% MC*

Samples	Deflection at beam center			Strains in compression			Strains in Tension		
	δ_{exp} (in)	δ_{FE} (in)	δ_{FE}/δ_{exp}	ε_{exp} ($\mu\varepsilon$)	ε_{FE} ($\mu\varepsilon$)	$\varepsilon_{FE}/\varepsilon_{exp}$	ε_{exp} ($\mu\varepsilon$)	ε_{FE} ($\mu\varepsilon$)	$\varepsilon_{exp}/\varepsilon_{FE}$
wood	0.0852	0.0869	1.0203	677.4	770.7	1.1382	683.1	770.7	1.1282
FRP-wood	0.0649	0.0643	0.9906	575.5	581.5	1.0103	563.5	581.5	1.0319

*All data shown are obtained from the average values of the samples under 2 x 100 lb loading.

TABLE 5. MOISTURE CONTENTS ON DEFLECTIONS OF WOOD AND WOOD-FRP SAMPLES*

δ_{wood} at 12% MC (Lot 1)	$\delta_{FRP-wood}$ at 12% MC (Lot 3)	Difference (%)
0.0852	0.0649	23.8
δ_{wood} at 31% MC (Lot 2)	$\delta_{FRP-wood}$ at Saturation (Lot 4)	Difference (%)
0.0945	0.0776	17.9

*All data shown are obtained from the average values of the samples under 2 x 100 lb loading.

Table 6. ULTIMATE LOADS FOR WOOD AND WOOD-FRP SAMPLES

Wood at 12% MC	FRP-Wood at 12% MC	Wood Saturation	FRP-Wood soaked for 60 days
2,193 lb	2,809 lb*	1,453 lb	2,472 lb**
Difference = 28%		Difference = 70%	

*Data for 5 samples only; **Estimated MC 15 to 20%.

POTENTIAL IMPLEMENTATION OF FRP-WOOD CROSSTIE

The design concept of an FRP-wood crosstie developed in this study provides an innovative approach for rehabilitating wood ties in service and for increasing the performance capabilities of new ties. The composite reinforcement increases the stiffness and ultimate load capacity of a wood crosstie, while decreasing stresses and providing a tough surface to resist plate cutting and ballast abrasion. The manufacturing of FRP-wood ties by the filament winding process using low cost glass fiber rovings and phenolic/vinylester matrix appears to be an efficient and economical way of producing this new product. Future work with prototype full-size FRP-wood crossties will involve long term laboratory and in-service testing, particularly for the effects of chemical environments and fatigue. Several prototype FRP-wood crossties will be installed on an actual railroad track bed to observe the effects of actual train travel on performance of FRP-wood crosstie. In the field, other adverse conditions, such as wood decay and degradation of the FRP wrap, will be addressed.

CONCLUSIONS

In this paper, a combined modeling, design optimization and prototype evaluation study for the potential development of actual FRP-wood crossties is described. A design for a wood crosstie wrapped by or encased in glass fiber-reinforced composite is developed through finite element modeling and experimental testing. A 3-D finite element (FE) model of a beam on elastic foundation developed in this paper can accurately predict the linear response of the wood-composite crosstie, and it can be applied in the design optimization of the composite reinforcement. The final recommended design is to use a two-thickness geometry of the composite and a fiber wrap-angle of +/- 45° (Figure 2). To verify the results of the model, prototype samples of 3 ft length are manufactured and tested. The test results demonstrate an improved performance of wood crossties reinforced with FRP. It is significant that through this work, a new optimal FRP-wood crosstie is proposed for railroad applications, and the present approach is successfully used for the development of this new potential product, from the conceptual design to the final manufacturing and testing of prototype samples of new FRP-wood crossties.

ACKNOWLEDGMENTS

This study is partially funded by the Federal Railroad Administration, WV Department of Highways, and Reichhold Chemicals. We thank Hexcel-Fyfe Co., PPG Industries, and Indspec for supplying materials. Our special appreciation to Industrial Fiberglass Specialties, Inc. for producing the FRP-wood samples by filament winding. We thank Harve Dailey of Indspec for his valuable cooperation in this study.

REFERENCE

Arora, J. S. 1989. "IDESIGN USER'S MANUAL," Optimal Design Laboratory, The University of Iowa, Technical Report No. ODL-89.7.

Barbero, E. J., J. F. Davalos, and U. Munipalle. 1994. "Bond strength of FRP-wood interface," *J. of Reinforced Plastics and Composites*, 13(9):835-854.

Dailey, T. H., S. S. Sonti, V. H. S. GangaRao, and D. R. Talakanti. 1996. "Using phenolics for wood composite hybrid members," Composites Institute's 51th Annual Conference and Expo'96, SPI, Cincinnati, OH.

Davalos, J. F., E. J. Barbero, U. M. Munipalle, H. A. Salim. 1992. "Interface bond strength of laminated wood-FRP composite beams," 25th Int. SAMPE Technical Conf., Philadelphia, PA.

Davalos, J. F., H. A. Salim, P. Qiao, R. Lopez-Anido, and E. J. Barbero. 1995. "Analysis and Design of Pultruded FRP Shapes under Bending," *Composites: Part B, Engineering J.*, 27(3-4): 295-305.

Gardener, D., J. F. Davalos, and U. Munipalle. 1994. "Adhesive bonding of pultruded fiber-reinforced plastic (FRP) to wood," *Forest Products J.*, 44(5):62-66.

Hetenyi, M. 1946. "Beams on Elastic Foundation," University of Michigan Press, Ann Arbor, MI.

Qiao, P., J. F. Davalos, and E. J. Barbero. 1996. "Design optimization of fiber reinforced plastic composite shapes." submitted to J. of Composite Materials.

Sonti, S. S., J. F. Davalos, M. G. Zipfel, and H. V. S. GangaRao. 1995. "A review of wood crosstie performance," *Forest Products J.*, 45(9):55-58.

Sonti, S. S. and H. V. S. GangaRao. 1995. "Strength and stiffness evaluation of wood laminates with composite wraps," Composites Institute's 50th Annual Conference and Expo'95, SPI, Cincinnati, OH.

Zipfel, M. G. 1996. "Development and Evaluation of a Composite-reinforced Wood Railroad Crosstie," Master of Science Thesis, West Virginia University, Morgantown, WV.

BIOGRAPHIES

Dr. Julio F. Davalos is Associate Professor in the Department of Civil and Environmental Engineering (CEE) of West Virginia University (WVU) and a member of the Constructed Facilities Center (CFC) conducting research on material modeling, characterization of wood and fiber-reinforced plastic composites, and analysis and design of structures. He has received several research, technology transfer, and teaching awards, including Researcher of the Year (1995), Teacher of the Year (1995), and Outstanding WVU Teacher - the highest teaching award in the university.

Michael G. Zipfel is an Aerospace Engineer with Lockheed Martin Astronautics, Denver, CO. He is a former Graduate Research Assistant of CFC/WVU, and his research emphasis was on the development and evaluation of composite-reinforced wood railroad crosstie.

Pizhong Qiao is a Research Assistant and Ph.D. Candidate at CEE/CFC/WVU. He is currently conducting research on analysis, design optimization and infrastructure applications of composite materials.

Kira L. Kaleps is Manager of Applications Engineering at Industrial Fiberglass Specialties, Inc., Dayton, OH. Her work involves several projects related to the filament winding process.

Composite Wicket Gate Development at the Olmsted Prototype Dam

MOSTAFIZ CHOWDHURY, ROBERT HALL, PETER HOFFMAN,
BYRON McCLELLAN AND JAMES MUNDLOCH

ABSTRACT

The specification, design, fabrication and testing of a composite wicket gate for a prototype of the Olmsted Lock and Dam will be presented. Since this type of gate remains in a retracted position on the floor of the river 60 percent of its life, it is exposed to not only the corrosive effects of river water but also the abrasive effects of the continuously moving river silt. A single composite wicket gate was specified for the prototype dam to allow its comparison to the performance of its steel counterparts. The project was administrated by the Louisville District of the US Army Corps of Engineers. The gate was designed by McDonnell Douglas Aerospace (St. Louis, MO) under subcontract to Massman Construction (Kansas City, MO). Assembly took place at Production Products Manufacturing and Sales (St. Louis, MO). Several design constraints were imposed to assure compatibility and interchangeability of the composite gate in any of five gate stations on the prototype dam. These constraints included utilization of the same hinge and load introduction fittings used on the steel gates. In addition, a specific gravity of three was required and the deflection performance of the composite gate had to equal its steel counterparts. A bolted/bonded design produced from off–the–shelf pultruded E–glass/vinylester flat plates and I–beams was developed. A steel strongback supplied hard points for bolted attachment of the hinge and load introduction fitting and added the mass required to meet the specific gravity requirements. The composite gate was delivered and installed on the prototype dam in July 1995. Testing initiated in December 1995 and was successfully completed in August 1996.

INTRODUCTION

The composite wicket gate was developed and tested to demonstrate the feasibility of utilizing low cost composite materials in US Army Corps of Engineers navigation structures. Operation of the dam type which utilizes wicket gates is illustrated in Figure 1. When the river levels are high enough to allow safe navigation over the dam, the wicket gates remain in a retracted position. When the river level drops below safe limits, the wicket gates are lifted into position to raise the upper pool. With the wicket gates in the up position, a traditional lock is utilized to allow river traffic to pass. At the Olmsted site, the gates will remain in the retracted position approximately 60 percent of the year. Each wicket gate is approximately 10' wide and 25' tall. Originally, 220 hydraulically operated wicket gates were slated to comprise the dam portion of the Olmsted project. However, current plans call for 140 boat operated gates to be used in combination with a more traditional stationary dam. The Olmsted Lock and Dam will replace Locks and Dam 52 and 53 on the lower Ohio River (Soast 1994). The dams of these existing projects consist in part of boat operated wooden wicket gates.

Mostafiz R. Chowdhury, US Army Corps of Engineers Waterways Experiment Station, 3909 Halls Ferry Rd., Vicksburg, MS 39108 (601-634-2567)

Robert L. Hall, US Army Corps of Engineers Waterways Experiment Station, 3909 Halls Ferry Rd., Vicksburg, MS 39108 (601-634-2567)

Peter L. Hoffman, McDonnell Douglas Corporation, Box 516, St. Louis, MO 63166 (314-233-3863)

Byron K. McClellan, US Army Corps of Engineers, Louisville District, P.O. Box 59, Louisville, KY 40201 (502-582-5783)

James D. Mundloch, McDonnell Douglas Corporation, Box 516, St. Louis, MO 63166 (314-234-1509)

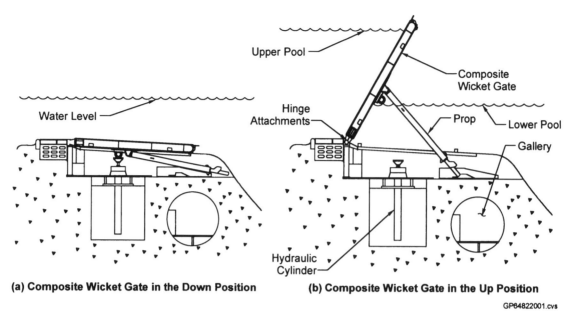

Figure 1. Prototype Dam Components and Wicket Gate Operation

A highly abrasive and corrosive environment is present for the wicket gate in this reach of the Ohio River. As such, the application was considered ideally suited for composite materials. The prototype composite gate is installed in a prototype dam which was constructed at the Smithland Lock and Dam on the Ohio River. The prototype dam is full scale but only five gates wide (approximately 50'). It is located parallel to the existing Smithland dam with river water access supplied by a 2000' channel. The prototype dam was constructed to test hardware, gate, and coating options for the Olmsted Lock and Dam. The facility provides the capability to refine components and select the best materials for use.

DESIGN AND ANALYSIS

Realizing that very high localized loads occur on the wicket gate and stiffness requirements similar to the current steel design must be achieved, a hybrid design was employed. This hybrid design consisted of a welded steel center strongback to react the localized loads and provide the required stiffness. In addition, it allowed the composite facesheets reinforced by composite I–beams to react the distributed pressure loading. The steel strongback was designed to be similar to the existing steel wicket gate design to ensure interchangeability and ease of installation. The weight of the steel also allowed the composite wicket gate to achieve the required specific gravity of 3.0. Since the steel strongback was sandwiched by composite facesheets its steel surface was not exposed to the flow of the water and silt. Elimination of abrasion on the steel structure greatly reduces the aggressive cycle of corrosion followed by abrasion.

Standard shaped pultruded composite I–beams and flat panels were used for the composite portion of the wicket gate. Pultruded structural shapes represent one of the lowest cost composite material forms available. In addition, the use of standard off–the–shelf material and shapes eliminated the need for any special prototype tooling and allowed a very aggressive schedule to be met. The composite structure consisted of Morrison Molded FiberGlass (MMFG) Extren 625 glass/vinylester pultruded sheets (0.75" thick) stiffened transversely and longitudinally by MMFG Extren 625 I–beams (18" x 3/8"x 4–1/2" x 1/2"). This composite structure was attached to an ASTM A572 Gr. 50 steel strongback weldment as illustrated in Figure 2. The interior of the strongback was metallized per US Army Corps of Engineers requirements and seal welded water tight. The remaining portions of the steel strongback were coated with a corrosion resistant paint system similar to the type used on the steel wicket gates. The pultruded composite sheets and I–beams were bolted and bonded to the steel weldment utilizing Huck (Huck Manufacturing Company) fasteners and Pliogrip 660 adhesive. Areas of the steel strongback targeted for bonding to composite components remained unpainted.

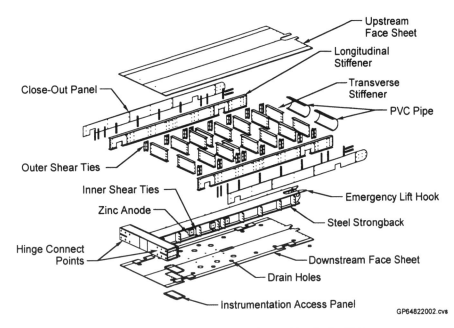

GP64822002.cvs

Figure 2. Composite Wicket Gate: Exploded View

The composite gate had to serve as a direct replacement of any steel wicket gate with no modifications to the dam or gate. As such the geometry, strength, and stiffness criteria for the composite wicket gate was required to match the steel wicket gates. Table 1 details the strength and stiffness requirements for the composite wicket gate. The pultruded Extren 625 E–glass/vinylester composite material allowables were based on minimum ultimate coupon properties published by MMFG. The strength properties were reduced by 15 percent to account for fully saturated conditions in a wet environment. Test coupons were machined from Extren 625 E–glass/vinylester flat panels and tested to verify the assumed material allowables.

TABLE 1. STRENGTH AND STIFFNESS REQUIREMENTS FOR THE COMPOSITE WICKET GATE

Strength - The Composite Wicket Gate Shall Be Capable of Withstanding the Following Loads:	

Hinge Connection:
    Horizontal Shear Key Load
    Bolt Tension Load
	80 kips/Hinge
	76 kips/Hinge

Cylinder and Prop Connection:
    Maximum Perpendicular Compression Load
    Maximum Parallel Load
	269 kips
	203 kips

Emergency Lift Tip Load	160 kips @ 45° to Horizontal

Moments and Shears
    Maximum + Moment
    Maximum − Moment
    Maximum Shear
	958,000 ft-lbs
	1,219,000 ft-lbs
	149.6 kips

Stiffness - The Composite Wicket Gate Shall Have Adequate Stiffness to Limit Deflection to the Following as Measured at the Center of the Top Edge When Subjected to Nominal Hydrostatic Load and Supported Rigidly at the Hinges and Cylinder Locations:	

Bending
    65° Cylinder Supported
    65° Prop Supported
    Horizontal Cylinder Lift
	1.5 in.
	1.0 in.
	6.0 in.

Torsion (Water Head on Half of the Wicket) Torque of 273,000 ft-lbs	4.0 in.

GP64822003.cv

Detail analysis was performed to demonstrate the structural adequacy of the composite wicket gate. The analysis focused on the composite and steel structural components, bolted/bonded joints, welded joints, and load introduction areas such at the hinge attachments, cylinder shaft, and prop. The analysis consisted of both hand calculations and finite element methods. A global finite element model (FEM) of the wicket gate was created to determine deflections of the structure.

Half symmetric FEM models were utilized for three of the four stiffness critical load conditions due to symmetry about the center beam. A full FEM model was created for the torsional load case due to unsymmetrical loading on the gate. Both models were created from shell elements in PATRAN 2.5. The hinge elements, prop, and cylinder shaft solid beam elements, boundary conditions, loading conditions, and material properties were added to the model by means of the input file for ABAQUS 5.4-1. The model was pinned at the hinge points and at the base of the prop or cylinder shaft (depending on the loading condition). The half model was also reacted for symmetry at the structure's centerline.

The materials in the gate were assumed to be linear elastic. The composite elements were assumed to exhibit lamina orthotropic plane stress behavior. The transverse stiffness of the composite facesheets was increased to include the presence of the caps of the composite I-beam stiffeners. In the areas where the facesheets are covering a steel plate, such as the center beam and end beam, the material was assumed to be steel. The contribution of the facesheets, however, were added into the thickness of the elements as a proportion of Young's modulus. The prop and cylinder connections to the gate were modeled as solid beam elements. These elements, however, were constrained as rigid beams so they would not rotate or change length. The areas of the beams were calculated from the volume of the actual connection and the length of the beam elements.

FABRICATION

Fabrication of the composite wicket gate was performed by Production Products Manufacturing and Sales (PPMS) (St. Louis, MO). The pultruded composite I-beams were pre-drilled at MMFG due to the limited flange area available for locating the holes. The face sheets were dry fit and match drilled utilizing the pre-drilled I-beam holes as locators. Once all the holes were drilled, the bonded surfaces were cleaned with an alcohol wipe. The two part Pliogrip adhesive was then mixed and applied in an even layer. After the adhesive was applied, the components were assembled and the Huck fasteners were installed.

The steel strongback was fabricated at B&D (St. Louis, MO) and delivered to PPMS for assembly of the composite components. Due to the size and weight of the steel strongback, it was elevated on wooden blocks to a convenient working height and the composite components were assembled to it. The assembly sequence started with installation of the downstream face sheet. Next, the transverse I-beam stiffeners were installed by bolting and bonding them to the lower face sheet and shear ties of the metal strongback. Once the transverse stiffeners were installed, dual composite angles were assembled to their outboard webs to serve as shear ties to the outer longitudinal composite I-beams. The upstream face sheet was installed next followed by the longitudinal I-beams. Finally, the composite wicket gate assembly was completed by installing non-structural panels on the longitudinal I-beams to close out this structure and make it more conducive to water flow. In addition, a half round standard PVC pipe was installed on either side of the steel lift point to improve the flow characteristics of the end of the gate.

INSTALLATION

The composite wicket gate was shipped to the Smithland prototype dam site where the hinges and load introduction fitting were installed. These components were supplied by the same company that produced the components for the steel gates. This approach ensured compatibility between the composite and steel gates for these critical components and also proved to be a very efficient approach since one company produced all the hardware. After the hinge and load introduction fitting were installed, the gate was lifted into position and hinge pins installed.

INSTRUMENTATION AND EXPERIMENTATION

The Corps of Engineers used several physical and analytical models of hydraulically operated wicket gates for studying the dynamic forces exerted on the wicket gates for various flow and operating conditions. The results of these models were used by the Corps of Engineers, Louisville District to design a hydraulically operated wicket gate for the Olmsted Lock and Dam.

The first wicket gate studies were performed with a 1:25 scale model of a curved gate. These initial studies reported loads which indicated potential problems with flow induced vibrations. These dynamic loads were caused by flow conditions which resulted in the formation of a low pressure region on the downstream section of the gate. Once the gate penetrates the water stream over the gate, air was allowed to release the low pressure bulb. This aspiration of the nappe resulted in oscillations of the gate and a loading condition which could possibly result in damage to the prototype structure. Further studies in procedures to aspirate the nappe resulted in redesigning the wicket gates from curved to straight.

Further studies were conducted on 1:25 scaled models of a straight gate. The redesigned gate resulted in loading conditions minimizing the vibrations resulting from the aspiration of the nappe (Chowdhury Hall and Pesantes 1995). Studies continued with a 1:5 scale model. The 1:5 scaled model was designed to be a structural similitude model which could produce scaled results of structural static and dynamic loads on the prototype steel gate (Chowdhury and Hall 1996).

The principal objectives of the prototype experiments at the Smithland prototype facility was to validate the functionality of the hydraulically controlled wicket gates to be placed at Olmsted Lock and Dam. The results of the forces from the prototype gates were compared to 1:5 scaled model studies to validate the use of a scaled model to produce both static and dynamic loads on a prototype gate (Chowdhury Ross and Hall 1996). Also, a composite gate was placed in the prototype facility to investigate the effectiveness of composite materials in this harsh environment.

A non-contact Scanning Laser Doppler Vibrometer (SLDV) response measurement system was used to perform shaker excited modal analysis of the prototype steel and composite wicket gates. Both gates were prop-supported in a dry configuration. I-DEAS Master Series CAE/CAM software was used to determine the modal vibrational characteristics from the data acquired with the Lazon system. The laser head was positioned 62' downstream from the test gate and a MB Model 50 Exciter with two added inertia blocks (133.4N each) were added to excite the wicket using a burst random signal (Chowdhury Ross and Hall 1996). The first seven mode shapes of each gate are shown in Figure 3. The mode shapes of the two different gates were surprisingly similar with the exception that modes 4 and 5 for the composite gate have been transposed. The results indicate that the two gate systems have similar dynamic characteristics.

The steel gate also had six pressure cells. Three of these pressure cells measured the upstream pressure while the remaining three cells measured the downstream pressures on the back of the gate. The pins connecting the gate to the sill were instrumented to determine the reactions of the steel gate. The composite gate also had nine uniaxial accelerometers for recording upstream and downstream vibrations. Two pressure cells, one at the top surface and another mounted on the back of the gate, measured the pressure, and a tiltmeter measured the inclination of the composite wicket gate.

The flow induced loading experienced by a wicket gate depends upon the position (up or down) of the gates to its immediate right and left. Fifteen flow experiments were conducted to include all critical gate configurations. Figure 4 shows both gates in the raised position. The composite wicket gate is the number four gate and the instrumented steel gate is number three.

a) Composite Gate Mode Shapes

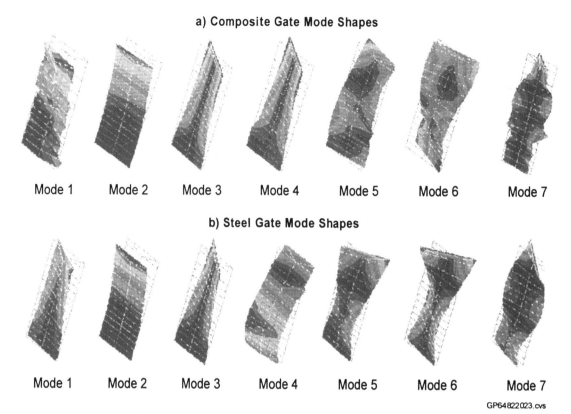

Mode 1   Mode 2   Mode 3   Mode 4   Mode 5   Mode 6   Mode 7

b) Steel Gate Mode Shapes

Mode 1   Mode 2   Mode 3   Mode 4   Mode 5   Mode 6   Mode 7

GP64822023.cvs

Figure 3. Composite and Steel Gate Mode Shapes

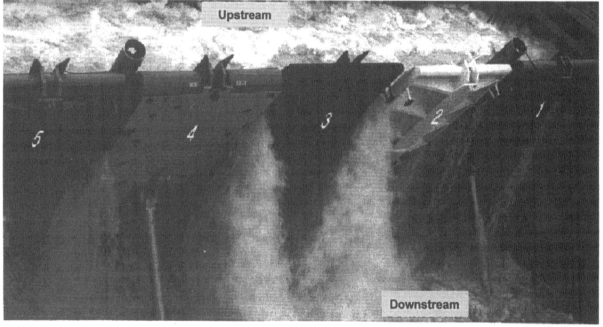

GP64822024.cvs

**Figure 4. Downstream View of Prototype Gates in the Engaged Position
(Gate #4 is the Composite Wicket Gate)**

Figure 5 provides comparisons of the upstream and downstream pressure gages for the three gate–gap flow configurations. A three gate–gap configuration resulted in the highest static and dynamic loads on the gates. As seen in the figures, upstream pressure for both gates were very similar to each other. Downstream pressure for the submerged gates were similar for both cases. After the breaking of the air pocket under the gate, the downstream pressure, however, showed some variation. This difference resulted due to the effect of the water flow beneath the gates and the variations of water levels.

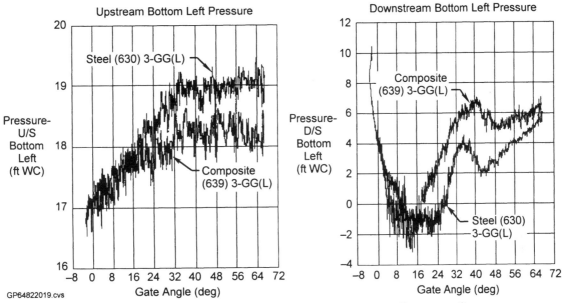

GP64822019.cvs

Figure 5. Comparison of Upstream and Downstream Pressure Gages

Accelerometer power–spectra densities (PSD) and time domain accelerometer responses indicate more vibrations in the composite gate than in the steel gate. Peaks in the composite gate PSDs were more energized than those of the steel gate. The second and third resonant frequencies of the dry composite wicket gate coincided with the corresponding peaks in the frequency domain response of the operational case. The resonant frequency content of the operational steel gate, however, does not coincide with any of the major structural modes and the anticipated flow-induced vibration does not impose any immediate threat to the structural performance of the wicket gate. Further studies need to be completed to insure that these vibrations will not cause any localized fatigue problems.

POST TEST INSPECTION AND LESSONS LEARNED

The prototype dam was de–watered for inspection on 12 August 1996. The composite wicket gate was removed from the dam for evaluation. In general, both the upstream and downstream faces of the gate remained in good condition. However, the non–structural close out panels added to the edges of the composite wicket gate for hydrodynamic improvements experienced varying levels of damage which resulted in loss of the structure in some areas. Similar damage was witnessed on both edges of the gate ruling out the damage as random. The damage mechanism appears to be a deep gouging caused by foreign objects flowing down river. The edges of the gate may be particularly susceptible to this type of damage since the gaps between the gates have constant water flowing between them in the up position which can act as traps for foreign objects. In addition, the damage was primarily restricted to the non structural close–out panels in part due to their light-weight construction and limited attachment to the main structural members of the gate. An exposed edge of the downstream facesheet appeared to have experienced impact damage. The damage could have been caused by a foreign object being locally trapped and smashed during retraction of the gate.

Many of the lessons learned and suggestions for an improved application of composite material can be attributed to the constraints imposed on the wicket gate design to ensure its compatibility with the steel gates. Lessons learned and suggestions for improved performance of composites in this type of application follow:

- Elimination of single point lifting of the gate with a hydraulic ram. Highly localized out–of–plane loading of this type is difficult to introduce into a composite structure without the aid of local metal details.
- Utilization of unitized composite structure versus built–up bolted/bonded structure to reduce overall costs. Emerging resin infusion techniques would be ideal candidates to produce the composite wicket gate as a unitized structure.
- Elimination of secondary composite structure. In general, if a composite component is designed to carry minimal loading it is also prone to unpredictable damage in navigation structures.
- Addition of stainless steel rub strips on the edge of the gate to offer improved resistance to foreign object damage.
- Elimination of all free edges in damage prone areas.
- Modal behavior of the composite and steel gates were very similar except for the 3rd and 4th modes which have been transposed from one to another.
- Upstream pressure for both gates were very similar, while the downstream pressure after the breaking of the air pocket beneath the gate varied between the wickets.
- The composite gate has a higher vibratory response than the steel gate but no evidence of damaging vibration was determined during these experiments.

REFERENCES

Chowdhury, M. R., Hall, R. L., and Pesantes, E. (1995). *"Flow–Induced Vibrational Test Results for a 1:25 Flat Wicket Gate,"* A Report Prepared for the US Army Engineer District, Louisville, Louisville, KY.

Chowdhury, M. R., Hall, R. L. (1996), *"Flow–Induced Structural Response of a 1:5–Scale Olmsted Wicket Model,"* Waterways Experiment Station, Vicksburg, MS (In progress).

Chowdhury, M. R., Ross, D. G., and Hall, R. L. (1996), *"Experimental Comparison of Prototype and 1:5 Wicket Gates,"* Waterways Experiment Station, Vicksburg, MS (lab review).

Soast, A. (1994). *Engineering News Record COVER STORY,* "Record Size Lock and Dam Project Gets a Soggy Launch as Rains Swell the Lower Ohio River for Months", pp. 24–26.

CONVERSION FACTORS

1 in. = 25.4 mm, 1 lbf = 4.4484 N, 1 psi = 6,895 Pa

ACKNOWLEDGMENTS

The research presented in this paper was sponsored by the US Army Engineer District, Louisville, Kentucky. We appreciate the cooperation of the authorities at the US Army Engineer Waterways Experiment Station, McDonnell Douglas Corporation, and the Office, Chief of Engineers, US Army Corps of Engineers that permitted us to prepare and present this paper for publication.

FRP Composite Rebars Come of Age: A Rational Strategy for Commercialization

CHARLES R. McCLASKEY

ABSTRACT:

FRP composite technology utilizing C-bar™ introduces to the construction industry a relatively low cost noncorrosive reinforcement alternative for commercialization consideration worldwide.

FRP COMPOSITE REBARS COME OF AGE: A RATIONAL STRATEGY FOR COMMERCIALIZATION

Corrosion of reinforcing steel in Portland cement concrete structures is a serious problem around the world. The quest for more effective solutions is leading many civil engineers to consider Fiber Reinforced Polymer (FRP) composite technology. The development of a new second-generation manufacturing process for FRP composite rebars offers the construction industry a viable noncorrosive reinforcement alternative. However, if the construction industry is to reap the desired benefits of lower cost, longer structure life and reduced maintenance under adverse conditions, careful attention must be paid to how and where this technology is applied.

In most instances FRP rebars cannot simply be substituted for steel in a given design. While we in the composites industry like to think of composites as being "better," to the design/engineering community FRP composites are unknown and "different." It should surprise no one that a material should not be considered a direct substitute for steel if it is anisotropic, and has roughly twice the tensile strength, one fifth the modulus of elasticity and one quarter the weight.

Clear patterns have emerged over the last forty years as FRP composites have replaced traditional materials in a host of applications. Typical sequential stages of development are: substitution, optimization and enabling. The first composite replacements usually mimic a design optimized for a different material. C-bar™ composite rebar features U. S. steel deformation patterns largely for reasons of market acceptance. These deformations were developed in the 1940's around the requirements of steel rolling mills. We expect the next generation of composite rebars to feature "non-steel" deformations, which will improve bond strength. Fortunately, the manufacturing process by which C-bar™ composite rebar is made

Charles R. McClaskey, Reichhold Chemicals, Inc., P. O. Box 13582, Research Triangle Park, NC 27709-3582

lends itself to deformation design flexibility at relatively low cost. Users of this product should not assume functional equivalency to steel just because it is dimensionally similar A second major difference between traditional steel reinforcing bars and the newer FRP composite products is that the durability of the latter is not absolutely knowable in the short term. The failure mode and useful life of structures reinforced with steel bars subjected to chloride ion attack have been well researched. It is possible to model the expected service life of a structure with reasonable certainty given critical variables such as chloride concentration, depth of cover, etc. This predictability in performance has not been evident with epoxy coated steel rebar, which is one reason alternate technologies are being investigated.

For cost reasons "E-glass" is the fiber of choice in current FRP composite rebars. (Other fibers can be processed and are being investigated concurrently.) Bare E-glass fibers exhibit rapid degradation when exposed directly to high levels of alkalinity. However, there is a substantial body of existing science developed in "corrosion-resistant" applications where E-glass composites have given successful service in excess of thirty years in alkaline environments. It is the role of the matrix resin to protect the fiber in an FRP composite structure. Many of these well documented cases involve vessels, piping and structures for the chemical processing industry. The loads on the walls of a large FRP composite storage tank containing hot caustic soda solution, for example, are not identical to the loads "seen" by reinforcing bars. No direct correlation for the purposes of defining FRP rebar service life is possible.

It is well understood how resin type, fiberglass surface treatment, as well as design (laminate architecture) and processing conditions all influence the corrosion resistance of FRP composites. Several generic lessons have been learned. First, matrix resins must be chemically compatible with the expected service. Major resin producers publish manuals which specify products suitable for service in given chemical environments. Glass producers can recommend products with surface treatments (sizings) that are compatible with the intended use. Usually, a resin rich surface devoid of glass fibers is created to ensure durability. The specially chosen resin in the corrosion-resistant laminate must be adequately cross-linked (cured) to provide the desired performance. A variety of standard tests exist to measure goodness of cure during production and on site. Finished products must be performance tested since the listed influencing factors are necessary but not sufficient to ensure quality.

Work recently published by researchers at the West Virginia University Constructed Facilities Center (Altizer, et al. 1996) on the durability of pultruded E-glass rods showed superior performance for urethane modified vinyl ester resin under stressed pH 13 conditions. It is no accident that this same technology is employed in the outer corrosion layer of the C-bar™ composite rebar. Urethane modified vinyl ester resins exhibit superior fracture toughness compared to conventional vinyl esters and provide better protection for glass fibers in reinforcing service. The specification of proven resin technology will be particularly important before FRP rebar specifications and standards are available.

The record is clear. E-glass FRP composites have given long-term service in structural applications under conditions of high alkalinity. Some commercially available composite reinforcing bar products have been shown to fail rapidly under load at pH 13. The puzzle of durability cannot be solved by rebar manufacturer's claims or scientific studies that test the performance of one product (N=1) and report generic conclusions without understanding or even listing critical parameters.

If we accept that FRP composite rebars perform differently than steel, that at present no guarantees which define long-term durability can be made, and that the durability of various E-glass FRP composite rebar products are in question, how can it be determined if this technology holds value for the construction industry? Fortunately, there appears to be an international consensus on adopting an inclusive cooperative development model.

All constituencies from materials suppliers to end users are participating: industry, academia and government interests are represented. This concept is championed by the Civil Engineering Research Foundation and The Society of the Plastics Industry (SPI), Composites Institute in the USA, the Advanced Composite Materials in Bridges and Structures network in Canada, and the members of the Eurocrete project in Europe. A global network of trade and professional associations is now in communication. One observation that can be made is that there is not yet enough participation by the practicing engineers/designers or contractors.

Interested parties can choose from a growing number of technical venues such as the Non-metallic Reinforcement for Concrete Structures symposia (Vancouver 1993, Ghent 1995, Sapporo 1997), the Advanced Composite Materials in Bridges and Structures Conferences (Sherbrooke 1992, Montreal 1996), and the International Conference on Composites in Infrastructure series (First Annual, Tucson 1996).

Thanks to the leadership of Dr. John "Jack" Scalzi of the U. S. National Science Foundation, the (International) Research on Advance Composites in Construction (IRACC) program is coordinating North American researchers and their counterparts elsewhere. IRACC has identified critical gaps in the scientific knowledge base. The "Reinforced Concrete" subcommittee is responsible for FRP composite rebar issues.

What are the tasks this network of interested parties should undertake? We can identify both commercial and scientific objectives. First, researchers must propose common durability test methods for adoption by the various testing bodies, ASTM, RILEM, etc. These test methods should reflect the need for accelerated results and could be used to judge the relative performance of FRP composite rebar products. Later these test results can be correlated with actual field experience to generate the credible durability predictions we now lack.

On the commercial side, as with most emerging technologies, some form of self policing must evolve to keep unsafe products from use. Voluntary (minimum) performance standards should be adopted. In industrialized nations the code bodies will not allow general use until consensus standards and design guidelines have been developed by professional engineering associations. However, there are many applications where noncorrosive reinforcing is desired for which no codes exist. Manufacturers need to exercise control over how these products are used. We simply do not have enough information to use FRP composite rebars alone in primary load bearing structures at this time (with the exception of demonstration). End users must be made aware of the potential danger of substituting FRP composite rebars in structures designed for steel. For these reasons, Marshall Industries Composites (inventors/producers of C-bar™ composite rebar) has established the criteria for its distributors that they provide engineering design services to customers and individually screen proposed applications.

Another issue FRP composite rebar manufacturers must face is the conflict between proprietary technology and standardization. Composite manufacturers want the freedom to employ all of the material's inherent design flexibility in the pursuit of lower cost and higher efficiency. They expect to be compensated for the value of unique products that deliver greater value. The civil engineering community sees the bewildering and seemingly endless combination of resins, fibers, designs and manufacturing processes as an impossible barrier to understanding and trusting this promising technology. A contract delivery system based on specifications and low bids (in the U. S.) has little patience for single sources and unique solutions.

Edward DiThomas, Chief Engineer of the Turner Construction Company, addressed this topic during his comments at the 1996 Annual Business Meeting of the SPI Composites Institute. He used Sony's Betamax™ and Apple's Macintosh™ as examples of superior technology that failed commercially because it was not made available to the mass market. The emerging FRP composite reinforcements must provide common performance if we are to develop the comprehensive design systems available for traditional materials. Performance standards for FRP composite rebars could meet the needs of both civil engineer users and composite product designers. Marshall Industries Composites' strategy to license their technology worldwide heeds Mr. DiThomas' advice.

Researchers are now developing design guidelines for FRP composite reinforcements around the world. The Japanese Society of Civil Engineers expects to complete their guidelines in the fall of 1996. A joint Canada/U.S. group is coordinating work on separate design specifications that are to be adopted by respective national codes such as the American Concrete Institute 318 and the Canadian Building Design Code. Design guidelines are a main thrust of the Eurocrete project.

One activity critical to the commercialization of composite technology has already begun: demonstration projects. The centerpiece for the U. S. is the 54 meter, 3-span McKinleyville Bridge in West Virginia. It is the first bridge in the world to use FRP rebars in the concrete deck. Owned by the West Virginia Department of Transportation, partners in the project include the Federal Highway Administration and the Constructed Facilities Center of West Virginia University. FRP rebars will be included in a variety of other bridge deck projects currently in the design stage. In Canada bridge demonstrations including carbon FRP tendons and advanced "smart" sensing technology predate U. S. work. The Headingley Bridge in Manitoba uses (glass) FRP composite rebars for one side of the bridge curb. The Eurocrete project has similarly sponsored case study structures which are being monitored for evaluation. It is vital that data be collected from as many structures as possible to provide calibration for the theoretical models being developed to describe composite reinforcement behavior.

One responsible approach to using FRP composite rebars now involves a powerful hybrid design concept advanced separately by UK (Arya, Ojori-Donko, and Pirathapa 1955) and Sweden (Mogahadam and Sentler 1955) based researchers. They propose using FRP composite rebars in combination with steel rebars to provide long-term system performance with low risk. The "Supercover" concrete concept patented by South Bank University (Arya and Piranthapan 1996), London, uses thick concrete cover (nominally 10 mm) to delay corrosion of the steel, and includes FRP composite rebars to prevent the thick cover from cracking. The structure is designed to code requirements based on the steel reinforcements, minimizing risk. It is possible to model the increase in service life and to calculate whether the increase in materials cost is justified in a particular application where salt exposures are known. This concept, if widely adopted, could deliver cost savings in harsh environments, as well as provide multiple opportunities for data collection, speeding the day when we have enough confidence to build, with even lower costs, all-composite reinforced systems.

In conclusion, there is a global need for cost-effective, noncorrosive concrete reinforcements to replace traditional materials in harsh environments. We can expect the use of FRP composite reinforcements to become significant first in the third world in those geographies where steel fails rapidly. The risk of using relatively unproven technology is small where steel fails in less than ten years due to salt contamination during concrete mixing. As scientific data is collected from demonstrations and non-code secondary applications in the industrialized nations, we will learn how to design and install composites in the third world. By the time code adoption is achieved in the U. S., leading edge composite reinforcement systems will integrate stay-in-place forms and reinforcements to make steel-like rebar technology obsolete. This enabling technology will reduce installed costs to the degree that composite systems will begin to compete with steel even in some noncorrosive applications. Additional benefits will accrue to the construction industry as suppliers of traditional reinforcements improve the cost/benefit performance of their products in reaction to the composite threat. I am proud to be part of a dynamic industry that intends to make this vision a self-fulfilling prophecy.

REFERENCES

Altizer , S. D., P.V. Vijay, H. S. GangaRao, N. Douglass, R. Pauer 1996 "Thermoset Polymer Performance under Harsh Environments to Evaluate Glass Composite Rebars for Infrastructure Applications,", *Composite Institute's 51st Annual Conference & Expo '96,* Session 3-C: 1-19.

Arya, C. F.K. Ojori-Donko and G. Pirathapa, 1955 "FRP Rebars and the Elimination of the Reinforcement Corrosion in Concrete Structures, Non-metallic (FRP) Reinforcement for Concrete Structure," *E&FN* : 227-234.

Arya, C. and G. Piranthapan, 1996. "SUPERCOVER CONCRETE: A new method for preventing reinforcement corrosion in concrete structures using GFRP rebars," South Bank University, *SUPERCOVER Information Pack.*

Mogahadam, M., L. Sentler 1955. " GFRP as Crack Control Reinforcement, Non-metallic (FRP) Reinforcement for Concrete Structures," E&FN: 243-250.

BIOGRAPHY

Charles R. "Charlie" McClaskey is Vice President, Business Development for Reichhold Chemicals, Inc. Reichhold, with headquarters in Research Triangle Park, North Carolina, is a leading manufacturer of resins, adhesives and polymers for the composites, coatings, adhesives, textile, carpet, construction and graphic arts markets.

Mr. McClaskey began his career in the composites industry with Owens Corning Fiberglas Corporation in 1977, serving in line sales and marketing management positions. In 1981, he joined the Norac Company, supplier of organic peroxides and metallic stearates where he rose to the position of Vice President of Marketing and Sales. Mr. McClaskey joined Reichhold's

Reactive Polymers Division as National Sales Manager in 1988 and was named Vice President soon after. In 1995, he assumed responsibility for business development for the Corporation.

Mr. McClaskey has served on the Board of Directors of the SPI Composites Institute and has chaired its Market Development Alliance. He is past chairman of the Western Composites Institute. In 1989, he received a Distinguished Service Award from the Western SPI for his leadership in negotiating the precedent-setting SCAQMD Rule 1162.

Mr. McClaskey has been active in the Composites Fabrication Association as a speaker and member of its Government Affairs Committee. He received CFA's highest accolade, the President's Award, in 1995 for his contributions to the association and the industry.

A long-time America Concrete Institute member, Mr. McClaskey is active on Committee 440 (Fiber Reinforced Plastic Reinforcement). He represents Reichhold on the Corporate Advisory Board of the Civil Engineering Research Foundation, the research arm of the American Society of Civil Engineers.

Mr. McClaskey received a BA in International Relations from Brown in 1967. Following graduation, he served in staff and fleet assignments in the U. S. Navy, separating later from the Reserves as a Lieutenant Commander. In 1977, he received an MBA from Dartmouth's Amos Tuck School of Business Administration.

Innovation of Pultrusion Structural Shapes for Infrastructure

GLENN BAREFOOT, DAVE SITTON, CLINT SMITH AND DAN WITCHER

ABSTRACT

MMFG has developed a new hybrid reinforced pultruded structural shape under an ATP Award from NIST; Georgia Institute of Technology has provided some design advice and testing support for this profile. This shape incorporates innovations in composite design and shape design to correct problems normally experienced with using a mat/roving glass reinforced I or W shape. An 8" section using these concepts has been successfully pultruded at the Bristol Division of MMFG.

This paper will present the shape, some comments on processing, and the test results. The testing will demonstrate that this shape is more efficient in its performance than other commercially available pultruded structural shapes and can be processed using different reinforcement and resin matrix combinations.

BACKGROUND

It would be redundant for MMFG to review the importance of infrastructure repair to the composites industry and to the nation. The Department of Commerce through NIST established a series of ATP (Advanced Technology Program) programs for composite materials, and MMFG submitted a project directly related to infrastructure. The MMFG project title was "Development of Innovative Manufacturing Techniques to Produce a Large Phenolic Composite Shape." The MMFG project team added Dr. Abdul Zureick of Georgia Institute of Technology whose function was to advise MMFG for the design of the structural shape and composite from a mechanical performance viewpoint and to perform the testing of the pultruded product to determine the performance and design parameters. MMFG made the final decisions based on processing. A summary of the principal objectives of this program is:

Glenn Barefoot, Morrison Molded Fiber Glass Company, Box 580, Bristol, Virginia 24203-0580
Dave Sitton, Morrison Molded Fiber Glass Company, Box 580, Bristol, Virginia 24203-0580
Clint Smith, Morrison Molded Fiber Glass Company, Box 580, Bristol, Virginia 24203-0580
Dan Witcher, Morrison Molded Fiber Glass Company, Box 580, Bristol, Virginia 24203-0580

1. Adjust the modulus of elasticity from the typical 2,500 ksi for the mat/roving type of composite to 6,000 ksi for the new innovative shape.

2. Pultrude a shape significantly larger than the EXTREN® 24" I shape, the current largest pultruded structural shape by MMFG.

3. Optimize the shape of the pultruded section to provide a more structurally efficient member (stiffness and buckling) than currently available commercial shapes.

4. Pultrude this innovative large structural shape in the phenolic resin system.

CONCEPTUAL SHAPE DESIGN

The innovative structural shape design was determined to resolve two critical issues: what is required to improve the structural performance of the shape and what is required in the shape configuration that will process successfully in phenolic pultrusion. At the time this program was initiated with NIST, only a 1-1/2" x 1/8" square tube had been processed in the phenolic resin system used at MMFG.

Figure 1 displays several different design concepts that were considered by the MMFG team which included Dr. Abdul Zureick. The sketches in Figure 1 are presented only for a conceptual discussion. Shapes such as (a), (b), and (d) contain a flange tip which offers stiffness but some problems in pultrusion to properly control the fiber reinforcement placement in the tip area. The flanges were anticipated to be thicker than the web, and an all-roving flange tip or a high level of roving concentration in the flange tip might be difficult to maintain its structural integrity in aggressive applications; the tip may break easily. Shape (a) was also rejected because it would have a definite positive and negative orientation which might accidentally not be installed correctly in the field. The flange tip models also have a problem in that they could become areas where water could easily collect within the beams.

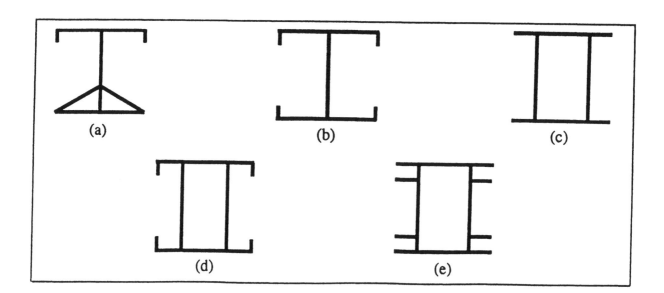

Figure 1

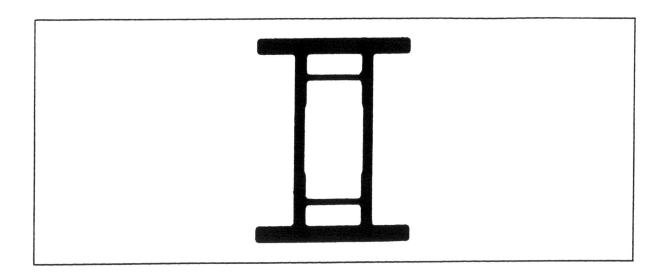

Figure 2

Shape (c) appears to offer more inherent stability, but the box shape is not a dramatic innovation. Shape (e) offers two supplemental flanges on the exterior which may aid in some connections and will offer additional stiffness to the shape design; however, these exterior flanges might also be somewhat difficult to pultrude and maintain its integrity on the structural shape. The final design is conceptually presented in Figure 2 in which the exterior flanges were moved to the interior to enhance processing; these interior flanges are much less fragile than the exterior flanges. These interior flanges also serve to tie the two webs together and reduce the effect of column lengths of the web; this adds improved stability to the structural shape. The improved processibility of the Figure 2 design also has the most cost effective tooling cost.

CONCEPTUAL COMPOSITE DESIGN

The conceptual composite design is relatively straightforward. The glass has an inherent modulus of elasticity of approximately 10,000 ksi, and a 6,000 ksi modulus on the composite shape can only be achieved with a 60% glass fiber volume applied unidirectionally. However, a structural shape intended for infrastructure usage must contain a significant amount of off axis material to resist buckling and twisting of the member. Hence, carbon fiber is required to provide the necessary addition to the composite modulus of elasticity to adjust for the significant amount of reinforcement applied in the off axis orientations. The maximum structural and cost benefits from the carbon fiber are obtained by adding the carbon fiber to the outside of the flanges.

The torsional performance will be driven by the composite structure in the web, and this is accomplished by adding 0°/90° stitched fabric and ± 45° fabric primarily to the web of the structural shape. The complexity of developing the phenolic structural shape was the additional requirement that all of the reinforcement had to be compatible with the phenolic resin matrix used at MMFG. Close cooperation with the raw material suppliers provided was required to select a roving compatible with the MMFG phenolic resin matrix that was also compatible with the stitching process. A glass surfacing veil was also used on this composite to enhance the visual appearance, although one of the challenges for future phenolic production is the development of a synthetic veil that will have the same FST (flame, smoke, and toxicity) performance as the phenolic resin matrix itself.

It should also be noted that while theoretical models could be developed to determine the amount of carbon fiber to be placed on the structural shape, MMFG ultimately worked with Georgia Institute of Technology to develop an empirical model. MMFG pultruded a 4" x 2" x 1/4" I shape on the laboratory pultruder and also separately pultruded carbon fiber strips. These strips were adhered to the surface of the I shape in different thicknesses, and the I shape was subsequently tested with and without the carbon fiber to make an estimate of the amount of carbon fiber required to change the modulus of elasticity on the shape.

PRELIMINARY TEST RESULTS

MMFG pultruded an 8" x 6" subscale prototype shape (the full section will be 36") in both the vinyl ester and phenolic resin matrices. The vinyl ester resin matrix was DERAKANE® 411 and the phenolic resin matrix was a custom blend from Georgia-Pacific. Dr. Abdul Zureick of Georgia Institute of Technology is in the process of performing an extensive amount of testing on both the vinyl ester and phenolic composites pultruded, but the purpose of this paper will be to present a few of the preliminary conceptual findings. Dr. Zureick's work will be extremely thorough and complete, and the information presented here only indicates MMFG's current progress towards achieving the objectives.

Figure 3 contains a measurement of the flexural modulus which shows that the 6,000 ksi value has been exceeded. Figure 4 presents the rotation of the beam under deflection and indicates an extremely stable structure when compared to the normal I shape; this is expected of a box beam and is indicative of having a relatively uniform raw material placement. Figure 5 presents confirmation testing performed by Dr. Jack Lesko of Virginia Polytechnic Institute and State University on the modulus of elasticity, again confirming Dr. Zureick's initial results of a modulus of elasticity exceeding 6,000 ksi for this hybrid beam.

THE FUTURE

MMFG will pultrude a 36" version of the 8" x 6" subscale prototype during the Spring of 1997. This 36" structural pultrusion will be made only with the phenolic resin matrix because MMFG no longer requires a preliminary trial in vinyl ester to verify the processing. The pultrusion of the 36" optimized beam will require a significant innovation in pultrusion machinery and tool design. This machine will require an estimated 120,000 lbs. of pull force and 140,000 watts of heating capacity.

MMFG has an opportunity to demonstrate the suitability of the 8" x 6" subscale prototype working with Virginia Polytechnic Institute and State University. A consortium of Virginia Polytechnic Institute and State University, Morrison Molded Fiber Glass Company, the Virginia Transportation Research Council, the Virginia Department of Transportation, and the town of Blacksburg, Virginia, are collaborating on a project to replace a small bridge over Tom's Creek which requires rehabilitation. This bridge was initially built in 1932 and was then reconstructed in 1964. The bridge is 24" wide and spans 17' 6" with twelve 20' - 10WF x 21 steel stringers,

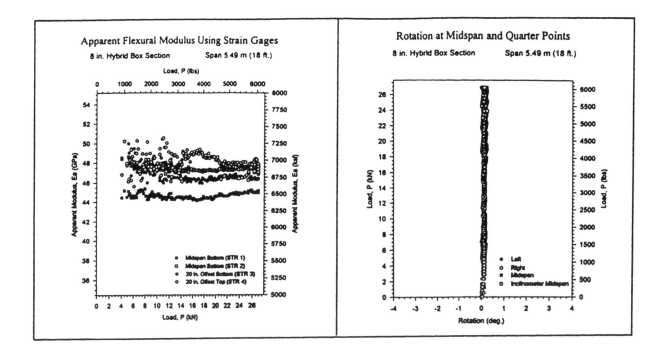

Figure 3   Figure 4

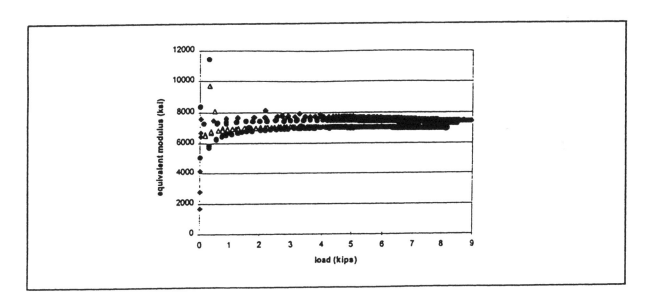

Figure 5

supporting a deck composed of 4" x 8" transverse wood beams and two to three inch asphalt.
A recent inspection of the bridge identified a significant corrosion problem with a steel stringer.
This consortium will replace the steel members with the 8" x 6" subscale prototype product. This
application will provide an opportunity to track the material behavior of the composite structural
members in service, and this monitoring will be performed by Virginia Polytechnic Institute and

State University. Virginia Polytechnic Institute and State University has performed a significant amount of testing already in support of this program which will be the topic of a future paper.

MMFG utilized a spin-off with this ATP technology by pultruding the standard EXTREN® 24" I shape with carbon fiber in the flanges. This shape was pultruded in support of a program initiated by Dr. Issam Harik of the University of Kentucky who is working with the Kentucky Department of Transportation to install a pedestrian bridge in the Daniel Boone National Forest (Figure 6). Dr. Harik will monitor the continued durability of this bridge as part of the program. This shape is not the double webbed I shape developed under the ATP program but is an application of ATP type of technology in the placement of the carbon fiber.

CONCLUSION

The second generation of composite structural shapes is here in that "smart" composites can be made taking advantage of the various types of reinforcements available for optimizing the pultruded composite, in addition to the equally important activity of optimizing the structural shape. A relatively straightforward addition of carbon fiber to the flanges of the composites and off axis stitched reinforcements to the web of the beam significantly improves the composite's performance for more advanced applications such as infrastructure. There is an apparent need for this advanced composite structural shape as MMFG has already had two applications, one from the Virginia Polytechnic Institute and State University and one from the University of Kentucky, using the technology developed during this program. The 36" structural shape to be pultruded during 1997 will complete the ATP program with technology that can be capitalized on other types of pedestrian and highway bridges.

Rendering of the hybrid fiber reinforced composite pedestrian bridge in the Daniel Boone National Forest

Figure 6

Composites in Action

The format of this session was a hands-on demonstration; therefore, there were no formal papers.

Subject Index

NOTES: Glass Fiber is not normally an index term because it is common to so many of the papers; other reinforcements are generally indexed by name. When the word "glass" is used, it always means glass fiber. Similarly, Polyester and Epoxy are indexed only when particular emphasis is given them. FRP refers to fiber reinforced plastics generally, while GRP indicates glass fiber specifically. The following abbreviations are used: ATH = Alumina Trihydrate; Dvlpt = Development; Envm = Environment; FEA = Finite Element Analysis; Flex = Flexural; Lam = Laminate; Matl = Material; Mech Props = Mechanical Properties; Mfg = Manufacturing; Perf = Performance; Props = Properties; Req't = Requirement; Struct = Structural; Temp = Temperature; TP = Thermoplastic; TS = Thermoset.

Polymer abbreviations follow ASTM 1600-92: HDPE = High Density Polyethylene; PF = Phenol Formaldehyde, PVAC = Polyvinyl Acetate; PPS = Polyphenylene; PUR = Polyurethane; SP = Saturated Polyester; UP = Unsaturated Polyester; and VE = Vinyl Ester.

Credit for developing this index is given to *Harry E. Pebly*. He is a consultant in writing and editing in plastics, composites and adhesives, located at 198 Center Grove Road, Randolph, New Jersey 07869.

AUTHOR INDEX